William Vincent Legge

A History of the birds of Ceylon

William Vincent Legge

A History of the birds of Ceylon

ISBN/EAN: 9783744740500

Printed in Europe, USA, Canada, Australia, Japan

Cover: Foto ©berggeist007 / pixelio.de

More available books at **www.hansebooks.com**

A HISTORY

OF THE

BIRDS OF CEYLON.

BY

Captain W. VINCENT LEGGE, R.A.,

FELLOW OF THE LINNEAN SOCIETY.
FELLOW OF THE ZOOLOGICAL SOCIETY
MEMBER OF THE BRITISH ORNITHOLOGISTS' UNION.
SECRETARY LATE ROY. AS. SOC. (C. B.), CORR. MEM. ROY. SOC. TASMANIA,
ETC., ETC.

LONDON:
PUBLISHED BY THE AUTHOR.
1880.

ALERE   FLAMMAM.

PRINTED BY TAYLOR AND FRANCIS.
RED LION COURT, FLEET STREET.

TO

HIS ROYAL HIGHNESS

ALFRED ERNEST ALBERT, DUKE OF EDINBURGH,

K.G., K.T., G.C.M.G., P.C., etc.,

THIS WORK IS DEDICATED,

BY SPECIAL PERMISSION,

IN MEMORY OF THE VISIT OF HIS ROYAL HIGHNESS TO CEYLON

IN THE YEAR 1870.

BY HIS OBEDIENT HUMBLE SERVANT,

THE AUTHOR.

2

PREFACE.

Of late years Ornithology has made more rapid strides than perhaps any other branch of zoological research. In Oriental regions more particularly many naturalists have, within the last quarter of a century, prosecuted their studies with the greatest vigour ; enormous collections have been made, entirely new regions explored, and their avifauna investigated with all that energy which collectors of the 19th century bring to bear on their work and doings in the forests of the tropics. The pens of Blyth, Jerdon, Wallace, the Marquis of Tweeddale, Swinhoe, Père David, and Allan Hume have brought our knowledge of the avifauna of India and the countries to the eastward of it to a high degree of perfection. At the time of the author's arrival in Ceylon much had been done by Layard, and the results of his labours were being largely added to by the researches of Mr. Holdsworth ; but nevertheless, up to that period, no complete treatise on the birds of the island had been written. As a rising British colony, with fast-developing resources and wealth, an increasing European community, and an educated element in the native population, the production of a book on its avifauna which should take a place in the series of zoological works which are invariably the outcome of civilization seemed to the author a positive necessity.

This idea was combined with a strong desire to create a taste for natural history in the minds more particularly of the educated native community, and the hope of founding an ornithological school in Ceylon, such as had been the effect of the labours of Jerdon in the Indian empire.

With this view, therefore, the author devoted his entire spare time during an 8½ years' residence in the island to the study of its ornis and the amassing of a large collection of specimens. Towards the close of his work he received no little encouragement in a promise of help from the Government made to him by the late Governor, Sir William Gregory, who, during his term of office in Ceylon, did so much for the advancement of science in all its branches, and to whom the author is much indebted for his recent exertions with the existing Government on his

b

behalf. On his return to England in 1877 it only remained for the author to combine his acquired knowledge of the life-history of the birds of Ceylon with a comparison of his collection (the largest ever made by one individual in the island) with series of specimens and skins in the British Museum and the collections of brother ornithologists in London, illustrative of the ornis of adjacent countries; and after three years of *incessant* labour the work has been brought to a conclusion.

A non-residence in London, within daily reach of the libraries, with their stores of ornithological literature, and the collections with which that great civilizing centre teems, has been a serious disadvantage to the author. Furthermore the vast amount of correspondence and supervision which the publication of the work entailed on him was much increased by his residence at a distance from those engaged in its printing and illustration. The scientific reader will therefore, it is to be hoped, pardon the various shortcomings which the author feels must, on this account, exist throughout the work.

Its mission, however, is not to impart knowledge to the scientific ornithologist in Europe, for it cannot pretend to any such degree of merit; it is intended purely as a text-book for the local student and collector in Ceylon; and though the author has as yet met with comparatively little support among the class for which he has worked so hard, yet if he succeeds in inculcating in the minds of only a few of the inhabitants of Ceylon a taste for the study of birds, which he apprehends must always rank foremost among the wonderful creations of an all-wise and bountiful Providence, his labour of love will not have been in vain. On the other hand, while his sincerest gratitude is evoked by the patronage which the Royal Family have been graciously pleased to bestow upon his humble labours, the author cannot but tender his best thanks to his friends and the general public in England for the cordial manner in which they have supported him.

W. V. LEGGE,
Captain R.A.

Aberystwith, September 2, 1880.

INTRODUCTION.

The island of Ceylon, although it contains none of those remarkable forms which characterize the birds of some of the Malay islands, undoubtedly possesses a rich avifauna ; and, considering its geographical area (about five sixths that of Ireland), the number of species is very large. The tropical position of Ceylon, coupled with its location in the path of the monsoon winds and rains, fosters the growth of luxuriant vegetation and verdant forests, which, as a matter of course, teem with all that wonderful insect-life necessary for the sustenance of birds, and hence the large number of resident species inhabiting it ; whilst the fact of its being situated at the extreme south of an immense peninsula makes it the finishing point of the stream of Waders and Water-birds which annually pass down the coasts of India ; and, lastly, the prevalence of a northerly wind at the time of the migration of weak-flying Warblers brings these little birds in numbers to its shores.

The abundance of the commoner species inhabiting the cultivated country near the towns on the west coast, and the semicultivated interior traversed by the railway and the highroads leading to the principal towns, at once strikes the traveller on his arrival in the island ; and the wonderful variety of bird-sounds heard during the course of a morning stroll, though they cannot vie in sweetness with the notes of the denizens of English groves, are, notwithstanding, quite as attractive. The laughing voice of the larger Kingfishers, the extraordinary booming call of the "Jungle-Crows" (*Centropus rufipennis* and *C. chlororhynchus*), and the energetic shouts of the Barbets when first heard fill the European traveller with astonishment, and more than compensate for the absence of the mellifluous voice of the Thrush and Blackbird.

As regards brilliancy of plumage, when we consider the tropical nature of their abode, the birds of Ceylon are decidedly mediocral. We find but little of that conspicuous beauty which characterizes the avifauna of many of the islands of the Austro-Malayan region, or even some of the birds of the Himalayas, nor do we meet with the gorgeous plumage of those of tropical America, or even the handsome dress worn by so many of the feathered inhabitants of African forests. When the naturalist has made the acquaintance of the Sun-birds, Pittas, and King-fishers there is not very much left in the way of brilliant plumage to attract him. Notwith-standing, many species are conspicuous for grace and elegance of form combined with an attractive coloration ; and if we except the above-mentioned families, the *peculiar* birds of the island number among their ranks some of the most beautiful species inhabiting it.

Before proceeding to the consideration of the ornithological features of the island, it will be well to notice briefly the labours of those naturalists who have heretofore interested themselves in the birds of Ceylon.

b 2

Labours of former Writers.—In 1743 George Edwards, Library Keeper to the Royal College of Physicians, published a work entitled 'A Natural History of Uncommon Birds,' and in it figured several species inhabiting India and Ceylon, among which were "The Black Indian Cuckow" (*Eudynamys honorata*), "The small Red-and-green Parrakeet" (*Loriculus indicus*), "The Black-and-white Kingfisher" (*Ceryle rudis*), "The Indian Bee-eater" (*Merops viridis*), "The Black-hended Indian Icterus" (*Oriolus melanocephalus*), "The Crested Red or Russet Butcher-bird" (*Lanius cristatus*), "The Pyed Bird of Paradise" (*Terpsiphone paradisi*), "The Purple Indian Creeper" (*Cinnyris asiaticus*), "The Cowry Grosbeak" (*Munia punctulata*), "The Short-tailed Pye" (*Pitta coronata*), "The Minor" (*Eulabes religiosa*), and "The Emerald Dove" (*Chalcophaps indica*). Of these it will be observed that but one species, the Lorikeet, is peculiar to the island.

During the latter half of the eighteenth century Gideon Loten was nominated Governor of Ceylon by the Dutch, and, happening to be a great lover of birds, collected and employed people to procure specimens of species which attracted his notice; and from his labours we first learn something of the *peculiar* birds of the island. He had drawings prepared of many species, which he lent to an English naturalist named Peter Brown, who published in London, in 1776, a quarto work styled 'Illustrations of Zoology.' His descriptions of the birds he figured were given in French and English, and related to the following species named by him thus:—"The Brown Hawk" (*Astur badius*), "Great Ceylonese Eared Owl" (*Ketupa ceylonensis*), "Red-crowned Barbet" (*Xantholæma rubricapilla*), "Yellow-cheeked Barbet" (*Megalæma flavifrons*), "Ceylon Black-cap" (*Iora typhia*), "Spotted Curucui" (*Cuculus maculatus*), "Red-vented Warbler" (*Pycnonotus hæmorrhous*), "Yellow-breasted Flycatcher" (*Rubigula melanictera*), "The Green Wagtail" (*Budytes viridis*), "The Rail" (*Rallina euryzonoides*), "The Pompadour Pigeon" (*Osmotreron pompadora*). The artist who delineated these species was Mr. Khuleelooddeen. Some of the drawings are fairly accurate; but others are grotesque and unnatural, showing the poor state of perfection to which the illustration of books had up to that time been brought.

We pass on now to a man of a different stamp, Johann Reinhold Forster, who gave Latin names to several of the peculiar Ceylonese forms which now stand, having been published after the Linnean period (1776). This author was likewise indebted to Governor Loten, of whom he speaks in his Introduction that he found a great field for his tastes in the science of natural history, and to assist him in his researches taught several slaves drawing. Forster's work, entitled "Indische Zoologie," was published at Halle, in Germany, in 1781, and is written in German and Latin, purporting to be a "systematic description of rare and unknown Indian animals." The following species are figured and described:—*Circus melanoleucus*, *Strix bakkamuna*, *Trogon fasciatus*, *Cuculus pyrrhocephalus*, *Rallus phœnicurus*, *Tantalus leucocephalus*, *Anser melanonotus*, *Anhinga (Plotus) melanogaster*, *Anas pæcilorhyncha*, and *Perdix bicalcarata*. Through Loten's instrumentality, therefore, 10 species were described by Forster, in addition to those which Brown figured, and which were afterwards named by Linnæus, Gmelin, and others. Prior to the advent of Templeton and Layard he did more for Ceylon ornithology than any other naturalist. One or two species were made known by Latham in his 'Synopsis,' such as the "Ceylonese Crested

Falcon " (*Spizaetus ceylonensis*) and the " Ceylonese Creeper " (*Cinnyris zeylonicus*) ; but these were afterwards found to inhabit India ; and Levaillant figured two Barbets in his ' Histoire Naturelle des Barbes,' one of which (the Yellow-fronted Barbet) is peculiar to the island.

A long gap now occurs, when little or nothing was done to elucidate the avifauna of the island ; and we hear nothing of the birds of Ceylon until Dr. Templeton, R.A., went out there to be stationed. Taking a great interest in the natural history of his temporary home, and at the same time not being a sportsman himself, he depended on his friends for specimens, which he forwarded to Blyth, then curator of the Asiatic Society's Museum, Calcutta, for identification. Fortunately for ornithology one of these friends was Mr. Edgar Leopold Layard, the now well-known ornithologist, and at present Her Majesty's Consul at Noumea. This gentleman, on his arrival in the island, set about collecting for Dr. Templeton, and, in his capacity as an officer in Government service, had ample opportunity for travel and exploration of the jungle.

The same zeal and untiring energy which has throughout life characterized Layard's career was brought to bear upon the study of the Birds of Ceylon ; and in a few years his great exertions in collecting bore fruit in a series of papers called "Notes on the Ornithology of Ceylon," published in the ' Annals and Magazine of Natural History,' which demonstrated to the scientific world that Ceylon was far richer in birds than any one had supposed. The account of his important labours is best given in his own words, contained in his kind notice of this work in a late number of ' The Ibis ' :—" I arrived in Ceylon in March 1846, and for some time, having no employment, amused my leisure in collecting for my more than friend, Dr. Templeton, who had nursed me through a dangerous illness, and in whom I found a congenial spirit. My chief attraction there was the glorious Lepidoptera of the island ; but I always carried a light single-barrelled gun in a strap on my back to shoot specimens for the Doctor. He himself, like Dr. Kelaart, never shot, but depended on his friends for specimens. I, of course, soon became interested in the ' ornis ; ' and on Templeton's leaving at the end of 1847 or beginning of 1848, he begged me to take up his correspondence with the late Edward Blyth, then curator of the R. A. S. Calcutta Museum. He left me his list of the species then known to exist in the island, numbering 183, and Blyth's last letter to answer. From that day almost monthly letters passed between the latter and myself, till I left Ceylon in 1853. The list and the correspondence are still in my possession.

" When I left I had brought up the list to 315 ; deduct from this the novelties added by Kelaart, and some which I think he has wrongly identified (but which are included in my list in the ' Annals and Magazine of Natural History '), 22 in number, and it leaves me the contributor of 110 species to the Ceylonese ornis, examples of most of which fell to my own gun.

" My collecting-trips never extended to those hill-parts where Dr. Kelaart collected, Nuwara Elliya, &c. I was twice in Kandy, once at ' Carolina,' an estate near Ambegamoa, and once as far as Gillymally, *viâ* Ratnapura."

Besides this, Layard, as he informs me *in epist.*, collected from Colombo to Jaffna, *viâ* Puttalam, Jaffna to Kandy on the Central Road, Colombo to Galle, and round to Hambantota,

Pt. Pedro to Mullaittivu, and thence back to the Central Road. The specimens procured on all these trips, as well as during Layard's residence at Pt. Pedro and other parts of the island, were sent to Blyth for identification, which resulted in the names given by the latter to not a few of the peculiar forms. He published papers from time to time in the 'Journal of the Asiatic Society of Bengal,' and also in 'The Ibis' so late as 1867.

Blyth, however, received specimens from another source, namely, from Dr. Kelaart, a native of Ceylon, and who went out from England in 1849 as Staff-Surgeon to the Forces. This gentleman, though he did not shoot himself, obtained specimens of many of the hill-birds inhabiting the vicinity of Nuwara Elliya, where he resided, and furnished Blyth with skins and notes for some of his papers, one of the most important of which is a "Report on the Mammalia and more remarkable Species of Birds inhabiting Ceylon," published in the 'Journ. Asiat. Soc. Bengal' for 1851. In 1852 Dr. Kelaart published his 'Prodromus Faunæ Zeylanicæ' in Ceylon, chiefly noted for the outline account of the mammals and reptiles of the island, with which he was better acquainted than with its birds. For this work, Layard, as he writes in 'The Ibis' for the current year, supplied him with all his "lists and numerous specimens, not only of birds, but of many mammals and reptiles new to him; and it was arranged that we should bring out a second part of the 'Prodromus' (then in MS. only), which should consist of the Birds, to be written by me." It appears, however, that Kelaart broke faith with him, and issued his 'Prodromus' with the notice of the birds (Part II.) compiled by himself. Thus "left out in the cold," Layard, on his return to England, published the valuable notes referred to above. He also compiled a considerable portion of the notice of the birds of the island contained in Emerson Tennent's 'Natural History of Ceylon,' and furnished the author with voluminous notes, whilst his large collection supplied the materials for the "List of Birds" printed in the work. This was published in 1868, and besides describing the habits and instincts of the mammalia, birds, reptiles, fishes, and insects of the island, includes an interesting monograph of the elephant. During the interval between the last-mentioned date and the year 1854 scarcely any thing was published concerning the ornis of Ceylon, with the exception of a stray paper now and then contributed to the 'Journal of the Asiatic Society,' some of which emanated from the pen of Mr. Hugh Nevill, C.C.S., who recorded the occurrence of the Wood-Snipe in the Part for 1867, which was not published till 1870. At this time Mr. Holdsworth was devoting his attention to the ornithology of the island, and a co-worker, the author, who arrived on the island a year later (Oct. 1868), had likewise commenced to collect vigorously. Mr. Holdsworth, who landed in the island in September 1875, was sent out from England to study the habits of the Pearl-Oyster, and find out the cause of the failure of the Pearl-fisheries, with a view of advising the Government what should be done for their better management. His appointment necessitated his residence, off and on, at Aripu, which is adjacent to the Pearl-banks, and while there he devoted his spare time to a study of the birds in the vicinity of the station. He also collected at Colombo and at Nuwara Eliya during both monsoons. The outcome of his labours during seven years' residence in Ceylon was his "Catalogue of the Birds found in Ceylon," published in the 'Proceedings of the Zoological Society of London,' 1872, and by far the most complete treatise which had ever been compiled on the avifauna of the island. The author devoted particular attention to the synonymy of the birds, which, up to that time, was in a very confused state; and the result was the working-out of the correct title of each species, which constituted a most valuable addition to the literature of the Ceylonese ornis. The catalogue numbers 326 species.

In this list 24 species were added by the author, which are published under the following titles:—*Hypotriorchis severus, Picus æruginosus, Pandion haliaëtus, Buteo desertorum, Huhua pectoralis, Brachypternus puncticollis, Prionochilus vincens, Erythrosterna hyperythra, Arrenga blighi, Geocichla layardi, Zosterops ceylonensis, Estrelda amandava, Chrysocolaptes festivus, Francolinus pictus, Chettusia gregaria, Terekia cinerea, Tringa salina, Sterna leucoptera, Sterna gracilis, Phaethon rubricauda, Sula fiber, Taccocua leschenaulti, Drymoipus jerdoni, Gallinago nemoricola.* Of the above, *Zosterops ceylonensis* was not an additional species, but a new name for *Zosterops annulosus* included in Layard's list. The *Butalis muttui* of Layard's list appeared also under another title (*Alseonax terricolor*), though this identification afterwards proved to be erroneous. *Arrenga blighi* was a new species described by Mr. Holdsworth from specimens procured by Mr. Bligh and himself.

A few were omitted, which the author considered had been species wrongly identified, or had been recorded by Kelaart on doubtful evidence; these were—*Ephialtes scops, Malacocercus griseus, Cisticola homalura, Phyllopneuste montanus, Phyllornis aurifrons, Hetærornis malabaricus, H. cristatella, Picus macei, Alauda malabarica, Cuculus bartletti, Turtur humilis,* and *Branta rufina.* Among this number *Cuculus bartletti* appears to be the Small Cuckoo (*Cuculus polio-cephalus*, p. 231); and *Turtur humilis* seems to have been accidentally omitted, as no mention whatever is made of the species. The *Zosterops annulosus* of Layard's notes, as we have seen, was discriminated as a new species. Figures are given of *Arrenga blighi, Zosterops ceylonensis* and *Z. palpebrosus, Brachypteryx palliseri,* and *Erythrosterna hyperythra.*

Review of present Work.—The total number of species included in the present work is 371, of which two are introduced birds, viz. *Padda oryzivora* and *Estrelda amandava.* Of the remainder, 18 species, besides the two just mentioned, are treated of in footnote articles, or noticed in the Appendix; of these, *Falco chicquera, Accipiter nisus, Scops malabaricus* (App.), *Palæornis columboides* (App.), *Lanius lucionensis, Siphia nigrorufa, Munia rubronigra, Fuligula rufina, Phalacrocorax fuscicollis,* and *Fregata aquila* are considered as doubtfully occurring. The following, *Cotyle obsoleta, Oceanites oceanicus, Phaethon indicus, Coturnix communis,* are looked upon as doubtfully identified; two species, *Schœnicola platyura* and *Brachypternus intermedius* (red race of *B. puncticollis*), are doubtfully determined; *Alauda parkeri* is perhaps not a good species; whilst one bird (*Stercorarius antarcticus*) may have, perhaps, been conveyed to the island in the form of the single example of the species noticed. In addition to the above, *Fuligula ferina, Turnix sykesi,* and a species of *Anser* are referred to in "Notes" as likely to occur. The following 24 species have been added by the author to Mr. Holdsworth's list :—*Baza ceylonensis*, sp. n., *Scops minutus*, sp. n., *Glaucidium radiatum*, *Cuculus poliocephalus*, *Brachypternus intermedius*, ? *Schœnicola platyura*, *Locustella certhiola*, ? *Cotyle obsoleta*, *Prinia hodgsoni*, *Turtur tranquebarica*, *Coturnix communis*, *Machetes pugnax*, *Calidris arenaria*, *Ægialitis geoffroyi*, *Æ. jerdoni*, *Glareola orientalis*, *G. lactea*, *Tringa temmincki*, *Ciconia alba*, *Tadorna casarca*, *Sterna saundersi*, *Sterna fuliginosa*, *Anous stolidus*, *Sula cyanops*, *Phaethon flavirostris*. Four species have been renamed—*Spizaetus nipalensis*, *Pyctorhis sinensis*, *Prinia socialis*, and *Acridotheres tristis*, which appear in this work as *Spizaetus kelaarti*, *Pyctorhis nasalis*, *Prinia brevicauda* (App.), and *Acridotheres melanosternus*.

In the subjoined Table will be found all the species which are recognized in the work as peculiar to the island; among them are included two birds about which there are doubts as to their not being found in India. These are *Drymœca insularis* and *Brachypternus intermedius*, the former of which may perhaps be the same as a South-Indian Wren-Warbler (*D. inornatus*). The birds here tabulated are all figured, with the exception of *Prinia brevicauda* and *Turdus kinnisi*, the reasons for the omission of which will be found in the Appendices.

TABLE OF BIRDS PECULIAR TO CEYLON.

Families.	Number of species.	Name.	Hill-district (under 5000 feet).	Low country.	Nuwara-Eliya plateau and over 5000 feet.
Falconidæ	2	Spizaetus kelaarti .... Spizaetus ceylonensis	* *	.. *	* *
Bubonidæ	3	Athene castanonota .... Scops minutus................ Phodilus assimilis	* * *	* ? *	*
Psittacidæ	1	Palæornis calthropæ	*	!	
Trichoglossidæ	1	Loriculus indicus	*	*	
Picidæ	3	Chrysocolaptes stricklandi .... Brachypternus ceylonus .... Brachypternus intermedius(App.II.)	* * ..	* * *	*
Capitonidæ	3	Megalæma zeylanica .... Megalæma flavifrons .... Xantholæma rubricapilla	* * *	* * *	
Cuculidæ	2	Centropus chlororhynchus .... Phœnicophaës pyrrhocephalus	* *	* *	
Dicruridæ	2	Buchanga leucopygialis .... Dissemurus lophorhinus..........	* †	* *	
Corvidæ	1	Cissa ornata................	*	¶	*
Muscicapidæ	3	Stoparola sordida .... Alseonax muttui............... Hypothymis ceylonensis..........	* .. *	.. * *	

TABLE OF BIRDS PECULIAR TO CEYLON (*continued*).

Families.	Number of species.	Name.	Hill-district (under 5000 feet).	Low country.	Nuwara-Eliya plateau and over 5000 feet.
Turdidæ	4	Myiophoneus blighi	*	..	*
		Turdus kinnisi (App. II.)	*	..	*
		Turdus spiloptera	*	*	*
		Oreocincla imbricata	*	..	*
Brachypodidæ	2	Rubigula melanictera	*	*	†
		Kelaartia penicillata	*	..	*
Timaliidæ	10	Malacocercus rufescens	*	*	*
		Garrulax cinereifrons	*	‡	..
		Pomatorhinus melanurus	*	*	*
		Alcippe nigrifrons	*	*	‡
		Pellorneum fuscicapillum	*	*	†
		Pyctorhis nasalis	*	*	
		Prinia brevicauda	*	*	
		Elaphrornis pallisori	†	..	*
		Drymœca valida	*	*	
		?Drymœca insularis	..	*	
Dicæidæ	2	Pachyglossa vincens	*	*	
		Zosterops ceylonensis	*	..	*
Hirundinidæ	1	Hirundo hyperythra	*	*	§
Ploceidæ	1	Munia kelaarti	*	*	
Sturnidæ	3	Acridotheres melanosternus	..	‖	
		Eulabes ptilogenys	*	¶	?*
		Sturnornis senex	*	‖	
Columbidæ	1	Palumbus torringtoniæ	*	*	
Phasianidæ	2	Gallus lafayettii	*	*	*
		Galloperdix bicalcarata	*	*	*

† Not common. ‡ Certain forests of Western Province in N.E. monsoon.
§ Occasional. ‖ Spreading into the forests at the base of the hills, particularly in the W. Province.
¶ In the forests of the Passodun Korale, down to 600 feet near Moropitiya.

It will be seen that this Table comprises 47 species. One *peculiar genus* (*Elaphrornis*) inhabits the island, its nearest ally being the Malayan and Himalayan *Brachypteryx*; and a subgenus (*Sturnornis*) is likewise recognized.

Affinities of the Ceylonese Avifauna.—We now come to the important point of the relationship of the Ceylonese ornis to that of adjacent regions; and this, as might be expected from the geographical position of the island and its separation from the mainland merely by a

c

shallow strait, is closer to that of South India than to the avifauna of any other part of the peninsula. Wallace, in his great work on the Distribution of Animals, considers the island of Ceylon and the entire south of India as far north as the Deccan as forming a subdivision of the great " Oriental Region." It is, however, in the hills of the two districts, which possess the important element of a similar rainfall, where we find the nearest affinities both as regards birds and mammals ; and this is exemplified by the fact of some of the members of the Brachypodidæ and Turdidæ (families well represented in both districts) being the same in the Nilghiris and the mountains of Ceylon, while many of the Timaliidæ and Turdidæ in one region have near allies in the other. For example, *Malacocercus (Layardia) rufescens, Pomatorhinus melanurus, Alcippe nigrifrons, Garrulax cinereifrons, Myiophoneus blighi, Oreocincla imbricata, Turdus kinnisi,* and *Palumbus torringtoniæ* in Ceylon are respectively represented in the hills of South India by *Layardia subrufa, Pomatorhinus horsfieldi, Alcippe atriceps, Garrulax delesserti, Myiophoneus horsfieldi, Oreocincla nilgherriensis, Turdus simillima,* and *Palumbus elphinstonii.*

But though this strong similarity in the avifauna of the mountains in question, as well as their geological characters, indicate a contemporaneous upheaval and enrichment with animal life of their surfaces, a similar connexion is found between the northern parts of the island and the low country of the Carnatic. Here, again, we have in the fossiliferous limestones of the two regions an undoubted connexion, and also an affinity in their avifauna, which differs totally from the mountain-districts on either side of the straits. The northern parts of Ceylon, as well as the south-eastern, both of which I shall speak of in my remarks on the geographical features of the island, may be considered to constitute an Indo-Ceylonese subregion, and are inhabited by the same species as the south-east coast-districts of the peninsula. *Brachypternus puncticollis, Anthracoceros coronatus, Malacocercus striatus, Pycnonotus hæmorrhous, Merops viridis, Pyrrhulauda grisea, Mirafra affinis, Turtur risorius, Buchanga atra,* and perhaps *Cursorius coromandelicus* are species characteristic of the north of Ceylon and of Ramisserum Island and the plains of Tanjore, but which are not inhabitants of the damp Malabar district. On the other hand it is noteworthy that *Gallus sonnerati* and the Lesser Florrikin (*Sypheotides aurita*), common in the Carnatic, have not yet been detected in North Ceylon. It is by way of the low-lying country of the Carnatic (the fauna of which, it may be remarked in passing, is allied to that of Central India) that the cool-season migrants enter the island of Ceylon, leaving numbers of their fellows in Southern India; and this forms an additional ornithological bond between the two districts. Some of these migrants come from the regions at the foot of the Himalayas, and tend to the supposition that there is a Himalayan element in the avifauna of Ceylon; but this is but very slight, if, indeed, it should at all be recognized, for migratory species, such as *Scolopax rusticula* and *Gallinago nemoricola* (which only inhabit the upper ranges and the high mountains of Southern India, and whose *locale* depends solely on climate), cannot be taken into consideration. One genus (*Pachyglossa*) certainly does constitute a bond of affinity. The distinctness of the avifauna of the Southern-Indian and Ceylonese mountains from that of the Himalayas may be shown by the fact that most of the Himalayan typical Timaline genera, *Suthora, Stachyris, Trochalopteron,*

Actinodura, are wholly absent from Ceylon, and but poorly represented in the hills of South India, there being only three species of the numerous genus *Trochalopteron* in the Nilghiris and Palani hills and not any of the others. Again, there is only one species of *Garrulax* in South India and one in Ceylon. Of the widely spread genus *Pomatorhinus*, found in the Himalayas, Burmah, and Java, there is only one species in each of the southern hill-regions in question. The genus *Alcippe* is about equally represented in both regions. These data show that though there is a connexion between the ornis of the Himalayas and that of Ceylon it is but slight, and only what one would expect in mountain-districts of adjacent ornithological regions. It is noteworthy that the Liotrichidæ, or Hill-Tits (one of the three peculiar families of the Oriental Region, and which are abundant in the Himalayas), are absent from Ceylon.

Certain Indian families are entirely absent from Ceylon, either as residents or migrants; they are the Eurylaimidæ (Broadbills)—a Himalayan and Malayan form,—the Pteroclidæ (Sand-Grouse), the Otididæ (Bustards), Gruidæ (Cranes), and Mergidæ (Mergansers). Among these families it is remarkable that some member of the Gruidæ has not yet been found in the cool season in North Ceylon; for, though the country is not thoroughly suited to their habits, the members of this family being migratory (and one of them, the Demoiselle Crane, extending to South India), it is singular that they do not extend their migration a little further south and reach the shores of Ceylon. I have heard a vague rumour of a Crane being seen near Mullaittivu; and it is not wholly improbable that the above-mentioned species (*Anthropoides virgo*) will some day be added to the occasional migrants during the N.E. monsoon. Another family, Vulturidæ, has a place in the Ceylonese avifauna, owing to a straggler having recently appeared in the island. Here, again, is an instance of species which, one would think, ought to occur as visitants in the N.E. monsoon; for I am informed that Vultures are not unfrequently seen in the Tanjore district; and *Gyps indicus* breeds in the Nilghiris.

Besides the widely distributed Grallatorial and Natatorial forms common to both India and Ceylon, certain Indian genera of western distribution are represented in the island. They are *Cuculus, Ceryle, Halcyon, Cypselus, Caprimulgus, Corone, Lanius, Turdus, Phylloscopus, Cinnyris, Hirundo, Motacilla, Corydalla, Turtur, Francolinus.* Of these the Cuckoos are remarkably numerous.

If we turn now towards the Malayan region we find, in spite of its more remote geographical position, quite as close an affinity as with the Himalayas—which may perhaps be accounted for on the theory held by some that there was at one time a connexion between the two regions. It may, however, be remarked, in passing, that if this did occur it must have been, in all probability, by way of the Andamans and Malacca, as we find the 15,000 feet contour of ocean-depth passes up near the east coast of the island into the Bay of Bengal to lat. 10° N. This Malayan affinity is shown in the existence in Ceylon of a Malayan form, *Phœnicophaës*, and a member of a typical Sunda-Island genus, *Myiophoneus*. It is also worthy of note that the island is visited by a Malaccan emigrant, *Gorsachius*, which has rarely been met with in India. This is remarkable, as, in all probability, before the submergence took place which altered the Malayan region, the hills of South India were just as much connected with Malacca as those of Ceylon. A closely allied Swallow to our "peculiar" *Hirundo hyperythra* is found in Malacca; and Malayan genera of Pigeons (*Carpophaga, Osmotreron,* and *Chalcophaps*) are also found in Ceylon, and perhaps to a greater extent, when we look at its small geographical area, than in India. Certain Australian and Malayan birds, such as *Haliaetus leucogaster, Coturnix chinensis* (found also in China), *Mycteria australis,* extend into Ceylon, not to mention the Waders (*Limicolœ*), which range from Asia thence to the Australian continent, taking in Ceylon in their path.

The island, however, is not dependent on these latter for its migratorial Waders, in which, as also in some water-birds (Anatidœ), it is very rich. It forms, in fact, the southernmost Asiatic limit of the flight of many European and Asiatic Grallatorial and Natatorial forms; and hence the large numbers of these birds which are found in the cool season along its shores. Of these the following species are noteworthy:—*Scolopax rusticula, Gallinago nemoricola, Machetes pugnax, Tringa minuta, Totanus ochropus, Totanus fuscus, Tringa minuta, Limosa œgocephala, Himantopus candidus, Recurvirostra avocetta, Œdicnemus scolopax, Hæmatopus ostralegus, Anas acuta, Anas circia, Anas crecca,* and *Phœnicopterus roseus.*

Geographical Features and Inland Distribution.—Having now considered the important question of the affinities of the Ceylonese avifauna it is necessary to notice the geographical features of the island as bearing upon the inland distribution of the birds inhabiting it. Ceylon is an island of about 270 miles in length and 138 in breadth, lying between lat. 5° 50' and 9° 50' N., and between long. 79° 40' and 81° 50' E.; it is separated from the mainland of India by a shallow strait 35 miles wide, which is traversed by a chain of islands, between which lies a long sandy shoal called Adam's Bridge, which is alternately raised and lowered on the north and south by the action of N.E. and S.W. monsoons. For ornithological purposes the island may be divided into four regions or districts—the dry forests of the entire north and south-east, the arid maritime belts of the north-west and south-east coasts, the damp Western-Province region, and the hill-zones of the Central and Southern Provinces. The northern part of the island consists of a vast plain covered with forest, except near the sea, where, particularly on the north-west coast, there are open tracts studded with low thorny jungle. This region is called in the present work the "northern forest-tract," and is here and there studded with very rocky abrupt hills, rising suddenly out of the forest-clad plain. Sigiri, Rittagalla, and Mahintale rock are some of the most notable among these acclivities. This region, which lies to the north of the high land intercepting the moisture brought up from the ocean by the S.W. and N.E. monsoons, is alternately swept by a dry westerly and easterly wind, and is covered with tolerably luxuriant forest and wild secondary jungle, inhabited chiefly by members of those Indian families which are most strongly represented in the island, the Flycatchers, Drongos, Barbets, Bulbuls, Babblers (Timaliidœ), and Cuckoos, but also contains many of the forest-loving "peculiar" forms, which have their stronghold further south. The northern forest-tract likewise is the home of many of the larger

Water-birds and " Waders" which affect the numerous tanks* in the heart of the jungle. The most luxuriant vegetation in this part of the island is to be found on the banks of the rivers, where the Koombook (*Terminalia glabra*) is one of the most characteristic trees. In the drier parts the forest is sprinkled plentifully with the iron-wood (*Mimusops indica*), the fruit of which is the favourite food of many birds. The open scrubby belt of land bordering the N.W. coast, as also the island of Mannar and parts of the peninsula of Jaffna, are characterized by a very different flora. Here almost every tree is of a thorny nature, and the low and almost impenetrable masses of brushwood are filled with Euphorbia-trees (*Euphorbia antiquorum*), which is the characteristic plant of the district. This region is the home of plain-loving birds, such as *Pyrrhulauda grisea*, *Merops viridis*, *Munia malabarica*, and is the almost exclusive habitat of *Buchanga atra*, *Lanius caniceps*, *Turtur risorius*, *Ortygornis pondiceriana*, and *Cursorius coromandelicus*, which appear to have extended their range from the Carnatic hither and not passed beyond the forests which hem in the district. Here, too, is the great haunt of the migratory Waders, which swarm on the muddy flats between Jaffna and Mannar, and also congregate round the salt lagoons of the N.E. coast. These latter are surrounded with heavy jungle, inhabited by the same birds as further inland, but which stands back at some distance from their grass-begirt shores.

Southward of the region just considered we have on the west coast the damp, luxuriant, typically Ceylonese region, cultivated with rice in some parts and in others clothed with tall forest, of which the characteristic trees are the gigantic Hora (*Dipterocarpus zeylonensis*), the Doon (*Doona affinis* and *Doona congestiflora*), the stately Keena (*Calycophyllum tomentosum*), and the lofty Dawata (*Carallia integerrima*). This tract, which comprises the Western Province and "South-western Hill-district," is intersected with ranges and groups of hills heavily timbered in some parts

---

* Many of these large irrigation-works claim a place among the most gigantic monuments of ancient enterprise and labour; they literally astonish the traveller and fill his mind with wonder as he stands on the vast bunds and looks down on the wild and lonely scene, pondering on the means and appliances which the engineers of those distant times must have used to get the great stones in their places. Whole valleys have been dammed up, and sometimes the strong floods of three rivers thrown back and spread out into a great lake, the waters of which must have irrigated thousands of square miles. The bund of the great Padewiya tank extends for 11 miles across a valley, and in olden times, before this enormous embankment was broken down by the rush of mighty floods, the water was, as Emerson Tennent tells us, thrown back for 15 miles along the valley. I regret to say I never visited this tank; but I have seen other bunds of great size, of which perhaps that which holds back the waters of Kanthelai tank is the finest. This tank, which has been lately restored, was built by King Maha Sin, A.D. 275; and the following details kindly furnished me by Mr. E. Scott Barber, C.E., who repaired the bund, may not be uninteresting to my readers:—" When up to 'spill-level' (22 feet). the tank contains 3580 acres, and is 17 miles round. The bund is 60 feet high and 200 feet in width at the bottom; it is 6800 feet in length, and contains 10,121,296 cubic yards of material. It is 'pitched' with large boulders from bottom to 60 feet up the slope and from 3 to 4 courses deep. The outlet was by two culverts 4 feet by 2 feet, situated at either end of the bund; the stones forming them average 1½ to 2 tons in weight, and are 'tongued' together in the centre." The top of this mighty embankment was about 60 yards wide and covered with jungle and large trees. As it was, it gave one the impression, when walking along it, of standing on a natural ridge or long low hill!

and covered with bamboo-cheena in others; the valleys, constantly rained on during the south-west monsoon, and likewise receiving a heavy downfall in the north-east monsoon, are the dampest spots in the island, and harbour numbers of Timaliidæ (*Malacocercus rufescens, Garrulax cinerei-frons, Alcippe nigrifrons, Pellorneum fuscicapillum*), also Brachypodidæ (*Hypsipetes ganeesa, Criniger ictericus, Rubigula melanictera*). The cultivated districts are conspicuous for the numbers of the common Bulbuls, Barbets, Doves (*Turtur suratensis*), smaller Timaliidæ (*Cisticola, Prinia, Drymæca,* &c.), as well as some numbers of the Heron family, which are seen about the paddy-fields. A considerable portion of the uncultivated soil in the Western Province and also in the lower hills is overgrown with a dense bramble (*Lantana mixta*), popularly known as "Lady Horton's wood," and which was introduced (unfortunately) into the island about the year 1830. It thrives on gravelly soil, and especially on land which has once been cultivated, sometimes clothing more than an acre without a single break. The fruit of this pest is eagerly sought after by many birds, particularly Bulbuls (*Rubigula, Pycnonotus, Ixos*); and to this fact the wonderful manner in which it has been propagated is due. The damp, heavy forests of the Adam's-Peak range descend continuously into the low country of Saffragam, and through them several true hill species (*Eulabes ptilogenys, Palæornis calthropæ, Garrulax cinereifrons*) range to a lower level than anywhere else, being quite common in portions of the Kuruwite and Three Korales.

We now come to the consideration of the fourth ornithological district, the lofty hills of the Southern Province, rising up on the north of the valley of Saffragam, of which Ratnapura is the chief town. The first-named region is entirely occupied by a group of high mountains and elevated valleys, forming a perfect mountain-zone, inside of the base of which there is scarcely any land of less elevation than 1500 or 1700 feet. This lofty district culminates in the high Pedrotallagala range (8200 feet), just on the north of the plain of Nuwara Eliya, from which extends an elevated plateau, intersected by forest-clad ridges, and dotted here and there with the curious natural fields called patnas, for some 20 miles south to the Horton plains (7000 feet), whence the lofty Haputale range stretches to the east and the Adam's-Peak range round to the west as far north as the Four Korales, the slopes of both dropping at once into the low country. The coffee-districts of Dimbula and Dickoya are enclosed by the latter on the east of the Nuwara-Eliya plateau, each with its dividing range; while the Uva patna-basin (a curious tract of grass-covered or patna-hills) forms its eastern flank, and slopes out into the Bintenne country through the valley of Badulla, being bounded on the extreme east by the lofty ridges of Madulsima. On the north of the Pedro mountain high ranges jut out towards the upland valley of Dumbara, beyond which the Knuckles and Ambokka ranges, running on each side to the north-west and north respectively, complete the Kandyan mountain-system. The southern hill-ranges bound the south side of Saffragam, and are comprised of the Kukkul, Morowak, and Kolonna Korales, the highest point being Gongalla, a little over 4400 feet in altitude. Of late years the forest has been felled for the planting of coffee, as in the Central Province; but there are still large tracts of forest in the Kukkul Korale in which Central-Province birds (*Cissa ornata, Eulabes ptilogenys, Sturnornis senex,*

Palæornis calthropæ, Zosterops ceylonensis, Culicicapa ceylonensis) abound, and in which both *Gallus lafayettii* and *Galloperdix bicalcarata* are plentiful. The northern portion of this korale, lying between the Karawita hills and the hilly forests of the Passedun Korale, consists partly of semicultivated land and partly of a curious and little-known tract of open grassy hills with wood-dotted dingles, resembling the patnas of the Kandyan country, and on the open parts of which Grass-Warblers, Wren-Warblers, and Munias are common, while Babblers (*Pomatorhinus*) are found in the groves; but otherwise an absence of bird-life is decidedly noticeable.

It is in the coffee-districts and valleys lying beneath the estates which are dotted with patna-grasses, particularly "Maana-grass" (*Andropogon martini*), and patched here and there with groves of luxuriant trees lining the courses of the streams, where the hill-species, both "peculiar" and Indian, intermingled with not a few low-country forms, abound; but it is also in these spots where the original ornithological features of the country are being gradually changed by the disappearance before the woodman's axe of such a vast area of forest, and species such as *Palumbus torringtoniæ, Merula kinnisi, Eulabes ptilogenys, Stoparola sordida*, and *Culicicapa ceylonensis* (true hill-species) are being driven into the upper forests, or are locating themselves to a considerable extent about the open estates where once their forest-home stood.

In the upper forests and in the Nuwara-Eliya plateau we lose the stately trees of the genera *Doona, Dipterocarpus*, &c., and find stunted, though thick-trunked, arboreal forms, for the most part profusely clothed with handsome mosses; and these woods, with their circumscribed patnas, are the favourite haunts of the peculiar birds enumerated in my table, as well as many Indian species, both permanent and migratory. Of the former may be mentioned *Merula kinnisi, Culicicapa ceylonensis, Parus atriceps, Cisticola schœnicola, Pericrocotus flammeus, Pericrocotus peregrinus, Hypsipetes ganeesa, Pratincola bicolor, Orthotomus sutorius, Corydalla rufula*; of the latter, *Turdus wardi, Erythrosterna hyperythra, Larvivora brunnea, Hierococcyx varius, Phylloscopus nitidus, Phylloscopus magnirostris* are noticeable.

The eastern subdivision of Southern Ceylon, which is shut off from the influence of the south-west monsoon by the eastern slopes of the Kolonna and Morowak-Korale mountains and their spurs, which run south towards Matara, presents one of the most remarkable instances of a sudden change in physical aspect and floral character that can, perhaps, anywhere be met with in such a small island. Possessing a totally different climate, and consequently a distinct flora, the avifauna of this region has little relation to that of the damp south-western division. The birds of the vast forest which stretches southwards from the Haputale mountains to the confines of the scrubby maritime district are the same as those of the northern forests; and the ornis of the coast-region is precisely the same as that of the north-west coast, except that it includes several species, such as *Prinia hodgsoni, Taccocua leschenaulti*, and *Pyctorhis nasalis*, which seem to have their head-quarters here, and are not found (in such abundance, at any rate) in that part. Characteristic species of the two regions are *Xantholæma hæmacephala, Pyrrhulauda grisea, Merops viridis, Picus mahrattensis, Upupa ceylonensis*, and *Cittocincla macrura*, none of which, with the exception of the latter bird, are found in the adjoining damp district. The numerous shallow salt lagoons and leways are the resort of Waders, Terns, Herons, Flamingoes, and Water-birds, all of which are characteristic of the north-west of the island.   The north-eastern part of the sub-division in question is called the Park country, the borders only of which, I much regret to say, are known to me. This tract consists of open glades and small plains covered with long grass and surrounded by heavy jungle, in which there are numbers of birds, the prevalence of Woodpeckers being noticeable. As regards the open country, it is not unlikely that some new Timaline species may be found in it.

Lastly, with regard to the great families of Scolopacidæ and Charadriidæ, which form such a large proportion of the Ceylonese ornis, and which migrate to the island in vast numbers at the commencement (October and November) of the cool season, as will be seen on a perusal of this work, their great haunts are the lagoons, tidal flats, marshes, and tanks near the coast along the northern shores of both sides of the island.   On the west coast these cease to the southward of Negombo, and the sea-board is only interseeted with deep mangrove-lined lagoons and lakes, which are quite destitute of "Wader"-life, save that of one or two species, as the ubiquitous *Tringoides hypoleucos* and the very abundant *Totanus glareola*. The entire east coast, however, is more or less inhabited by Sandpipers, Stints, Shore-Plovers, and other members of these families. From the Virgel down to Battical on the sea-board is not so favourably suited to their habits as further south, where they again become very abundant, and occupy the coast-line, with its numerous estuaries, leways, and lagoons, down to Hatagala.   Nowhere, however, do these interesting birds muster in such force as from the Jaffna peninsula, with its inland salt lagoon and large salt lake, down the west coast to the immense tidal flats at the embouchure of the Manaar channel. The entire coast of this region is shallow, the tide receding some distance, and leaving exposed an oozy shore, covered in places with green weed. On these flats myriads of small Waders congregate, and species (such as the Turnstone and that anomalous bird the Crab-Plover) which are not plentiful on the east coast are here found in abundance.   In this district are of course included the islands of Palk's Straits, on which these birds are likewise equally abundant.

Monsoons and Seasons.—There are, roughly speaking, two seasons in Ceylon, which are ushered in by the advent of two monsoons, the south-west and north-east.   The former com-mences to blow in April, after the termination of the hottest time of the year, the sultry weather of March.   For about a fortnight violent squalls, accompanied by downpours of rain, drive in from the sea on the west coast; and along the western slopes of the mountain-ranges, where the moisture resulting from this wind collects, the rain is just as heavy and more continuous.   This weather, which is called the "little monsoon," is, though unpleasant, preferable to that which preceded, when there was an absence of wind and the nights were very sultry.   It is the signal for the commencement of the spring migration. Inscssorial birds (Warblers &c.) immediately move northwards, and the Waders, which throng the salt lagoons and estuaries on the northern and eastern coasts, commence their long flight towards northern regions.   After the cessation of the little monsoon there is a lull, when the weather is again unpleasantly hot and "steamy," until the end of May, when the south-west wind again blows with greater violence than before, for in some years the "little monsoon" is not by any means strong. The rain at this period is also much more continuous, and sometimes very heavy downfalls are experienced, as in 1876, when 11 inches fell at Colombo in twenty-four hours. At this time of the year perfectly different weather is experienced on the east coast, when the same south-west wind, deprived of its moisture by its passage over great tracts of forest, has become intensely dry and almost warm. After the burst of the monsoon is over the wind gradually lessens throughout the months of July, August, September, and beginning of October, when the weather again becomes sultry. The great autumn migration is now setting in: myriads of Sandpipers, Stints, and shore-birds in general are now travelling southward from Northern Asia, and some species, as the Pintailed Snipe and the Golden Plover, arrive on the north coast, and even reach the south-western district (Galle) as early as the middle of September; at the same time Warblers and Wagtails arrive in the island and rapidly spread over the country.

About the middle of October, and sometimes as early as the first week in that month, the first signs of the N.E. monsoon may be looked for on the east coast. Heavy thunderstorms coming from the land every afternoon betoken the breaking up of the S.W. monsoon; they continue for about a fortnight, and then the wind, with rain, sets in from the north-east; at the same time on the west coast heavy thunderstorms are experienced every evening, which, in the same manner as those which preceded them on the east coast, take place later each consecutive evening until they cease. During this time migrants from India continue to arrive, and a local movement of birds towards the west coast takes place. The north-east wind, which is not so strong as the south-west, reaches the west coast only in the form of a land-breeze at night, which is scarcely felt until about Christmas. In the meantime, at the end of November, a strong northerly breeze sets in down the west coast; this is locally styled the "long-shore wind," and is mainly conducive in adding to the ranks of migrants of all classes, but particularly to those of the Grallatorial order. Snipe now come in great numbers, and by the middle of December large bags may be made in almost any good district.

Internal Migrations.—It is natural that the prevalence of two winds blowing at different seasons from opposite quarters across the island should cause a movement of species inhabiting the coast districts on each side of it. This is most observable on the coast of the Western Province, south of Negombo, as here the wind is damp, and, as we have just seen, accompanied by heavy rains, which induce certain species to leave the sea-board and retire inland in order to obtain shelter from the force of the monsoon. It would appear to any one studying the avifauna of a coast-district, like that of Colombo for example, that all these birds had left that side of the island; but this is not the case, as they are mostly to be found after the rains of June in the sheltered districts of the interior, not far from the coast. On the other hand, however, various species which are not resident on the west coast visit it when the S.W. monsoon has died away and the N.E. monsoon has commenced to blow on the east coast, tending to carry them towards

d

the south-west. Instances of such birds are to be found in the Paradise Flycatcher (*Terpsiphone paradisi*) and the Indian Sky-Lark (*Alauda gulgula*), which latter bird is found during the south-west monsoon in numbers at the tank-meadows in the northern forests, while the former (in the red stage) inhabits both the northern and south-eastern forest-tracts. Species that move away from the *immediate* western sea-board are *Dendrochelidon coronata, Eudynamys honorata, Thamnobia fulicata, Tephrodornis pondicerianus*, and *Parus cinereus* ; but a few miles inland, in sheltered spots, these birds may be found all the year through, except perhaps the latter, which must be classed as an uncertain N.E. monsoon visitant to the maritime districts of the Western Province.

In the mountains the movements of the hill species are very noticeable in those districts west of Nuwara Eliya which are exposed to the violent winds and rain which accompany the incoming of the monsoon in May. The Hill-Myna (*Eulabes ptilogenys*), the Blue Tit (*Parus atriceps*), the handsome Torrington Wood-Pigeon (*Palumbus torringtoniæ*), the large Bulbul (*Hypsipetes ganeesa*), the Orange Minivet (*Pericrocotus flammeus*), the Jay (*Cissa ornata*), the Hill-Barbet (*Megalæma flavifrons*), the Jungle-fowl (*Gallus lafayettii*), and the Spur-fowl (*Galloperdix bicalcarata*) are among the more prominent species which appear in the upper ranges (from 5000 to 8000 feet) as soon as the calm weather of the N.E. monsoon has set in in November.

At this season of the year also low-country birds, which, as a rule, only range into the hill-zones to an inconsiderable elevation, ascend to the upper hills. *Artamus fuscus, Oriolus melanocephalus, Upupa ceylonensis, Pycnonotus hæmorrhous, Layardia rufescens, Terpsiphone paradisi*, and *Hypothymis ceylonensis* are species which either occasionally ascend to altitudes above 5000 feet, or are found yearly in the upper zone during the N.E. monsoon.

True Migrants.—The arrival of the migratory species, which takes place, as already mentioned, at the termination of the S.W. monsoon, gently adds to the avifauna of the island. The Insessorial migrants consist chiefly of Muscicapidæ, Laniidæ, Motacillidæ, and Sylviidæ, while the Grallatorial are made up of Scolopacidæ and Charadriidæ. The members of the first-mentioned order are wholly migratory ; but certain species of the two latter remain to some extent as non-breeding loiterers throughout the year. The following is a table of migrants :—

ACCIPITRES.	PICARIÆ.	INSESSORES.
Baza lophotes.	*Cuculus canorus.	Oriolus indicus.
Falco peregrinus.	Cuculus micropterus.	Lanius cristatus.
Cerchneis amurensis.	Cuculus poliocephalus.	Buchanga longicaudata.
Circus æruginosus.	Cuculus passerinus.	Alseonax latirostris.
Circus cinerascus.	*Cuculus maculatus.	†Siphia ruboculoides.
Circus melanoleucus.	Hierococcyx varius.	Muscicapa hyperythra.
	Coccystes coromandus.	*Cyanecula suecica.
	Merops philippinus.	Larvivora brunnea.

Turdus wardi.
Geocichla citrina.
*Mouticola cyana.
*Sylvia affinis.
Acrocephalus dumetorum.
Locustella certhiola.
Phylloscopus nitidus.
Phylloscopus magnirostris.
Phylloscopus viridanus.
Hirundo rustica.
*Hirundo erythropygia.
*Motacilla maderaspatensis.
Motacilla melanope.
Budytes viridis.
Corydalla richardi.
Corydalla striolata.
†Pitta coronata.

COLUMBÆ.

*Turtur pulchratus.

GRALLÆ.

*Portana bailloni.
Portana fusca.
Rallina euryzonoides.
? Hypotænidia striata.
*Rallus indicus.
*Scolopax rusticula.
*Gallinago scolopacina.
Gallinago stenura.
*Gallinago gallinula.
‡Limosa ægocephala.
‡Terekia cinerea.
†Totanus glottis.
†Totanus stagnatilis.
‡Totanus fuscus.
Totanus calidris.
Totanus glareola.
Totanus ochropus.
†Tringoides hypoleucus.
*Machetes pugnax.
†Tringa subarquata.
†Tringa minuta.
†Tringa subminuta.

*Tringa tommincki.
‡Limicola platyrhyncha.
*Calidris arenaria.
§Strepsilas interpres.
Numenius lineatus.
Numenius phæopus.
‡Recurvirostra avocetta.
Squatarola helvetica.
Charadrius fulvus.
†Ægialitis geoffroyi.
†Ægialitis mongolica.
*Chettusia gregaria.
Hæmatopus ostralegus.
†Sterna caspia.
Larus brunneicephalus.
Tadorna casarca.
Anas acuta.
Anas circia.
Anas crecca.
Spatula clypeata.
?§ Phœnicopterus roseus.
*Ardea goliath.
Gorsachius melanolophus.

* Rare stragglers to the island in N.E. monsoon, or irregular migrants in small numbers.
† Migratory for the most part, non-breeding birds remaining throughout the year.
‡ Possibly a regular migrant in small numbers.
§ Rarely a loiterer in Ceylon in S.W. monsoon.

In this list the families Cuculidæ and Sylviidæ muster strongest among land-birds, but do not, it will be observed, furnish as many representatives as the Grallæ (Waders). Among the latter it is noteworthy how many species "loiter" or remain behind in the breeding-season. A knowledge of this fact is all the more interesting, as, until very recently, it was not known that members of the Gralline order, such as *Totanus, Tringa*, and *Ægialitis*, ever remained in the tropics throughout the year; now, however, the researches of Mr. Hume in the Andamans, and of myself in Ceylon, have fully proved this to be the case. Stragglers to Ceylon at uncertain times of the year have not been included in the list, as they cannot be looked upon in any way as migrants. Among these may be mentioned *Neophron ginginianus, Nisaetus pennatus, N. bonelli, Baza ceylonensis, Buteo desertorum, Pastor roseus, Alsocomus puniceus, Sterna dougalli, Anous stolidus, Sula leucogastra, S. cyanops, Stercorarius antarcticus, Phaethon flavirostris, P. indicus,* and *Fregata minor.* Of these, *Pastor roseus* and *Sterna dougalli* are the only species which, when they do visit the island, appear in numbers.

Breeding-season.—The majority of Ceylon birds breed during the first half of the year, the exact times varying according to locality and climate. In the Western Province the height of the breeding-season is, as in India, during the rains of April, May, and June. At this time the

d 2

jungles teem with insect-life, and all forest-birds are busy rearing their young. In very moist districts, such as Ratnapura and the Passedun Korale, eggs may be found in August and even September. Among early breeders in the Western Province may be cited the Barbets and Wood-peckers. On the eastern side of the island many birds commence to breed in November and December, while the heavy rains are falling ; but the season continues, nevertheless, throughout the first three or four months of the year, and many birds may be found nesting, as on the western side, in May and June. In the hills, and more particularly in the upper ranges, where the nights are cold and frosty in January and February, the nesting-season commences at the end of March or beginning of April, and continues until June and July, corresponding in this respect with the breeding-time in temperate climates. In the north of Ceylon the larger Waders (Ardeidæ), and the Water-birds that breed with them, commence to nest in November; but on the south-east coast the season is later, the Heronries not being resorted to as a rule, I think, before January.

Remarks on the plan of the Work. 1st. *Classification.*—The classification followed in this work is totally different from that used by Jerdon, principally taken from Gray, and which continues still in vogue among some Indian ornithologists. This is, I must confess, inconvenient for Indian field-naturalists and collectors ; but as, in my opinion, it was not possible to follow the above-mentioned system, and as the main object of this work is to endeavour to inculcate a taste for ornithology among *local students* of the science in Ceylon, it behoved me to adopt that system which appeared to me to accord best with the generally recognized affinities of the various orders into which the Ceylonese ornis divides itself, and at the *same time coincided best with the classification employed by Jerdon*, and which I am aware many who have taken up the study of ornithology in Ceylon are familiar with. The divisions adopted have been Orders (in one case also a Suborder), Families, and Subfamilies, and, in the great Order Passeres, Sections have also been made use of. The Accipitres, or Birds of Prey, have been granted precedence simply as a very favourite and specialized order, and because it has until recently been the practice among English ornithologists to follow Gray and place them first. The Psittaci, or Parrots in the possession of a cere and a very high degree of intelligence, seem to occupy a place not far distant from the Hawks. The interesting order Picariæ, in which the posterior margin of the sternum has a double notch, inasmuch as many of its groups possess zygodactyle feet, comes next the Parrots. The satisfactory arrangement of the vast order Passeres presents great difficulties ; and here the system adopted by Mr. Wallace in classifying according to wing-structure has been adopted. The Columbæ (Pigeons) are a highly specialized order, and in preceding the Gallinæ, or Game-birds (aptly called Rasores, or "Scratchers," by some systematists), must of necessity come next the Passeres. In the arrangement of the remaining orders in the work (Grallæ, Gaviæ, Anseres, Pygopodes, Herodiones, and Steganopodes) I have followed the bent of my own views on the subject, considering these six orders as naturally divisible into two great classes—1st, those with *autophagous* or independent young ; 2nd, those with *heterophagous* or dependent young. It is impossible to follow a linear arrangement ; but nevertheless there *are* forms in each of the orders composing these two divisions which possess affinities for one another, and consequently tend to group them in the rotation which they take in this work. The same rule has been followed, as much as possible, in considering the order in which the various families composing these orders should be arranged. It will not be necessary to enter into any disquisition in this Introduction on the much-disputed subject of classification, or to explain further my reasons for not following the more modern systems of Professors Parker and Huxley, or, still better, the modification of these systems by Messrs. Sclater and Salvin, as they have been sufficiently set forward in testifying above my desire to adopt a system best suited to the requirements of the local student, at the same time avoiding a total reversal of Gray's classification.

2nd. *Plan of the Articles.*—It has been thought best to define the characters of the various orders, families, subfamilies, genera, and species in accordance with their external characteristics, in order to simplify their comprehension to beginners. Reference is, however, made frequently to the sternum, a generally important, though not in some families (Scolopacidæ, for instance) always a reliable character.

The accompanying woodcut represents the sternum of the Malay Bittern (*Gorsachius melanolophus*), together with the bones attached to it. It has been selected as an example of a sternum with a single notch in the posterior margin. The various parts are named beneath.

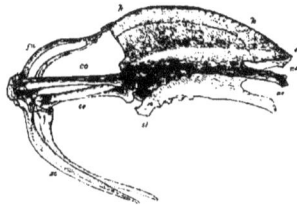

st, sternum ; *k*, keel of sternum ; *no*, notch in posterior margin ; *fu*, furculum ; *co*, coracoid bones ; *sc*, scapula.

In the great division Carinatæ, which comprises all living birds but the Ostrich family and its allies, the "carina" or keel is more or less deep so as to hold the powerful pectoral muscles which lie in the angle between it and the body of the sternum. In the latter (Ratitæ), however, the keel is slightly developed only, the sternum being flat, inasmuch as the same development of muscle is not required for non-flying birds. The furculum is in most birds a single bone, but in some Parrots, Pigeons, and Owls consists of two separate clavicles. In some genera of the Steganopodes it is anchylosed to the keel, and this latter is not produced to the posterior edge of the sternum.

The synonymy at the head of the articles is not supposed, by any means, to be complete. Besides local references, only those of a leading nature, as also relating to the recent writings of Indian ornithologists, more particularly contributors to 'Stray Feathers,' have been given, as these were all that were necessary to the local student. Towards the close of the work I have been obliged to curtail the synonymy, even in its reduced form, and many Indian references have been omitted which did not relate to notes of much interest on the species in question. Mr. Ramsay's distribution list of Australian birds has been of much service to me as regards Australian distribution ; but, owing to want of space, I have been unable to quote, except in one or two instances, this important contribution to Australian ornithology. In respect to Ceylon references, I have not quoted my paper on the "Distribution of the Birds in the Asiatic Society's Museum," contained in the local journal for 1874, as it was printed in mistake during my temporary absence from the island, and contained many errors in distribution, which, owing to the result of subsequent experience, I had intended to correct.

In regard to the local names for the birds of the island preference has been given to those used in Asiatic and Malayan countries, and, in the case of Waders and Water-birds, Heuglin's Egyptian names have been quoted. Sinhalese names have been supplied from Layard's catalogue and from a list furnished me by Mr. MacVicar, of the Survey Office, as well as from information obtained myself from the natives. This gentleman also supplied me with a list of Tamil and Ceylon-Portuguese names, which I have used throughout the work.

The measurements of specimens, with regard to which I have been particular, all relate to Ceylonese specimens in the flesh, except when the contrary is stated (as in the case of Waders and sea-birds particularly) in brackets. My system of wing-measurement, it is well to remark, consisted in straightening the metacarpal joint by pressure in the hand, or on the table in the case of large birds, and then measuring on the upperside of the wing. The dimensions attained in this manner exceed those taken of dried specimens, when the metacarpal joint has stiffened in the usual convex form, by from 0·1 to 0·3 of an inch. Contrary to the usage of most writers, I have placed the measurements before the description, simply because it is in accordance with the practice of field-naturalists to measure their specimens first. In the description of the plumage I have endeavoured to follow a uniform system throughout: beginning with the head and back, the wings and tail are then described, thus completing the upper surface; the lores and face are then mentioned, and *ensuite* the under surface, the under wing coming last.

It is hoped that the figure of a bird which has been engraved to show the various portions of the plumage in terms of scientific nomenclature will be of service to those who are not ornithologists, should they have occasion to peruse the description of the plumage of any species in which they may be interested.

The observation (*Obs.*) on each species has been given for the benefit of the *local student*, in order to furnish him with as much information as possible of allied species inhabiting India, and, in fact, the entire Oriental Region. Many of my observations on kindred species and genera may seem superfluous to the ornithologist in England, with numerous libraries at his command ; but it is to be hoped that, as far as the naturalist in Ceylon is concerned, they will be of some use. Likewise with a view of assisting the local student, an *outline* of the entire *geographical distribution* of each species has been sketched out ; this matter, again, may seem, to European readers, superfluous in a work of local nature.

As the system of spelling has recently been changed, I have followed, to the best of my knowledge, the new method, but which, however, I am bound to remark, is subject to variation* at the hands of those who conform to it. For instance, the names of some places are spelt differently in the road-maps of the Surveyor-General and in that published by the editor of the 'Observer'; for example, the name of a celebrated tank is spelt "Kantalay" in the one and " Kantaleyi " in the other, whereas, after the old spelling " Kandelay " was abolished, the word used to be spelt by some civil servants " Kanthelai," and as such it appears in this work. My readers will therefore, I trust, bear with the somewhat variable orthography of Ceylonese names in the ' Birds of Ceylon.'

In the early part of the work the name of the territorial division "Pattuwa" will be found, in some instances, incorrectly spelt " Pattu ;" but in the map, compiled from road-maps of Provinces, kindly furnished me by Col. Fyers, R.E., I have followed in all instances the new method of spelling. The figures indicating the rainfall are taken from tables likewise furnished me by the Surveyor-General.

As regards the *nidification* paragraph, I regret to say, as far as local students are concerned, that I have been compelled largely to quote from Mr. Hume's ' Nests and Eggs,' owing to the difficulty in obtaining information about, or finding one's self, the nests of birds in Ceylon. Yet the admirable notes contained in that work are perhaps better than those which I could have obtained in the island. If, however, the Appendix be consulted much interesting additional information will be found supplied by my valued correspondent Mr. Parker, who has done more in Ceylonese oology than any recent collector.

It now remains for me to return my grateful thanks to the many ornithologists, naturalists, and collectors who have furnished me with assistance and information, and placed their valuable collections at my disposal during the time I have been compiling this work. I am much indebted, first and foremost, to Dr. Günther, Director of the Zoological Department of the British Museum, and to Mr. Bowdler Sharpe, Senior Assistant of the same; for, through the kind permission of the former, the vast collections, both mounted and in the skin, were placed at my disposal for purposes of comparison with my own; while the latter, under whose care these collections are placed, rendered me every assistance in the procuring and examination of the large series of specimens that it was necessary to examine, and was always ready and willing to impart information on difficult points with which his great experience and unexceptionally central position enabled him successfully to deal. Again, to Mr. Seebohm I am highly indebted for having placed at my disposal his large collections, the extensive Chinese series of skins collected by the late Mr. Swinhoe being of great service for purposes of comparison ; also to Mr. Howard Saunders, who, as regards his particular group (the Laridæ), furnished me with much assistance. To Messrs. Gurney, Harting, Dresser, Sclater, Salvin, and Godman my thanks are likewise

---

* Letters sent me from *Mannar*, spelt thus correctly by the writer, are impressed with the post-mark *Manaar*!

due for aid rendered as regards the several groups which they have made their study. I must not forget to acknowledge the assistance rendered to me by Mr. F. H. Waterhouse, Librarian of the Zoological Society, in answering my frequent queries as to references and data from the many scientific works required to be consulted, and which, from time to time, I omitted to collect while prosecuting my studies in London. Mr. Holdsworth's kindness in giving me access to his valuable collection of Ceylon birds, and also benefiting me by his opinion on matters connected with island distribution &c., has been of much service to me. The premature death of the late Marquis of Tweeddale, and the consequent closing to the scientific world for the time being of his collection, was no small loss to the author, who was at the time just entering on the study of the Passerine birds, and reaping the advantage of that correspondence which this distinguished ornithologist was always ready to enter into with his brother naturalists. By this untoward event an anticipated visit to the magnificent collection at Yester, which, on a former trip I had only time to glance at, was also put aside. On his return to England from Afghanistan, Captain Wardlaw Ramsay, into whose possession the collection passed, kindly lent me such specimens connected with the Third Part of the work as I required. To Canon Tristram, also, I am indebted for the loan of eggs and skins of several interesting species. I have likewise to acknowledge, with thanks, the receipt of information on various points from Herr Meyer, of the Royal Museum at Dresden, Herr Von Pelzeln of the Imperial Museum at Vienna, and Mr. Edward Nolan, Secretary of the Academy of Natural Sciences, Philadelphia. From a still more distant region, New Caledonia, I have had the advantage of correspondence with my enthusiastic forerunner in the field of Ceylon ornithology, Edgar Layard, who from time to time supplied me with details of his old experiences in the island.

Last, but not least, I must acknowledge with gratitude the aid I have received from my correspondents in India and Ceylon. Of the former I must mention particularly Mr. Allan Hume, C.B., and likewise not omit the names of Mr. Blanford, F.R.S., President of the Asiatic Society of Bengal, Captain Butler, 83rd Regt., and Dr. Edie, of the Madras Museum. In Ceylon my valued correspondents Messrs. Bligh and Parker, Ceylon Public Works Department, kept me constantly supplied with new material concerning the habits and nidification of many species : the former furnished me with copious notes on hill-birds, while the latter worked hard on the little-known districts of the north-west, and, being a most enthusiastic lover of birds and a close observer of Nature, the information supplied by him has been most valuable. In point of fact the better part of the Appendices is made up of material supplied by this gentleman from the Mannar district, where he has recently gone to be stationed. To Messrs. H. MacVicar, Forbes Laurie, R. Wickham, L. Holden, E. Cobbold, Captain Wade-Dalton, and other gentlemen now or formerly resident in the island, I am indebted for notes on the habits and local distribution of several interesting species. In conclusion, I am constrained to remark that had others among my Subscribers corresponded as vigorously with me during the progress of the work as Messrs. Bligh, Parker, and MacVicar, much more local information would have been contained in it.

W. V. L.

SYSTEMATIC INDEX.

Order **ACCIPITRES.**

Suborder FALCONES.

Family VULTURIDÆ (**1 species**).

e

Order PICARIÆ.

---

* Incorrect title at head of article. † *Vide* description of " Red Race."

e 2

SYSTEMATIC INDEX.

* Incorrect title at head of article.

Order **PASSERES.**

* Incorrect title at head of article.

SYSTEMATIC INDEX.

* Incorrect title at head of article.

f

Order GRALLÆ.

f 2

LIST OF PLATES.

MAP OF DISTRIBUTION *to face* Titlepage.

LIST OF WOODCUTS.

xliv

ERRATA ET CORRIGENDA.

Page 85, 2nd line of synonymy, for *vociferans* read *vociferus*.

110, line 21 from bottom, *after* cere, *eliminate brackets, and read* cere, all but the tip of bill.

174, 2nd line of synonymy, *for* 1786 *read* 1788.

186, at head of article, *for* ΟΥΜΝΟΡΗΤΠΑLΜΟΒ *read* ΟΥΜΝΟΡΗΤΠΑLΜΙΒ.

224, 5th line of *Nidification*, *for* it is *read* they are.

273, line 8 from bottom, *for* Nikerawettiya *read* Nikaweratiya.

310, in 4th line of *Observation*, the semicolon *should precede* "in."

463, at head of article, ΜΥΙΟΡΗΟΝΤΒ *should be more correctly* ΜΥΙΟΡΗΟΝΕΒ.

647, line 18, for *H. hypoxanthus* read *P. hypoxanthus*.

674, line 25 from bottom, *for* 1856 *read* 1866.

DATES OF PUBLICATION

AND

CONTENTS OF PARTS.

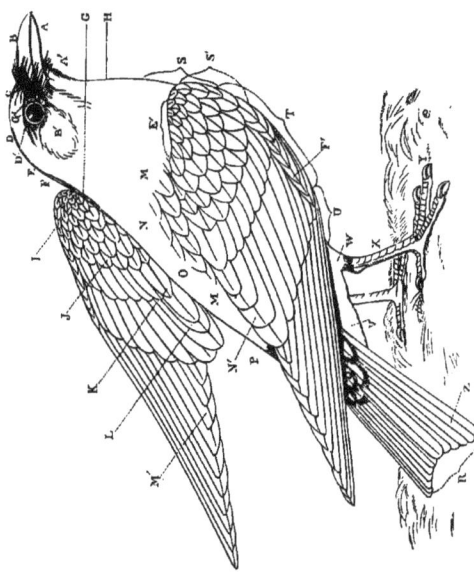

A. Gonys.
A'. Chin.
B. Culmen.
B'. Ear-coverts.
C. Forehead.
D. Crown.
D'. Supercilium.

E. Occiput.
F. Ulna.
F'. Nape.
F''. Winglet.
G. Hind neck.
G'. Lores.
H. Fore neck.

I. Lesser wing-coverts.
J. Median wing-coverts.
K. Greater wing-coverts.
L. Secondaries.
M, M'. Scapulars.
M'. Primaries.
N. Interscapular region.

N'. Tertials.
O. Back.
P. Rump.
Q. Upper tail-coverts.
R. Rectrices.
S. Chest.
S'. Metacarpus.

T. Breast.
U. Abdomen.
V. Under tail-coverts.
W. Thin.
X. Tarsus.
Y. Middle toe.
Z. Penultimate rectrix.

BIRDS OF CEYLON.

Order ACCIPITRES.

Bill short, strong, stout at the base, the upper mandible longer than the lower, the culmen strongly curved, the direction of the tip perpendicular; nostrils placed in a cere or soft membrane. Wings with ten primaries. Feet strong, armed with powerful talons of an elongated conical shape, curved, sharp, and rather smooth. Talons capable of being bent under the foot, the inner one stronger than the others. (*Sundevall* in part.)

Suborder FALCONES.

Eyes placed laterally in the head; no facial disk. Tail generally with twelve feathers, in some with fourteen. Outer toe not reversible; toes bare. Plumage compact.

Fam. VULTURIDÆ.

" Head naked, or clothed with down ; no true feathers on crown of head ; nostrils not perforated, rounded, perpendicular, or horizontal." (*Sharpe*, Cat. Birds, i. p. 2.)

B

VULTURIDÆ.

VULTURINÆ.

Genus NEOPHRON.

Differs from the other genera of its family in having the bill long and slender, and the tip much curved; in the cere being more than half the length of the bill, with the nostrils placed horizontally in it; the head is bare only to the occiput; the wings much pointed, the 3rd quill being the longest, the tail wedge-shaped, and the membranes uniting the toes ample.

NEOPHRON GINGINIANUS.

(THE LESSER SCAVENGER-VULTURE.)

Vultur ginginianus, Lath. Ind. Orn. i. p. 7 (1790); Daud. Traité, ii. p. 20 (1800).

Neophron percnopterus, Blyth, Ann. Nat. Hist. 1844. xiii. p. 115; Jerdon, B. of Ind. i. p. 12; id. Ibis, 1871, p. 236.

Neophron ginginianus, Gray, Hand-l. of B. i. p. 4; Hume, Rough Notes on Indian Raptores. i. p. 81; id. Nests and Eggs (Rough Draft), i. p. 9; Sharpe, Cat. Birds, i. p. 18; Legge, Str. Feath. 1876, p. 195 (first record from Ceylon); Hume, Str. Feath. i. p. 150.

Gingi Vulture of Latham; *White Scavenger-Vulture*, Jerdon, Birds of India.

"*Pharaoh's Chicken*," "*Pharaoh's Hen*," "*Dirt-bird*," popularly in India and Egypt.

Kal-Murgh, Hind. Shikarees; *Manju-Tiridi*, Tam., lit. "Turmeric-stealer," also *Pittri-gedda*. lit. "Dung-Kite."

Adult male and female. Length to front of cere 21·0 to 23·5 inches; culmen from cere 1·35; wing 18 to 19, reaching to tip of tail; tail 8·5 to 10·5; tarsus 3; mid toe 2·4 to 2·5, its claw (straight) 0·85; bill, gape to tip, 2·4.

Iris brown; naked skin of the head, face, and throat yellow; cere yellow, bill yellowish horny; legs and feet fleshy yellow.

Above, the neck, back, scapulars, wing-coverts, including the lesser primary-coverts, tail, entire under surface with the under wing-coverts white; primaries and winglet green-black, the outer webs of the long quills, from the notch to the base, and the entire web of the shorter quills pervaded with greyish; secondaries dusky greenish black, the outer webs towards the tips silvery whitish; tertials brown. changing into whitish towards the tips.

Young. Birds of the year have the head partly clothed with short rudimentary feathers; a broad blackish stripe passes from the forehead over the centre of the crown, and spreads out on the occiput; the lores are divided by a narrow blackish stripe, running forward to the cere; upper surface blackish brown, deepest on the back and sides of the neck and the chest, and paling into dark brown on the scapulars and wing-coverts; feathers of the back, scapulars, and wing-coverts more or less broadly tipped with fulvous; upper tail-coverts fulvous brown; tail pale brown, tipped with fulvous grey; quills black, as in the adult, but the outer webs washed with brownish grey at the tip;

breast and abdomen brownish, the feathers tipped with fulvous; under tail-coverts fulvous; under wing brown, with fulvous tippings along the edge.

With age the whole of the upper surface pales, the median wing-coverts remaining darker than the rest of the wing and the back; the sides of the neck likewise remain dark, while the rest of the under surface becomes "light brownish:" the upper tail-coverts are paler than the rump during the transition stage. Examples, however, vary in their mode of acquiring the adult plumage, the back of the neck in some being quite blackish, while the back and wing-coverts are almost white.

Obs. This species is distinguished by Mr. Sharpe, in his 'Catalogue of Birds,' from that common to Egypt and the countries surrounding the Mediterranean and Red Seas, on account of its smaller size and yellow bill. Mr. Hume. however, is unable as yet to determine whether there is any constant difference in size or colour of bill to be depended on. He remarks (Str. Feath. i. p. 151) that he has "procured and measured numerous specimens in many different parts of India, both of black- and yellow-billed birds, and with and without more or less of slender white feathers on the throat," and that he is unable to detect any marked distinction as regards size in the two. When a very large series from different localities can be got together, this point may perhaps be satisfactorily determined.

Distribution.—The Scavenger-Vulture of India can only take its place in the avifauna of Ceylon as the veriest straggler. An immature example made its appearance at Nuwara Elliya in March 1874, and was shot by Mr. Grinlinton, of the P. W. Department, while roaming about the bazaar in search of food. Its occurrence at that season of the year in the highlands of Ceylon proves it to have been driven to the south by the north-east monsoon, a wind which often brings Indian Raptores, not usually found in such low latitudes, to the island. It is therefore not improbable that, under similar circumstances, it may again find its way to Ceylon.

On the continent the White Scavenger-Vulture, if it be considered distinct from the Egyptian bird, is, according to Jerdon, "abundant throughout the greater part of India, being more rare in Central and Northern India, and unknown in lower Bengal." Subsequent observers record it as being numerous in stated localities, such as the Nilghiris, Northern Guzerat, in Sindh, Rajpootana, Khandala, and even in the subsidiary ranges of the Himalayas, where, according to Mr. Hume, it breeds up to 8500 feet. Mr. Brooks, in a paper on the birds of the Sulimán hills, in 'Stray Feathers,' 1876, remarks that he found it more abundant in that locality than anywhere in India, owing probably to the fact of there being no other Vultures there to dispute the territory with it and rob it of its easily-earned and noxious food.

Habits.—This Vulture, which, from its unclean propensities, is perhaps the least interesting of its family, is nevertheless an important support to the somewhat deficient sanitary customs which usually obtain about native villages and bazaars in India. By reason of its weak bill, it is unable to tear the flesh of carrion in company with other Vultures, of which it, moreover, is said to stand in considerable fear, and it therefore subsists by devouring all sorts of offal and other disgusting substances.

It is a denizen of most towns and villages in India, and, in common with the Grey Crow, displays an utter fearlessness of man, frequenting the dirtiest native quarters, or hovering round the abattoirs, where it appeases its ravenous appetite on the refuse thrown out during the night. It does not, however, confine itself to the vicinity of human habitations, being often found about open country, both flat and hilly, and likewise on the borders of such large sheets of water as the Sambhur lake, &c.

Like its near ally, the Egyptian Vulture, which I have seen easily advancing with almost motionless wings against a strong wind, this species has considerable powers of flight. Its usual mode of progression is with heavy and rather measured flappings of the wing; but when collected in flocks near some tempting spot it soars to a considerable height, and takes a quiet survey of the ground beneath it. It passes much of its time on the ground after feeding, and stands with an erect deportment. Jerdon remarks that it walks with ease, lifting its legs very high.

Nidification.—The spots chosen by this bird to nest in are the tops of walls, buildings, temples, &c., and in the upper branches of large trees in the vicinity of houses. The nests are described by various writers as untidy, rather loosely-put-together structures of sticks and large twigs, with but a slight depression in the centre, which is lined with rags, pieces of cloth, wool, and the many suitable substances to be found about

B 2

human dwellings. Mr. Hume mentions having found nests entirely lined with human hair, while others had nothing but green leaves to protect the eggs. These are usually two in number, but sometimes three, broad oval in shape, of a greyish-white or reddish ground-colour, and covered with very variable markings. Mr. Hume remarks, in his 'Nests and Eggs,' that "every possible shade of brownish red and reddish brown is met with, and every degree of marking, from a few distinct scattered specks to streaks and blotches, nearly confluent over the greater portion of the egg's surface." They average 2·6 inches in length by 1·98 in breadth.

Fam. FALCONIDÆ.

"Crown of the head always clothed with feathers, though the sides of the face are often more or less bare. Outer toe (except in the *Polyborinæ*) only connected to the middle toe by interdigital membrane." (*Sharpe*, Cat. Birds, i. p. 30.)

Subfam. ACCIPITRINÆ.

"Outer toe connected to middle toe by an interdigital membrane; tibia and tarsus to all intents equal in length, the difference between them not so great as the length of hind claw." (*Sharpe*, Cat. Birds, i. p. 46.)

FALCONIDÆ.

ACCIPITRINÆ.

Genus CIRCUS.

Bill accipitrine, short, moderately robust, compressed, high at the base, the culmen curved gradually from the base of the cere to the hooked tip; margin slightly festooned. Nostrils large, oval, placed forward in the cere, and protected by the bristles of the lores. Wings long and pointed, the 3rd and 4th quills subequal and longest. Tail long, even or rounded at the tip. Tarsus long, subequal with the tibia, slender, covered in front with transverse, behind with large reticulated scutæ, plumed a little below the knee. Toes slender, the outer and the middle connected at the base by a membrane; the middle toe about half the length of the tarsus; inner toe short; claws much curved and very acute. Lower part of face surrounded by a ruff of thick-set feathers, forming a partial disk.

CIRCUS ÆRUGINOSUS.

(THE MARSH-HARRIER.)

Falco æruginosus, Linn. S. N. i. p. 130 (1766).

Falco rufus, Gm. S. N. i. p. 266; Yarr. Brit. B. i. p. 90.

Circus æruginosus, Savign. Syst. Ois. Egypte, p. 90 (1809); Gray, Gen. B. i. p. 32; Schl. Vog. Nederl. pls. 20–22; Jerdon, B. of Ind. i. p. 99; Gould, B. Gt. Br. pt. xiii.; Hume. Rough Notes, ii. p. 314; Holdsworth, P. Z. S. 1872, p. 414 (first record from Ceylon): Sharpe, Cat. Birds, i. p. 69; Legge, Ibis, 1874, p. 10, 1875, p. 278, 1876, p. 126; Scully, Str. Feath. 1876, p. 126.

Le Busard roux, Brisson, also *Le Busard de Marais.*

Moor-Buzzard, Albin, Birds, i. pl. 3 (1731), also popularly in England; "*Harpy*" of some writers; *Swamp-Hawk*, *Paddy-field Hawk*, Sportsmen in Ceylon.

Mot-chil, Beng., lit. "Meadow-Kite"; *Sufeid Sira* of Mussulmen in Bengal (*apud* Jerdon). *Akbash-Sā*, Turkestan, lit. "White-headed Kite."

Kurula-goya, Sinhalese; *Prāndu*, Tam.

Adult male and female. Length to front of cere 20·0 to 21·25 inches; culmen from cere 1·0 to 1·21; wing 15·5 to 16·6 : tail 8·5 to 9·5; tarsus 3·2 to 3·9; mid toe 1·9 to 2·0, claw (straight) 0·8 to 0·9; height of bill at cere 0·43 to 0·45.

Obs. There is no constant difference in the size of males and females, some of the former equalling if not exceeding the largest of the latter.

Iris golden yellow, cere yellow; bill black; base of under mandible, legs, and feet yellow; claws black.

Fully matured plumage. Head and nape buff-white, deepening into rufescent buff on the hind neck; the feathers of the head with clear, blackish-brown mesial stripes, increasing in width on the hind neck, on the lower part of which they spread over the feather into the deep glossy brown of the back, scapulars, median wing-coverts, and longer tertials: in some examples, probably the oldest, the head-streaks are reduced to narrow shaft-lines; least wing-coverts above the flexure and along the ulna, in the female, buff, with dark central streaks overcoming the feathers on the lower series; the median wing-coverts and the scapulars margined with indistinct rufous; upper tail-coverts pale grey, often shaded with tawny patches, and the basal portion of the feathers white; greater wing-coverts, secondaries, primaries (with the exception of the four longer quills), their coverts, and the winglet dull silver-grey, with dark shafts; longer primaries black; basal portion of the inner webs of all the quills, edge of the wing, and under wing-coverts pure white; tail paler grey than the wings, with a whitish tip and a brownish hue near it; the shafts white.

Lores and round the eye slaty blackish, with the bases of the feathers white; ear-coverts brownish, edged with tawny; ruff blackish brown, margined broadly with buff; throat, chest, and breast buff; the chin with narrow dark shaft-lines, and the remainder regularly marked with broad, pointed, sepia-brown streaks, paling on the lower parts into dull rufous, and spreading over the feathers, which are often pale-margined, or with buff bases showing here and there on the surface; under surface of tail whitish.

In such fully matured birds the lower parts vary much, the feathers in some being as pale-margined as the breast.

A younger stage, but one in which the bird is adult, and which is more frequently met with than the above, has the head and hind neck rufescent buff, the feathers with broad mesial brown stripes; the forehead is not so pale as the crown, and the ear-coverts are conspicuously brown; the shorter primaries are dusky, or not so grey as the coverts; the fore neck and chest, and sometimes the better part of the breast, are rufous-buff, with rufous-brown stripes, while the whole of the lower parts, including the under tail-coverts, are dark rufous, with dark stripes on the breast: under wing-coverts rufescent.

Young. Iris brown: cere, legs, and feet greenish yellow, the bill sometimes greenish about the base of lower mandible. Whole upper surface, wings, and tail uniform dark brown, while the entire under surface from the throat down is chocolate-brown: the forehead, crown, and chin buff, with narrow brown shaft-stripes; the tail is tipped with buff, and the feathers of the lower parts, in some examples, very finely margined with the same. Occasionally the forehead and crown are both brown and the buff confined to the nape, while very rarely the entire bird is a very dark brown.

Progress with age. The brown iris becomes mottled with yellow, and the cere becomes yellowish above, the legs losing at the same time their greenish hue.

The buff of the head spreads down the hind neck, increases on the throat, and a patch of the same appears on the chest : in females the lesser wing-coverts become rufescent buff, with dark central streaks : the under wing-coverts pale into rufous, but the quills remain as in the nestling plumage. Examples killed at the end of the season in Ceylon are usually in this dress, which is probably acquired by a change in the feather itself.

At the next moult, the buff continues to spread chiefly on the fore neck, uniting in some cases with the pale space on the chest : the lower parts become dark rufous : the primary-coverts, secondaries, and their coverts are pervaded with grey ; the upper tail-coverts are rufous, the lower feathers tipped with ashy, and the tail is brownish ashy.

Obs. The amount of yellow on the upper surface varies much in all these adolescent stages, some examples having the feathers of the lower back even broadly margined with it : it varies, in females, on the wing-coverts, and in all males I have ever examined is absent from that part.

Distribution.—This large Harrier (or the Moor-Buzzard, as it is sometimes called in England) arrives in Ceylon on its annual migration southwards through India in November, and remains in the island until the usual month of departure, the following April. It confines itself chiefly to the sea-coast, and is even there somewhat local in its distribution. Although tolerably numerous on the open plains of the Jaffna peninsula and about the vast rush-beds at the lower end of the great Jaffna *lagoon*, as well as on the coasts of both sides of the island as far as Mannar and the delta of the Mahawelliganga, it is equally so, during some seasons, in the extreme south of the island, and makes its appearance there as early, if not earlier, than in the north.

There can, I think, be no doubt that our seasonal migrants arrive from the north in two separate streams—the one from the north-east driven across the Bay of Bengal from Burmah and the eastward-trending coast to the north of the Godavery ; the other making its way down with what is called the "long-shore wind" of October and November from the southernmost point of the Carnatic or the region about Cape Comorin, and landing

its components on the south-western shores of Ceylon. In the case of more species than one, to be hereafter
noticed, I have observed migrants in the extreme south at an earlier date than in the very north of the island.
The Marsh-Harrier is more numerous some seasons than others; and this irregularity in its numbers was
particularly noticeable at Galle in 1871 and 1872, in the first of which years it was so common that it now
and then frequented, one or two at a time, the open and public esplanade without the fort walls, coming into
the "camp" and sitting on the ground near the barracks; it was at the same time to be found in the marshes
all through the district. In the following year, however, I noticed very few examples anywhere in that part of
the island. It frequents the paddy-lands and swamps far up the Gindurah, and is likewise found in the interior
of the country to the north of Hambantota, as well as in swampy districts along the south-east coast as far
as the irrigated plains below the Batticaloa lake, the largest tract of paddy-land in the island and a favourite
locality for all marsh-loving birds. I have not unfrequently seen it on the swamps between Colombo and Kotté.

As regards its geographical range, the Marsh-Harrier is one of the most widely diffused of its genus. It
may be said to have its permanent head-quarters in Europe and Siberia, south of 60° N. lat., and in Western
Asia as far as the region immediately north of the Himalayas. In the non-breeding season, however, its
wandering propensities carry it over an immense portion of the Old World. It migrates through all India
and into China and Japan, spreading southward even into the Philippines. In Africa it spreads over Egypt
and Abyssinia, Algiers, and Eastern Morocco, and reaches the Canary Islands, where Professor Newton, in his
edition of Yarrell, says that Ledru obtained it in the island of Teneriffe. It occurs likewise in South Africa,
where Mr. Ayres procured it in the Transvaal Republic. It is most abundant in the marshy districts of
Europe, being very common in Turkey, and swarming, according to Mr. Howard Saunders, in the marshes of
the Guadilquivir. Since the draining of the fens and marshes in England, it has become, according to Professor
Newton, almost entirely banished.

Habits.—The Marsh-Harrier, as its name implies, is a denizen of swamps, fens, damp moor-land, marshes,
wet pasture-lands, and, in the East, of tracts of rice or "paddy" cultivation, which supply it with the same
kind of food as the first-named localities. It is a bird of powerful but heavy flight, traversing considerable
distances with a few strokes of its long wings, followed by onward sweeps, in the course of which it guides
itself along just above the ground, ready to drop on the first prey which it espies. It is by no means a shy
bird, either when seated or on the wing, and in the course of its beating round or crossing a piece of ground
will fly close to the sportsman.

It is the most predatory of all the Harriers, not contenting itself with living on reptiles, frogs, rats, and
other small mammals, but seizing wounded Snipe and other birds without fear of the gun, and capturing fish
with as much skill as the Fish-Hawk. I have killed it with a large Lulu*, weighing nearly two pounds, in
its talons, and have likewise detected the remains of young Pipits in the stomach of one shot in the marshes
of Jaffna. On seizing a lizard or snake, these birds usually devour it there and then, fixing it to the ground
with the talons, in the same manner that any ordinary Hawk pins its prey to a branch.

The Moor-Buzzard sometimes soars to a great height, circling round and round above swamps and
marshes, and on account of its large size has much the appearance of an Eagle in the distance, until its
long tail be observed, this feature at once ensuring its identification. It perches on the ground like its
congeners, but not unfrequently rests on dead trees at the borders of marshes, and is the only Harrier I have
seen thus perched in Ceylon.

In his interesting paper on the birds of Turkestan, Dr. Scully remarks that besides feeding on frogs,
rats, and lizards, the Marsh-Harrier kills the Reedling (*Calamophilus biarmicus*), this little bird no doubt
coming constantly beneath its notice as it hovers round the reed-beds of swamps in that country.

Nidification.—This species, it appears, has been known to breed in India, Mr. Hume having received
a pair of eggs taken near the Kistna river. The natives of Oudh have also informed that gentleman that it
breeds in their province; and as it has been shot in other parts of the country during the breeding-season, it
seems certain that a few birds breed within the Indian limits. The nest is said to be placed on the ground,
among sedge or reeds, and to be made of sticks, rushes, or coarse grass.

---

* A common freshwater fish in Ceylon.

Mr. A. B. Brooke, in his notes on the ornithology of Sardinia ('Ibis,' 1873, p. 154), writes as follows concerning a nest in the neighbourhood of Oristano, where these Harriers swarm:—"A nest I found in the end of April was built in the middle of a reedy, marshy lake, placed halfway up the stems of the reeds, *just clear of the water*; the bottom was formed of rough coarse sticks, and the interior of dried matted rushes, in some cases with their roots attached, the egg lying carelessly in the middle." "The eggs are usually three in number, white, with a pale greenish tinge, and sometimes slightly spotted with bright reddish brown. They measure from 2·08 to 1·84 inch in length, by 1·58 to 1·44 inch in breadth"*. Mr. Hume describes some eggs in his collection as having a good number of markings, consisting, in one instance, of specks and spots chiefly at one end, and in another of large blotches and smears of pale brown.

Mr. Hewitson remarks that the eggs are "most commonly white," though they are sometimes spotted. This variation in their character accounts for the difference of opinion expressed by Montagu, Latham, and Selby on the subject. The figure in Mr. Hewitson's plate represents a slightly-marked egg, there being a few small spots of pale reddish scattered pretty evenly over the surface and intermingled with some pale blotches of bluish grey. As many as five eggs are sometimes laid, though four is the usual number.

* Newton's ed. Yarr. Brit. Birds, p. 130.

CIRCUS MELANOLEUCUS.

(THE PIED HARRIER.)

Falco melanoleucus, Forster, Ind. Zool. p. 12, pl. 11 (1781).
Circus melanoleucus, Vieill. N. Dict. d'Hist. Nat. iv. p. 465 (1816); Gray, Gen. Birds, i. p. 32;
    Kelaart's Prodromus, Cat. p. 115; Layard, Ann. & Mag. Nat. Hist. 1853, xii. p. 105;
    Jerd. B. of Ind. i. p. 98; Hume, Rough Notes, ii. p. 307; Holdsworth, P. Z. S. 1872,
    p. 414; Sharpe, Cat. Birds, i. p. 61 (1874); Hume, Stray Feath. vol. iii. p. 33; Swinhoe,
    Ibis, 1874, p. 266, pl. 10; Gurney, Ibis, 1875, pp. 226-7, and 1876, p. 130; Hume,
    Str. Feath. vol. v. p. 11 (1877).
The Black-and-White Falcon, Pennant, Ind. Zool. p. 33, pl. 2 (1790); Kelaart, Prodromus.
Pahatai, Hind.; *Ablak Petaha,* Nepalese *(apud* Jerdon).
Kurula-goya, Sinhalese.

Adult male. Length to front of cere 16·5 to 18 inches; culmen from cere 0·75; wing, of 5 examples from different
parts of India, 13·7, 13·0, 14·2, 14·2, 14·5; tail 8·2 to 9; tarsus 2·9 to 3·25; mid toe 1·2 to 1·4, claw (straight)
0·55; height of bill at cere 0·35.

Obs. In Mr. Hume's table of measurements of 34 old males (Str. Feath. vol. v. p. 12) the wings range from 13·2 to
14·34 inches, and the tarsi from 2·8 to 3·25.

Iris bright golden yellow; cere varying from grey to greenish yellow; bill black, paling into leaden at the base; legs
and feet chrome-yellow.

Entire head, neck, chest, back, upper scapular feathers, and median wing-coverts black, glossy on the upper parts and
dull on the fore neck and chest; least and greater wing-coverts, point of the wing, shorter primaries, secondaries,
rump, and upper tail-coverts pale silvery grey, the quills brownish at the tips; longer primaries blackish on the
terminal half, with the bases of the inner webs white; tertials brownish near the tips, much darker in some
examples than in others; tail light sullied grey, paler on the lateral feathers; shafts of all but the latter feathers
brownish; beneath the chest, together with the under wing, pure unmarked white.

Young. Iris "ochreous yellow" (Swinhoe): cere greenish grey or greenish yellow; gape and loreal skin yellowish;
    bill pale at the base.
I subjoin here the description of Mr. Swinhoe's specimen, figured in 'The Ibis,' 1874, inasmuch as it appears, according
to Mr. Gurney's judgment ('Ibis,' 1875, p. 226), to be, in all probability, the first plumage of the bird :—"Upper
parts light brown, the feathers on the back dark-stemmed. Crown, nape, and scapulars blackish brown in centre
of feathers, with broad yellowish-red margins. Underparts light buff, with yellowish-brown streaks, broad and
darker on the breast; tibials and vent chestnut-buff, with darker stems to feathers. Quills brown, tipped light,
with lightish stems, and barred across inner webs, more obscurely towards their tips; axillaries reddish cream,
with reddish-brown spots; under wing whitish cream, with conspicuous bars. Upper tail-coverts greyish white;
tail whitish brown, with three broad bars; a fourth, indistinct bar crosses near base of tail."

Obs. This example appears to be a male, as it has a wing of 13·0 inches only, although it is worthy of remark that
in some Harriers immature females are sometimes smaller than the other sex. The plumage of the specimen, as
described, is much like that of an adult female to be noticed hereafter; and the presence of three "broad bars" on
the tail instead of a greater number of narrow ones, as ought to be the case in a young bird, is singular.
Mr. Gurney remarks, in the same article, that "the progress towards maturity is marked in all cases by the spreading
of a conspicuous grey tint over the greater and middle wing-coverts, and over the outer webs of the secondaries and
of the upper portion of the primaries." This is doubtless the case up to a certain point in the bird's change of
plumage; but it appears evident that the entire adult plumage, as is to be expected in an attire so marked in

c

its character, is put on at one final moult. Adult males are always to be found in the perfect pied dress without any intermingling of immature characteristics pointing to a gradual assumption of the black-and-white livery. There is, however, much to be learnt concerning the plumage of this species, particularly in respect to the females, and a thorough knowledge of it can only be attained by means of the acquisition of a large series of carefully sexed and dated specimens.

Adult female. The wing, in 10 examples measured by Mr. Hume, varied from 13·7 to 15·1 inches *, and the tarsus from 3·05 to 3·3.

It is a matter of difficulty to determine in this species which type really represents the fully adult female. The following are the dimensions and description of a female shot by myself near Trincomalie, which I have compared with examples in the British Museum, and which Mr. Sharpe considers to be fully mature :—

a. Length to front of cere 17·8 inches ; culmen from cere 0·8 ; wing 14 ; tail 9·5 ; tarsus 3·1 ; mid toe 1·6, claw (straight) 0·67.

Iris citron-yellow ; cere gamboge-yellow ; bill dark horn, bluish at the gape and the base beneath ; legs and feet gamboge-yellow.

Head and upper surface, with the wing-coverts and tertials, a subdued though glossy sepia-brown ; the longer scapulars with a greyish bloom ; the crown-feathers margined with rufous, and the hind neck with dull whitish, not extending to the tips ; edge of forehead, above the eye, and the face whitish ; the lesser coverts, from the shoulder along the flexure of the wing, pure white, with brown mesial stripes, gradually extending over the feathers on the succeeding series : winglet, primary and greater coverts, shorter primaries, and the secondaries silvery grey, barred with brown, the subterminal band broad, and the tips of the feathers dull white : longer primaries darker brown, barred with the hue of the tips, and the interspaces of the outer webs greyish : inner edge of all the quills towards the base white ; upper tail-coverts almost unmarked white ; tail above greyish, with four dark bars, the subterminal one some distance from the tip, which is pale ; the interspaces of the two outer feathers towards the base white, and the bars on that part rufous.

Chin and gorge whitish, striped from the gape round to the ear-coverts with rufous brown ; ruff white, with broad brown central stripes : under surface and under wing white, the fore neck and chest with bold dashes of brown, almost confluent on the sides of the neck, and diminishing to mesial stripes of a more rufescent hue on the breast, the lower parts having shaft-lines of the same : lower series of the under wing-coverts with rufescent brown bars, the rest with rufous shaft-lines : lower surface of tail dull whitish, the bars showing indistinctly.

b. An example in the British Museum, from the collection of Capt. Pinwell, is marked as a female and is in the following plumage :—

Mantle glossy dark clove-brown, much deeper than in the above ; centres of frontal, occipital, and hind-neck feathers blackish brown, those of the first-named parts edged with rufous, of the latter with a paler or fulvescent hue ; the outermost series of greater wing-coverts silvery white, crossed with broad bands of dark clove-brown ; secondaries, shorter primaries, and their coverts of the same ground-colour, with blackish bars : 1st, 2nd, and 3rd quills with the terminal portions brown, barred with a darker hue on both webs ; internal portion of the inner webs of all the quills white ; tail dusky silvery grey, crossed with five clove-brown bars, those on the lateral feathers gradually changing into rufous.

Sides of the throat, together with the posterior part of face and ear-coverts, rufescent, with dark shaft-stripes ; ruff whitish, striped with dark brown ; chest fulvescent whitish, the feathers with broad rufous-brown centres ; beneath, from the chest pure white, the breast with light rufous-brown stripes, decreasing in width to lines on the abdomen, lower flanks, and under tail-coverts.

Obs. This example differs from the Tamblegam bird in being darker as regards the brown plumage, and paler as regards the grey colouring of the wing-coverts ; while the rufous edgings of the head and throat-feathers are more brought out, which latter characteristic savours of youth, in spite of the apparently more adult coloration of the back and wing-coverts.

It is in much the same dress as an "adult" female described by Mr. Hume in his excellent and exhaustive article already referred to. Another obtained by Col. Godwin-Austin in Assam, and described by Mr. Gurney (Ibis, 1876, p. 130), is darker than either of these—"the entire mantle being blackish brown, increasing in intensity as it approaches the tips of the lower scapulars, which are almost black ; the wings show a remarkable approach to the plumage of the adult male, but the band which extends across the wing-coverts, instead of being black, is dark chocolate-brown, varied by some of the brown feathers passing, in part, into a decided black."

It is probable that each of the above examples were sufficiently mature to breed ; but it does not follow that the darkest birds were the oldest. My bird had the ova developing, and would have bred in the succeeding June, and

---

* Colonel Godwin-Austin's bird measures, according to Mr. Gurney, 15·8 ('Ibis,' 1876, p. 131).

was in a paler phase than any of the others. It follows, however, from what has been made known by various writers of late, that, as in other Harriers, the female of this species has no fixed character of adult plumage, but that as the bird gradually grows older it inclines towards the melanistic dress of the male, never actually acquiring it, and always retaining the striped under surface peculiar to the sex. The length of tarsus will likewise serve to distinguish an adult female from an immature brown-plumaged male.

Young. Iris "light brownish yellow: cere slaty greenish grey" (*Armstrong*).
Nestling plumage as in young male.

Distribution.—This handsome Harrier, which, in common with the other three species in our list, is a migrant to Ceylon in the cool season, is undoubtedly a rare species in the island. On the few occasions on which it has occurred it has been a straggler no doubt, from the numbers which visit, during the N.E. monsoon, parts of the eastern coast of India and Burmah. Layard, with his usual good fortune, while investigating the ornithology of the island, shot a specimen on the north-west coast near Mantotte, an excellent district for Harriers; he also mentions having seen a drawing of another example made by Mr. Mitford, District Judge at Ratnapura, from a bird brought to him by a native, and captured near that place. In the early part of 1860, I observed a bird in the black-and-white plumage in the cinnamon-gardens at Colombo, and in March 1875 I shot the female above described on the shores of Tamblegam Bay. It is possible that immature birds, in a dress in which they may be mistaken at a distance for other members of the genus, may visit the northern shores of Ceylon; but the old, pied birds can very rarely do so, for during an interval of more than eight years' collecting, always on the look-out for Raptores of all kinds, and two of which were passed in the north of the island, I never succeeded in detecting but the one adult bird above mentioned.

The Pied Harrier is, during the season of its wanderings, more abundant in Assam and Burmah than elsewhere, and radiates outwards from that region down the eastern parts of the Indian peninsula to Ceylon. Mr. James Inglis records it in 'Stray Feathers' (vol. v. p. 11) as extremely common, from September until April, in North-eastern Cachar. Dr. Jerdon writes that it is found in abundance in districts where rice-cultivation is carried on, "as on the Malabar coast, in parts of the Carnatic, and in Mysore," but that it is rare in the Deccan and Central India, though common in Bengal. To the east it spreads from its head quarters, which are evidently the Mongolian territory to the north of Burmah, into China and the Amoor Land, from which regions Mr. Swinhoe records it.

Habits.—The Pied Harrier is said to prefer grassy jungles to swampy land. I have seen it both in marshy places and low scrubby jungle; and the district in which Layard obtained his specimen is one of open plains, studded here and there with clumps of low bushy growth, or dotted with scattered trees. Jerdon says that, in India, it is common in districts which are cultivated with rice; and it therefore does not appear to confine itself to one particular description of country; but, like its congeners, to traverse such open tracts as abound in the food on which it subsists. Being a bird of slender frame and long wing, its flight is particularly easy and graceful: it glides over wide fields impelled by a few slow, though powerful strokes of its ample pinions; and when hunting for its prey it "quarters" a tract of ground with the greatest regularity; starting at one end, it sweeps across from side to side, backwards and forwards, with a graceful turn at the end of its course, and while rising and falling, so as to skim just above the top of the long grass, it is enabled to drop like a stone on its prey.

Its diet consists of small reptiles, lizards, and no doubt small birds, or young ones taken from the nest when its more favourite food is not procurable. It alights and rests on small eminences on the ground, banks, or stones, and roosts, like its congeners, on *terra firma*, thus falling a prey not unfrequently to nocturnal animals. Mr. Oates writes that near Poungday, in Pegu, it is often found on the large plains of mixed jungle and paddy-land, and that it prefers inundated paddy-land to any other.

Nidification.—Where this Harrier breeds is still a matter of conjecture with Indian writers, and consequently nothing is known of its nidification. The late Mr. Swinhoe could obtain no information concerning its nesting in China; and the inference therefore is, that it retires in the breeding-season to the region between the Himalaya and the east of China. Mr. Hume is of opinion that it breeds in part of this district, namely Assam; and Dr. Jerdon remarks that he saw several birds at Purneah in July, at which time they ought to have been nesting somewhere. In the female I killed in March the ova were commencing to develop largely, and she was evidently about to breed at no great date from that time.

CIRCUS CINERACEUS.

(MONTAGU'S HARRIER.)

Falco cineraceus, Mont. Orn. Dict. vol. i. (1802); Temm. Man. i. p. 76 (1820).

Circus cinerarius, Leach, Syst. Cat. Mamm. &c. Brit. Mus. p. 9 (1816).

Circus montagui, Vieill. N. Dict. xxxi. p. 411 (1819).

Circus cinerasceus, Steph. Gen. Zool. xiii. p. 41 (1825); Kelaart, Prodromus, Cat. p. 115;
    Layard, Ann. & Mag. N. H. 1853, xii. p. 105; Schlegel, Vog. Nederl. pls. 18, 19 (1854).

Circus cineraceus, Cuv. Règ. An. i. p. 338 (1829); Gould, B. of Europe, i. pl. 35 (1837);
    Jerd. B. of Ind. i. p. 97; Gould, B. of Gt. Britain, pt. xii.; Hume, Rough Notes, ii.
    p. 303; Newt. ed. Yarr. Brit. B. i. p. 138; Holdsworth, P. Z. S. 1872, p. 413; Shelley,
    B. of Egypt, p. 184; Legge, Ibis, 1875, p. 278.

Circus pygargus, Sharpe, Cat. Birds, i. p. 64 (1874).

The Ash-coloured Harrier, Montagu; *The Ashy Falcon* of Kelaart; *Swamp-Hawk*, Sports-
    men in Ceylon.

Rétu üli, Transylvania. *Cenizo*, Spanish.

Pilli-gedda, Tel.; *Puna-Prändu*, Tam.

Kurula-goya, *Rajaliya*, Sinhalese.

Adult male (from European, Indian, and Ceylonese examples). Length to front of cere 16·5 to 17·5 inches; culmen
from cere 0·69 to 0·71; wing 13·6 to 15·5 (sometimes reaching to the tip or even beyond the tip of tail); tail 8·6
to 10; tarsus 2·2 to 2·3; mid toe 1·1, claw (straight) 0·45; height of bill at cere 0·32 to 0·34.
The following are some measurements of old birds from examples in the British Museum exemplifying the above
variation :—

	Wing.	Tail.	Tarsus.
	in.	in.	in.
♂ . Bengal	14·9	9·9	2·2
♂ . N. Bengal	15·5	10·0	2·3
♂ . Seville	13·6	8·6	2·2
♂ . Seville	14·2	9·1	2·3

N.B. In this species the second primary-covert does not reach within ⅜ inch of the notch in the second primary, falling
short of it, in females, by as much as 2¾ inches.

Iris bright yellow : cere, loreal skin, and base of lower mandible yellow, top of cere tinged with greenish : bill blackish
at the tip, paling into bluish horn-colour at the base; legs and feet chrome-yellow, claws black.

Head, upper surface, and wing-coverts dark bluish ashen, amalgamating with the paler bluish of the throat, fore neck,
and chest, a darkish tint usually prevailing across the back and scapulars; 1st to the 5th primary blackish
slate-colour, the rest, together with their coverts, silver greyish with black shafts; secondaries duller silver-grey,
crossed by two dark brown bands; upper tail-coverts white, banded broadly with slate-grey; two central tail-
feathers slate-grey, the next two paler grey, barred with brown, the remainder with the ground-colour white,
more or less tinged with rufous towards the base and barred with dark-edged rufous bands.

Beneath, from the chest, white, striped with narrow streaks of rufous down to the under tail-coverts; axillary plume
and under wing-coverts barred with rufous, but not extending to the wing-lining beneath the ulna and carpus.
In some examples the ashen hue extends much further down the breast than in others.

Obs. A very remarkable melanistic variety of the adult form exists, some fine examples of which, from Mr. Howard
Saunders's collection, are in the British Museum. The whole bird is dark sooty brown, with the cheeks, back,

belly, and lower flanks blackish brown; tail brownish grey; quills and secondaries blackish brown, and the under surface of the tail pale greyish.

Young. The chick is first clothed with white down, which changes in about ten days to fawn-colour on the upper surface: in a fortnight more, according to Mr. H. Saunders's observations, the breast and flanks become clothed with chestnut feathers, and the quills come out blackish brown with a rich rufous border.

Male bird of the year. Wing from 13 to 14 inches; females not exceeding the males at that age.
Iris brownish yellow; cere, bill, and legs much as in the adult.
Above sepia-brown: nape and upper tail-coverts white, the former with the centres of the feathers brown, and the latter with terminal spots and occasionally bars of the same; occiput and hind neck edged with rufous; wing-coverts margined with fulvous; primaries blackish brown, the longer feathers washed on the outer webs with greyish, and the inner webs white towards the base and mottled with brown: tail with the six central feathers brownish grey, barred with brown, the latter becoming broader than the grey ground on the outer of these feathers; the remainder brown, barred with rufescent white.
Cheeks and a broad eye-streak whitish; a gular band of dark rufous-brown, and below it a ruff of paler, dark-centred feathers, *not contrasting*, however, with the band, or *setting it off*, as in *C. macrurus*; chin and gorge rufescent whitish; throat and chest dull brownish rufous, with distinct dark shafts to the feathers, and gradually melting into the yellowish rufous of the breast and lower parts, which are striated with broad stripes of rufous; axillary plume dark rufous, with light marginal spots; median under wing-coverts rufous, with pale margins, the major series brownish.

Obs. The above is a description of *one* example, as presenting a fair type of the young male. The under surface, however, varies much, though it is always darker than that of *C. macrurus*, and differs from that species in the more conspicuously streaked lower parts, as well as in the duller gular band and less conspicuous ruff below it.

Progress towards maturity. The change from this to the adult phase is gradual but *systematic*. The upper surface becomes cinereous brown, the upper tail-coverts sometimes coming out in the adult form (white, with blue-grey bands): the tail becomes grey, the bars vanishing on the central tail-feathers, and the interspaces on the laterals are white in some and rufous-white in others; the chest and fore neck are rufescent, mingled frequently with ashen feathers, and the breast and lower parts pale fulvescent, streaked with rufous stripes; the lower surface of the primaries and the bases of the inner webs are white; under wing-coverts with more white than in the first stage.

After the next moult the lower parts become white with tawny streaks, as in the adult, and the chest is often ashy with cinereous-brown strie; at the same time the head usually retains its brown dress, and the tail has the lateral feathers as darkly barred and as much tinged with rufous as in the younger stage. The gular band is usually dark brownish, contrasting with the pale whitish ruff assumed at this age.

Young female [*]. In the first year, females do not exceed males in size, measuring sometimes quite as low in the wing as the smallest of the latter.
Iris, in some brown, in others yellow, mottled with brown: bill, legs, and foot as in male.
Much resembles the male in plumage, but usually not so dark a rufous beneath, and with the strie not so strongly pronounced; these are, however, variable in extent, being mostly confined to the chest in some, and extending in others to the lower parts; the primaries are barred on both webs with narrow bands of brown, and the secondaries are crossed on their inner webs with broader bars of the same; the wing-coverts vary, being sometimes almost uniform, and occasionally very deeply edged with rufous, the brown hue being confined to the centre of the feather.

In the next stage the rufous ground-colour of the under surface disappears from the edges of the feathers, and the mesial stripes contrast markedly with the lighter hue of the rest of the web; the head continues to be edged with rufous as before, and the margins of the hind-neck feathers are the same as in the yearling plumage: the upper tail-coverts are scantily barred or pointed with rufous, and the quills more pervaded with ashy than in the first plumage.

---

[*] The adult plumage in this sex varying so much, I have considered it advisable to commence with the young, and follow the changes to the old bird.

The under surface continues to alter until the bird is fully matured; but the adult dress, after it is acquired, varies not a little in different individuals. The following is a description of an example in the British Museum, which, judging by the regular alteration of character during adolescence, appears to be a fully matured bird.

Adult female. Above sepia-brown, pervaded on the back with greyish; the head margined with rufous, and the hind neck with fulvous, the centres of the feathers being blackish brown; median wing-coverts broadly margined with rufous-buff as in the younger bird; terminal portion of primaries and the secondaries deep brown, with a purple lustre; the outer webs of the longer primaries greyish, and both webs barred with narrow bands of blackish brown; inner webs near the base isabelline grey; upper tail-coverts white, with greyish-brown bars near the tips of the longer feathers: central tail-feathers drab-grey, with four narrow bars and a broad subterminal band of deep brown, the remainder crossed with the same number of wider bands, the interspaces paling to white on the lateral feathers, where the bars are narrower again, and tinged with rufous at the base.

Face and a small space above the ears white; the gular band deep brown, margined with rufous and tinged with ashy; ruff blending with the throat and fore neck, which are rufescent, with broad cinereous-brown stripes: beneath, from the chest downwards, fulvescent whitish, with bold central stripes of rufous-brown on the chest, and of rufous on the lower parts; under wing-coverts rufescent white, boldly dashed with rufous; edge of wing-lining whitish. Soft parts as in the adult male.

Length* to front of cere 18·5 inches, culmen 0·7; wing 14·0 to 15·3; tail 10·0; tarsus 2·3; mid toe 1·13, claw (straight) 0·58; expanse 43·0.

The following are measurements of several European and Indian examples of adult females, which are all exceeded by those of a male from N. Bengal:—

Wing.	Tail.	Tarsus.
in.	in.	in.
14·3	9·1	2·5
14·3	9·0	2·4
14·6	9·0	2·4
14·6	9·3	2·3

Distribution.—This widely dispersed Harrier is, as might be expected, a winter or cool-weather visitant to India and Ceylon, arriving in the latter place about October and departing again in April. After concentrating itself in considerable force in the Jaffna peninsula, the adjacent isles, and on the coast of the Northern Province, it spreads down both sides of the island, but does not apparently wander into the interior after the manner of the last species. On the west coast it is chiefly confined to such open localities as the cinnamon-gardens of Negombo, Colombo, Morotuwa, &c., and the almost impenetrable swamp called the Mutturajawella. In the Galle district it never came under my notice as an identified bird, but may have figured among the many observed on the wing between the port and Baddegama; in the south-east, however, it occurs, but not so frequently as in the north. During a visit, in March 1876, to Jaffna and the neighbourhood, I found it at several islands in Palk's Straits, among which were the twin islets of Erinativoe, on which several were seen in the course of a day's excursion. In the island of Manaar and at Aripu I likewise observed and procured it.

Montagu's Harrier has a very similar geographical range to the next species. In Europe it is perhaps more generally distributed, as it extends in the summer to the British Isles, and is also common in Spain, but chiefly during the winter, whereas the Pale Harrier does not move westward of 8° E. long. It does not confine itself to the south of the continent alone, for it has been recorded from both Heligoland and Sardinia. It is found in Scandinavia, but does not appear to range into Northern Russia, although it inhabits the south of that country. From the Caucasus it extends, like the last bird, through Palestine, to the elevated region of Turkestan, from which Severtzoff records it. It is abundant in India in the cool season; but though Jerdon remarks that he found it in all parts of the empire, the experience of recent observers, as appearing in 'Stray Feathers,' tends to show that it is more local than either the Marsh- or the Pale Harrier. Mr. Hume does not record it from Sindh, and Mr. Ball states that it is not common in Chota Nagpur. In the Deccan, Mr. Fairbank says it is common; and it occurs, but not abundantly, in the Khandala district. It is found in Burmah, and has been obtained as far east as the Yangtsze river in China.

---

* From the flesh in Ceylonese examples.

Habits.—Montagu's Harrier delights in swamps, marshes, and open country, more or less studded with low jungle and copse, over which it sweeps at a considerable height, rising and falling in its rapid progress, and appearing to take in a more extended view of its ground than the Pale Harrier can do in its low-directed flight. I have seen it, however, in the great swamp of Mutturajawella, flying steadily from end to end, with a slow beating of its long wings, keeping just above the tangled vegetation, and now and then dropping out of sight in the sudden manner peculiar to its family. It is crepuscular in its habits, flying about its hunting-grounds so late that it cannot be discerned when a little way off; and sharp indeed must be its eyesight to enable it to capture the small prey that it lives on, among grass and herbage, with so little light. Layard, whose observations tended to show that it fed much upon snakes, has the following well-written description of its flight in the Ann. & Mag. Nat. Hist. :—"Nothing can exceed in gracefulness the flight of this bird when beating over the ground in search of its quarry. Its long pointed wings smoothly and silently cut the air; now raised high over its back, as the bird glides along the furrows; now drawn to its sides, as it darts rapidly between the rows of standing paddy; now the wings beat the air with long and even strokes, and now extended, they support their possessor in his survey of the marsh over which he is passing. Suddenly he drops, and after a momentary halt speeds away, with a snake dangling in his talons, to some well-remembered stone or clod of earth, and commences his repast."

I have found the bones of small mammals, probably mice, as well as grasshoppers in this Harrier's stomach; but in Ceylon, according to my experience, its chief food consists of lizards. In countries where reptiles do not abound, such as England and other parts of Europe, it preys to a certain extent on small birds; and Mr. Howard Saunders, in his very interesting account of the nesting of this Harrier in the Isle of Wight, published in the ' Field ' of the 2nd September, 1875, found amongst the food brought to the young in the nest, " the remains of several small birds—skylark, titlark, stonechat, and yellow hammer." It will also kill snakes, as appears from the above extract from Layard's writings, and no doubt very frequently preys on them in the fetid swamps of the East. Professor Newton, in his edition of Yarrell, speaks of one "which was observed to hover about a trap, baited with a rabbit, without pouncing, but on a viper being substituted for the rabbit, the bird was immediately caught." The same writer likewise speaks of its swallowing birds' eggs whole.

Montagu's Harrier, like the Moor-Buzzard, seems to prefer perching on level ground to settling on little knolls and elevations. It roosts also on the ground, and is probably often captured in the East by the stealthy jackal, or in northern climes by the still more clever fox. This Harrier does not appear to have strong powers of vision, when they are subjected to the force of the sun's rays. I once observed three birds alight, one after the other, on the bare soil, and stand with erect carriage, all looking in the same direction, after the manner of Gulls; and being between their position and the rays of the setting sun, I appeared not to be noticed by them, for I was enabled to creep steadily forward towards them in the open, and thus secured, from among the trio, one of the finest female specimens in my collection.

I have heard this Harrier make a weak squealing note, but can say nothing further as to its voice; in fact the Harriers, as a group, seem to be among the most silent of raptorial birds, little or nothing concerning their notes having been placed on record by the numerous observers of their otherwise interesting habits. When viewed on a glorious tropical morning, there is something very striking in the noiseless course of this and other Harriers as they glide silently over the misty paddy-swamps of the interior, while the luxuriant forest surrounding these, to the lover of nature, most interesting spots re-echoes with the voice of hundreds of the smaller bird creation.

This species thrives in confinement; and Mr. Saunders, in his article above referred to, records that the young bird in question, when it had acquired the free use of its wings, flew " round the lumber-room in which it had been placed in a buoyant manner, and took great pleasure in a bath, in which it would stand knee deep, enjoying being sprinkled with water, after which it would spread its wings and bask in the sun."

Nidification.—The Ashy Harrier does not breed within the Indian limits, but in northern climes, where it propagates its species, it nests in May and June. In Europe and Great Britain its nest is built, as elsewhere, on the ground, and is made of small sticks, rushes, grass, roots, &c., the latter composing the interior or lining. It is more slightly built, as a rule, than the nests of other Harriers; but its size must necessarily depend on the situation in which it is placed, for if this should be in damp ground, where the water is liable to rise, instinct teaches the bird to raise the body of the nest until above the level at which its eggs might be destroyed. In 'The Ibis,' 1875, Messrs. Danford and Harvie Brown remark that at Mezőség, in Transylvania, they "found them nesting among reeds, the nest being sometimes considerably above the ground." The nest found by Mr. Saunders in the Isle of Wight was a "mere bottom, lined with dry grass, with an outside border of fine heather-twigs." The eggs are four to six in number, generally white and unspotted, but sometimes bluish white; the specimen figured by Mr. Hewitson is of a very pale blue: they measure 1·72 to 1·51 inch in length by 1·30 to 1·25 in breadth.

While sitting, the female is said by some writers to be attended by the male, who brings food to her; but I note that Mr. Saunders writes that, in the course of many hours' watching at different times, he "never observed the male approach the nest as if to bring food or take his turn at incubation." It is possible, however, that some birds display different propensities to others. I conclude this article with quoting still further from the interesting account of the nesting of this species in the Isle of Wight, as touching one of the most interesting features in a bird's economy, viz. its manner of returning to its nest in order not to betray its whereabouts. The writer remarks as follows:—"It was most interesting to watch the movements of the Harrier when returning to her nest; the wide circles which enabled her to take in the position of any large object on the downs gradually narrowed; then quartering would begin again, to be succeeded by more circles, till every one might be expected to be the last. Then, perhaps, she would change her mind, and go off for another series of wide flights; but when the moment came there was no hesitation or hovering, but a sudden closing of the wings as she swept over the spot, and she was down in so stealthy a manner, that if the eye were taken off her for a second, it was impossible to say whether she had settled or merely gone over the brow of the hill again."

CIRCUS MACRURUS.

(THE PALE HARRIER.)

Accipiter macrourus, S. G. Gmelin, N. Comm. Petrop. xv. p. 439, pls. viii. & ix. (1771).
Circus swainsonii, Smith, S. Afr. Q. Journ. i. p. 384 (1830); Gray, Gen. B. i. p. 32; Kelaart's
    Prodromus, Cat. p. 114; Layard, Ann. & Mag. N. H. 1853, xii. p. 104; Jerdon, Birds of
    Ind. i. p. 96; Hume, Rough Notes, ii. p. 298; Holdsworth, P. Z. S. 1872, p. 413; Legge,
    Ibis, 1874, p. 10, et 1875, p. 278; Dresser, Ibis, 1875, p. 109 (Severtzoff's Fauna of
    Turkestan).
Falco herbæcola, Tickell, J. A. S. B. 1833, p. 570.
Circus pallidus, Sykes, P. Z. S. 1832, p. 80; Gould, B. of Europe, i. pl. 34; Hume, Stray
    Feath. i. p. 160.
Circus macrurus, Sharpe, Cat. Birds, i. p. 67 (1874).
The Pallid Harrier of some writers; *Swainson's Harrier* of others.
Pale-chested Harrier in India.
White Hawk, Paddy-field Hawk, in Ceylon.
Dastmal, Hind.; *Puna-Prāndu*, Tam., lit. "Cat-Kite;" *Pilli-gedda*, Tel., also "Cat-Kite."
Boz-Sä, Turkestan, lit. "Grey Kite."
Kurula-goya, Ukussa, Sinhalese.

Adult male. Length to front of cere 17·0 to 17·8 inches; culmen from cere 0·72; wing 13 to 14, averaging about
13·7; expanse 42; tail 9·0 to 9·5, exceeding the closed wings from 1·5 to 3·0; tarsus 2·5 to 2·8; middle toe 1·2
to 1·3, claw (straight) 0·6 to 0·65; height of bill at cere 0·38.

Female. Wing from 14·2 to 14·9 inches, averaging about 14·6.

Male. Iris golden yellow, very rich in the oldest birds; cere yellow, tinged with green above; gape greenish yellow:
bill blackish at top, paling to blue at cere; legs and feet chrome-yellow; claws black.
Head and entire upper surface, including the wing-coverts, bluish ashy, the upper tail-coverts barred with white; in
most examples, except the very oldest, there is a brownish wash on the nape and mantle; primaries ashy grey,
the 3rd, 4th, and 5th more or less black, according to age, from the notch to the tip, and white from that part
along the inner web to the base; secondaries ashy grey, tipped pale, and with their inner edges white; tail pale
grey, the three lateral feathers white, and all but the central pair banded with narrow grey bars; forehead and
an ill-defined supercilium, chin, throat, under surface of body, and wings pure white; throat, sides of neck, and
chest very pale bluish grey, blending into the white of the lower parts, and with the shafts darker than the webs:
cheeks faintly striated with grey.

Young male. Wing in the first year varying from 12·8 to 14·0 inches, quite equal to the adult.
Iris greenish, or less bright yellow than in the adult; cere and eyelid greenish yellow; bill blackish horn-colour, bluish
near the cere; legs and feet citron-yellow.
Above chocolate-brown, with an angular white nuchal patch, and with the upper tail-coverts white, with terminal
rufous-brown spots: the entire upper surface edged with rufous, narrowly on the head, more broadly on the hind
neck and lesser wing-coverts, and on the back and scapulars confined chiefly to the tips of the feathers; quills
brown, the inner webs barred with darker colour, and the interspaces from the notch upwards buff; a grey wash
on the outer webs of the primaries, and the tips, as well as those of the secondaries, pale fulvous; tail drab-brown
on the centre feathers, changing on the three laterals to buff, the tip buffy white, and the whole, with the exception

D

of the outer feather, crossed by 4 or 5 bars of dark brown; a narrow supercilium and a patch below the eye whitish; lores brown; a broad brown gular band succeeded by a fulvous, dark-striped ruff; under surface with the under wing pale uniform rufescent, the shafts of the feathers slightly darker rufous.

Between this and the adult grey stage birds are found in great variety of change to the pale plumage; after the first moult they become ashen brown above, with generally some rufous feathers about the nape; the upper tail-coverts become barred with greyish brown, the central tail-feathers ashy, the three lateral ones whitish, barred with rufous-brown; the quills pale ashy brown at the base; the long primaries black on the terminal half, with their inner edges whitish; the facial markings become very pale, the throat, fore neck, and chest bluish ashy, the ruff with darkish streaks, and the chest striated with pale brown; beneath, very pale bluish ashy in some, quite white in others, many specimens, likewise, having rufous shaft-stripes, while others are completely unmarked: the under tail-coverts pure white. It is in the chest and under surface that the greatest variation takes place.

In an example from the Deccan in the British Museum, in the ashen-brown upper plumage of the second year, the entire under surface from the chin, including the under wing-coverts, is *pure white*, with a few shaft-lines of rufous on the chest and throat: the tail almost as pale as in the oldest specimens: the basal part of the web of the inner primary webs partakes of the same albescent character as the under surface, being quite white.

Young female. Wing in the first year averaging about 13·5 inches, but frequently no longer than that of young males. Differs from the young male in being usually of a deeper brown, the wing-coverts very broadly edged with rufous, the upper tail-coverts with brown mesial *stripes*, and the under surface *much darker, of a rich uniform rufous tawny*: the gular band of a very dark brown, *contrasting strongly* with the whitish cheek-patch. The white nuchal patch varies, but is, I think, stronger, as a rule, than in the young male.

Progress towards maturity. In the next stage, the upper surface loses the conspicuous character of the edgings; the head and hind neck contrast with the back, the latter becoming paler; the cheeks and gular band remain the same; but the under surface undergoes a gradual change, commencing with the fading out of the rufous, particularly on the lower parts, leaving this colour confined to the centre of the feather, the fore neck and chest being heavily streaked, and the lower breast and flanks lightly so.

The lower parts continue to pale with age until, in the oldest birds I have been able to examine in a large series, they become fulvescent white, and are, with the under tail-coverts, unmarked, save with a few light streaks of pale rufescent; the throat is marked with brownish mesial lines in such examples, and the fore neck and chest with dark brownish streaks on a rufescent ground; the ruff is greyish, with darker longitudinal spots; the upper parts are glossy cinereous brown, and the wing-coverts rather darker brown, the broad fulvescent yellow edgings showing more conspicuously even than in young birds; the nape is light, and the feathers of the head and hind neck edged with rufous; the scapulars and tertials are tipped with a paler hue; the central tail-feathers are ashen grey, with six brown bars, the ground-colour of the three lateral feathers remaining buff; the forehead and eye-streak are whitish, and the gular patch greyish brown.

Obs. This species may be distinguished at all ages from *C. cinerarius* by its having the tip of the second primary-covert reaching to, or even overlapping, the notch on the second primary, by its closed wings not reaching within 1·3 inch of the tip of the tail, and sometimes falling short of it by 3 inches, and by its longer middle toe, this latter not exceeding 1·1 inch in the last species. In addition to these characteristics, the young, in which alone mistakes are likely to be made, may be recognized at a glance from *C. cinerarius* by the lighter-coloured ruff contrasting with the dark cheek- and ear-patch.

Distribution.—This handsome Harrier visits Ceylon, on its southward migration through India, about the commencement of October, and spreads in considerable numbers over the whole island, including the mountain-zone at its highest parts. Unlike its congeners, however, it remains behind in the island to a limited extent, those which do not leave being young birds, and they confine themselves in the wet season to the upper regions and the north coasts. Mr. Holdsworth has seen them in Nuwara Elliya in July and August, and I have met with specimens shot at the Elephant plains about the same time. I cannot but think, however, that such occurrences are rare exceptions, its remaining in Ceylon at this season being a most remarkable feature in this Harrier's economy. Mr. Holdsworth is an authority for its existence, out of season, in the north, as he observed it at Aripo throughout the year. On the opposite side of the island it is not seen during the S.W. monsoon; and I imagine that it is limited at that time to the north-west coast, on the plains of which, both species, this and the last, abound, attracted thither, no doubt, by the myriads of lizards which overrun these open wastes.

Besides the above locality, in the north, I have found the Pale Harrier numerous in the Jaffna peninsula and adjacent islands, at Mannar in the open pasture-lands and plains, in the great delta of the Mahawelliganga, and on the south bank of the Virgel, along the seaboard of the Eastern Province, and about the salt lakes or "leways"* of the Hambantota district. In the Western Province it is mostly confined to the paddy-lands and the marshes, round the large brackish lakes on the sea-coast, but in the northern half of the island it is found at all the large tanks of the interior. In the Kandyan Province it frequents the patna-hills and "plains" of the upper ranges, wanders over the open country in the coffee-districts, and is not unfrequently found in Dumbara.

The Pale Harrier is a bird of wide geographical range during the cold season of the northern hemisphere; and though it perhaps does not cover as much ground as its near ally the Hen-Harrier, its southern limits are more extended. It is common in some parts of Europe, and absent from other portions of that continent, not visiting, for instance, the British Isles. Lord Lilford does not seem to have noticed it in Spain, and Messrs. Harvie-Brown and Danford found it rare in Transylvania. In the small island of Heligoland Mr. Gätke records its occurrence. From Europe it extends southwards through Egypt and Eastern Africa to Cape Colony, and eastwards through Palestine, where Canon Tristram found it common, to Persia and Turkestan, in which latter highland Mr. Severtzoff and Dr. Scully observed it. Throughout India it is more or less abundant in the cold season, extending into Burmah and thence into China, where it has been procured on the Yangtsze river. In the peninsula of India it does not appear to remain in the breeding-season, Messrs. Adam, Butler, Ball, and others recording it only during the cool weather; and this makes its occurrence in Ceylon, the most southerly limit of its range, all the more strange during the south-west monsoon.

Habits.—Swainson's Harrier does not frequent openly-timbered plains or scrubby land, the favourite haunts of the last species, so much as swamps, marshes, rice-fields, and pasture-land, more particularly those in the vicinity of water. It passes most of its time on the wing, and rarely perches on any thing higher than a fence, preferring to rest from its labours on *terra firma*. Few, if any, of the Harriers exceed it in grace and ease of movement, and none are so skilful in sailing along close to the ground, or gliding with motionless wings just above the tops of the reeds or long swamp-grass. On espying its prey it suddenly closes its wings, or makes a quick turn, and drops like a stone upon the ground. By the margin of the extensive salt lakes on the Magam Puttu I have witnessed its powers of flight to great advantage. It would suddenly come into view above the top of the surrounding jungle, and sweep instantly down to the surface of the plain, along which it would skim for several hundred yards without any movement of its wings, and as easily rise over an intervening strip of wood, again to descend with rapid swoop, and glide along the shore of the glistening salt-pan, until, with a sudden but easy turn, it would commence to quarter backwards and forwards in search of food. Its favourite diet consists of the lizards (*Calotes*) which swarm on the open land in Ceylon; but it likewise captures mice in long grass, and frogs or beetles in the marshes which it frequents.

I once shot one at Jaffna in the act of swooping down on a wounded Gull-billed Tern; but its movement was most likely one of curiosity, as it would have had some difficulty in disposing of such large quarry. It must nevertheless frequently have the opportunity of picking up wounded or sickly birds of small size. It roosts on the ground; and Jerdon remarks that it is sometimes surprised at night by a jackal or fox. It has the same silent habit as other Harriers.

Nidification.—It has been clearly ascertained that this species does not breed within the limits of the Indian empire, in which very few specimens are seen after the month of April; and this fact renders its remaining in Ceylon during its regular nesting-season all the more singular. The birds that frequent the Indian region may no doubt breed in Kashgaria or in the steppes of Siberia; but I do not observe any account of its nidification in the writings of those who have visited the Central-Asian region. Its only known breeding-haunts are the steppes of South-eastern Russia, whence Dr. Bree figures two interesting specimens of its eggs in his 'Birds of Europe.' These are:—(1) pure white, unspotted; (2) white, with a few pale reddish blotches of moderate size, some of which are confluent round the small end. They measure 1·75 by 1·3 inch, and 1·8 by 1·35 inch. The nest is placed most likely on the ground among bushes and the stunted growth dotting the barren Russian steppes.

_____

* Shallow lagoons in which the annual salt formations take place.

D 2

ACCIPITRES.

FALCONIDÆ.

ACCIPITRINÆ.

Genus ASTUR.

Bill stouter than in the last genus, with the culmen not descending so suddenly from the base; festoon tolerably pronounced; cere large; nostrils oval, unprotected by bristles; lores scantily plumed. Wings short and rounded, the 4th and 5th quills subequal and longest, the first a little longer than half the fifth. Tarsus short, moderately stout, covered in front and behind with large transverse scutæ, or with a smooth plate in front as in the subgenus *Scelospizias*. Toes short, the inner toe reaching to the last joint of the middle one, the outer one slightly longer; claws well curved and acute.

ASTUR TRIVIRGATUS.

(THE CRESTED GOSHAWK.)

Falco trivirgatus, Temm. Pl. Col. i. pl. 303 (1824).
Astur trivirgatus, Cuv. Règ. An. i. p. 332 (1829); Gray, Gen. B. i. p. 27; Kelaart's Prodromus, Cat. p. 105; Layard, Ann. & Mag. N. H. 1853, xii. p. 104; Jerdon, B. of Ind. i. p. 47; Schl. Mus. P.-B. *Astures*, p. 22; id. Vog. Nederl. Ind., Valkv. pp. 18, 57, pl. 10; Holdsworth, P. Z. S. 1872, p. 410; Sharpe, Cat. Birds, i. p. 105 (1874).
Astur palumbarius, Jerd. Madr. Journ. x. p. 85 (1839).
Lophospizia trivirgatus, Hume, Rough Notes, i. p. 116; Gurney, Ibis, 1875, p. 35.
Lophospiza trivirgata, David & Oustalet, Ois. de la Chine, p. 22 (1877).
Sparrow-Hawk, Europeans in Ceylon.
Three-streaked Kestrel, Kelaart.
Gor-Besra, H., lit. " Mountain Besra ;" *Kokila dega*, Tel., lit. " Cuckoo Hawk " *(apud* Jerdon).
Ukussa, Sinhalese.

Adult male. Length to front of cere 14·25 to 14·8 inches; culmen from cere 0·7 to 0·78; wing 7·5 to 8·3; tail 6·25 to 7·0; tarsus 2·0 to 2·2; mid toe 1·1 to 1·2, its claw (straight) 0·5 to 0·58; hind claw (straight) 0·77; height of bill at cere 0·45.

Adult female. Length to front of cere 14·8 to 15·0 inches; culmen from cere 0·8; wing 8·0 to 8·5; tail 6·5 to 7·2; tarsus 2·2 to 2·4; mid toe 1·35, claw (straight) 0·6; hind claw 0·85.

Obs. The above measurements are from a series of Coylonese and Indian birds, including several examples from Malacca. Some birds from Malaya (Borneo, for instance) have the wing more than 9 inches in the female.

Iris golden yellow, in some examples beautifully pencilled with brown at the exterior edge; cere, gape, and eyelid greenish yellow, in some yellow; bill dark bluish brown, pale at the base, and with the tip blackish; legs and feet sickly or, sometimes, gamboge-yellow.

Hind neck, back, and wings glossy brown, in some old specimens with an ashen hue pervading these parts; forehead, crown, nape, crest, and face cinereous brown, generally with an ashen hue, particularly above and behind the eye: the crest, which is usually from 1½ to 1¾ inch in length, springs from the nape; upper tail-coverts deep brown, with the longer feathers broadly tipped with white; primaries and secondaries barred with dark brown, the under surface of the brown interspaces whitish; tail light drab-brown above and pale grey below, with a pale tip and four dark-brown bars: in the female there is usually a fifth bar concealed beneath the coverts.

Chin, throat, and under surface, from the chest downwards, white; a bold dark-brown chin-stripe, and the lower edge of the cheeks equally dark, generally forming a gular stripe; chest brownish rufous, the centres of many of the feathers darker than the edges; breast and flanks rather closely barred with deep brown; the thighs more closely barred with narrower bands of the same; under tail-coverts, in the female, with a few terminal bars of brown: under wing-coverts white, spotted with brown.

In a younger but still mature phase (in which I have found birds paired) the feathers of the lower part of the throat and centre of chest have broad white edges and bold central drops of dark brown, which pale off into rufous towards the sides of the chest, and there spread over the entire feather; the breast and flanks openly barred with broad bands of sepia-brown, and the thighs narrowly barred, generally, with a darker hue. This appears to be the commonest phase of what may be called the mature dress, the uniform-chested birds being rarely met with.

Young. Iris greenish yellow, sometimes mottled with brown; cere and eyelid greener than in the adult; legs and feet greenish yellow.

The nestling in first plumage is light smoky brown above, the bases and edges of the feathers very pale; head and crest very dark, the bases of the feathers tawny; quills barred much as in the adult, but the inner edges and interspaces white, shaded with tawny grey; tips of the secondaries and their coverts and those of the upper tail-coverts pale; tail light drab-brown, with either three or four visible bars across the centre, and an additional one at the base of the two lateral feathers; beneath white (some examples are much coloured with a rufous line), the fore neck and chest boldly streaked with dark brown, and the rest marked with oval, lighter brown spots; thigh-coverts *barred* with darker brown than the breast-spots.

The change towards the adult dress takes place by the darkening of the upper surface and the tips of the upper tail-coverts gradually becoming whiter; the sides of the chest at the same time become uniform rufous-brown, this colour spreading by degrees over the entire feather, except at the inner edge: the breast and flank-markings turn into bars, at first broad and far between, and then narrower, darker, and closer together.

Obs. The larger Nepal race, originally described as *Spizætus ruftinctus* by M'Clelland, has the wing varying, according to Mr. Hume, from 9·3 to 10·6 inches, while Jerdon gives that of a female as 11·5. Several examples I have measured in the British Museum exceed 10, and differ in the character of their plumage as well. Mr. Sharpe now considers this a good species, and Mr. Hume has always accepted it as such. Above, these birds are a more ruddy brown than the smaller species: there is no ashy tint; the upper tail-coverts and tail are tipped with a more subdued colour; the neck and chest are marked with very broad rufescent brown drops on a buff-white ground, and the markings of the under surface have a more rufescent character than in the small bird. Some Malayan specimens of the latter race which I have examined exhibit a marked similitude to these Himalayan birds in their coloration: and on the whole the South-Indian and Ceylonese races are the darkest, and more nearly resemble each other than those from any other two localities. The Formosan bird is evidently a larger race than ours, as Mr. Swinhoe records a female with a wing of 9·0 (Ibis, 1866, p. 395).

Distribution.—The Crested Goshawk is widely dispersed through the low country, inhabiting those parts which are covered with forest or heavy jungle. It is found pretty generally all through the jungles to the north of the Dedaru Oya; but I do not think it occurs in the Jaffna peninsula. In the wilds of the Eastern Province, and the thickly wooded country to the south of Haputale, it is tolerably frequent, but difficult of observation on account of its sylvan propensities. It is liable to be met with in most of the isolated forests or reserves in the Western Province, such as the Ambepussa Hills, the Ikkade Barawe forest near Hanwella, but chiefly, according to my own observations, during the north-east monsoon; the same may be said of the south-western corner of the island, where, from November until May, I have known it to occur about Amblangoda, Baddegama, and as near Galle as the Government reserve at Kottowe. Further inland, in the jungles of the Pasdun Korale and the district of Saffragam, it is doubtless resident and breeds. In the Peak Forests it is likewise not uncommon. As regards its range into the mountain-zone I do not know of its having been found above 3500 feet. About the neighbourhood of Kandy, and at Nilambe and Deltota, it is frequently shot, there being in general one or two examples in Messrs. Whyte and Co.'s establishment.

The *Gor-Besra*, as it is called in India, is spread over the peninsular portion of the empire, inhabiting the Nilghiris perhaps more commonly than other wooded regions. It does not appear to be an abundant species, as but few instances of its occurrence are recorded in 'Stray Feathers,' whereas frequent mention is made of its northern ally from the Himalayas, Nepal, Kumaon, and Assam. Our bird appears to be found in Pegu, as it is included in Mr. Oates's list, and to the south-east of Burmah it seems to have a very extended range, inhabiting Malacca, Java, Sumatra, portions of Borneo and the Philippine Islands, together with Formosa. From the island of Sumatra it seems to have been first known, Cuvier giving that island as its sole habitat. It does not extend eastwards from Burmah towards China, which is a singular feature in its distribution, seeing that it has such an extensive south-easterly range. Père David did not meet with it anywhere in the latter country; nor did Mr. Swinhoe in all his experience on the coast of the Celestial Empire.

Habits.—This bold bird is almost entirely a denizen of the forest, in the tallest trees of which I have usually met with it, giving out its shrill monosyllabic scream (or, more properly speaking, whistle) as a call-note, perhaps, to its mate, or in defiance of the group of small birds which very frequently are found haranguing it at a respectful distance. In this latter respect it much resembles its smaller cousin, the Besra (*Accipiter virgatus*); for I have more than once found it surrounded by a host of angry White-eyebrowed and Forest Bulbuls*, accompanied by one or two equally energetic Kingcrows, darting and flying round in the highest state of excitement, while the Goshawk, with an air of injured innocence, sat stolidly on the capacious limb of some enormous Koombook tree, screaming at its tormentors to the utmost of its powers. This habit of the small birds, I must here state, carries with it some amount of injustice; for though this hawk is frequently given a bad character for not respecting the life of his feathered friends, and appropriating for his larder sundry small chickens, pigeons, and that ilk, I have invariably found his food to consist of lizards, to none of which is he so partial as to the Green Calotes (*Calotes viridis*). I have shot him in the forests of the Vanui, screaming with delight over a brilliantly green Lizard which hung, pinned by his talons, to a branch, while his stomach was crammed with just such another. Layard, in his 'Notes on Ceylon Ornithology,' says that it swoops down to the poultry-yard from "some towering tree or butting rock, and, despite the fury and resistance of the faithful mother, rendered fiercer by despair, the foe generally carries off one, if not two, of her family."

Jerdon also remarks, in the 'Birds of India,' that "it is not very rare in the Neilgherries, and occasionally commits depredations on pigeons and chickens, making a pounce on them from a considerable height. It generally keeps to the woods or their skirts, dashing on birds sometimes from a perch on a tree, but generally circling over the woods, and making a sudden pounce on any suitable prey that offers itself."

Layard says that they are used by native falconers in Ceylon for hunting, and mentions that he saw one at Anaradjapura, which had been hoodwinked by having its eyelids sewn up, "the thread running through them so as to draw the edges together at pleasure." I have seldom seen it fly any distance, nor observed it far away from the outskirts of woods; but its progression from point to point in the forest is swift and performed with quick beatings of the wings.

It was formerly, according to Jerdon, used for falconry in India, and was taught to strike Partridges.

Nidification.—The nest of the Crested Goshawk does not appear to have ever been found. Mr. Hume has not succeeded in eliciting any information from his numerous correspondents concerning its nidification; and all we know concerning its breeding is what Layard tells us—that it nests in the "holes and crevices of precipitous rocks."

---

* *Ixus luteolus* and *Criniger ictericus*.

ASTUR BADIUS,

(THE INDIAN GOSHAWK.)

Falco badius, Gm. S. N. i. p. 280 (1788).

Accipiter dukhunensis, Sykes, P. Z. S. 1832, p. 79.

Accipiter badius, Strickl. Ann. N. H. xiii. p. 33 (1844); Kelaart's Prodromus, Cat. p. 115;
Layard, Ann. & Mag. N. H. 1853, xii. p. 104.

Micronisus badius, Bp. Consp. i. p. 33; Jerdon, B. of Ind. i. p. 48; Hume, Rough Notes, i.
p. 117; Holdsworth, P. Z. S. 1872, p. 411; Hume, Nests and Eggs, i. p. 24; Legge,
Ibis, 1875, p. 276.

Astur badius, Sharpe, Cat. Birds, i. p. 109 (1874); David and Oustalet, Ois. de la Chine, p. 24.

Scelospizias badius, Gurney, Ibis, 1875, p. 360.

The Brown Hawk, Brown, Ill. Zool. p. 6, pl. 3 (1776).

The Shikra, Jerdon.   *Indian Sparrow-Hawk*, popularly in India.

Shikra (female), *Chipka* (male), Hind.; *Chinna-Wallur*, Tam. (*apud* Jerdon).

Brown's Sparrow-Hawk, Kelaart.

Ukussa, Sinhalese south of Ceylon; *Kurula-goya* in north.

Adult male. Length to front of cere 11·5 to 12·8 inches; culmen from cere 0·6 to 0·63; wing 6·9 to 7·0; tail 5·5 to
0·2; tarsus 1·75 to 1·9; middle toe 1·0 to 1·12, its claw (straight) 0·4 to 0·45; hind toe 0·8, its claw (straight)
0·6; height of bill at cere 0·31 to 0·36.

The largest examples do not equal those from Northern India; the average length of wing of Ceylonese birds is about 7·3.

Iris usually light crimson or orange-red, in very old examples fine crimson; cere and orbitar skin greenish yellow.
the top of the generally greenish bill bluish, darkening at the tip; tarsi and foot yellow, the front of the tarsus
streaked with greenish.

Above bluish ashy, palest on the rump and upper tail-coverts; top of the head and the nape shaded with brownish,
and a ruddy tinge generally on the hind neck; quills ashy brown, the inner webs for two thirds of their length
from the base edged and barred with white, the brown interspaces being darker than the rest of the feather;
beyond the notch there are indications of darkish bars; tertials and scapulars with a large concealed white patch
down their centres; tail bluish grey, tipped with whitish; central feathers unbarred, but slightly darker towards
the tip; the outer feathers with faint brown bars towards the base of the inner web, the next with five bars on
the same web; the two adjacent feathers extend to the outer web; the barring of the outer
tail-feathers varies in extent even in birds which are similarly pale throughout their plumage; lores greyish; chin
and gorge white; cheeks, ear-coverts, and a narrow chin-stripe cinereous grey; chest, breast, and flanks pale
sienna-colour, narrowly barred with white, which in no two specimens is alike*, being in some open and in others
very close, particularly on the chest; belly, thighs, and under tail-coverts, with the sides of the upper coverts,
white, the bars gradually fading out on the lower breast; under wing and lower surface of quills rufescent white.

In a slightly younger stage of the adult plumage the upper surface is darker and pervaded with a cinereous hue; the
bars on the inner web of the outer tail-feather extend nearly to the tip, and on the adjacent one there are as
many as in the young bird.

Adult female. Length to front of cere 12·6 to 13·8; culmen from cere 0·63 to 0·65; wing 7·7 to 8·2; tail 6·5 to 6·8;
tarsus 2·0 to 2·2; mid toe 1·25.

Females, except perhaps those that are very old, are browner on the upper surface than males; the barring of the under

---

* In one remarkable specimen from Uva the entire under surface, from the throat to the lower breast, is openly
barred, the width of the white bands being the same throughout.

surface is bolder, and the brown bands have a more perceptible dark edging : they are also variable in hue, and are continued more to the lower parts than in the other sex, some examples having the thigh-coverts barred like the flanks : the outer tail-feathers, as demonstrated by Ceylonese examples at any rate, are seldom without very narrow bars on the inner webs.

Obs. A very marked difference exists between fully-aged birds and those that have just assumed the barred phase ; in the latter the upper surface is very brown, and the bands of the lower surface are far apart and conspicuously edged with brown, giving the whole an umber rather than a sienna appearance. In this stage Ceylonese examples of this Hawk resemble, on their under surface, the race characterized by Mr. Hume, in 'Stray Feathers' (vol. ii. p. 325), under the name of *Micronisus poliopsis*, and to which he contends the Pegu birds belong. The diagnosis of this species is, " Very similar to *M. badius*, Gmel., but larger, the adult males a paler and purer grey, wanting the nuchal rufescent collar and the central throat-stripe, and with the cheeks and ear-coverts unicolorous with the crown." The young birds are also described as having no more than four bands on the central tail-feathers, instead of five or six as in *M. badius*, and " in both sexes the barring of the lower surface seems on the whole broader and more strongly marked than in any specimens of true *badius*." As regards the latter feature Mr. Sharpe remarks (Cat. Birds, i. p. 110) that it is " banded with broader and brighter vinous bands than its near ally." The absence of the throat-stripe and the few caudal bars are valuable characteristics in differentiating it from Ceylonese *M. badius*, but not so the vinous bands ; in this respect *M. badius* appears to vary to a considerable extent, particularly as regards birds not fully aged ; and this inclined Mr. Gurney to consider the Ceylonese example spoken of (*l. c.*) to belong to the *poliopsis* race. I have, however, shown him specimens in my collection exhibiting this peculiarity, and he now considers it to be of no specific value.

Young. These attain the full dimensions in the first year.

Iris, at first greenish yellow, changing to saffron-yellow with age, and passing then through various shades of orange to the red hue of old birds ; cere and orbitar skin greenish, changing to yellow : legs and feet greenish yellow, the feet changing first to the adult yellow, and then the posterior part of the tarsus ; bill dark brown, with the base only bluish.

Head, upper surface, with the wing-coverts rich brown, pervaded with an ashy hue, and conspicuously edged with brownish rufous, which, on the scapulars and tertials, is fulvescent, and across the hind neck often pales into whitish ; the basal portion of the feathers is white, which shows more on the latter part than elsewhere : forehead, face, and above the eye buff-white, striated with brown, which coalesces on the ear-coverts with the rufous-brown of the sides of the head : quills brown, crossed by narrow dark bars (faint towards the tips), which show as blackish brown on the buff under-wing. Tail brownish grey, crossed on both webs of all but the lateral feathers with brown bars ; these are usually five in number on the central feathers in the male, and six in the female, the basal bar lying beneath the coverts ; on the remaining feathers the number varies, the penultimate in some examples having no more than the central feather, while others have six or seven according to sex.

Throat, fore neck, and under surface buff-white, the first-named part with a brown centre-stripe, and the rest of the feathers down to the belly with large umber-brown " drops " and dark shafts ; these vary much in individuals—pale and narrow in some, dark and heavy in others, particularly on the chest : on the thighs and lower parts they narrow almost into stripes ; under tail-coverts, in pale examples pure white, in dark, heavily-marked birds with narrow mesial stripes of brown.

In the younger stages the drops have a constant tendency to turn into bars, these latter being most prevalent on the flanks ; many birds, I believe, have a tendency to the bar-like form of marking from the first, although this does not lead to any quicker or more gradual assumption of the sienna barring, peculiar to the adult plumage ; for this is put on by a moult at once throughout the whole under surface, which takes place in some birds while the upper surface is still in the immature dress, and is sometimes mingled with the chest-drops and bold dark bars of the flanks. At this first moult to the adult dress the cheeks are generally streaked with brown on a pale ground.

Distribution.—The Shikra is distributed throughout the island, extending into and resident in most parts of the Kandyan Province. On the Nuwara Elliya plateau I have not observed it ; but it is no doubt a visitant to that elevated region during the dry season. It is not uncommon on the Fort MacDonald patnas, and I have procured it on Namtooni-Kuli Mountain, near Badulla, which has an elevation of more than 6000 feet ; it is also met with in Dimbulla and the Knuckles district, so that it may be said generally to affect the mountain-zone. In the interior of the lowlands it is resident ; and during the north-east monsoon it is common in the cultivated districts round the sea-coast, taking up its abode in the vicinity of human habitations. It is fond of establishing itself on cliffs, such as those at Trincomalie, and is frequently seen about the ramparts at Galle and

Jaffna. In the early part of May it retires into the interior to breed, and is not seen about its maritime haunts until October. In spite of this local migration to the sea-coast, the Shikra may be found throughout the year, in spots suitable to its habits, in most of the inland districts. In the Eastern Province I found it tolerably frequent in October; but scarcely met with it at all during two trips to the south-eastern forest districts. In the Western Province it is an inhabitant of the cocoa-nut districts bordering the sea-coast, retiring for the most part into the interior, as is the case on the east coast, during the south-west monsoon.

The Shikra is found pretty well all over the plains of India from the extreme south to the Himalayas, into which it ascends to an elevation of 5000 feet. It is a bird of local distribution, notwithstanding its extensive habitat. Mr. Hume speaks of it in Sindh as being not uncommon in the cultivated portions, but not found in the "desert or rocky tracts." Mr. V. Ball, again, says that it has a somewhat local distribution "in the large district of Chota Nagpur." It extends into Burmah and Malayana, and thence, according to Père David, into China, that is, if all the birds found in these regions belong to the true *badius* race; westward of Sindh it is found as far as Afghanistan; but this, I believe, is its furthest limit.

Habits.—This interesting little Hawk may be observed in every variety of situation but heavy forest. Cliffs on the sea-coast, rocky eminences in the interior, isolated groves of trees, cocoa-nut compounds surrounding native villages, the borders of paddy-fields and cinnamon-plantations dotted with large trees, are among the localities which it frequents. In the wilder parts of the country it is partial to "cheenas"* and new clearings in the forest, where it may be seen flying rapidly from tree to tree, or seated on a blackened stump discussing the remains of some lusty lizard. It affects coffee-plantations in the hills and bushy patnas, and is often seen in the vicinity of the bungalows, on the look-out, perhaps, for stray chickens. Its favourite diet is the ubiquitous lizard (*Calotes*), the remains of which I have found in every example dissected. It feeds also on mice and large beetles; and I once shot one on the Fort-MacDonald patnas in the act of darting at a Bulbul. It no doubt captures birds when pressed with hunger, but small reptiles and insects form the better part of its sustenance. It is commonly trained in India, and is taught to catch small game-birds; but its courageous disposition prompts it to attack (according to Jerdon) even "young Pea-fowl and small Herons." It is a persistent tormentor of both the Common and the Carrion-Crow in Ceylon, and may be often seen pursuing them high in the air, darting at them from above and beneath, much to the discomfiture of the "Corbies," who usually escape by a sudden swoop into the trees below. Its flight is a steady, straight-on-end movement, performed with quick beatings of the wings; but it sometimes soars to a considerable height, making quick circles, and then suddenly swoops down, alighting in an adjacent tree. It is a very noisy bird, making its shrill two-note whistle or scream heard for some distance, and furnishing a capital sound for the clever imitative powers of the Green Bulbul (*Phyllornis jerdoni*).

Mr. Ball remarks of it, in his avifauna of Chota Nagpur ('Stray Feathers,' 1874), "that at the season of the jungle-fires numbers of these birds assemble to hunt the grasshoppers and other orthopterous insects which are compelled to take flight before the advancing flames." Another writer, Mr. Thompson, says that they are very fond of frogs.

Nidification.—I have never succeeded in getting the eggs of this Hawk in Ceylon, though it must breed freely in the interior and not very far from the sea-coast. The nesting-season, I have ascertained from dissection of many examples, is from April to June; and it retires to sequestered jungles to rear its young, as I have met with it in the wilds of the interior at this season in a state of breeding. In India it breeds in April and May, and, in some parts, in June. The nest, writes Mr. Hume, "is usually placed in a fork high up and near the top of the tree. It is but loosely built of twigs and smaller sticks, lined with fine grass-roots, and averages about 10 inches in diameter." As architects he does not attribute to them much talent, remarking that they take "a full month in preparing their nest, only putting in two or three twigs a day, which they place and replace as if they were very particular and had a great eye for a handsome nest; whereas, after all their fuss and bother, the nest is a loose ragged-looking affair, that no respectable crow would condescend to lay in!" The eggs are usually three, but sometimes four; they are oval in shape and smooth in texture; they are delicate, pale bluish white, either devoid of markings or sprinkled openly throughout with faint greyish specks and spots. They average 1·55 by 1·22 inch.

---

* Land cleared by the natives for the purposes of cultivation.

E

FALCONIDÆ.

ACCIPITRINÆ.

Genus ACCIPITER.

Bill slightly shorter and more feeble than in *Astur*; festoon equally prominent. Nostrils large, oval, protected by the loral plumes. Wings similar to *Astur*, but the first quill longer; tail longer. Tarsi long, slender, the scutæ less pronounced; middle toe long, the inner reaching only to the first joint. Structurally of slender form.

ACCIPITER VIRGATUS.

(THE JUNGLE SPARROW-HAWK.)

Falco virgatus, Temm. Pl. Col. i. pl. 109 (1823).

Accipiter virgatus, Vig. Zool. Journ. i. p. 338 (1824); Blyth, Cat. B. Mus. A. S. B. p. 22 (1849); Jerdon, B. of Ind. i. p. 52; Hume, Rough Notes, i. p. 132; Jerd. Ibis, 1871, p. 243; Holdsworth, P. Z. S. 1872, p. 411 (first record from Ceylon); Hume, Stray Feathers, 1874, p. 141; Legge, Ibis, 1874, p. 10, and 1875, p. 276; Sharpe, Cat. Birds, i. p. 150; Gurney, Ibis, 1875, pp. 480–83; David and Oustalet, Ois. de la Chine, p. 26 (1877).

Nisus virgatus, Less. Man. d'Orn. i. p. 97 (1828); Schlegel, Vog. Nederl. Ind., Valkv. pp. 20, 59, pl. 12. figs. 1–4 (1866).

Accipiter stevensoni, Gurney, Ibis, 1863, p. 447, pl. xi.

Accipiter besra, Jerd. Madr. Journal, x. p. 84 (1839); id. Ill. Ind. Orn. pl. 4 (1847).

The Besra Sparrow-Hawk, Jerdon; *Besra*, popularly in India.

Besra (female), *Dhoti* (male), Hind. (*apud* Jerdon).

Jungle-Hawk, Europeans in Ceylon.

Yao, Chinese at Pekin (Père David).

Ukussa, Sinhalese.

Adult male (Ceylon). Length to front of cere 10·0 to 10·3; culmen from cere 0·5; wing 6·0 to 6·4; tail 4·6 to 5·0; tarsus 1·9 to 2·05; middle toe 1·2 to 1·25, its claw (straight) 0·35 to 0·4; height of bill at cere 0·27. Iris yellow; cere, loral skin, and eyelid yellow; the top of the cere sometimes greenish; bill dark horn, base and near the gape bluish; front of tarsus greenish yellow; posterior part with the sides of the toes and the soles lemon-yellow.

In the fully aged bird the head, hind neck, back, and wings are very dark ashen, the head deeper than the rest and concolorous with the cheeks and ear-coverts; frequently a brownish wash is perceptible on the back; the hind neck often with a rufous hue; quills brown, with a series of faint lighter bars, which show whitish beneath on the terminal half and buff towards the base; secondaries and tertials barred near the inner edge with white; tail brownish ashy above, tipped pale and crossed with four dark bands, the terminal one at the tip and the basal sometimes concealed beneath the coverts; on the inner web of the lateral feathers there are five indistinct bands, which fade out entirely in very old birds.

Chin and throat buff-white, with a broad central stripe of dark slate-colour; chest, sides of breast, and flanks uniform rufous, or sometimes with a few white streaks, caused by the edges of underlying brown-centred feathers; centre of the breast and the belly barred with rufous on a white ground; thighs rufous, barred slightly with pale grey: under tail-coverts pure white; lower surface of the tail grey: under wing-coverts buff-white, spotted with brown.

Obs. The under surface varies considerably in birds which have not quite reached the above fully-matured dress; these have usually an ashen hue on the sides of chest, the edges of centre chest-feathers white, and the middle of the lower breast and belly plainly barred with rufous ashou. Other examples have the white barring continued across the whole breast to the flanks, and in these the thighs are boldly barred.

Adult female. Length to front of cere 13·0 to 13·3; culmen from cere 0·52: wing 7·5 to 8·0; tail 0·0 to 6·5; tarsus 2·2 to 2·4: extent of wing 24·0 to 25·5.

Ceylon females do not often exceed 7·6 in the wing, the above limit applying to a fine specimen from Northern India in the British Museum.

Iris yellow, in some orange-yellow, with a dark outer rim occasionally; bill, legs, and feet as in male.

Hind neck, back, and wings smoky brown, but the head and nape similar to those of the male; the cheeks and ear-coverts paler than the crown: the light portions of the tail have the same smoky hue instead of being ashy as in the male: throat, fore neck, and centre of the chest white, the latter part boldly dashed with dark brown, running into the broad chin-stripe above; sides of the fore neck and chest rufous, the latter, together with the breast, flanks, and lower parts, boldly barred with white, the interspaces being rufous on the upper parts, and rufous ashen on the belly and thighs; under tail-coverts white, in some with terminal streaks of brown; under wing-coverts as in the male.

The female appears never to acquire the uniform rufous breast of the male; and the above description represents, I

ACCIPITER NISUS.

(THE SPARROW-HAWK.)

Falco nisus, Linn. S. N. i. p. 130 (1766).
Accipiter nisus, Pall. Zoogr. Rosso-As. i. p. 370; Gray, Gen. Birds, i. p. 29, pl. 10. fig. 4; Kelaart's Prodromus, Cat. p. 115; Layard, Ann. & Mag. Nat. Hist. 1853, xii. p. 104; Jerd. B. of Ind. i. p. 51; Hume, Rough Notes, i. p. 124; Sharpe and Dresser, B. of Europe, pt. ix.; Sharpe, Cat. Birds, i. p. 132 (1874).
Basha (female), *Bashin* (male), Hind., *apud* Jerdon; *Karghai,* Turki (Scully).

Adult female (India). Length to front of cere 14·0 to 16·0 inches; culmen 0·55; wing 8·0 to 10·0; tail 7·5 to 8·0; tarsus 2·3 to 2·5; middle toe 1·6 to 1·8, claw (straight) 0·5; height of bill at cere 0·3.

Adult male (India). Length to front of cere 11·8 to 12·2; culmen from cere 0·5; wing 8·0 to 8·3; tail 6·0 to 6·4; tarsus 2·2 to 2·3; middle toe 1·35.

Iris varying from saffron-yellow to orange-yellow; cere yellow; bill dark horn, bluish at the base; legs and feet gamboge-yellow; claws black.

Male. Above dusky slate-colour, darkest on the head, and more so on the upper back than on the rump; the feathers at the sides of the hind neck edged with rufous, and those at the back with white bases; quills ashy brown, the terminal portions of the primary outer webs greyish, the inner webs barred widely with brown and white internally towards the base; tail greyish brown, with four or five brown bars, the subterminal one the broadest.

A lightish space just above the lores; cheeks and ear-coverts more or less rufous; throat whitish, washed with rufous

E 2

think, the limit of this coloration. It is taken from a bird shot near Trincomalie, containing an egg ready for expulsion ; and Mr. Sharpe, with his wide experience of the Accipitrinæ, remarked of this specimen that it was one of the oldest he had ever seen.

Young. Iris greenish yellow, sometimes mottled with brown specks ; cere dull brownish green or greenish yellow : eyelid yellowish green ; legs and feet greener in front than in adults ; bill duskier.

The bird of the year has the head and nape deep brown, tinged with ashen ; a whitish eye-stripe or supercilium ; the bases of the nape-feathers white, showing on the surface more or less ; the upper surface is chocolate-brown, edged with brownish rufous, brightest on the hind neck (and deeper throughout in the female) ; tips of the secondaries and tertials paler than those of the back feathers ; quills barred with dark brown, the interspaces whitish at the inner edges ; tail pale smoky, crossed by four bands, as broad as the interspaces, the terminal one at the tip ; the inner web of the lateral feathers with 5 or 6 narrow light bars.

Throat and entire under surface buff-white, the chest and upper breast-feathers edged with rufescent buff or yellowish buff (in the female) ; a broad throat-stripe and long oval drops on the neck and chest of sepia-brown ; the sides of the latter part brownish rufous ; the lower parts with rounder spots of a lighter hue ; flanks barred with rufous-brown ; thighs with bold spots of brown ; under tail-coverts narrowly streaked with the same ; under wing-coverts buff, handsomely spotted with dark brown.

Obs. In this species a great variety of coloration in the plumage of the male is met with between the youngest phase and that noticed above of moderately mature birds, but notwithstanding the rufous character of the chest commences directly to assert itself, and serves to distinguish it from the opposite sex. By a change of feather in the first year the sides of the chest become rufous, the centre of the breast assumes a bar-like form of marking, while the flanks and thigh-coverts become regularly banded with rufous-brown. After the next moult the white centre of the chest becomes dashed with rufous and ashen streaks, and the flanks and sides of the breast assume their rufous covering, and present the appearance described above in not fully matured males ; this is accompanied by the assumption of the cinereous upper surface and the consequent disappearance of the rufous edgings. In some birds of the second year the chest is striped with rufous, and the surrounding white portions of the feathers washed with the same. Malabar specimens are identical in size and character with Ceylonese ; and an example from Chofoo in the British Museum corresponds as regards size with birds from Ceylon.

Distribution.—The Besra was first recorded as a Ceylonese bird by Mr. Holdsworth (*l. c.*). It is, however, a common species in the island, and, as Mr. Holdsworth remarks, may have been the bird referred to by

---

and with the shafts dusky ; chest, breast, flanks, and lower parts whitish, barred somewhat narrowly with rufous-brown bars edged with rufous ; on the sides of the chest the bars are broadest, and on the abdomen they are wide apart ; thighs narrowly barred, the insides more or less tinged with rufous, and a patch of the same on the lower part of each flank ; under tail-coverts whitish or rufescent white, banded with narrow pointed bars of brown. Examples with marked rufous cheeks have the rufous portions of the lower parts and the under wing-coverts of a corresponding intensity.

Female. Less ashy above, the head and hind neck dark as in the male, and the latter part much edged with rufous in some examples ; tail with an additional bar, there being always five on the central feathers ; the markings of the under surface are browner, the darker hue predominating on the bars, which are only edged with rufous, and which are likewise more pointed than in the male ; the chin and throat fulvous, with dark shafts to the feathers : under wing-coverts white, barred with dark brown.

Young (nestling). " Clothed with white down ; the feathers of the back deep sepia-brown, with rufous margins ; breast fulvous fawn, the chest longitudinally streaked with brown, inclining to arrow-head markings on the abdomen and to bars on the flanks." (*Sharpe,* Cat. Birds.)

Bird of the year. Iris paler yellow than in the adult ; bill paler and yellowish at the base beneath.
Above brown, the feathers edged with rufous, and the nape marked with white, arising from the exposed basal portions of the webs ; crown darker than the hind neck ; quills rufescent white on the inner webs from the notch to base, both webs conspicuously barred with dark brown ; tail brown, with five or six broadish bands of a darker hue, the lateral feathers with an additional bar and the inner webs pale.
Cheeks and ear-coverts brown, striated with whitish ; throat white, with broad mesial brownish stripes ; under surface

Kelaart as the European Sparrow-Hawk. It is possible, nevertheless, that the Doctor's identification may have been correct; and in support of the idea that the European Sparrow-Hawk may have occurred in Ceylon, I would here remark that I have lately received a specimen of the European Hobby from Ramisserum, auguring favourably for the occurrence of other northern Hawks in the latitude of Ceylon. As to the present species it is widely distributed in the low country and a frequent bird on the hills, ranging into the jungles of the main range, whence I possess an example killed at Nuwara Elliya. It is not uncommon in the northern forests, in the Eastern Province, and in the south-western hill district. I have obtained it at Baddegama and in Saffragam, and have met with it in other forests on the west side of the island. It is frequently obtained in the Kandy district and in the surrounding ranges, whence it figures now and then in the collections of Messrs. Whyte and Co., of Kandy.

This species is, according to Jerdon, found in all the large forests of India, inhabiting the Nilghiris, the Eastern Ghats in places, the Malabar and Central-Indian forests, and the slopes of the Himalayas. It is likewise an inhabitant of Burmah, the Malaccan peninsula, the Andaman Islands, Java, Borneo, Timor, and the Philippine Islands, and to the eastward of Burmah extends into China, Siberia, and Japan, if the birds from the latter country do not all belong to Mr. Gurney's larger, short-legged race *A. stevensoni*. It is not a very common bird in India, for most of the writers in 'Stray Feathers' speak of it as being local in the regions they treat of. Captain Feilden appears to be the only one who has procured it of late years in Burmah; and in the north-east of India, in the Mount Aboo district, but few specimens have been obtained.

In China, however, Père David says it arrives in the spring at Pekin in great numbers, and breeds in the mountains of the provinces.

Habits.—This little Sparrow-Hawk is a denizen of the jungle, rarely coming into the open country at any distance from its sylvan haunts. I have frequently met with it in pairs, both old and young, and have always found it a noisy bird, haranguing its feathered companions of the woods, who oftentimes collect in excited mobs and annoy it with their incessant clatterings. It generally perches on the large limbs of trees and flies from one to another, uttering its loud and shrill squeal, which somewhat resembles that of the little Goshawk (*Astur badius*). Its cries must often lead to its discovery in the jungles which it frequents, as on all occasions on which I have either met with it or shot it I have traced it by its note, which can be heard at some distance in the stillness of the primeval forest. It is shy, as a rule; but on one occasion, finding three immature birds

white, boldly banded with spear-shaped broadish bands of brown and rufous, the latter hue mostly confined to the centre of the bar; the markings of the flanks darker than the rest; thighs barred with brown; under wing-coverts buff, with arrow-headed spots of brown; under tail-coverts whitish, unmarked.

Obs. This plumage is acquired by the dissolving of the longitudinal streaks of brown into the bold bar-shaped markings: this is well shown in the series of feathers given by Mr. Sharpe in an article on the subject (P. Z. S. 1873, pl. 39), and from which it can be observed that the longitudinal "drop" in the process of its dissolution expands at various points into bars, the white portion of the web advancing as an interspace to the shaft, leaving, however, at this first stage, pointed projections at the lower side of the bar, these being in reality the remains of the stripe.

Note.—I can neither include this, nor one or two other species to be noticed further on, as undoubted Ceylonese birds, as their occurrence in the island is, as in the present case, matter of uncertainty, or has been accepted from mere visual testimony. It appears advisable, however, to include them in footnotes, in order that sufficient information may be given to enable my Ceylon readers to identify the species should they occur hereafter in the island.

Distribution.—The evidence as to the occurrence of this species in Ceylon is summed up in the following sentence (Kelaart, Prodromus, p. 90):—"*Accipiter nisus* is very rare; we have only seen one *live* specimen." It is possible, as I have remarked in my article on *Accipiter virgatus*, that Kelaart may have been correct in his identification; but it must be remembered that, though a naturalist, he did not make ornithology his study, and those birds which were collected for him (he never used a gun himself) were identified by others, chiefly by Blyth, I believe. In the case of this bird, the specimen was a living one, so that I incline to the belief that it was a Jungle Sparrow-Hawk and not the European species. The latter is a cold-weather visitant to India, and is spread during that season, sparingly, over the whole of the empire. It is always to be found in the Nilghiris, on the Eastern Ghauts, and other hilly portions of Central India. It together, I shot one, and the others were so tame as to fly out of the tree and immediately return to it again, one of them thereby following his companion into my collection. The diet of the Besra consists of small reptiles, coleoptera, and other large insects, the lizard (*Calotes*) being its favourite food in Ceylon. In India it is, says Jerdon, highly esteemed among native falconers, and is caught by means of a trap called there " Do Guz." This is a small, dark-coloured net, fixed to two thin bamboos lightly stuck in the ground, and which give way on the bird striking the net while it is dashing at a decoy picketed in front of it. On this happening the meshes instantly fold round the hawk and effectually prevent its escape. It is flown at partridges, snipes, and doves, and " is particularly active and clever in the jungle." The male is, however, according to the same writer, rarely trained. I do not think it is in the habit of soaring as much as its European ally. I have on one occasion seen it taking a few small circles in the air; but they were quickly over, and it again dashed off to its sylvan haunts.

Nidification.—In Ceylon this Hawk breeds about the month of May, during which I once procured a female containing an egg almost ready for expulsion, but which was unfortunately broken by my shot. It was of a pale green colour and unspotted, but would have most likely received some markings had the bird lived to lay it. In India nothing seems to be known concerning its nidification, and I never heard of its nest being found in Ceylon.

---

is common on the southern slopes of the Himalayas, where also Mr. Hume's larger race (*A. melanochistus*) has its home. It is doubtfully recorded from Tenasserim; and in 1870 Mr. Hume received a specimen from the Andamans, which is its most south-easterly limit.

Dr. Jerdon remarks that " it comes in very regularly about the beginning of October, and leaves again about the end of February or March according to the locality."

It is spread over the whole of Europe, including Great Britain, and extends through Central Asia to China, and southwards from the Mediterranean into Algeria and north-east Africa.

Habits.—The Sparrow-Hawk frequents wooded country, and preys on small birds and quadrupeds. It is a bird of powerful flight, but not so active as its Indian congener, the Besra; but it is nevertheless trained for falconry in some parts of India. It is a bird of predatory disposition, and consequently it is under a ban in a game-preserving country like England. It is described in Yarrell's ' British Birds ' as being so " daring during the season in which its own nestlings require to be provided with food as frequently to venture among the out-buildings of the farmhouse, where it has been observed to rapidly skim over the poultry-yard, snatch up a chick, and got off with it in an instant."

Nidification.—This species breeds sparingly in the Himalayas up to 8000 feet, building, as it does in Europe, in trees. In England it often takes possession of the nest of a crow, and repairs the lining for the reception of its own eggs. These are four or five in number, of a bluish-white or greenish-white ground, handsomely blotched and spotted with rich reddish brown or brownish crimson, the markings being sometimes collected in a zone near one end. Dimensions 1·7 by 1·3 inch. The beautiful specimens figured by Mr. Hewitson (plate vii. figs. 2 & 3) represent in a very interesting manner the variety in the eggs of the Sparrow-Hawk. The first is openly and handsomely blotched throughout with rich sepia, softened at the edges over other markings of light brownish, while the second has the obtuse end covered with confluent clouds of sepia-brown, overlying rather small and somewhat lineated blots, which are scattered rather thickly over the entire surface.

ACCIPITRES.

FALCONIDÆ.

BUTEONINÆ.

Bill weak, the festoon usually less developed than in the last subfamily; wings longer; tail moderate; tibia longer in proportion to the tarsus than in *Accipitrinæ*, the difference being " more than the length of the hind claw "[*]; outer and middle toe connected at the base as in *Accipitrinæ*.

Genus BUTEO.

Bill small and short, sloping from the base, the cere considerably advanced, and the comissure nearly straight, the festoon being only slightly developed. Wings long, pointed, the 4th quill the longest, or subequal with the 3rd. Tail compact, moderate in length, the feathers rigid; tip reaching beyond the closed wings and cuneate in shape. Tarsus moderately stout, the upper portion plumed more or less below the knee, the rest protected by broad transverse scales in front and *behind*. Toes shortish, the inner reaching to the last joint of the middle; claws short and moderately curved.

BUTEO PLUMIPES.

(THE INDIAN BUZZARD.)

Circus plumipes, Hodgson, Gray's Zool. Misc. p. 81 (1844).
Buteo plumipes, Hodgs. P. Z. S. 1845, p. 37; Jerd. B. of Ind. i. p. 91; Hume, Rough Notes, ii.
    p. 285; Jerdon, Ibis, 1871, p. 340; Hume, Str. Feath. vol. iii. p. 358, et vol. v. p. 347;
    Gurney, Str. Feath. vol. v. p. 65; Sharpe, Cat. Birds, i. p. 180, pl. vii. fig. 1 (1874).
Buteo vulgaris japonicus, Temm. and Schlegel, Faun. Jap. pls. vi. & vi. n.
Buteo japonicus, Bp. Consp. i. p. 18 (1850); Schl. Mus. P.-B. *Buteones*, p. 7 (1862); Blakist.
    Ibis, 1862, p. 314; Swinhoe, Ibis, 1870, p. 87; Jerdon, Ibis, 1871, p. 337; David and
    Oustalet, Ois. de la Chine, p. 19 (1877).
Buteo desertorum, Jerdon, Ibis, 1871, p. 338.
La Buse commune du Japon, Schl. Faun. Jap. p. 16.
The Harrier-Buzzard, Jerdon, B. of Ind.
Kara-Sä (in common with other Buzzards), Turkestan (Dr. Scully).
Kurula-goya, Sinhalese.

[*] Sharpe, Cat. Birds, i. p. 158.

Adult female. Length to front of cere 18·5 to 19·0 inches; culmen from cere 0·85; wing 15·0 to 16·5; tail 7·5 to 9·0; tarsus 2·6 to 3·2, bare front of tarsus 0·9 to 1·35; mid toe 1·5, its claw (straight) 0·7; height of bill at cere 0·4.

Adult male. Length to front of cere 18·0 to 18·5 inches; wing 13·5 to 15·6; tarsus 2·7 to 2·9, bare front of tarsus 0·9 to 1·5.

Obs. In the series from which the latter measurements are taken is included what appears to be an immature though not a very young bird in my own collection, from the south of Ceylon. Its detailed dimensions are:—Length to tip of bill 18·25 inches; wing 13·5; tail 7·25; tarsus 3·0, bare portion of front of tarsus 1·5; mid toe 1·5.

Iris dull yellowish mingled with brown or light hazel; cere varying from greenish yellow to yellow; gape yellow; bill blackish; legs and feet citron-yellow, claws black.

Above sepia-brown, dark and uniform on the forehead and back, and pale on the hind neck and greater wing-coverts, the feathers more or less margined with rufous mostly on the hind neck, scapulars, and wing-coverts, on the first of which the white bases of the feathers show considerably, and there is a dark nuchal patch; primaries and their coverts dark brown, the outer webs of the longer quills washed with greyish, the inner webs white internally and crossed with narrow bars of brown; secondaries paler, dark near the tips and with both webs barred, the white portions of the inner webs washed, in some, with rufous; the lateral feathers of the upper tail-coverts broadly margined with rufous and some barred with the same; tail rufous or brownish rufous, more or less shaded with brown, and washed at the margins of the rectrices with greyish, tipped with dull buff, and with a softened subterminal band and a number of narrow bars (incomplete in old birds towards the base) of brown; lateral feathers white internally.

Lores and a superciliary line blackish, a postorbital and moustachial streak dark brown, the feathers, as on the ear-coverts, pale-edged; throat whitish striped with brown; sides of the neck and chest rufous, in some brown, the shafts dark, and the margins of the feathers indented with rufescent whitish, which in some examples is conspicuous on the centre of the chest; breast and belly whitish or rufescent white, the feathers dark-shafted and barred with brown, in some on the lower breast, while other examples have the breast crossed with a wash of dark brown; lower flanks cinereous brown, greyish in old birds; thighs rufous, more or less cross-marked with brown; under tail-coverts fulvous, barred with rufous-brown; under wing whitish, painted down the centre with rufous and barred with brown.

Obs. The above description is taken from a number of examples in the British Museum, and is intended to embody as much as possible the characteristics of the very variable plumage in this species. Scarcely any two specimens are alike on the under surface; the older the bird the more covered with rufous-brown are the lower parts, and the less conspicuously barred is the tail. Many individuals exhibit a fuliginous phase, which is thought to be the result of old age, and which I will notice here as such, remarking, first of all, that such an example formed the type of Hodgson's species, which has been figured in Mr. Sharpe's admirable catalogue of the Accipitres.

Dark phase in old bird. In this the head, hind neck, and back, together with the wings, are uniform brown; tail dark brown, with the bands crossing the feathers completely, the subterminal one much marked; beneath, almost uniform brown, the centre of the breast alone being crossed with paler bands. In an example from Etawah the under surface is very dark, but the feathers have paler lateral margins, and the under tail-coverts are brownish buff, banded like ordinary adult birds, showing thus a remnant of the usual mature plumage, and demonstrating the fact that the fuliginous coloration has been a further advance beyond that stage and is the result of old age.

Young. Similar to mature birds described above, although scarcely any two specimens are alike. The primaries are paler brown, and have not the outer webs washed with ashy; the ground-colour of the upper tail-coverts not so pervaded with ashy; tail very variable, sandy brown, brownish grey, or greyish rufous, plainly barred on the central feathers with rather wavy bands of brown, uniting with the darker margin of the feathers, and the inner webs of the lateral feathers not so white as in adults.

Edge of forehead whitish, cheeks whitish striped with brown, the moustachial stripe streaked with white; throat and all beneath white or whitish buff; the chest and fore neck more or less broadly striped with brown, the markings coalescing down the sides of the fore neck in some; sides of the lower breast generally brown, uniting with the dark flanks; thighs fulvous, with brownish-rufous markings, in some showing indications of bars; abdomen and under tail-coverts buff, spotted with rufous; under wing whiter than in the adult, and the primary under-coverts with less brown on the terminal portions; basal half of primaries beneath pure white.

With age the thighs and flanks commence first to darken and the central rectrices lose their plainly defined bars, the brown hue gradually diminishing at the edge of the feather.

The following is a description of the Southern Ceylon example above alluded to :—

Head, hind neck, back, and wings sepia-brown ; the mantle, wing-coverts, and rump with moderately deep rufous margins ; the concealed edges of the scapulars and wing-coverts indented with whitish bars ; the margins of the head and hind-neck feathers fulvescent whitish ; the nuchal feathers dark brown at the tips, elongated as a rudimentary crest and showing much white at the base ; primaries very dark brown, washed with grey on the outer webs, particularly about the notch ; the inner webs almost entirely white from the notch upwards ; secondaries paler brown, the internal portions white, crossed with narrow incomplete bars of dark brown ; lateral upper tail-coverts rufous at the edges, and the concealed portions barred with rufescent white ; general hue of tail rufous ashy, crossed with numerous narrow bars of dark brown, tipped with fulvous, the subterminal bar broader than the rest, the internal portions of all the lateral feathers white ; inner webs of the central pair paling into white near the shaft.

Loral plumes dark with white bases ; a narrow blackish line beneath the eye and a brownish postorbital stripe, beneath which the ear-coverts are whitish, narrowly lineated with rufous-brown ; chin and throat buff-white, openly striated with narrow lines of brown and bounded on either side by a plainly marked brown moustachial streak ; chest and under surface whitish buff, the former with large rufous-brown terminal patches almost covering the feather ; the breast with smaller and indented central patches of the same ; lower flanks well covered with brown, and the sides of the abdomen marked with pointed bars of rufous-brown ; thighs in front and at the sides brown, with indistinct bars of rufous ; interiorly fulvous, patched with brown ; under tail-coverts with a few terminal spots of brown ; under wing-coverts rufous, tipped paler and centred with brown ; greater series uniform dark brown.

Obs. The African Buzzard (*Buteo desertorum*), with which Indian examples of the present species have until lately been confounded, is a smaller bird, the limit of the length of wing in the male being, according to Mr. Gurney's investigations, 13·5 to 15·4 inches, and in the female 14·3 to 16·85 inches. In their plumage, however, some specimens of this species so closely resemble the Indian bird that it is difficult to define the differences which exist by a mere description. It is not my province here to go into this matter, as the African bird is not likely to find its way to Ceylon. I will remark, however, that the dimensions of my bird from Southern Ceylon are low enough to relegate it to the ranks of the African species ; but the locality in which it was shot, coupled with the fact of Lord Tweeddale possessing an unmistakable example from Ceylon, makes it necessary to refer my bird (in spite of its diminutive size and comparative large amount of bare tarsi) to the Asiatic form. Mr. Gurney, who carefully examined the specimen, supports this view, and informs me that he has never heard of a true *B. desertorum* having been procured to the eastward of Erzeroom. Furthermore Mr. Hume, in his exhaustive notice of the various Indian Buzzards ('Stray Feathers,' vol. iii. p. 58), removes *B. desertorum* from the Indian avifauna, assigning the specimens from the Nilghiris, formerly referred by him to this species, to the subject of the present article.

Distribution.—This interesting Buzzard, the Asiatic representative of the European *B. vulgaris*, is a very rare visitor to Ceylon, which island forms the most southerly limit of its wanderings in the cool season. Not more than two instances of its capture are known to me—the first of which is that of a large female in the museum of the Marquis of Tweeddale, and the second that of the example above alluded to in my own collection. The former was procured about the year 1865 by Mr. Spencer Chapman, but from what exact locality his Lordship is unable to inform me. I understand that the major portion of Mr. Chapman's collections was made in the west and north-west of the island, in one of which districts the Buzzard was probably met with in its passage from the Malabar coast to Ceylon. The specimen in my possession was shot in October 1871 at Maha Modera, a few miles to the north of the port of Galle, by Mr. Wylde, a gentleman for some time resident at the latter place. It had been haunting the vicinity of the bungalow for several days, having made its appearance there after the prevalence of high northerly winds, which usually bring down many of the Ceylonese migrants from the coast of India[*].

Dr. Jerdon ('Ibis,' 1871, p. 338) writes, under the head of *Buteo desertorum*, that this species has been sent from Ceylon ; but he probably refers to the specimen above mentioned as procured by Mr. Chapman, and which Mr. Holdsworth alludes to in his catalogue (*loc. cit.*).

---

[*] This bird was referred to by me ('Stray Feathers,' vol. i. p. 488) as *Butastur teesa* ; this, however, was a mistake, as the latter is a much smaller bird, and is now removed to a different subfamily, chiefly on account of the character of the scales on the hinder part of the tarsus.

In the south of the Indian peninsula the Harrier-Buzzard is found, during the cool season, in the Travancore and Nilghiri hills.

With regard to the former locality, Mr. Bourdillon, as quoted by Mr. Hume in his "First List of the Birds from the Travancore Hills" ('Stray Feathers,' 1876), says:—"This bird, a winter visitor, seems not to be uncommon during December, January, and February." From the Nilghiris Mr. Hume himself records it.

In the north of India it is found in Nepal, whence Mr. Hodgson's original specimen of *Buteo plumipes* came, along the southern slopes of the Himalayas to Sikhim, and thence into British Burmah, where Captain Feilden procured it in the province of Upper Pegu. On the north of the Snowy range it is found as a winter visitor in Kashgar, though Dr. Stoliczka, during his excursion to that remote region, met with it but rarely. Another observer, however, Dr. Scully, in his valuable "Contribution to the Ornithology of Eastern Turkestan" ('Stray Feathers,' 1876), mentions, at p. 125, the shooting of three examples at Yarkand in January, and this locality appears to form the westernmost limit of its range. He further remarks that it is common there during the winter, but was never met with in the plains after that season was fairly over, having moved away northwards about the 20th of April.

It is remarkable that when a movement of these birds *does* take place southwards in winter so many remain in the great upland of Turkestan, which, one would think, must possess quite as rigid a climate as the more northerly lower-lying regions, where they no doubt breed, and which may very likely be the mountainous country bounding the vast Mongolian empire on the north. Jerdon, however, in his note on *B. japonicus* ('Ibis,' 1871, p. 337), writes that he procured it "at Darjeeling, in Kumaon, and in Kashmir in *summer*, at a height of from 9000 to 10,000 feet," which savours much of its breeding in the higher parts of the outlying Himalayas.

The vast territory lying between the Himalayas and Eastern China has been but little explored, and therefore this Buzzard has not yet been recorded from it, though it doubtless inhabits, at one season or other of the year, the whole of this region. Père David, in his work on the 'Birds of China,' says that, although it is found in winter in the provinces of the S.E. of China, it penetrates rarely into the interior, and that he only got one example in the neighbourhood of Pekin. He remarks that Middendorf and Dybowski found it in East Siberia; so that its range would seem to lie in a more northerly track from Turkestan, probably through the north of Mongolia to Siberia and Japan, in which latter country it is the common Buzzard, and styled as such in the 'Fauna Japonica.' In the winter it moves in a southerly direction down the coast of China, where Mr. Swinhoe found it as far south as the island of Hainan. Captain Blakiston procured it in the island of Yesso, the most northerly of the Japanese group, and Col. Prejevalsky observed it during a voyage from Kiachta to Pekin.

Habits.—This species seems to prefer open country to forests and jungle, in which it exhibits much of the nature of a Harrier, hunting for its food over marshes and bare land with a steady flight. My specimen, Mr. Wylde informed me, took up its quarters in the cocoa-nut compounds and paddy-fields near his bungalow, about which it appeared to prowl as if intent on the capture of some of the poultry. When dissected, however, its stomach contained the remains of lizards. Its manners, however, on this occasion were evidently those of a new arrival by no means at home in its quarters; and after a few days it would evidently have betaken itself to some open upland district in the interior.

Captain Feilden remarks ('Stray Feathers,' 1875, p. 30):—"I found this bird at the edge of the parade-ground in tolerably thick tree-jungle with partially cleared underwood."

In Turkestan, Dr. Scully observed it, in company with *Buteo vulgaris* and *B. ferox,* hunting everywhere over the rush-grown frozen marshes, these birds being "so intent on the work they had in hand that they often seemed to disregard one's presence and approached so close as to be easily shot."

Further testimony as to its Harrier-like habits is afforded by Mr. Dourdillon's observations of it in the Travancore hills, "where two or three might be seen steadily quartering the ground, and occasionally pouncing on some mouse or lizard," and were noticed "to perch both on trees and on stones, and bent backwards and forwards over a field of young coffee."

Mr. Swinhoe writes:—"I fell in with this bird on the island of Naochow. He was resting at noon, after a meal off *Passer montanus,* in one of the bushy trees of a small grove. My appearance disturbed him, and he flew across heavily, when I secured him." ('Ibis,' 1870, p. 87.)

The testimony of various observers therefore goes to prove that this Buzzard is a bird of solitary habit, straying about alone, and usually so intent on securing the various prey on which it exists, that it is any thing but a shy bird.

Nidification.—I am unable to give my readers any information on the breeding of this Buzzard. In these days of ornithological research the day is doubtless not far distant when it will be discovered nesting in the Himalayas, or its breeding-haunts in the comparatively unknown regions of Central Asia penetrated by some adventurous explorer.

Subfam. AQUILINÆ.

Bill variable, usually lengthened and straight at the base; but in some (smaller genera) more curved and shorter, the margin festooned. Wings generally long; the 4th quill usually the longest, in some the 3rd and 4th, and in others the 4th and 5th. Tarsus less than the tibia by more than the length of the bind claw, but more than half its length; in some feathered entirely to the toes, in others partially, with the *hinder portion* always *reticulate.*

ACCIPITRES.

FALCONIDÆ.

AQUILINÆ.

Genus NISAETUS.

Bill strong, moderately lengthened, but not so much so as in *Aquila*, the culmen curving from the cere, its length not exceeding the hind toe; tip much hooked; the margin prominently festooned. Nostrils large, oval, and directed downwards. Wings with the 5th quill the longest, of moderate length, shorter than in *Aquila*. Tail moderate, even at the tip. Tarsus shorter than the tibia, stout, clothed with feathers to the toes, which are large and covered with three large scales at the tip. Claws large, much curved, the inner claw much larger than the middle.

NISAETUS FASCIATUS.

(BONELLI'S EAGLE.)

Aquila fasciata, Vieill. Mém. Linn. Soc. Paris, 1822, p. 152.
Falco bonellii, Temm. Pl. Col. i. pl. 288 (1824).
Aquila bonellii, Less. Man. Orn. i. p. 88 (1828); Gould, B. of Europe, i. pl. 7 (1837);
    Kelaart, Prodromus, Cat. p. 114 (1852); Layard, Ann. & Mag. Nat. Hist. 1853, xii.
    p. 98; Tristram, Ibis, 1865, p. 252; Shelley, B. of Egypt, p. 206.
Eutolmaetus bonellii, Blyth, J. A. S. B. xiv. p. 74 (1845); Hume, Rough Notes, i. p. 189.
Nisaetus grandis, Jerd. Ill. Ind. Orn. pl. 1 (1847).
Pseudaetus bonellii, Hume, Nests and Eggs, i. p. 33.
Nisaetus bonellii, Jerd. B. of Ind. i. p. 67 (1862); Holdsw. P. Z. S. 1872, p. 411.
Nisaetus fasciatus, Sharpe, Cat. of Birds, i. p. 250 (1874); Dresser, B. Eur. part xxxiv. (1874).
The Crestless Hawk-Eagle, Jerdon; *The Genoese Eagle*, Kelaart.
Perdicero; Aquila blanca, Spanish.
Mhorungi, lit. "Peacock-killer," Hind.; *Rajali*, Tam. (*apud* Jerdon).

Adult male. Length to front of cere 25·0 to 26·5 inches; culmen from cere 1·6; wing 18·5 to 19·5, expanse 62·0; tail
11·5; tarsus 3·5 to 3·7; mid toe 2·3 to 2·5, claw (straight) 1·2; hind toe 1·3 to 1·6, claw (straight) 1·6; height of
bill at cere 0·7.

Female. Length to front of cere 26·0 to 27·0; culmen from cere 1·6 to 1·7; wing 18·6 to 20·3; tail 11·2 to 12·0;
tarsus 3·6 to 4·0; mid toe 2·6, claw (straight) 1·3; hind toe 1·6.

Obs. Some adult females are quite as small as males. In Hume's 'Rough Notes' the dimensions of the wings of three
females are given at 20·0, 19·65, and 19·65, and the expanse of the largest 67·0.

Iris bright yellow, in some brownish yellow; cere yellowish; bill blackish brown, paling into bluish horn about the cere,
the gape yellowish; foot yellowish or whitish brown.

Above deep brown, very dark on the rump; the feathers of the head, neck, upper back, and wing-coverts with pale margins, and the concealed portions white; longer scapular feathers almost black near the tips; feathers of the nape elongated; edge of the wing from the flexure to the front white; median coverts paler brown than lesser, with a dark patch near the tips and the bases mottled; primaries and secondaries black-brown, the outer webs of the longer quills washed with grey, the inner webs of all whitish towards the base and crossed by narrow bars: inner webs of secondaries mottled with white; upper tail-coverts tipped with greyish white; tail brownish grey or cinereous brown, with a *broad terminal band* of blackish brown, and the basal part of the central feathers marked transversely near the shaft with wavy brown rays, which, on the more lateral feathers, develop into narrow irregular bars.

Loral plumes blackish; a blackish-brown moustachial patch; ear-coverts and the sides of the neck below them tawny brown, striped with a darker hue, and the space above them at the posterior corner of the eye whitish; under surface from the chin to the belly white; the throat with five mesial lines, and the fore neck, chest, and breast with blackish-brown central stripes, generally broadest at the sides of the breast and flanks, and in some specimens very wide on the chest; thighs variable, in some specimens dark brown with pale indentations, in others much paler, but with the same character of marking; abdomen and under tail-coverts lighter brown than the thighs, barred with whitish; axillary plumes brown, spotted with white; under wing-coverts blackish brown, much marked with white along the edge; tarsal feathers pale brownish.

Obs. Some examples incline from their youth to be darker on the thighs and abdomen than others, and consequently a considerable variation exists in these parts in adults. As a rule the older the bird (a sure characteristic being the tail) the narrower are the stripes of the under surface.

Occasionally it would appear that the tawny hue, to be noticed presently, continues to remain on the under surface, the stripes and the dark colouring of the underparts being as in the normally white birds. There is a beautiful example in this plumage in the British Museum, from Mr. Howard Saunders's Spanish collection.

Young. The bird of the year has the upper surface and wing-coverts of a medium or sandy brown, the feathers with dark shafts; the head and hind neck tawny, with the feathers dark-centred; primaries lighter than in the adult, and the outer webs similarly pervaded with grey; the bars of the inner webs more extensive; secondaries broadly tipped with dull white; these and the greater coverts have in some examples a strong purplish lustre; upper tail-coverts brownish, paling into white at the tips, and with dark shafts; tail light sandy brown, mottled on the central feathers, and with a deep pale tip, the whole crossed with seven or eight narrow wavy bars of dark brown, without any broad terminal band; in many examples the bars are undefined, and run into the mottlings of the interspaces; *no broad band* at the tip.

Face, ear-coverts, and sides of head concolorous with the adjacent brown parts; the ear-coverts striated with a darker hue; throat and entire under surface uniform brownish rufous, paler on the chin and with clearly defined shaft-lines, diminishing towards the lower parts; abdomen, thigh-coverts, and under tail-coverts unstriated, but with the centres of the concealed portions of the feathers brown, showing the tendency of these parts to become dark with age; under wing-coverts rufescent like the breast and striped with brown, the lower series dark brown.

With age the rufous of the under surface becomes white, the mesial lines expand at the tip into "drops," and thence into broadish stripes; the thighs and legs become brown with darker stripes, while the belly and under tail-coverts are heavily dashed with the same; the under wing-coverts become blackish brown at the same time.

Distribution.—This powerful Eagle, the finest of the short, curved-billed genus *Nisaetus*, and so well known in Southern Europe and Northern India, has been once procured in Ceylon. It can therefore only be looked upon as a straggler to the island, and takes its place in our lists as such, in common with the Scavenger-Vulture and the Amurian Kestrel.

Layard writes, in his notes on Ceylon ornithology (*l. c.*):—"This Eagle was procured by R. Templeton, Esq., R.A., several years ago, and I do not know from what part of the island it was obtained. It has not fallen under my notice*, nor has Dr. Kelaart enumerated it amongst his acquisitions at Nuwara Elliya." Many

---

* There, notwithstanding, is a faded specimen of this Eagle in the Poole collection. It has the wing 10·5, tail 10·0, tarsus 4·0 inches; the peculiar brown coloration of the exterior of the thighs is still visible, although the head and all the under surface are bleached. Mr. Layard writes me that he does not remember any thing about this specimen, and its presence, evidently as a Ceylon bird, in this collection is somewhat puzzling. Can it be that this is Dr. Templeton's specimen, afterwards presented to the collection while at Sir Ivor Guest's.

years have elapsed since this occurrence, which was prior to 1858; and since then I am unable to find any record of its having been met with in Ceylon. The specimen referred to was identified by Mr. Blyth, when he was Curator of the Calcutta Museum, so that there is no chance of the species having been mistaken for any other Eagle.

Bonelli's Eagle is, as far as the Indian peninsula is concerned, chiefly confined to the northern part of it. It is not uncommon in portions of Bengal, but not so in the lower districts of the Province.

In Madras and the south generally, it is rarer; and I notice that it is not included in the "First List of Birds from the Travancore Hills" ('Stray Feathers,' vol. iv.) even as a rare visitor. Dr. Jerdon, however, records it from the Nilghiris, whence it no doubt visited Ceylon when procured by Templeton.

It is an inhabitant of the slopes of the Himalayas; Mr. Brooks records it among the birds he observed between Mussoori and Gangaotri. In the north-east of India it is more common than elsewhere; for in Sindh, Mr. Hume says, "one, two, or more pairs are to be met with about every large lake, making terrible havoc amongst the smaller water-birds, and carrying off wounded fowl before one's eyes with the greatest impudence." In Southern Europe, Bonelli's Eagle is a well-known bird. Lord Lilford and Mr. Howard Saunders speak of its common occurrence in Spain, and on the northern coasts of Africa it is also pretty freely distributed.

Mr. Brooke has met with it in Sardinia; and Canon Tristram remarks that it is more common in Palestine than the next species, being generally found in the wooded hills about Carmel, Tabor, and the Lake of Galilee.

Habits.—Rocky wooded hills, mountainous jungles, and forests in the vicinity of high land are the habitat of this bold and daring Eagle. Jerdon says that "it is much on the wing, sailing at a great height, and making its appearance at certain spots, in the districts it frequents, always about the same hour." The latter propensity is noticeable in other birds of prey, for I have remarked it in the Sea-Eagle and Crested Hawk-Eagle of Ceylon. The present species is very powerful in the legs and feet, and is known to kill the smaller kinds of game and hares with ease. It is, however, so strong and active on the wing that it preys largely on various birds, such as Jungle-fowl, Partridges, Ducks, and Herons, and, according to Jerdon "even Peafowl."

It is very destructive among Fowls and Pigeons; and it is recorded, in the 'Birds of India,' that a pair committed great devastation among several pigeon-houses in the Nilghiris.

The following interesting account of the manner in which these robbers captured the Pigeons is given by Jerdon at page 69 of his first volume:—"On the Pigeons taking flight, one of the Eagles pounced down from a vast height on the flock, but directing its swoop rather under the Pigeons than directly at them. Its mate, watching the moment when, alarmed by the first swoop, the Pigeons rose in confusion, pounced unerringly on one of them and carried it off; and the other Eagle, having risen again, also made another and, this time, a fatal stoop." Such a bird as this would do much damage in the poultry-yards of many a pretty bungalow in the Kandyan province.

Concerning its economy in Palestine, Canon Tristram remarks as follows:—"It perches on some conspicuous point of rock looking out for its prey, and after a short circling excursion will again and again return to the same post of observation. I take it to be more truly a game-killing Raptor than any of the preceding Eagles" (the Golden, Imperial, Tawny, and Booted), "and less addicted to carrion-feeding than any of its congeners. The Rock-Pigeons are its favourite quarry in the winter, and it preys much on the Turtle-Doves in the Ghor and the plain of Genuesaret. I have also seen it pursue Kites, apparently with the intent of robbing them."

Its fondness for Pigeons was noticed by Mr. Hume, who killed the male of the pair which form the subject of his interesting article in 'Rough Notes,' returning to the nest with a Little Brown Dove (*Turtur cambayensis*) in its talons. This Eagle has a singular habit of packing in large flocks, one of which very unaquiline assemblies was witnessed by Lord Lilford in Spain, he being informed that such flights were not unfrequently seen. This was in May, during the breeding-season of the species; and, as is remarked, the bird being a permanent resident in the country, it is a difficult matter to account for such an assemblage. The note of this Hawk-Eagle is described as being a "shrill croaking cry."

Nidification.—In the plains of India Bonelli's Eagle breeds in December and January, and in the Himalayas and the district of Kumaon much later, commencing in April and continuing until June. In the

Nilghiris it breeds as early as December. The nest is usually placed in the ledge of a cliff, but it has been found fixed in the branches of large trees. It is a huge platform of sticks, containing in the centre a circle or layer of fresh green leaves, on which the eggs are laid, and which the bird covers them with on leaving the nest, in the same manner that I have myself seen the Grey-backed Sea-Eagle do.

Mr. Hume, in his interesting account of the taking of this Eagle's nest, given in 'The Ibis' for 1869 (p. 143), speaks of one nest visited as being five feet in diameter. The situation of this nest is thus described:—"About a mile above the confluence of the clear blue waters of the Chambal and the muddy stream of the Jumna, in a range of bold perpendicular clay cliffs that rise more than 100 feet above the cold-weather level of the former, I took my first nest of Bonelli's Eagle. In the rainy season, water trickling from above had (in a way trickling water often does) worn a deep recess into the face of the cliff, about a third of the way down. Above and below it had merely grooved the surface broadly, but here (finding a softer bed, I suppose) it had worn in a recess some 5 feet high and 3 feet deep and broad. The bottom of this recess sloped downwards; but the birds, by using branches with large twiggy extremities, had built up a level platform that projected some 2 feet beyond the face of the cliff. It was a great mass of sticks fully half a ton in weight, and on this platform (with only her head visible from where we stood at the water's edge) an old female Eagle sat in state."

The eggs are usually two in number; but some nests have been found with only one. They are described by Mr. Hume as "moderately broad ovals, varying slightly in size." They are whitish in colour, sometimes quite unmarked, but usually are faintly blotched with pale yellowish or reddish brown. The markings in others, as given by Mr. Brooks, are darker, or "bright reddish brown, sparingly intermixed with light reddish grey." They average in size 2·78 by 2·1 inches.

In the Holy Land, Canon Tristram found it nesting on the cliffs of the deep gorges characteristic of that country. He writes, in 'The Ibis,' 1865, p. 253:—"It does not appear to lay till the end of March, and then generally a single egg. These are either white or with the faintest russet spots. One nest, which contained two eggs both fairly coloured, baffled all our attempts at its capture. It was comfortably placed under an overhanging piece of rock near the top of the cliffs of Wady Hamam, in such a position that no rope could be thrown over to let down an adventurous climber; and yet from another point, which projected nearly parallel to it, we could look into the nest with longing eyes. The old birds seemed perfectly aware of the impregnability of their fortress."

NISAETUS PENNATUS.

(THE BOOTED EAGLE.)

Falco pennatus, Gm. S. N. i. p. 272 (1788); Temm. Pl. Col. i. pl. 33 (1824).

Aquila pennata, Vig. Zool. Journ. i. p. 337 (1824); Gould, B. of Europe, i. pl. 9 (1837); Fritsch, Vög. Eur. tab. 5. figs. 3, 4, 5 (1858); Kelaart's Prodromus, Cat. p. 114 (1852); Layard, Ann. & Mag. Nat. Hist. 1853, xii. p. 98; Jaub. et Barth. Rich. Orn. p. 36, pl. 3 (1859); Jerdon, B. of Ind. i. p. 63; Holdsworth, P. Z. S. 1872, p. 411; Shelley, B. of Egypt, p. 207 (1872); Legge, Str. Feath. vol. iv. p. 249; Dresser, B. Eur. pt. xxxii. (1874).

Butaetus pennatus, Blyth, Journ. Asiat. Soc. Beng. xiv. p. 174 (1845).

Hieraetus pennatus, Blyth, Journ. Asiat. Soc. Beng. xv. p. 7 (1846); Hume, Rough Notes, i. p. 182 (1869).

Nisaetus pennatus, Sharpe, Cat. Birds, i. p. 253 (1874).

Le Faucon patu, Briss. Orn. vi. App. p. 22, pl. 1 (1760).

The Dwarf Eagle, Sportsmen in India.

Bagati Jumiz, Hind., lit. "Garden Eagle;" also *Gilheri-mar*, lit. "Squirrel-killer;" *Oodatal Gedda*, Tel., lit. "Squirrel Kite" (*apud* Jerdon).

Punja-Prandu, Tam., lit. "Field-Kite."

Rajaliya, Sinhalese.

Adult male. Length to front of cere 10·5 to 21·0 inches; culmen from cere 1·02 to 1·2; wing 14·5 to 15·5; tail 8·2 to 8·5; tarsus 2·3 to 2·4; mid toe 1·5 to 1·7, claw (straight) 0·75 to 0·8.

Adult female. Wing 15·0 to 16·4; tail 8·5 to 9·5; tarsus 2·3 to 2·5; mid toe 1·5 to 1·7; culmen from cere 1·15 to 1·3.

This limit of wing is from a series of Bengal, Turkish, and Spanish examples. Mr. MacVicar's specimen, referred to below, which was a female and an Indian-bred bird, measured 15·7 in the wing; a male, in my own collection, purchased from Messrs. Whyte and Co., 15·2.

Iris varying from pale brown to chestnut-brown; cere yellow; bill black at the tip, paling into leaden or bluish at the base, and with the gape yellow; feet yellow; claws black.

Head and hind neck brownish tawny, darkest on the forehead and crown (in some paler or fulvous tawny), the shafts of the feathers dark and their margins pale: back, rump, scapulars, lesser and greater secondary wing-coverts dark earth-brown, with the edges of the feathers slightly paler; median wing-coverts, uppermost tertials, and some of the scapular feathers pale brownish, darkening towards the shaft of the feather; primaries and secondaries blackish brown, with obsolete bars on the light portions of the inner webs and the extreme tips whitish; upper tail-coverts sandy fulvous; tail blackish brown, lighter than the tips of the quills; traces of obsolete transverse marks exist in many specimens; the inner webs of the lateral feathers mottled with whitish.

Plumes of the lores and round the eye black; checks, ear-coverts, and a space below them dark tawny, with a narrow blackish-brown moustachial stripe; throat and fore neck buff, paling slightly on the whole under surface and under wing into buff-white, the throat marked with central stripes concolorous with the ear-coverts; these become narrower on the chest, and gradually change into shaft-lines on the breast and flanks and secondary under wing-coverts; primary under coverts spotted with dark brown. The amount of striation on the under surface varies much, and some examples have the stripes confluent across the throat.

Dark form. The plumage above has the same character as the foregoing, but is much darker throughout both as regards body and the wings and tail; the light portions of the wing-coverts and tail are very much darker than in the pale bird; the forehead and crown are well covered with black feathers, and the hind neck rufous instead of pale fulvous; the upper tail-coverts are darker than in the pale bird; the chin and cheeks are boldly dashed with blackish brown; and the entire under surface uniform wood-brown, the centres of the feathers black, blending with the ground-colour.

Young. The nestling has the iris brown, and the legs and feet yellow, like the adult.

Obs. In the splendid series possessed by the British Museum, many of which were collected by Mr. Howard Saunders in Spain, are two nestlings obtained from the nest, with the parents, by that gentleman. One is a light bird, and the other a very dark one, demonstrating the fact that *light* and *dark* birds exist from the very nest, and are the offspring of the same parents. This fact solves the problem as to the light and dark birds of both sexes, which has so long engaged the attention of naturalists. Mons. Bureau, in his paper on this Eagle, published in the ' Proceedings' of the Association Française pour l'avancement des Sciences, Nantes, goes very fully into this singular feature in the economy of the Booted Eagle, proving by his observations that sometimes birds of the light and dark type pair together, the union of similar-plumaged birds being of course the commoner; and he remarks, with reference to the progeny, " De l'une ou l'autre de ces unions naissent habituellement des jeunes d'un seul type : plus rarement on trouve dans une même nichée des jeunes de l'une ou de l'autre race." This conclusion is substantiated in the case of the two young birds now alluded to, the parents of whom belonged to the two phases. In the ' Birds of Europe,' Mr. Dresser cites several instances of light and dark birds breeding together in Russia and producing young of both descriptions. The description of the above-mentioned nestlings is as follows :—

Pale form. Head and hind neck light but rich sienna, the feathers of the crown with dark shafts : back, lesser secondary wing-coverts, and longer scapulars deep wood-brown, with a purplish lustre; the tail broadly tipped with whitish; scapulars, tertials and major wing-coverts, primaries, and secondaries blackish brown, the latter paling at the tips into the hue of the coverts; upper tail-coverts light fawn-brown with dark shafts; under surface very pale fawn, richest on the chest, where the feathers have dark shaft-stripes.

Dark form. Head and hind neck rich tawny, the forehead blackish, and the crown with dark shaft-lines; dark portions of the upper surface much the same as the pale bird, but the scapulars and wing-coverts darker; cheeks, fore neck, and entire under surface dark brown, quite as intense as in the full-grown dark bird.

With its advance towards maturity, the pale bird becomes lighter on the head and under surface. The head and hind neck are rich tawny, with the shaft-stripes narrower than in the adult, and the crown not so dark; the ear-coverts and sides of the neck are rich tawny brown, this part blending evenly into the paler fawn-colour of the chest : the moustachial streak is dark and unites with the surrounding tints; the wing-coverts and scapulars have a greater extent of pale tipping, which extends to the least coverts along the front of the wing : the upper tail-coverts are very pale, and the light tip of the tail deeper than in the adult : the entire under surface is pale fawn, blending into the darker hue of the chest, which is handsomely striated as in the adult, but the streaks not contrasting so much with the feather.

With age, in the dark form, the tawny hue of the head and hind neck gradually changes to the darker coloration of the adult; the crown and forehead become more uniformly brown, and the light edgings of the back feathers less conspicuous, finally darkening into the ground-colour.

Distribution.—This bold little Eagle, so well known in Southern Europe and India, appears to pay occasional visits to Ceylon, and has been obtained both in the maritime and moderately elevated hill-districts. It was first killed by Edgar Layard near Pt. Pedro, during his official residence at that place. The season of the year was that in which Asiatic Raptors usually visit the island, and at the same time, during the prevalence of the north-east monsoon in 1875–6, two additional examples were collected. The first, a fine female, was killed by Mr. H. MacVicar, of the Survey Department, in the cinnamon-gardens close to Colombo, and was presented by that gentleman to the Colonial Museum ; the second was shot in the district of Dumbara, near Kandy, was preserved by Messrs. Whyte and Co., of that town, and afterwards passed into my hands.

I am under the impression that I have seen this species myself in the north-eastern part of the island ; but I can no more speak with certainty concerning it than I can satisfy myself as to the identity of several Hawks not in our lists, which I have met with in the forests of Ceylon and failed to shoot.

In India this Eagle is pretty fairly distributed as far as the plains are concerned ; but its numbers are greater in the north than in the south. It is not found at any elevation in the mountains, and does not inhabit

G

Burmah in any quantity. It is recorded as being comparatively rare in Pegu, neither Mr. Oates nor Captain Feilden having procured many examples of it in that region.

From the west of India its range extends through Persia to Palestine, south-eastern, southern, and central Europe; whereas on the south of the Mediterranean it inhabits Egypt and Algeria, and thence extends, probably by way of the east coast, to the south of the continent, having been procured in Damara Land by Mr. Andersson. Lord Lilford found it common in Spain near Seville, and remarks that it inhabits many other parts of the Peninsula. Mons. L. Bureau records it as an inhabitant of the west of France, and Count Wodzicki of the Carpathians, while other naturalists, as quoted in Mr. Dresser's 'Birds of Europe,' have met with it in Central Germany and many parts of Russia. In Palestine Canon Tristram believes it to be confined to the north, and only observed it between the months of October and March.

Habits.—The Booted Eagle frequents hilly, wooded country, as well as open plains, cultivated land, and ground covered sparsely with small timber and scrub, where it finds an abundance of food in birds, small vermin, and perhaps some kinds of reptiles. It is partial to districts where woods and clumps of forest are intermingled with low jungle. It is a bold and daring bird and very active on the wing, in testimony of which Mr. Hume, in his 'Rough Notes,' quotes from the letters of Mr. R. Thompson, who observed one of these Eagles dash into a tree, and seize a bird out of a flock of Parakeets, while on another occasion he witnessed the attempted capture of a rat on the ground. Layard, in writing of the specimen he shot at Pt. Pedro, after narrating that he had mistaken it in the twilight of the morning for a Brahminy Kite, remarks, " it suddenly pounced upon a Bulbul roosting in an oleander bush : this at once undeceived me; and as it rose with its victim in its claws, I fired and brought it to the ground. It fought with determined spirit and kept a small terrier at bay, till I killed it with the butt-end of my gun."

Jerdon, in his 'Birds of India,' notes its destructive habits, and says that it pounces on doves, pigeons, and chickens, and that it forages about villages in company with Kites, who are often unjustly blamed for the depredations in reality committed by the " Dwarf Eagle." Although fierce in its nature it is at times sociably inclined, even towards other members of its order; for Mr. Brooks has seen it, several at a time, seated on the ground in company with the Common Kite. The note of the Booted Eagle is a wild scream, which is said to be different from that of most other Eagles. It was observed by Capt. Feilden in Burmah to perch much in thickly foliaged trees, a somewhat abnormal habit for the Eagle family.

In Spain it appears, says Lord Lilford, " to prefer open country and isolated groups of trees to large extents of forest," and is, according to the natives, "the scourge of the Quails in Andalucia." It arrives in the country in April, breeds there, and departs in October.

Nidification.—The Dwarf Eagle does not breed commonly within the Indian limits. Mr. Hume records, in 'Nests and Eggs,' a nest found at Hurroor, near Saloun. It was built in the branch of a high banyan tree, about 50 feet from the ground, and consisted of dry twigs, being a circular platform in shape, with a slight depression in the centre and devoid of lining. The eggs were two in number, of a dead white ground-colour, and one of them blotched and streaked with reddish brown. The egg measured 2·13 by 1·78 inch. In Spain, Lord Lilford, who found many nests, chiefly built in pine-groves, says that they are invariably lined with green leaves, which is a common practice with the Eagle tribe. These nests, when built in pines, were situated at the junction of a large lower branch with the trunk, and all, as well as others found by him, contained two eggs. The figures on pl. x. of 'The Ibis' for 1866 show the variation in the colouring, the one being dull white with a few faint reddish blotches about the centre, and the other clouded and dashed with two or more shades of light reddish. The lighter of the two measures 2·04 by 1·73 inch, and the larger and more handsomely coloured 2·26 by 1·83 inch.

Bill curved more suddenly from the base than in *Nisaetus*, less stout, and with the tip not so prolonged; margin not prominently festooned. Nostrils circular, rather small, and placed near the edge of the cere. Wings moderate, reaching, when closed, beyond the middle of the tail; the 4th and 5th quills subequal and longest, or the 5th shorter than the 4th. Tail moderate, broad at the base, rounded at the tip. Tarsus as in *Nisaetus*; middle toe long, with the claw rather short; lateral toes nearly equal, but with the inner claw nearly as long as the hind one. Head crested; the feathers short, broad at the base, and pointed at the tip, forming a wedge-shaped crest, which originates above the occiput.

LOPHOTRIORCHIS KIENERI.

(THE RUFOUS-BELLIED HAWK-EAGLE.)

Astur kieneri, G. S. *, Mag. Zool. 1835 (Aves), pl. 35.
Spizaetus albogularis, Tickell, J. A. S. B. xi. p. 456 (1842).
Limnaetus kieneri, Strickland, Ann. N. H. xiii. p. 33 (1844); Jerdon, B. of Ind. i. p. 74;
    Bligh, J. A. S. (C. B.) p. 64 (first record from Ceylon); Legge, Str. Feath. 1875, p. 198;
    Gurney, Ibis, 1877, p. 433.
Spizaetus kieneri, Gray, Gen. B. i. p. 33 (1845); Schl. Mus. P.-B. *Astures*, p. 11; Wall.
    Ibis, 1868, p. 14; Hume, Rough Notes, i. p. 216; Hume, Str. Feath. i. p. 310 (1873).
Nisaetus kieneri, Jerd. Ill. Ind. Orn. p. 33 (1847).
Lophotriorchis kieneri, Sharpe, Cat. Birds, i. p. 255 (1874).

Adult male. Length to front of cere 19·5 to 20·5 inches; culmen from cere 1·0 to 1·1; wing 14·2 to 15·5; tail 8·2 to 9·0; tarsus 2·7 to 3·0; mid toe 2·0 to 2·15, its claw (straight) 0·85 to 1·1; inner claw (straight) 1·3; height of bill at cere 0·5 to 0·55. Expanse (of one with wing of 14·5) 45·0; weight of the same 1¾ lb.

A great disparity in size exists between the sexes in this species, but males also differ much *inter se* in this respect. The above dimensions are taken from a fair series of Indian, Ceylonese, and Malaccan examples. The wings of four Ceylonese males examined measure 14·2, 14·6, 13·5, and 15·0.

Adult female. From Mr. Hume's Darjiling specimens ('Stray Feathers,' vol. i. p. 311). Length 24·0 to 29·0 inches:

---

* The article here referred to merely has those initials appended to it, and some doubt exists as to whether they refer to G. Sparre or Geoffroy St.-Hilaire. Mr. Sharpe has adopted the latter in his 'Catalogue of the Accipitres.' I observe that, throughout the 'Mag. Zool.,' St.-Hilaire either signs his name in full or uses the abbreviation " Geoffroy St.-H.;" and I think there is no reason to infer that had he been the author of the two descriptive articles (*Astur kieneri* and *Pica mystacalis*) in the volume for 1835, which are signed " G. S.," he would have used these initials instead of his usual signature. In the Roy. Soc. Catalogue, vol. v., these two identical articles are referred to as written by G. Sparre; and, in all probability, this is the correct determination of their authorship.

culmen 1·2; wing 17·0 to 17·5; tail 10·0 to 12·5; tarsus 3·0; mid toe 2·3, its claw (straight) 1·18; inner claw 1·5; height of bill at cere 0·65.   Expanse 50·0.

An example from Sarawak in the British Museum, marked ♀, has the wing 13·0 and the tail 7·5.

Iris dark brown; cere yellow, in some greenish yellow; bill black, plumbeous at base; feet yellow; eyelid greenish yellow.

Lores, head, crest, back and sides of neck, upper surface, and wings dark blackish brown, almost black; crest of three or four stiffish, ovate feathers from 2·2 to 2·5 inches in length; inner webs of primaries (in the longer ones to the notch) whitish, crossed with narrow blackish bars; inner webs of secondaries more dusky, similarly barred; tail blackish brown, crossed with six or seven narrow smoky-brown but indistinct bars; in some examples the bars on the central feathers are nearly obsolete.

Chin, throat, and chest white, changing on the upper breast into the deep ferruginous of the lower parts, including the legs and under tail-coverts, and striped everywhere but on the chin and throat with lanceolate black shaft-streaks; under surface of tail greyish; under surface of primaries white, from the notch to the tip greyish, showing narrow black bars; lesser under wing-coverts pale rufous with black mesial stripes; greater secondary series and the primary row black, with white tips and fulvous edges.

Obs. In very old specimens the rufous colouring is very deep, and spreads upwards to the throat, the feathers being either tipped with it or washed with a paler hue than that of the breast. The extent of the shaft-streaks on the upper parts varies, the throat and chest having them in old birds. Mr. Bligh's male example (the first procured in Ceylon, and now in the Norwich Museum) has the rufous colouring extending no higher than the breast, and therefore represents a mature, but not an aged bird.

Young. I have not had an opportunity of examining this Eagle in its nestling plumage, and I therefore transcribe here the description given by Mr. Sharpe at page 458 of the 'Catalogue of Accipitres,' from a young bird in Lord Tweeddale's collection, which is evidently in its first dress:—"Above dark brown, the feathers lighter on their margins; wing-coverts coloured like the back, but the greater series with narrow white margins; hind neck paler than back, rufous-brown, with dark brown longitudinal centres, causing a slightly streaked appearance; quills blackish, with whity brown shafts; the secondaries paler brown, like the scapularies, all the quills narrowly banded with black, nearly obsolete on the primaries, but more distinct on the secondaries, especially underneath, where the lining of the wing is whitish; tail dark brown, whitish at tip, and crossed with seven or eight rather narrow bands of black.

"Crown of head dark brown, with tiny cream-coloured tips to the feathers; the occipital crest black, and 1·9 inch long; forehead and eyebrow very broad, rich creamy buff; cheeks and entire underparts creamy white, as also the tarsal feathers and under wing- and tail-coverts, the greater under wing-coverts with a few indistinct blackish bars."

Wing 13·3 inches.

The tippings of the head-feathers, margins of the wing-coverts, and creamy colour of the under surface testify to this bird being in nestling plumage.

An immature bird, apparently of the second year, in my collection is in the following plumage:—Head and upper surface very dark brown, the terminal portions of the feathers being blackish, but the basal parts paler brown than the centres; forehead at the edge of the cere, a narrow streak above the eye, and the basal portions of the head and nuchal feathers whitish; crest fully developed; lesser coverts on the point of the wing and along its edge with pale terminal margins; primary and greater secondary coverts and also the secondaries pale tipped, the former most clearly so; inner webs of the quills much as in the adult, but with the ground less white, being mottled between the bars; tail smoky brown, tipped pale, with narrower bars than in the adult, the subterminal one scarcely broader than the rest.

Chin, face, ear-coverts, and entire under surface with the under wing white; ear-coverts and sides of neck below them with terminal dark shaft-stripes; feathers at the sides of the breast and one or two on the chest with lanceolate dark brown shaft-stripes, surrounded by a wash of rufous; longer feathers of the flank-plumes dark brown, forming a prominent dark patch; thighs, tarsi, and under tail-coverts with rufescent feathers here and there; major under wing-coverts with blackish terminal patches. The rufous hue on the under tail-coverts is taking place by a change of feather; but there are some new feathers on the thighs of a darker hue.   Wing 15·0 inches.

In the Norwich Museum are two young examples from "Java" and "Batchian" in this stage of plumage.

With age the darkening of the lower parts and the gradual advance of the rufous up towards the chest is very perceptible. An example from Malacca, in the British Museum, in the next stage to the above has the throat, chest, and most of the breast white, the rufous hue appearing on the lower breast and extending downwards, while the shaft-stripes do not extend above the breast.

Obs. This interesting genus of Eagles, though comprising very few species, is widely diffused, taking both the Old and the New Worlds into its range. Until lately but two were known, the present and the large *L. isidorii* from Columbia, South America; recently, however, a third, *L. lucani* (Sharpe and Bouvier), has been added from the Congo river, S.W. Africa.

Distribution.—This rare and handsome Eagle has only lately been discovered in Ceylon; and the gentleman who has the merit of adding it to the avifauna of the island is Mr. S. Bligh, of Lemastota. The first Ceylonese example was procured by him in Kotmalie, a district at the base of the Nuwara Elliya ranges, lying at an altitude of about 3500 feet. It was shot on the 20th of October, 1873, and was a male in adult plumage. The next example was killed near Kandy at the latter end of 1875, and taken to Messrs. Whyte and Co.'s establishment, whence it passed into the Colonial Museum at Colombo; about the same time a young bird (above described) was shot near Peradeniya by a native, and procured from him by Mr. Whyte. Mr. Bligh met with another, which was seen close to his bungalow, on the 6th of June 1875, but evaded his pursuit; and in January 1876 I was equally unsuccessful in procuring another at Nalanda, a district to the north of the Matale hills, which are celebrated for the variety of Raptors found in their vicinity.

It has as yet, therefore, proved quite a hill species, which is in accordance with its habits in the Himalayas and elsewhere in the hills of Borneo and Malacca.

This Eagle is an inhabitant of the northern parts of India; but has not yet been detected in the south, which is the more strange when viewed in conjunction with its not unfrequent occurrence of late years in Ceylon; this, however, only substantiates the theory of the strong affinities of the Ceylonese avifauna with that of Malayana, in which region this Eagle is rather widely distributed.

According to Jerdon it is found in Central India, and Tickell obtained it near Chaibassa; but it has not been procured from there of late years; and Mr. Hume doubts if these specimens really belonged to the true *kieneri*, which was described originally from the Himalayas by Sparre, from a specimen at that time in Prince Essling's collection.

Along the southern slopes of the Himalayas it has been occasionally met with, particularly in Darjiling, Sikhim, and the eastern portions of the range; and in the collections made by Mr. Inglis for Mr. Hume in Cachar one example is noted. It is, however, rare in that district as everywhere else. Mr. Inglis writes ('Stray Feathers,' vol. v. p. 9):—"I was lucky enough to secure the only specimen of this handsome bird that I ever met with; I got it while on a fishing excursion on the Cheerie, close to the Cacharee Degoon Ponjee, at an elevation of 2000 feet."

From North-east India it extends southwards into Malacca, and thence into the islands of the archipelago. It has been procured in Java and Borneo, in the latter by Mr. Wallace, and from the former it has been sent to the Norwich Museum. From the island of Batchian, one of the Moluccas, there is likewise a specimen at Norwich, this locality (which is in lat. $0° 40'$ S. and long. $127°$ E.) being at present the furthest known limit of its range into the Malay islands.

Habits.—The Rufous-bellied Hawk-Eagle inhabits forest-clad hills, frequenting, in search of its prey, open glades, valleys, clearings, and patnas. In Ceylon, it is therefore found about the coffee-estates, which are bordered by wood and studded with dead trees, the latter furnishing it with an advantageous post of observation. It is a bird of truly predatory disposition, and is as bold and courageous as it is handsome.

Mr. Bligh remarks, in his note on the capture of his bird, contained in the 'Journal of the Ceylon Asiatic Society' for 1874, that it was "sailing just above the trees in circles in a very buoyant and graceful manner, rarely flapping its wings. My little terrier," he says, "was frisking about some thirty yards off, and on arriving over the spot, the bold bird at once altered its flight, hovering in small circles with a heavy flapping of the wings, evidently with a view of examining the dog." He further remarks that when brought to the ground with a broken wing, "it put itself in an attitude of defence at once; and a formidable bird it looked, with beak open, head thrown back, wings spread, and talons ready for action, and its beautiful brown eyes looking so fierce."

Mr. Inglis, in his note on the shooting of a specimen on the hills of Cachar, bears the same testimony to its plucky nature, and says that it fought most fiercely while it was being secured. As observed by myself its flight was buoyant, but not very swift, resembling somewhat that of the Ceylon Crested Eagle (*Spizaetus ceylonensis*) ; its white chest, contrasted with the dark lower parts, is a conspicuous characteristic when the bird is flying overhead. This Eagle preys on birds and small mammals, being capable, however, of capturing an animal of no diminutive size, so strong are its talons and so bold its disposition.

FALCONIDÆ.

AQUILINÆ.

Genus NEOPUS.

Bill longer than in *Lophotriorchis*, more suddenly hooked at the tip, the festoon less pronounced; cere large, the nostrils oval and partially covered by the loral bristles. Wings very long and exceeding the tail when closed; the terminal portions of the longer primaries very concave beneath; 4th, 5th, and 6th quills subequal and longest. Tarsus slender, feathered to the toes, which are short, the inner nearly as long as the middle, the outer very short. Claws slightly curved; *the inner claw very long*, exceeding the hind; outer claw very short, not reaching to the tip of the middle toe.

NEOPUS MALAYENSIS.

(THE BLACK KITE-EAGLE.)

Falco malayensis, Temm. Pl. Col. i. pl. 117 (1824).

Aquila malayensis, Vig. Zool. Journ. i. p. 337 (1824); Schlegel, Vog. Nederl. Ind., Valkv. pp. 8, 49, pl. 3. figs. 1, 2 (1866).

Aquila malayana, Less. Traité, p. 39 (1831).

Ictinaetus malayensis, Blyth, J. A. S. B. xv. p. 7 (1846); Kelaart's Prodromus, Cat. p. 114; Layard, Ann. & Mag. N. H. 1853, xii. p. 99.

Neopus malayensis, Jerd. B. of Ind. i. p. 65 (1862); Hume, Rough Notes, i. p. 187; Wald. Tr. Z. S. viii. p. 34; Holdsworth, P. Z. S. 1872, p. 411; Legge, Ibis, 1874, p. 8; Sharpe, Cat. Birds, i. p. 257; Bourdillon, Str. Feath. 1876, p. 355.

Heteropus malayensis, Hume, Nests and Eggs, i. p. 32.

The Black Eagle, Kite-Eagle, in India.

Hengong, Bhot.; *Adavi nalla gedda*, Tel., lit. "Jungle Black Kite" (*apud* Jerdon).

Kalu-Rajaliya, lit. "Black Eagle," Sinhalese.

Adult male. Length to front of cere 25·0 to 27·2 inches; culmen from cere 1·35; wing 20·6 to 21·75, expanse 63·0 to 64·0; tail 12·2 to 13·5; tarsus 3·2; mid toe 1·6 to 1·7, its claw (straight) 1·1 to 1·2; inner claw (straight) 1·6, hind claw (straight) 1·45; height of bill at cere 0·55.

Adult female. Length to front of cere 28·0 to 29·5 inches; culmen from cere 1·37; wing 23·0 to 25·0, expanse 75·0; tail 13·5 to 14·5; tarsus 3·5 to 3·8; inner claw (straight) 1·7 to 1·9; height of bill at cere 0·6. Weight 3¼ lb.

Obs. The chief distinguishing characteristic of this peculiar Eagle is its remarkable foot and straight claws, the inner of which is the longest, exceeding the hind by about 0·1 inch, which latter is just twice the length of the outer.

Iris hazel-brown; bill brownish horn-colour, paling into greenish at the cere; cere, gape, and base of lower mandible citron-yellow; feet gamboge-yellow.

Head and entire upper surface sooty black, darkest on the head, lesser wing-coverts, and scapulars, and paling into brown on the upper tail-coverts; entire under surface and legs blackish brown, blending into the black of the cheeks and hind neck; feathers of the head with spinous glossy shafts, bases of the loral plumes and a small space above them white; scapulars and outer webs of quills with a green lustre; bases of the inner webs of the longer primaries barred with white; on the remainder and those of the secondaries there are indications of bars slightly lighter than the ground-colour; concealed portions of the upper tail-coverts crossed with narrow incomplete white bars; tail with four or five interrupted bars, slightly paler than the ground-colour, the terminal one about 2¼ inches from the tip; on the under surface these bars show whitish, and mostly so on the lateral feathers, where they increase to seven; under wing-coverts uniform brownish black.

The amount of white about the lores varies in individuals, and a specimen from Ceylon in my collection has a small tuft of white feathers below the cheeks.

Young. In the nestling-plumage, as figured by Schlegel (*loc. cit*), the head, neck, and entire under surface are fulvescent buff, each feather with a central stripe of brown, the pale ground-colour darkening on the back and wings into blackish brown, and having the margins of the feathers buff.

Immature bird. Wing of an example in the British Museum 18·5 inches.

In this plumage the back, wings, and tail are but little paler than in the adult; crown almost uniform black, the feathers tipped with fulvous, which on the nape, hind neck, and behind the ears increases in extent, and gives those parts a striated appearance; the forehead and lores whiter than in the adult; lesser and median wing-coverts tipped pale; primaries as black as in the adult, the inner webs with narrow mottled bars of white as far out as the notch; bars of the tail-feathers narrower, closer together, and more numerous than in the adult, the terminal one nearer the tip; upper tail-coverts as in the adult; throat and fore neck deep brown, the feathers tipped with fulvous; breast, flanks, and thighs mingled with rufous and streaked and mottled with the brown of the fore neck; lower part of tarsi streaked and mottled with fulvous; under tail-coverts barred with the same; under wing-coverts buff, closely barred with irregular marks of blackish brown.

With age, as the pale striations and tippings of the upper surface disappear, the bars on the inner webs of the primaries diminish near the tips: the tail-bars likewise alter in character; but they are always perceptible on the central feathers in the oldest birds, and the bases of the primaries are never, as far as I have been able to examine specimens, without a few white bars. Mr. Sharpe observes, in his 'Catalogue,' that while the change to the adult plumage on the upper surface takes place by a partial moult, the alteration on the lower parts is acquired by the brown edgings of the feathers gradually occupying the whole of the web.

Distribution.—The Black Eagle is found both in the lowlands of Ceylon and the mountain-zone up to the highest elevations. In the low country it confines itself chiefly to tracts of forest and retired valleys in the vicinity of some rocky eminence, on which, in all probability, it breeds. I have seen it on several occasions in the Kurunegala district and about the Ambepussa hills; further south, in the more wooded portions of the Pasdun Korale and Saffragam, it is more plentiful, and in the hilly jungle-clad country between Galle and the southern mountain-range I have often seen it soaring round the forest-covered hills on the southern bank of the Gindurah, or gliding over the secluded valleys at the base of the Morowak Korale coffee-districts. In these latter it is not uncommon too. The endless jungles of the eastern side of the island, teeming with bird-life, form a grand refuge for these sable robbers; and I have observed them from the base of the Ouvah hills to the Friar's Hood forests, between which latter and the sea, at about an hour's walk from the Battiealoa Lake, I once shot a fine specimen. In the northern half of the island I have met with it as far up as the neighbourhood of Haborcnna, near which the lofty cliffs of Rittagalla and the precipitous rock of Sigiri no doubt furnish it with a permanent residence.

In the Central Province it is tolerably common, confining itself to the higher peaks in the Kandy district and the high ranges surrounding the Nuwara-Elliya plateau. I have seen it at Horton Plains and at Kandapolla, near the sanatorium; but it is oftener met with on the Uva side between Nuwara Elliya and Madulsima than anywhere else in the hills.

The Black Eagle is found in most of the hilly wooded districts of India, but appears to visit certain localities for a time and then depart again, reappearing the following year. In the south it is found in the Travancore district and in the Malabar region generally, following the west coast to the district of Surat.

Mr. Fairbank says it is rare at Mahabaleshwar, and in the Deccan he has not observed it. In the Himalayas it ascends generally to an elevation of seven or eight thousand feet, and is more common there from September till April than during the hot season. Col. Irby states that he has procured it as high as 10,000 feet. Mr. Ball does not include it in his avifauna of Chota Nagpur, nor does it appear in the "First List of the Birds of Upper Pegu" ('Stray Feathers,' 1875). Mr. Brooks records it as rare above Mussoorie.

To the south-east of the Himalayas its numbers commence to diminish; it finds no place among the birds collected in North-east Cachar by Mr. James Inglis ('Stray Feathers,' 1877); and though it is recorded by Jerdon and other naturalists from Burmah proper, it does not appear to be common there. According to Schlegel it is found in Malacca, and Wallace notes its occurrence in Java, Sumatra, and Celebes; but in these islands it appears to be far from numerous.

Habits.—This fine, long-winged Eagle is, on account of the singular structure of its feet and its curious habits, one of the most interesting, but at the same time perhaps the most destructive of Raptors to bird-life in Ceylon. It subsists, as far as can be observed, entirely by bird-nesting, and is not content with the eggs and young birds which its keen sight espies among the branches of the forest-trees, but seizes the nest in its talons, decamps with it, and (as Mr. Bourdillon, in his article on the Travancore birds in 'Stray Feathers,' observes) often examines the contents as it sails lazily along. Furthermore, Mr. S. Bligh informs me that he once found the best part of a bird's nest in the stomach of one of these Eagles which he shot in the Central Province! Its flight is most easy and graceful. In the early morning it passes much of its time soaring round the high peaks or cliffs on which it has passed the night, and about 9 or 10 o'clock starts off on its daily foraging expedition; it launches itself with motionless wings from some dizzy precipice, and proceeding in a straight line till over some inviting-looking patna-woods, it quickly descends, with one or two rather sharp gyrations, through perhaps a thousand feet, and is in another moment gliding stealthily along, just above the tops of the trees: in and out among these, along the side of the wood, backwards and forwards over the top of the narrow strip, it quarters, its long wings outstretched and the tips of its pinions wide apart, with apparently no exertion; and luckless indeed is the Bulbul, Oriole, or Mountain-Finch whose carefully-built nest is discovered by the soaring robber.

Mr. Frank Bourdillon, in his "Notes on the Birds of the Travancore Hills" ('Stray Feathers,' 1875, p. 358), in which district this Eagle is not uncommon above 500 feet, remarks, "I have never seen it make any attempt to seize a full-grown bird, but have once or twice seen one carry off a nest in its claws, and examine the contents as it sailed lazily along. It is a very silent bird, and may be seen steadily quartering backwards and forwards along the side of a hill and in and out among the tree-tops."

It is, I think, worthy of remark that the long inner claws of this bird seem especially adapted for the work of carrying off loose and fragile masses, such as the nests of small birds, as they would naturally form its chief means of grasp when such an object was being held by both feet during the process of flight.

Concerning its habits in India, Jerdon writes the following account, which is confirmatory of what I have above stated:—"I never saw it perch, except for the purpose of feeding or on being wounded; and the Lepchas of Darjiling, when I saw this Eagle, said, 'This bird never sits down.' It lives almost exclusively, I believe, by robbing birds' nests, devouring both the eggs and the young ones. I dare say if it saw a young or sickly bird it might seize it; but it has neither the ability nor dash to enable it to seize a strong Pheasant on the wing, or even, I believe, a Partridge; and Hodgson, I fancy, must have trusted to a native partially ignorant of its habits, when he says 'that it preys on the Pheasants of the regions it frequents as well as their eggs.'

"I have examined several birds shot by myself, and invariably found that eggs and nestling birds had been alone their food. In these cases I found the eggs of the Hill-Quail (*Coturnix erythrorhyncha*), of *Malacocircus malabaricus*, and of some Doves (*Turtur*), with nestlings and the remains of some eggs that I did not know. I have seen it also, after circling several times over a small tree, alight on it and carry off the contents of a dove's nest. In India, doves, and perhaps some other birds, breed at all times in the year; and it may,

H

perhaps, obtain eggs or nestlings at all seasons, by shifting its quarters and varying the elevations; if not, it probably may eat reptiles; but of this I cannot speak from observation."

I have been assured by several gentlemen in the planting-districts that it attacks fowls, and carries them off from the poultry-yards; and Mr. Northway, of Deltota, has a fine pair stuffed by Messrs. Whyte and Co., which were killed in so doing. It is the opinion of some naturalists that it does not attack large birds; but this fact is conclusive, though it may only carry off poultry when much pressed by hunger. The voice of this species is a shrill, very long-drawn scream, resembling the cry of the Serpent-Eagle somewhat, but much more powerful, and when heard in the deep gorges of the mountain forests in the upper ranges is a wild and stirring note.

Nidification.—It is extremely difficult to obtain information about the breeding-habits of a species frequenting such wild haunts as the Black Eagle. My endeavours to trace even the whereabouts of an eyrie were futile, although, during the last year I was in Ceylon, I learnt that a pair were thought to nest in the high cliff above the Nuwara Elliya and Kandapolla road. In 1872 a pair frequented a ravine near the Galle and Akkuresse road; and I believe they were breeding in the neighbourhood, but I was unable to discover their nest.

Mr. Hume has received eggs from two nests, with their parent birds, and has no doubt that they were rightly identified. These eggs were taken in January in India, and, in all probability, our birds breed about the same time. The nests were situated on ledges on the face of cliffs, and contained respectively one and three eggs. They were nearly perfect ovals, devoid of gloss and rough in texture, and of a greyish-white ground; and the single egg was richly blotched and mottled with brownish red, while the other three contained only a few brownish specks at one end. They varied "from 2·5 to 2·68 inches in length, and from 1·88 to 2·02 inches in breadth."

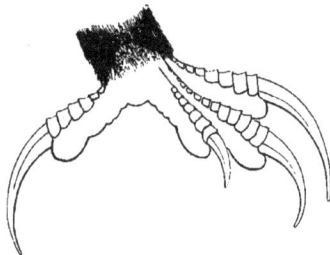

Foot of *Neopus malayensis.*

Genus SPIZAETUS.

Bill stouter, slightly shorter, and deeper than in *Neopus*; culmen curved much as in that genus, the festoon more pronounced; cere small. Nostrils large, oval, and directed obliquely upwards, and protected by the long loral bristles. Wings short and rounded; the 5th quill the longest, the 1st the shortest of all; tips of the secondaries falling short of those of the primaries by less than the length of tarsus. Tail long, rounded at the tip, exceeding the closed wings by more than the length of the tarsus. Tarsus long, but less than the tibia, feathered in some to the base of the toes, in others partly on the middle toe. Toes moderate, furnished at the tip with three transverse scales, the lateral toes subequal and slightly exceeding the hind toe; inner claw shorter than the hind. Head usually furnished with an elongated crest.

SPIZAETUS KELAARTI.

(THE CEYLON MOUNTAIN HAWK-EAGLE.)

(Peculiar to Ceylon.)

Spizaetus nipalensis (Blyth), Kelaart, Prodromus F. Zeyl. p. 96, and Cat. p. 114 (1852); Layard, Ann. & Mag. Nat. Hist. 1853, xii. p. 98; Blyth, Comm. Jerd. B. of Ind. Ibis, 1866, p. 242 (in part); Sharpe, Cat. Birds, i. p. 267 (1874) (in part).

Limnaetus nipalensis, Hodgson, Jerdon, B. of Ind. i. p. 73 (1862, pt.); Holdsworth, P. Z. S. 1872, p. 411.

Spizaetus kelaarti, Legge, Ibis, 1878, p. 201.

The Beautiful Crested Eagle, Kelaart, Prodromus.

Rajaliya, Sinhalese, Central Province.

Ad. similis *S. nipalensi*, sed piloo minùs nigricante, strigâ gulari et fasciis mystacalibus valdè angustioribus, pedibus robustioribus et unguibus validissimis, sed principuè corpore subtùs pallidiore brunneo et fasciis transversalibus omnino albis, rachide quoque albâ distinguendus.

Adult female. Length to front of cere 29·5 to 31·0 inches; culmen from cere 2·0; wing 18·0 to 20·0; tail 19·0 to 19·0; tarsus 4·4 to 4·6; mid toe 2·7 to 2·8, its claw (straight) 1·3; inner claw (straight) 1·7; hind toe 2·0, its claw (straight) 2·05 to 2·1, circumference 1·4 to 1·5; height of bill at cere 0·81. Weight 6 lb.
Iris yellow; cere blackish; bill black, paling to blackish leaden at the base; feet citron-yellow, claws black.

Mature female. Back, scapulars, lesser wing-coverts, rump, and upper tail-coverts blackish brown, the scapulars and upper tail-coverts tipped with white; forehead, crown, crest, and ovate centres to the feathers at the sides of the occiput and hind neck black, the latter very broadly margined with light sienna, diminishing gradually towards the lower part of the hind neck; crest of 5 or 6 feathers, 3½ inches in length and tipped with white; median and greater wing-coverts pale brown, darker near the tips, which are finely edged with whitish, except those of the inner feathers of the latter, which are rather deeply so; primaries and secondaries black, the latter tipped with white, and the whole crossed with obscure smoky-brown bars, which are white towards the base at their inner edges; tail blackish, tipped pale, with three pale smoky-brownish bands, and a fourth beneath the coverts, the subterminal one about 2 inches from the tip and about 1½ inch in width.

H 2

Chin, throat, and fore neck creamy white, with a very broad, mesial, black stripe, and with two others, less clearly defined, passing from the gape down the sides of the throat, and spreading out over its lower part ; cheeks and ear-coverts boldly striped with black, the edges of the feathers concolorous with the sides and back of neck ; chest, breast, flanks, and all the lower parts, including the legs and under tail-coverts, sienna-brown, darkest on the flanks, thighs, and under tail-coverts ; the feathers of the chest with wide and deep marginal indentations of white, and the breast, flanks, thighs, abdomen, and under tail-coverts barred with straight, complete bands of white, the *shaft* being of the same colour ; bars on the thighs narrow, but everywhere else broad, the brown interspaces on the sides of the breast and on the under tail-coverts with their lower edges darker than the rest ; tarsi pale brown, with whitish tips to the feathers ; lesser and median under wing-coverts concolorous with the chest and narrowly barred with white ; the greater series white, crossed with blackish-brown bars ; under surface of the light portions of the quills and tail-feathers greyish white.

Obs. The above description is combined from the examination of several fully-sized females, exhibiting each a different amount of intensity in the colour of the crown and hind neck, but none of them possessing the extremely dark features characteristic of adult Nepaul birds, or any inclination to the very broad chin-stripe of these latter, though this character is variable in that species. The older the Ceylonese birds become, no doubt the darker would be the head, and the bolder the chin and moustachial stripes, although I do not think they would ever acquire the same degree of melanism as the Indian species (*Spizaetus nipalensis*).

I have unfortunately no data of the dimensions of any *ascertained* adult males ; but the following of an immature bird, shot by Mr. Bligh, and the subject of the background figure in my Plate, will give some idea of the size attained by that sex.

Young male, apparently at the outset of the 2nd year :—Wing 16·3 inches ; tail 11·75 ; tarsus 4·5 ; mid toe 2·3, its claw (straight) 1·4 ; hind claw 1·7. (Two presumed males, in the British Museum, of *Spizaetus nipalensis,* have the wings 17·0 and 17·3 respectively ; and an ascertained male, recorded at p. 219 of ' Rough Notes,' measures 17·8, which, in view of the respective sizes of the females in the two races, will fairly represent that of adult males of *Spizaetus Kelaarti.*)

Above brown, the back, scapulars, and wing-coverts conspicuously margined with white as in the smaller species (*Spizaetus ceylonensis*) ; crown with the centres of the feathers dark brown, paling into fulvous at the margins ; rest of the head and hind neck paler, the edges of the feathers pale fulvescent ; crest well developed, the feathers black, deeply tipped with white ; greater wing-coverts pale brown, with much white on the inner webs and at the tips ; primaries and secondaries blackish brown, with paler smoky-brown bars than in the adult ; the inner webs white towards the base ; tail blackish brown, crossed with four pale brownish bands ; the black interspaces and terminal band narrower than in the adult ; tip whitish.

Chin, throat, and fore neck white ; the chin unstriped, a few blackish-brown drop-shaped marks on the throat, spreading laterally over the fore neck ; chest-feathers pale sienna-brown, indented at the sides with bar-like spots of white ; breast, flanks, abdomen, and under tail-coverts pale brownish, barred with complete white bands, wider than the brown interspaces, which are darker on the flanks than on the centre of the breast ; thighs barred more narrowly than the breast, the brown hue concolorous with that of the sides ; tarsi pale brownish, the feathers tipped with whitish ; under wing-coverts white, spotted with sepia-brown.

Obs. I discriminated (*loc. cit.*) this Hawk-Eagle from the Indian species (*Spizaetus nipalensis*), having made a careful examination of all the examples to hand in the British, Indian, and Norwich Museums, to aid me in my conclusions ; and the diagnosis of the distinctive characteristics of the two species, given in my article, will, I think, be sufficient to establish the Ceylonese bird as a good *subspecies* or local race, which I have named after Dr. Kelaart, who first brought to notice the existence of the species in Ceylon. For the benefit of my Ceylon readers and others who have not seen my remarks in ' The Ibis,' I now recapitulate in substance the remarks I there made.

The Ceylonese bird differs from the Indian in the peculiar barring of the entire under surface from the throat downwards, and in its very large feet and claws, the latter of which are especially noteworthy. Furthermore, it does not appear to acquire the black head and cheeks and the very broad black throat-stripe which are characteristic of *Spizaetus nipalensis.* In this latter bird the chest is usually dark brown, the centres of the feathers consisting of a broad dark brown " drop " or stripe, which pales off into an unbroken fulvous-brown margin, while in others the whole feather is sepia-brown, with slight marginal indentations of white ; this coloration is continued in most examples down to the breast, about the middle or upper half of which the barred feathers commence, and in which the white band is more or less irregular and interrupted at the shaft by the brown hue of the feather, the division varying from an exceedingly fine margin on each side of the *dark* shaft to a broad space of about $\frac{1}{16}$ inch. In many birds these bars do not even correspond or oppose one another on each side of the shaft, amounting in reality to nothing more than deep indentations of white. The thighs and under tail-coverts in the Nepaul bird are, however, barred in the same complete manner as the breast and flanks of the Ceylonese, but the perfect bar never seems to go to any higher than the tibials.

In the young of the Indian species the breast is marked with drop-shaped streaks, the bars being *confined to the flanks* and under tail-coverts: the markings are very dark as a rule, particularly on the chest and upper breast. It is, I may here remark, a very variable bird in its plumage, old birds differing *inter se* as much as young ones; and out of a score I have examined, no two were exactly alike. Five adult Ceylonese examples, which I have had the opportunity of examining, exhibited precisely the same character of barring over the whole under surface.

Lastly, as regards the massive foot and immense claws, which are characteristic of *S. kelaarti*, I have been unable (as will appear by a glance at my table of measurements, in 'The Ibis,' of seventeen examples of *S. nipalensis*) to find any Indian example of this latter species with the hind claw exceeding 1·0 inch; whereas in the Ceylonese bird it attains the great size of 2·1 inches, this measurement being taken, in accordance with my usual custom, across the arc from the tip to the exterior edge of the base.

Distribution.—This magnificent Eagle, the noblest representative of its tribe which Ceylon possesses, is peculiar to the island, and was first recorded by Dr. Kelaart from a bird procured by him near Badulla, mention of which is made at page 96 of his ' Prodromus,' as follows :—" This elegant crested Eagle is occasionally seen in the highest mountains. The only specimen we succeeded in procuring was shot on a mountain 4000 feet high, near Badulla." ' From that time until comparatively recently it does not appear to have been noticed by naturalists in the island; and so late as the year 1872, Mr. Holdsworth was unable to record any further instances of its capture since that of Kelaart's bird, although, doubtless, in the course of opening up the forests of the Central Province for the planting of coffee, the species may have been killed not unfrequently, and not recognized by its captors as any thing valuable.

It is entirely a mountain species, having its headquarters in the wild and little-trodden forests of the main range and other isolated lofty jungles, such as Haputale and the Knuckles, whence it descends to the neighbouring coffee-estates in pursuit or search of its quarry. In so doing it has lately been shot so frequently that it can no longer be considered one of our very rare Eagles. Not many years after the establishment of Messrs. Whyte and Co.'s business as naturalists and collectors, specimens began to find their way to them, and in 1875 I had the opportunity of examining two examples preserved in their collection. In March, 1876, a magnificent bird was shot by Mr. Bligh on the Catton Estate, Lemastota, and in the same year five examples were procured by Messrs. Whyte and Co., belonging to gentlemen in the surrounding planting districts. Three of the finest of these were obtained as follows :—(1) by Mr. A. Thom, on Oudasgeria Estate, Matale ; (2) by Mr. E. Nicol, Kitlamoola Estate, Deltota ; (3) by Mr. Gould, Maturata—all at elevations ranging from 2000 to 4500 feet. About the same time a sixth specimen was shot by Mr. Thurston near Nuwara Elliya, but unfortunately was not preserved.

Habits.—This fine Eagle frequents the retired recesses and forests of mountainous country, above an elevation of 3000 feet or thereabouts, probably not dwelling permanently or breeding below 4000 feet, although it may frequently be met with considerably beneath these altitudes when in search of food. Though bold and courageous in its disposition as a Raptor, it is very shy and wary of man, rarely coming beneath his notice, except when caught in the act of making a raid on the poultry-yards of the planters or seizing a hare on the mountain patnas. The first-named habit has on nearly all occasions led to its capture of late years in the planting districts. One of the finest examples above noticed was shot by Mr. Nicol after it had missed its mark at a fowl and settled on a tree near his bungalow ; and Mr. Bligh informs me that the magnificent example which he shot at Catton had its talons covered with the fur of a newly slaughtered hare.

It is occasionally seen about Nuwara Elliya, where the existence of isolated cottages and houses, with their accustomed live stock, is a weighty attraction for it ; it is quite powerful enough to be capable of carrying off the largest inmate of the poultry-yard, and, indeed, could make quick work with a moderately-sized lamb, were such to be found among the possessions of the fortunate owners of the many pretty bungalows which dot the plain of Nuwara Elliya. Its powers of flight and skill in catching game must be quite equal to those of its Himalayan relative, of whom Captain Hutton, as quoted by Mr. Hume in ' Rough Notes,' says, " it is most destructive to pigeons, fowls, and game." Mr. Thompson likewise writes of this bird :—" It feeds much on Pheasants, Hares, Black Partridge, Monaul and Cheer Pheasants, and sometimes on young deer."

Our bird may now and then be seen perched on the dead trees which stand in new coffee-plantations or upon the half-leafless ones peculiar to some of the higher patnas in the main range. Its flight is similar to that of the smaller low-country bird; and I have seen it quietly beating round the edges of the woods on the Horton Plains, probably on the look-out for the large black Squirrel (*Sciurus tennantii*), the "Kaloo Dando-leyna" of the Sinhalese, and which animal, I have no doubt, is often preyed upon by it. The note of this species is a loud scream, somewhat resembling that of *S. ceylonensis*.

[Since this article was sent to the press, I have received the following interesting note on this species from Mr. Bligh. Writing from Haputale, where the bird seems to be tolerably common, he says, "I often see the bird on the wing: now I know the species well, and I believe it to be nearly as common as *S. ceylonensis*; but they do not hawk for their prey so low down as the latter, which often skims through a valley of coffee within gun-shot of the ground, indeed often flies from one high stump to another, whereas the other would boldly sweep through the valley at a much greater elevation, and now and then, if really looking for prey, take a large sweeping circle. Lately I had the pleasure of seeing a pair of these birds on the wing together with a *S. ceylonensis*. I could easily distinguish the species; the small one, for some reason, kept above the others, and eventually soared away out of sight, as if he did not relish the neighbourhood of his powerful relations."

Nidification.—The nest of this species has never yet, to my knowledge, been found. The large tracts of forest which still clothe portions of the Nuwara-Elliya plateau, and stretch from the Horton Plains to the Peak, furnish it with a secure refuge in which to rear its young. It doubtless breeds on trees, nesting in a similar manner to the next species.

In the Plate accompanying this article, the figure in the foreground is taken from a magnificent female bird, mature, but not quite adult, for the possession of which I am indebted to Mr. Gould, of Maturata. The second figure is that of the young male described in this article, and for the loan of which I am indebted to the kindness of my friend Mr. Gurney, coupled with the civility of the authorities of the Norwich Museum, who loaned the specimen to me for the purpose of figuring. Mr. Keulemann's talented pencil has delineated this bird in the act of reposing on one leg, so characteristic of these Eagles.

Spizaetus kelaarti.   *Spizaetus nipalensis.*

The above woodcut of the *adult* breast-feathers of this Eagle and those of *Spizaetus nipalensis* shows the distinctive characters of marking in the two birds.

$\frac{2}{5}$

SPIZAETUS CEYLONENSIS.

SPIZAETUS CEYLONENSIS.

(THE CEYLON HAWK-EAGLE.)

(Peculiar to Ceylon ?)

Falco ceylanensis, Gmelin, S. N. i. p. 275 (1788).

Falco cristatellus, Temm. Pl. Col. i. pl. 282 (1824).

Spizaetus limnaetus, (Horsf.) *apud* Layard, Ann. & Mag. Nat. Hist. 1855, xii. p. 98 ; Kelaart, Prodromus, Cat. p. 114; Blyth, Cat. B. Mus. Asiat. Soc. Beng. (var. β) p. 25; id. Journ. Asiat. Soc. Beng. 1852, vol. xxi. p. 352.

Limnaetus cristatellus, Jerd. B. of Ind. p. 71 (in part); Holdsworth, P. Z. S. 1872, p. 411 ; Legge, Ibis, 1874, p. 9, and 1875, p. 277.

Spizaetus cirrhatus, Sharpe, Cat. Birds, i. p. 269 (in part).

Limnaetus ceylonensis, Gurney, Ibis, 1877, p. 431, et 1878, p. 85.

The Ceylonese Crested Falcon, Latham, Gen. Syn. i. p. 80 (1781).

Autour cristatelle, Temm. Pl. Col. 282.

The Crested Eagle and *The Hawk-Eagle* of Europeans in Ceylon.

Rajaliya, Sinhalese.

Ad. similis *S. cirrhato*, sed minor : alâ vix 15·2 unc. longâ : cristâ occipitali 3 vel 4 unc. longâ : pedibus flavis : iride flavâ.

Adult male and female. Length to front of cere 21·5 to 23·5 inches ; culmen from cere 1·1 to 1·25 ; wing 13·8 to 15·2, but rarely exceeding 14½ ; tail 9·0 to 10·5 ; tarsus 3·5 to 3·8 ; middle toe 1·8 to 2·0, its claw (straight) 0·85 to 1·0 ; height of bill at cere 0·5 to 0·56. Expanse 46 to 50. In the female I find no constant excess in size in the above measurements, taken from a series of fifteen examples ; one of that sex measures 14·2.

Iris leaden grey with a tinge of yellow, pale straw-colour or golden yellow ; cere dark leaden, in some with a greenish tint above : bill dark plumbeous, black at the tip, pale bluish at the gape and base ; feet lemon-yellow or greenish yellow ; claws black.

Obs. As will appear from the above, this Eagle is a bird of uncertain character in the coloration of its iris. It is likewise so in its plumage, there existing both a dark and a light phase, of which the latter, I think, contains the larger birds. To the dark form I will give precedence in this article, as I am able to furnish a more complete sequence of changes than in the pale.

1. *Dark form, old bird.* Head and hind neck dark tawny, the centres of the feathers blackish ; a crest of five or six elongated black feathers tipped with fulvous ; back, scapulars, and wing-coverts blackish brown, the feathers slightly paler at the margins, the coverts edged with tawny fulvous, blending gradually into the dark centres of the feathers and more conspicuous on the greater series than on the rest : lesser coverts pervaded with an ashen hue ; primaries and secondaries deep brown, with the terminal portions and a series of bars across both webs black, the basal portions of the inner webs white ; tertials paler brown than the secondaries ; rump and upper tail-coverts dark wood-brown ; tail dark ashen, crossed with three black bands, one at the coverts, another at the centre, and a third at the tip, about 1½ inch in width, having an interspace above it of about 2 inches wide.

Loral plumes and a superciliary streak blackish ; cheeks and moustache boldly streaked with black, passing into the blackish brown of the ear-coverts ; throat white, with a broad black chin-stripe, spreading over the fore neck and chest into a series of blackish "drops," paling into brownish at the margins of the feathers ; chest and under surface brownish rufescent, the bases and sides of the chest-feathers white, and each with a slaty-black central stripe vanishing on the lower parts into the dark smoky-brown ground-colour ; on the flanks, abdomen, and under tail-coverts the feathers have white bases, which show here and there, and disturb the uniformity of the ground-tint ; thighs paler than the abdomen and cross-rayed with obscure fulvous ; tarsi brownish fulvous ; under surface of tail greyish ; under wing-coverts whitish, dashed with tawny brown ; greater series white, with terminal black spots.

The above is a description of the example now at the Zoological Gardens, aged six years, which is by far the darkest bird, particularly as regards the under surface, which I have ever met with. Its iris is *very* pale straw-colour.

Mature bird. At about three or four years of age, in a stage of plumage in which most dark birds are met with, the head and hind neck are more or less sienna-brown, with the centres of the feathers blackish, least so on the hind neck; on the forehead and above the lores the narrow feathers are pale-edged; crest, which is sometimes 4½ inches in length, black, conspicuously tipped with white, the shorter feathers being blackish brown, paling into rufous at the white tips; back, scapulars, and wing-coverts deep glossy brown, paling off at the margins into a tawny hue, the greater coverts with less of the dark brown central hue, finely edged greyish, and with the concealed portions of the bases white; winglet and primary-coverts, the quills and secondaries dark brown, barred and terminated with black, much as in the above, but with more white on the inner webs, and with the tips of the secondaries whitish, a fulvous patch on the outer webs of the longer primaries opposite the notch; tertials wood-brown, paler than the scapulars; rump and upper tail-coverts of a similar hue; tail smoky brown, tipped white and crossed with four blackish bands, the subterminal one equal in width to the preceding interspace, the next two much narrower, and the basal one generally incomplete; on the lateral feathers there is an additional pale basal bar, and the interspaces are mottled with white.

Cheeks and the sides of the neck beneath them boldy streaked with blackish, the edges of the feathers being white: ear-coverts concolorous with the hind neck. Chin, throat, and under surface white, contracted at the centre of the fore neck between the tawny hue of its sides; a narrow blackish-brown chin-stripe passing down to the chest, from which to the abdomen each feather is centred with a broad drop-shaped dash of blackish brown; on the abdomen and flanks these expand until they cover the terminal portion of the feathers: the lower flank-plumes blackish brown, forming a large dark patch; under tail-coverts dark brown, usually tipped with white; thighs and upper part of tarsus a more rufous-brown, paling into buffy white at the feet; under surface of tail and of the quill-interspaces whitish; bases of quills beneath pure white; under wing-coverts white, dashed and striped in places with blackish brown, those beneath the ulna centred widely with rufous-brown.

In these birds I have invariably found the iris yellow, which is the normal colour, I imagine, of the eye in the adult.

Young*. Nestling clothed with white down, with the crest-feathers plainly indicated by three or four attenuated downy shafts; the wing-coverts, scapulars, and quill-feathers on first appearing are fulvous-brown, deeply tipped with white; the tail-feathers are similar, and the whole darken considerably in a short time, the hue of the interscapular feathers being deeper than that of the rest.

Nestling plumage at 3 months. Iris leaden grey; bill dusky plumbeous, blackish at the tip; feet light lemon-yellow. Head, back, and sides of neck with the ear-coverts light sienna-brown, edged with whitish: crest-feathers blackish, deeply tipped with white; back, scapulars, and lesser wing-coverts dark sepia-brown, the scapulars broadly tipped with white, and the back feathers margined, terminally, with rufous-grey, the bases being paler brown than the rest; median wing-coverts mostly white, with a longitudinal patch of brown; greater series broadly margined with white, the outer webs being a paler or fawn-brown; primaries and secondaries brown, the former the darker in hue, tipped with white and crossed with narrow bars of black, vanishing near the internal edges, which are white; first primary and terminal portion of the long ones almost uniform blackish: rump and upper tail-coverts fawn-brown; tail umber-brown, with a deep white tip and five or six narrow bars of blackish brown, the subterminal one slightly broader than the others, and the light interspaces, as in the quills, showing white beneath.

A thin white line from nostril over the lores; loral plumes blackish: lower part of cheeks, throat, and under surface pure white, dashed on the sides of the chest and breast with light sienna-brown "drops," those on the flanks being slightly darker and coalescing into a patch at the lower part; belly, thighs, and under tail-coverts dashed with pale brown and tipped with white; tarsi white. The extent to which the under surface is marked in this stage varies. The "drops," however, darken after the space of two months, as do also the feathers of the head and hind neck, which at the same time acquire darker mesial stripes: the brown of the back and wings also becomes more intense, and the bird is then in the normal plumage of the first year, with a long crest measuring from 3 to 4 inches.

At the second moult the example under consideration darkened on the head and hind neck, the crest remaining the same: the white of the wing-coverts diminished in extent, and the tail underwent a considerable change, the number of bars on the central feathers being reduced to four (of greater width than the last, especially the terminal one, which was preceded by an equally broad interspace), the chest "drops" increased in number and in intensity, and the lower parts became more covered with brown; the dark patches on the white under wing-coverts were also more numerous.

---

* These changes of plumage are described from observation, during youth, of the above living example, as well as from notes on other immature birds in my collection.

In the third year the upper surface continued to darken, the back became more uniform in hue, the white on the wing-coverts diminished, but the tail remained much the same, except that the brown was more cinereous in its tint; the crest, however, was almost entirely absent, but this was doubtless an abnormal characteristic; on the under surface a faint chin-stripe developed itself, and the coloration of the cheeks altered, becoming striated with dark shaft-lines, and each feather of the breast and under surface had a "drop" of umber-brown, those on the flanks completely covering the feather, while the abdomen, thighs, and under tail-coverts became uniform brown; under wing-coverts dashed with brown.

Iris during these years pale grey, without a sign of yellow in the coloration.

In the fourth year the "drops" on the under surface darkened, the marking of the tail altered, and the lower parts were more completely covered with brown; the crest was much shorter than it was in the second year, but otherwise the bird was in the plumage described above as mature, with the exception that the iris was still leaden grey.

No change took place after this until the autumn of the fifth year, when the bird commenced to moult several months after the time *, and assumed the fuliginous plumage in which it has been above described.

2. *Pale form.* Iris greenish grey; pale slate-grey; greenish grey faintly tinged with yellow. Head, back, and sides of neck tawny brown, the centres of the feathers black and broadest on the crown; crest as in the *dark form*, with the longer feathers boldly tipped with white and the shorter with rufescent greyish; major portion of the feathers of the back, scapulars, and lesser wing-coverts blackish, paling off at the margins and exposed bases into fulvescent brown, with the tips paler still; bases and most of the inner webs of the median and greater coverts white, showing most on the former, and the terminal portions blackish brown at the centre; rump, upper tail-coverts, and tertials pale brown, the tips of the coverts, in some, whitish; secondaries dark brown, barred and edged internally with white, as in the other phase; primaries and tail the same, but with a large fulvous patch at the quill-notches; cheeks and ear-coverts concolorous with the sides of the neck; the lower part of the face striated with dark brown; chin, throat, and entire under surface down to the abdomen white.

No chin-stripe; centres of the chest- and breast-feathers rufous-brown, many of them with dark shaft-lines, and on the flanks and sides of breast with patches of dark brown; a dark brown patch on the lower flank-plumes; under tail-coverts and thighs rufous-brown, the white bases on the latter giving them a chequered appearance; tarsi buffy white, dashed with the hue of the thighs; under wing-coverts white, the primary series with dark brown terminal patches; under surface of primaries white as far as the notch, that of the secondaries for two thirds of their length.

Obs. The above is a description of the oldest example in this phase of plumage that I have been able to procure. I obtained it on the shores of the Kantholai tank, and judging by the bars on the tail, which are three narrow ones, separated from a broadish terminal one by an interspace of equal width, it is only in the third year. Another example, of apparently similar age, has the cheeks whitish, streaked with brown lines, but no chin-stripe; there is a series of dark shaft-stripes on the chest, but the lower parts are less clothed in brown than in the aforementioned. It is in the above phase of plumage that by far the greatest number of examples are procured in Ceylon.

Young (bird of the year). Wing nearly equal to that of the adult. Iris leaden grey or pale slate-colour, sometimes tinged with greenish; in one example pearly white; cere and gape bluish leaden; bill blackish at the tip.

Forehead, crown, back, and sides of neck tawny buff, the feathers in some with dark shaft-lines, in others entirely without any dark coloration; crest as in the dark bird; on the lower hind neck the brown terminal centres gradually develop into the dark brown of the back, scapulars, and wing-coverts, which are margined with tawny and tipped with whitish: on the median coverts there is much more white than in the mature bird, in some examples the entire feathers being uncoloured, with the exception of the terminal portion of the outer webs, and form an extensive white patch across the wing; first primary uniform blackish brown as in the older bird; upper tail-coverts pale brown, tipped with white; tail smoke-brown, tipped with white and crossed with five narrow bars of blackish brown, the subterminal one broader than the rest; lateral feathers white internally and with an additional bar.

Entire under surface pure white: the pale rufescent feathers of the side neck encroaching on the throat, and a few dashes of the same hue on the sides of the chest, flanks, and belly; thighs and under tail-coverts shaded with pale rufescent brown, the feathers tipped with white; tarsi washed with the same in some, pure white in others;

---

* The moulting-time was a month later every year, a circumstance which apparently was caused by the natural want of vigour consequent on the captivity of the bird.

I

under wing-coverts white, the primary series spotted with dark brown, and the lining of the ulna washed with rufous-brown.

Obs. The Crested Hawk-Eagle of Ceylon is a miniature representative of the peninsular Indian species *S. cirrhatus.* Gmelin recognized Latham's Ceylonese Crested Falcon as a distinct form, and described it (*loc. cit.*) under the name of *Falco ceylonensis*; but subsequent naturalists, overlooking its smaller size, have treated it as one and the same with its large ally. Mr. Gurney refutes this idea with reason, as will be seen by reference to his remarks on the species ('Ibis,' 1877, p. 430). The maximum size which the insular bird attains in the wing is 15·3 inches, a measurement representing the minimum of *cirrhatus*, it being, however, at the same time about 2 inches below the average of the Indian species. I am not aware that the latter acquires the fuliginous plumage of *ceylonensis*: and the light phase of this is, moreover, paler than the immature dress of the Indian bird, which appears to partake somewhat of the characteristics of the *mature* form above described. I have examined a large series, and have found them all less pale on the head than Ceylonese young birds, and many of them possess the chin-stripe and striated cheeks unknown in our buff-plumaged young. It is possible that *ceylonensis* may prove not to be peculiar to Ceylon, Mr. Hume having described a small bird from Travancore as *Spizaetus sphinx*, which may, when a sufficient series is obtained, prove identical with it as a resident in S. India, or, should it turn out to have been a straggler, demonstrate the fact that *Spizaetus ceylonensis* strays over to the Indian coast from North Ceylon.

The dimensions of *Spizaetus sphinx* ('Stray Feathers,' vol. i. p. 321) are as follows :—" Length 22 to 23 inches ; wing 14·1 ; tail 10·2 ; tarsus 3·0 : mid toe and claw 2·5 (nearly)."

The upper plumage appears to bear a great resemblance to melanistic examples of the Ceylonese birds : " the whole back, top, and sides of the head (excluding the crest), back, and sides of the neck, a pale, slightly rufous-brown, each feather with a blackish-brown shaft-stripe." The lesser lower wing-coverts are " dull rufous, brown-shafted, more or less white-edged ; the rest white, very broadly *barred* * with deep brown." In this the species seems to differ from *S. ceylonensis*, as also in the coloration of the throat, which is described as follows :—" Chin and throat white, with one central and *two lateral blackish-brown streaks, which unite at the base of the throat* at the front of the neck ; below this for about an inch *dull rufous-brown, like the sides and the back of the neck* ; the breast white ; the feathers with huge dark brown drops, edged paler towards the tips ; sides, abdomen, lower tail-coverts, flanks, and exterior tibial plumes a nearly uniform, somewhat pale umber-brown, most of the feathers with inconspicuous very narrow whitish tips ; interior tibial plumes and tarsal feathers pale dingy yellowish brown, paling most towards the feet."

Distribution.—The small crested Eagle of Ceylon is chiefly a low-country bird, and is more or less dispersed throughout the maritime provinces and the interior jungles of the island. In the Eastern Province it is located in greatest force, and thence northwards it occurs principally along the coast, near salt-lakes and open tracts of land, to the delta of the Mahawelligunga and the district lying between Tamblegam and Kanthelai tank, where it is again more common than irremediately to the south of the Virgel. To the north of Trincomalie it is found in the open woods bordering the continuous salt-lakes of that part of the coast, and in the interior is met with generally in the vicinity of the tanks of the Vanni. Layard found it at Pt. Pedro ; but it is on the whole a scarce bird in the Jaffna peninsula. It occurs sparingly throughout the west of the island to the north of Negombo, but it is decidedly scarce between that place and Kalatara.

In the wooded districts interspersed with paddy-cultivation, which form the south-west corner of the island, it is more common than in the Western Province, and again further east, beyond the Morowak Korule ranges, it becomes more numerous still, frequenting the low-lying jungles between Hambantota and the Badulla mountains. In the Kandyan Province it is not unfrequent up to an elevation of 4000 feet, occurring chiefly in the Knuckles ranges, in Medamahanuwara, Dumbara, and southwards to Ambegamoa, as also round the eastern slopes of the Maturata district into Uva proper and Madulsima. Mr. Bligh has obtained it in Kotmalie and in the spurs of the Haputale range, and Mr. Holdsworth speaks of having seen it at Nuwara Elliya†. Layard mentions (*l. c.*) that Kelaart obtained it at Nuwara Elliya ; but the latter does not include it in his list of birds from that locality ('Prodromus,' p. xxix).

----

* The italics are mine.

† I have never seen any specimens of this bird from the Nuwara-Elliya plateau. Mr. Holdsworth speaks of the Eagle that he observed as soaring in " wide circles, with a squealing cry." This is a marked characteristic of the Serpent-Eagle (*Spilornis spilogaster*), whereas the Crested Eagle rarely soars, and seldom utters its cry on the wing. I think, therefore, that Mr. Holdsworth may have been mistaken in his identification.

Habits.—This noble little Eagle frequents open forest, the borders of heavy jungle, detached woods, cheenas, and scrubs interspersed with large trees. About such localities it prowls with a slow, though buoyant flight, being chiefly about in the mornings and afternoons, and searches the open ground for its favourite food, the large *Calotes* lizard. When satisfied with the result of its excursions it perches on solitary dead trees or exposed limbs of others in the forest, and enlivens the wilds with its complaining cry, which may be syllabized as *kre kre kre kreee, kre kre kre kreee*, quickly repeated, and continued to a wearisome extent. This is, however, the cry of the young or immature bird, and develops in the adult into a prolonged note in a different key, and in which the principal accent is laid on the second syllable, resembling the sounds *kre-kreēē-kre-kree*. This is as invariably the voice of the brown, dark-marked birds as the former is of the light-plumaged individuals.

Of the lonely cheenas of the Eastern Province, studded with blackened trees and stumps, and scantily covered with a straggling crop of " Kurrukkan " (*Eleusine indica*) or a few wild cucumbers, this Eagle forms a marked characteristic; perched motionless on the limb of a tall tree, it remains for a long time piping out its monotonous cry, which is perhaps answered from another cheena a little distance off. At such times it is seated bolt upright on one leg, with the other drawn up beneath its breast-plumes, its erect crest and its eyes staring proudly before it; and so regardless is it of all around it that it may easily be approached in the open from behind to within an easy shot.

It is a bold and courageous bird in its disposition, as is amply testified to when it is kept in confinement; but as regards its prey it captures nothing larger than jungle-fowl, squirrels, and other small mammals, and feeds more on lizards than any thing else. It is exceedingly active and quick-sighted, and rarely misses any thing upon which it pounces. It is quite capable of capturing a bird on the wing, and in the Kandy district it is often shot carrying off poultry from the planters' bungalows; in the villages of the Vanni it also commits considerable havoc in the same way about the houses of the natives. Layard, in his notes (*vide suprá*), speaks of one darting at a wounded Sparrow-Hawk which he had tied to a post in the verandah of his bungalow. Its flight is not, as a rule, swift, but performed with steady flappings of the wings; it rarely soars—and when it does, mounts in quick small circles for a short time and then flies off at a tangent.

The habits and disposition of birds of prey are well observed when they are in confinement; it may not, therefore, be out of place to subjoin here a short account of one of these Eagles which I reared from the nest and had five years in my possession, and which is now personified in the noble little representative of the species in the Zoological Gardens. When a chick he was fed upon lizards, which were first given him cut up; but as soon as he could stand up, he quickly learnt to devour them in the orthodox way, beginning at the head and finishing up at the tail, which he always swallowed whole. As it grew older, whenever food was thrown to it, and more particularly in the case of small birds or any thing which it was fond of, it seized the prey with both feet, squatted down on the tarsi, and spread forward its wings in a line with its head, at the same time expanding its tail and completely covering up its prize from view; it would then droop the head, looking at the coveted morsel, and commence uttering its querulous note, endeavouring to flap its wings when approached by any one, and altogether presenting a very singular appearance. This was its habit throughout life, and was more particularly practised when in company with other Raptors in the same aviary, being evidently its mode of shielding its prey from outward attack. He had the same method of standing on one leg and resting the other on the knee-joint, with the tarsus thrust out from the perch and the toes clenched, that I have observed in other Eagles, and which is no doubt a muscular exercise.

He would now and then seize a stone and fly round the aviary with it, or at other times endeavour by main force to tear up a clod from the floor of his aviary. During his first year he was a timid bird, sometimes retreating into a dark corner or " cot," inhabited by a Wood-Owl (*Syrnium ochrogenys*), and stretching himself out would remain there for hours; he likewise frequently allowed his nocturnal companion, who fed as much by day as by night, to rob him of his meat. Very different, however, was his nature after the first moult; he then developed both in muscular strength and courage, and became a bold and fierce little tyrant, commenced by attacking his companion, and finished by killing him outright. He displayed great agility and power of flight, one day darting up and seizing, through the bamboos of the aviary-roof, a Magpie-Robin that was perched upon it; at other times he would dart from his perch and catch, in the air, birds, rats, and other food thrown in to him. He was fond of bathing, and invariably stood out in heavy showers of rain, in which he

1 2

would expose himself to a thorough drenching, and then dry himself in the sun with his wings expanded. The most singular and interesting point in his disposition was his manifest display of anger and excitement, accompanied by a particular note of displeasure, consisting of a shrill scream, followed by a "champing".sound. This passion he exhibited, becoming quite furious when shown a stuffed bird of any size—a huge Pelican, which was his pet aversion, being usually subjected to the fiercest onslaughts when shown to him at the bars of his aviary; these were followed by a continued uttering of his note of anger until his passion died away. At about the age of twelve months he commenced to utter his adult note; but now and then, more particularly in the breeding-season, during the first three years, I heard the querulous cry peculiar to the young stage. When shown any object which excited his interest or curiosity, such as a tempting morsel of food, without the bars of his aviary, he had a singular habit of twisting his head round till it was completely turned upside down, all the time keeping his eyes fixed on the subject of his examination. At other times, when under the influence of excitement from any cause, he would throw his head back until it touched his back, and sway his head too and fro with a spasmodic outdarting of his wings, as if he were going to launch himself through the roof of his aviary. He made two voyages round the island with me, and one trip across country in a bullock-bandy, and during his life in Ceylon experienced several adventures, one of which well-nigh proved fatal. While at Trincomalie he narrowly escaped being killed by a wild cat, from whose clutches he must have escaped purely by dint of fierce struggles, and inflicting, no doubt, severe wounds on the animal with his talons. One morning, during my absence in the jungle, he was found to be missing, and on examining the aviary a large hole was discovered in the roof, through which he had evidently been dragged; search was made high and low throughout the whole premises, but not a sign of the eagle was anywhere to be seen. About midday, when the house-coolie went to draw water, the unfortunate bird was perceived floating on the surface, which was about 30 feet below the trap. On rescuing him from his perilous position he was found to be uninjured, with the exception of a wound at the point of the wing, evidently made by the teeth of a cat, which must have dragged him across the compound some 40 yards, with a view of taking him through an opening at the back of the wall, where the beast found the eagle's clutches too strong for him, and dropped him close to the trap, down which he had fallen in the darkness. Neither his mauling by the cat nor his five or six hours' cold bath in the darkness of the well had done much towards intimidating his eagleship; for the plucky little fellow fought vigorously while being secured, and it was only by dint of enveloping him in the coolie's cloth that he could be brought up again to *terra firma*. He was then tied to a stick and well dried in the sun, and then, much to my wife's satisfaction, was reinstated, undaunted by his adventures, in his aviary.

Nidification.—The Crested Eagle breeds in the south of Ceylon in February and March, but commences in the north somewhat earlier. In the neighbourhood of Trincomalie I twice found its nest during the course of its being built or repaired in January, but was unsuccessful in obtaining the eggs, for the birds deserted on both occasions. They were both large structures of sticks placed in the uppermost branches of banyan trees, and appear to have taken a long time to set in order, one nest being worked at for a month before I ventured to have it looked at, and then it seemed to have made but little advance. Only one young bird appears to be reared, for I am aware of two instances in which a solitary eaglet was taken from the nest.

The front figure in the Plate accompanying this article represents the dark bird now in the Zoological Gardens, and in his sixth year. The second is that of an immature *light* bird, which I shot with three others on the same day in the Batticaloa district.

Genus SPILORNIS.

Bill longer than in the last genus; festoon slightly pronounced, the culmen curved from the cere; the cere advanced. Nostrils oval, oblique, protected by the loral plumes. Eyelid furnished with long lashes. Wings short, rounded, the 4th and 5th quills the longest. Tail moderately long and ample. Tarsus slender, feathered slightly below the knee, protected with small hexagonal scales both in front and behind. Toes short, furnished at the tip with transverse scales. changing at their bases into the reticulated scales of the tarsus; claws short and rather straight.

Head furnished with a heavy rounded crest, extending entirely across the occiput.

SPILORNIS SPILOGASTER.

(THE CEYLONESE SERPENT-EAGLE.)

Buteo bacha, Vigors, Mem. Raffl. p. 650 (1830, nec Daud.).
Hæmatornis spilogaster, Blyth, Journ. As. Soc. Beng. vol. xvi. 1852, p. 351; Kelaart, Pro-
    dromus, Cat. p. 114 (1852); Layard, Ann. & Mag. Nat. Hist. 1853, xii. p. 100.
Hæmatornis cheela, Layard, Ann. & Mag. Nat. Hist. 1853, xii. p. 99.
Spilornis bacha, Holdsworth, Proc. Zool. Soc. 1872, p. 412; Legge, Ibis, 1875, p. 277.
Spilornis spilogaster, Walden, Ibis, 1873, p. 208; Gurney, Ibis, 1878, p. 100.
Spilornis cheela, Legge, Ibis, 1874, p. 9.
Spilornis melanotis, Sharpe, Cat. Birds, i. p. 289 (1874).
The Harrier-Eagle, Buzzard-Eagle, in India (pt.); The Cheela Eagle, Ceylon Eagle, "Cheela,"
    Kelaart and Layard; Serpent-Eagle, Europeans in Ceylon.
Rajaliya, Sinhalese; Cudoombien, Tam., in Ceylon (apud Layard).

♂ ad. similis S. bacha, sed ubique pallidior, gutture pallidè cinerascente nec nigricanti-brunneo: pectoris et abdo-
    minis maculis ocellatis minoribus, minùs rotundatis et saturatiore brunneo circumdatis: pedibus obscurè flavican-
    tibus: iride flavâ.

Adult female. Length to front of cere 22·0 to 23·5 inches: culmen from cere 1·3 to 1·4; wing 15·3 to 15·8; tail 9·0
    to 10·5; tarsus 3·3 to 3·4; middle toe 1·8, its claw (straight) 0·9; height of bill at cere 0·65.

Male. Length to front of cere 21·5 to 22·5 inches; wing 14·5 to 15·7 (average of seven examples 15·1); tarsus 3·2
    to 3·3.

Obs. The above measurements are taken from a series of examples shot in the low country or on its hill-borders,
    representing a small type of our Serpent-Eagle inhabiting these districts. The majority of examples from the
    Kandyan province are considerably larger, and may be fairly held to constitute a bigger race of the species, as will
    be seen from the following dimensions, taken from several specimens:—
Adult female. Length to front of cere 25·0 inches; culmen 1·5; wing 15·9 to 16·6; expanse 52·5; tail 11·0 to 11·5;
    tarsus 3·4 to 3·5; middle toe 1·9.
Male. Wing 1·5 to 15·8 inches.
Iris golden yellow, the external edge, in some, indented with black; cere and loral skin varying from greenish yellow

to "citron;" gape greenish yellow; bill bluish at cere and base, darkening at the tip to blackish; legs and feet citron-yellow (usually much stained).

Fully adult plumage. Forehead, crown, and elongated occipital feathers more or less overlying the entire hind neck jet-black, with the basal two thirds of the feathers white and concealed beneath the black portions, and the tips almost always faintly tipped with fulvous; cheeks and ear-coverts more or less blackish grey, according as the throat is light or dark, blending into the adjacent black; hind neck, back, rump, scapulars, wing-coverts, and tertials dark neutral brown, with a strong purplish lustre, particularly on the upper back and scapulars; lower part of hind neck paler than the back; upper tail-coverts tipped with white; least wing-coverts, including the winglet, with two terminal white spots on each feather: greater wing-coverts tipped white at the outer feathers; terminal portion of the primaries and secondaries (about $2\frac{1}{2}$ inches of the former), a narrow band across the centre of the feathers, and another near the base blackish brown, with a purple lustre, the interspaces smoky brown on the outer webs, gradually paling to white internally, the whole band showing whitish beneath; secondaries and shorter primaries tipped with white; winglet blackish brown, tipped with white, and barred near the base of the inner webs with the same; tail with a two-inch terminal band, and a second, nearly as broad, on the basal half purplish black; an equally broad interspace of dusky whitish, more or less clouded with light brown, and the space between the second band and the coverts paler, but not conspicuously lighter than the band; all the caudal feathers tipped with white.

Loral and rictal plumes black; chin and gorge iron-grey, more or less dark according to the individual, in some almost as pale as the fore neck, which, together with the chest and under surface, varies from a light earth-brown to a chocolate-colour, paling always slightly towards the belly and under tail-coverts; the feathers of the breast, flanks, belly, and thighs with a series of roundish, opposite white spots surrounded by a dark edge; on the under tail-coverts, and in some specimens on the thighs, the spots develop into bars, either continuous or interrupted at the shaft; under wing-coverts, as in the young stages, variable, the ground-colour of the lesser series usually more rufous than that of the breast and covered with large white spots, which, near the tip, predominate over the brown; greater series dark brown, spotted like the rest; edge of the wing unspotted white.

Mature plumage. In this phase the lower feathers of the crest are tipped with fulvous, and in some the basal portion washed with the same; the cheeks, ear-coverts, and chin are darker than in old birds; the scapulars, median wing-coverts, and rump-feathers are tipped with white; winglet more conspicuously tipped than in the above; markings of quills and caudal feathers much the same; sometimes the secondaries are terminated with brown, adjacent to the white tip, and occasionally there is a remnant of *white mottling* above the central tail-bar.

Under surface chocolate-brown, the fore neck darker, blending into the blackish-grey hue of the throat, and the feathers slightly edged with fulvous; lower parts darker than in the fully adult, the edging round the spots deeper, and those latter, therefore, more conspicuous; under wing-coverts dark like the breast, the spots on the lesser series smaller than in old birds; external edge of wing-lining sometimes unmarked white, at others striped or barred, like the rest of the feathers, with brown; under surface of tip of tail showing more white than the upper.

Obs. Great variation exists, particularly in this latter stage, in the markings of the under surface, although there is, as in the coloration of the throat, a certain similarity of type, which distinguishes the species from some of the more eastern forms. As regards the spots, in birds of the same age and with similar upper-surface plumage, they are in some examples very large and darkly bordered, in others small, and then, of course, more of them on each feather, the edge being sometimes scarcely darker than the ground of the feathers; in others, again, they are more bar-shaped than circular. There is, in this mature stage, sometimes an indication of fulvous cross-marking near the tips of the chest-feathers; but I have never seen it in a fully old, completely black-crested bird. In some specimens much more of the hind neck is blacker than in others, the sides of this part being black right round to the throat. Hill examples of the larger race are blacker on the chin and throat than the small birds of a like age. This is a peculiarity I have observed in 4 specimens at Norwich, 4 in Mr. Bligh's collection, and 2 in my own.

Young (bird of the year). Iris greenish yellow, sometimes with a brown inner circle, in one specimen I have seen (as also in that referred to by Mr. Blyth, *l. c.*, as drawn by a Mr. Moognart) white; cere, gape, and loral skin greenish yellow; bill dark horn-colour, the tip blackish; legs and feet pale yellow.

The white eye appears to be an abnormal, though doubtless not an unfrequent feature.

Head and elongated occipital feathers dusky fulvous, the feathers with an extensive subterminal blackish-brown patch and deep tips of buffy white, the bases white; interscapular region, wing-coverts, and scapulars rich sepia-brown, with a purplish lustre, becoming paler on the rump and upper tail-coverts, the feathers tipped with fulvous grey, least so on the lesser wing-coverts, which are darker than the median series; those and the greater coverts tipped generally with white and spotted on both webs with the same, the markings in some taking the form of bars; primaries and secondaries dark earth-brown (this colour corresponding to the light interspaces in the adult), terminal portions blackish, and the rest of the feather crossed with three bars of the same; the brown interspaces mottled with white at the inner edge, and the tips of all the quills deeply marked with the same; upper tail-coverts tipped with white; tail light brown, mottled transversely with white, with a subterminal band, one about the middle, and one at the base, of purplish black; lateral feathers with three bars and the tips of all white, the inner webs white at the edge.

Cheeks, ear-coverts, chin, and upper throat black, the feathers white at the base, and those at the lower edge of the gorge tipped with fulvous; in some birds the throat is marked up to the chin with this hue; fore neck, chest, and under surface chocolate-brown, light in some, dark in others, the breast marked inconspicuously with fulvous-brown cross rays; the lower parts, thighs, and under tail-coverts spotted with white as in the adult, the spots with less dark surroundings, those on the under tail-coverts developing into bars; under wing-coverts and axillary plumes concolorous with the chest, except the greater series, which are blackish brown, the whole with large round spots of white, but covering the ground-colour less than in the adult; external edge of wing-lining white: base of quills beneath white.

The change from this plumage, which is that of Blyth's *S. spilogaster*, to the mature dress following the adult plumage in this article takes place in the gradual increase of the black on the head-feathers, the decrease of the pale terminal margins of the upper surface, the widening of the subterminal black caudal band and its adjacent light interspace, causing a moving up of the centre dark band, and in the gradual lessening of the light interspace above it; the light transverse rays across the chest (in those birds which possess them) grow fainter, and the ocelli of the lower parts extend, as a rule, more up the breast. These changes take place in the second moult.

1[1]] birds in their first plumage often have the white wing-covert tippings, in the form of deep terminal margins, extending up the webs, on which the bar-like markings are more extensive than in low-country birds. In one specimen the basal portions of the mantle-feathers are fulvescent buff, showing on the surface very markedly; the head-feathers do not show much of the fulvous centres, and the throat is a very dark brown, blending into the chocolate of the fore neck.

Light phase of young plumage[a]. A pale or albescent form of plumage exists occasionally in the young of this species, which is analogous to the same feature in the Booted Eagle (see remarks in my article on this latter bird).

Such an example, in my collection, shot near Kadugannawa in October 1876, has the upper surface similar to other birds, except that the least wing-coverts are very lightly tipped with fulvous, whereas the greater and median coverts have much white about them, the inner webs of the latter being almost entirely of this colour and the outer margined with it; cheeks, ear-coverts, and chin black; throat, chest, and entire under surface buff-white: the feathers of the chest, sides of the breast, and flanks with large oval patches of deep brown on their terminal portions, which diminish on the lower parts into oval central streaks: the long flank-feathers covering the thighs barred widely with a paler brown, with which the thighs are closely banded; coverts immediately beneath the ulna spotted with brown; remainder of the wing-lining white, a few of the feathers with a single dark spot near the tip; base of primaries white beneath.

Obs. This Serpent- or Harrier-Eagle, as the genus is perhaps more generally styled in India, was referred to by Layard in his notes as *Hæmatornis cheela*, it having been identified for him as such by Blyth in the days when it was not discriminated as distinct from this latter northern form. Adult specimens are still extant in the faded collection at Poole. Subsequent (evidently) to his acquaintance with the bird in its adult character, immature specimens were procured by Dr. Kelaart and himself, and sent by both gentlemen, under the impression that they belonged to a new species, to Blyth, by whom they were described as such under the title of *H. spilogaster*, by which name it must now stand as a Ceylonese and Sumatran bird.

From the Northern-Indian *Spilornis cheela* it differs widely, inasmuch as it is a much smaller bird, has a paler throat, wants the yellowish-brown cross markings on the chest which are characteristic of the mature birds in that species, and differs in the character of the ocelli of the lower parts.

As regards the South-Indian form, whether the *Sp. melanotis* of Jerdon, or perhaps, more properly speaking, *Sp. albidus* of Temm., represents the common species of that part, I am unable to say; but if it does, the Ceylon bird is its inferior in size. The wing in Jerdon's type, from the foot of the Nilghiris, is 16 inches; but it is probably a male, as other examples in the British Museum, referred by Mr. Sharpe to that species, exceed 17 inches in the

---

[a] This occurs in *S. cheela*; there is a similar specimen in the British Museum.

wing, and I have lately examined a bird, not fully adult, from Malabar, which has a wing of 18·2. It may be well to remark that this specimen, which has since passed into the Norwich Museum, and was noticed by Mr. Gurney ('Ibis,' 1878, p. 145), does not differ much from immature examples of *S. cheela*, tho breast and under surface being isabelline brown, and the white ocelli surrounded each by a bold dark margin, in addition to which the axillaries and under wing-coverts are differently coloured. The nearest affinities of *S. spilogaster* are with the Malayan races, to which it approaches closely in size. It is, however, distinguishable from *S. davisoni* from the Audamans, which species has the ocelli small, very round, and more confined to the lower parts.

From the Javan Serpent-Eagle, also (the *S. bacha* of Mr. Sharpe's Catalogue, and to which it has of late been referred), it differs in a marked manner, inasmuch as the latter species is very dark above and beneath, and possesses in its adult stage an almost black throat, the contrast of *S. spilogaster* at that period ; the under wing is likewise different, the lesser coverts being concolorous with the greater, and not palor as in the Ceylonese form.

To the Sumatran bird, however, it approaches very closely, so much so that Mr. Gurney thinks ('Ibis,' 1878, p. 100) the two races are identical. I have carefully examined the series from Sumatra in the Norwich Museum, in company with that gentleman, and though slight points of difference exist, they do not appear sufficient to rank as specific, in which case the species should bear the same name as the Ceylonese, as it is not identical with *Sp. bacha* of Java, with which it has been hitherto united. The differences referred to are the darker throat of the adult, and the lighter, less clouded pale tail-band, resembling somewhat that of *S. pallidus* from Borneo. The four examples which constitute the series seem as a whole to be smaller than Ceylonese, although the wing in one attains 16·5. The race from Singapore is also not separable from the Sumatran ; but in its young stage it differs from Ceylonese immature birds (there are none forthcoming from Sumatra) in the subterminal pale band being considerably broader. It is therefore possible that when a larger series is got together from Sumatra, containing old and young, the race may be found, as in the case of the Singapore bird, to differ from the subject of this article in its young plumage.

I subjoin a synopsis of the several species of *Spilornis* referred to in the above observations, in order that the respective characteristics may be seen at a glance :—

    a. *Spilornis cheela*. *Hab.* Himalayan region, Burmah, China, Formosa.
        Large size : wing 18·0 to 19·5 inches.
        Ad. Chest almost always crossed with narrow transverse strim ; throat and cheeks iron-grey ; ocelli of the lower parts bar-shaped, with a brown border.
        Juv. Head black as in adult, but the dark hue separated from the white base of the feather by a fulvous patch ; throat and cheeks black.

    b. *Spilornis melanotis*. *Hab.* Peninsular India.
        Smaller : wing 16·5 to 17·8 inches.
        Very similar to *S. cheela* in plumage.

    c. *Spilornis spilogaster*. *Hab.* Ceylon, Sumatra, and Straits Settlements ?
        Smaller still : wing 15·3 to 16·6 inches.
        Ad. Chest uniform brown, without any transverse strim ; throat and cheeks pale iron-grey ; under-surface spots variable in shape and size, surrounded by a dark edge, which is also variable in intensity ; median under wing-coverts concolorous with the chest.
        Juv. Head-feathers conspicuously tipped with white ; throat and cheeks blackish.

    d. *Spilornis bacha*. *Hab.* Java.
        Similar to *S. spilogaster* in size.
        Ad. Very dark above and beneath ; throat and cheeks black-brown ; ocelli large, rounded, the edge scarcely darker than the ground-colour of the feather.

    e. *Spilornis davisoni*. *Hab.* Andaman Islands.
        Smaller than *S. spilogaster* : wing 15·0 inches (Hume).
        Ad. Ocelli small, very round, and not extending much above the abdomen.

Distribution.—The Serpent-Eagle is widely distributed over the whole island, but is much more numerous in the dry forest-clad tracts of country than in the humid and more cultivated portions. It is a common bird in the continuous jungles of the south-eastern low country, parts of the " Park," the Eastern Province, and the entire northern half of the island. In all these districts it is chiefly to be found in the vicinity of village tanks or on the banks of the forest-lined rivers.

In the Western Province it is a scarcer bird, and is mostly confined to the wild country commencing near Avisawella and stretching through Saffragam, and thence along both banks of the Kaluganga to the maritime districts at its mouth. In the hilly country between Galle and the Morowak Korale it is likewise an uncommon bird, being now and then met with on the outskirts of damp paddy-land and on the banks of the Gindurah and other streams. As regards the Kandyan province, it is found generally throughout the coffee-districts, extending even to the Nuwara-Elliya plateau; but it is chiefly noticeable about Kadugunawa, in parts of Dumbara and the Knuckles district, in Dolosbage, and thence into Ambegamoa. In Haputale it is not uncommon, Mr. Illigh having procured many specimens in that district. It is found as near Colombo as Atturugeria and Kaduwella; but going northwards of the capital it is not very frequently met with until the Maha Oya is passed and the drier districts near Kurunegala reached.

Layard, who thought it to be migratory, remarks of its distribution, under the head of *Hæmatornis cheela*, "Abundantly and widely distributed throughout the island;" and in speaking of the immature phase (*H. spilogaster*) says, " the Doctor " (Kelaart) " procured his specimen at Trincomalie, whilst I killed mine in the Vanni. I afterwards shot another pair at Pt. Pedro." From his observations it appears that the species visits the north (the Jaffna peninsula is probably meant) in March, and remains until July. It is very probable that a partial movement to the peninsula does take place at that season, which led to the belief that the species was migratory.

Beyond the confines of Ceylon, this species reappears in the island of Sumatra and extends thence to the Straits Settlements. It was first made known from Sumatra by Sir S. Raffles.

Habits.—This small Eagle, whose serpent-destroying propensities make it a useful bird, is a denizen of forest, frequenting the banks of streams and rivers and the borders of tanks, more especially the smaller class known as the village " Kulam." Every such sheet of water possesses its pair of Snake-Eagles, which haunt the heavy jungle and huge trees clothing the bunds or dams, and patiently watch throughout the day from some huge outstretching limb for the various snakes and frogs which disport themselves from time to time on the banks of the stagnant pool. On espying its prey, the yellow-eyed bird raises his massive topknot, and with glistening orbs darts noiselessly down with dangling feet, and sweeping off the luckless reptile, mounts to the nearest perch and there devours it, resuming there and then his patient watch. This sluggish existence is, however, varied by a daily soar above his accustomed haunt, in the blaze of the noonday sun, when he mounts to a great height in wide circles, and with loud screams proclaims his freedom and success.

Equally at home by the sandy beds of the dried-up rivers in the northern and eastern forests, one of these Eagles may be encountered, at every mile or so, during a ramble down these romantic watercourses. They are invariably seated on the overhanging limb of an immense Kumbook tree, and when disturbed skim noiselessly on before the intruder and take up their post again on the nearest inviting perch.

The last specimen of this Eagle procured by me in Ceylon fell to my gun at one of these riverine haunts under rather interesting circumstances. It was about 4 o'clock on the evening of a scorching day in the Seven Korales when I arrived at the banks of the dry course of the Kimbulana-Oya; and leaving my jaded bullocks to enjoy the welcome shade of the grand umbrageous trees overhanging the crossing-place, I started for a tramp down the heavy sandy bed of the river. Here, as in most rivers in the north and east, which in the wet season are mighty torrents, not a drop of water was now to be seen, save in some more than ordinary deep holes under the denuded roots of the great trees which grow on the bank or in the hollows of the large rock-masses which stood up here and there from the sandy bed. Above most of these tiny pools sat a solitary Little Blue Kingfisher, eagerly eyeing the water, round the edge of which ran quietly one or two Green Sandpipers or a Common Snipe, reduced by scarcity of food to a rare degree of tameness. On rounding one of the rocky barriers a huge Owl glided noiselessly from the branches of an overhanging tree, and immediately fell to my first barrel with a broken wing. As the wounded bird waddled off a Serpent-Eagle, evidently taken aback at the sight of his companion trailing his wing along the sand, swooped down on him, doubtless out of mere curiosity, and quickly followed him to ground with a fractured pinion. This brought the Owl to a sudden

K

halt; and the two birds now presented a most singular spectacle, standing almost side by side and glaring with manifest amazement first at me and then at each other—the Owl with his long aigrettes erected and his immense yellow orbs staring from beneath them as he angrily snapped his bill; while the Eagle stood in his most defiant attitude, his amber eyes glaring fiercely, and his bushy topknot rising and falling as I approached him! It was a fit tableau for an artist; but, alas! was soon spoiled; and ere many minutes the interesting birds were dangling dead from the roof of my bullock-cart.

The Harrier Eagle may often be met with by the sides of tracks in the northern and eastern jungles, and is usually found with a snake dangling in its talons, which has been killed at the open side of the rudely-made road. When wounded, it is a very handsome object, placing itself on the defensive with its glaring yellow eyes and huge uplifted topknot.

Serpents are killed by these birds before being carried off, a bite on the neck soon depriving them of life; they may often be seen dangling from their grasp in the air, or hanging dead from their talons when perched. The food I have always found on dissection to be torn in pieces; but it is sometimes devoured whole—Mr. Holdsworth recording an instance of a Tree-Snake (*Passerita*), which is the favourite quarry with the Serpent-Eagle, being disgorged whole from the stomach of a wounded bird. Lizards and frogs are likewise eaten, but not so commonly as snakes. The note of this species is a prolonged and clear scream of three syllables, with the accent on the first, and is not unlike that of the Kite-Eagle.

Layard refers to the "doleful moanings" of this bird "scaring the herd-boy from the tank side, or the lonely native threading his way through the jungle." I myself have never heard these sounds, although the species was constantly under my observation for two years in the jungles of the north, haunting sometimes the vicinity of my camp from morning to night; I infer, therefore, that they may be the utterances peculiar to the breeding-time, which I was not fortunate enough to hear.

With his accustomed keen powers of observation he marked the habits of the Harrier-Eagle well, and has the following descriptive paragraph of them in the 'Annals and Magazine of Natural History':—"Concealed in the dark foliage of some overhanging tree, it heedlessly marks the smaller frogs approach the grassy margin of the pool. Suddenly the large green Bull-Frog (*Rana malabarica*) uplifts its head and utters its booming call. The *Cheela* is now all attention; with outstretched neck it fixes its glaring eyeballs on its desired prey; lower and lower it bends, for the frog, which has now reached the sedges with a croak of triumph, gains a log. But a shadow glides over him—in vain he crouches—and his colour becomes a dull brown, so closely resembling the log, that human eyes would take him for a knot in the decaying timber; with noiseless rapidity the barred wings pass on, and the log is untenanted. Fast clutched in the talons of his merciless foe, the frog is borne to the well-known perch, and a sharp blow on the back of the neck from the bill of the bird deprives it of life."

In the dry season I have known it to take up its quarters permanently by the side of a small water-hole a few square yards in extent, so that it might live on the frogs and snakes which frequented the muddy little spot.

Nidification.—The nest of this Eagle has very seldom been found; and the eggs I have never been able to procure. It breeds in the Western Province in March and April, Mr. MacVicar, of the Ceylon Public Works Department, having received a young bird taken from a nest in the Hewagam Korale in the latter month. The nest was described to me as being a large structure of sticks placed in the fork of a tree.

Layard, who was very fortunate in finding the nests of rare birds, remarks that "it builds in the recesses of the forest on lofty trees. The structure is a mass of sticks piled together and added to year by year. The eggs, generally two in number, are 3 inches in length by 2 in diameter, of a dirty chalk-white, minutely freckled at the obtuse end with black dots."

FALCONIDÆ.

AQUILINÆ.

Genus HALIAETUS.

Bill very stout and long, the cere and base of culmen straight, tip suddenly hooked; festoon well developed. Nostrils round and directed straight backwards and quite exposed. Eyelid devoid of lashes. Wings long, the 3rd quill the longest, the 1st slightly exceeding the 7th; the outer webs of the longer primaries abruptly notched. Tail short, scarcely exceeding the closed wings, cuneate at the tip. Tarsus very stout and longer than the middle toe; its upper third feathered in front, the remainder covered with broad transverse sculæ, the hinder part shielded with narrower transverse scales. Toes protected by rectangular scales to the base; outer toe only slightly longer than the inner; claws short, rather small and well curved, trenchant beneath.

HALIAETUS LEUCOGASTER.

(THE WHITE-BELLIED SEA-EAGLE.)

———

Falco leucogaster, Gm. S. N. i. p. 257 (1788, *ex* Lath.); Temm. Pl. Col. i. pl. 49 (1823).

Falco blagrus, Daud. Traité, ii. p. 70 (1800).

Haliaetus blagrus, Cuv. Règne An. i. p. 316 (1817).

Falco dimidiatus, Raffles, Trans. Linn. Soc. xiii. p. 277 (1822).

Haliaetus leucogaster, Vig. Zool. Journ. i. p. 336 (1824); Gould, B. Aust. pt. 3, pl. 37. fig. 1 (1838); Schlegel, Mus. P.-B. *Aquila*, p. 14, 1862; Jerd. B. of Ind. i. p. 85 (1862); Schlegel, Vog. Nederl. Ind., Valkv. pp. 9, 50, pl. 4. figs. 1, 2 (1866); Hume, Rough Notes, i. p. 259 (1869); Holdsworth, P. Z. S. 1872, p. 412; Ball, Journ. Asiat. Soc. Beng. 1872, p. 276; Sharpe, Cat. Birds, i. p. 307 (1874).

Ichthyaetus leucogaster, Gould, B. Austr. i. pl. 3 (1837); Gray, Cat. Accipitr. 1844, p. 13.

Pontoaetus leucogaster, Gray, Gen. Birds, i. p. 18 (1845); Kelaart, Prodromus, Cat. p. 114 (1852); Layard, Ann. & Mag. Nat. Hist. 1853, xii. p. 100.

Cuncuma leucogaster, Gray, Cat. Accip. 1848, p. 24; Wall. Ibis, 1868, p. 15; Walden, Trans. Zool. Soc. viii. p. 35 (1872); Hume, Str. Feath. 1874, p. 149; id. Nest and Eggs, i. p. 48; Legge, Ibis, 1875, p. 278; Hume, Str. Feath. 1876, p. 461.

Blagrus leucogaster, Blyth, Cat. B. Mus. A. S. B. p. 30 (1849); Swinh. Ibis, 1870, p. 86.

Polionetus leucogaster, Gould, Handb. B. of Austr. i. p. 13.

White-bellied Eagle, Lath. Gen. Syn. i. p. 33 (1781).

Le Blagre, Levaill. Ois. d'Afr. i. pl. 5 (1797).

The Grey-backed Sea-Eagle of some writers; *The "Fish-Eagle," "Fish-Hawk,"* Europeans in Australia; *The "Sea-Eagle," "Osprey,"* Europeans in Ceylon. *Duck-Eagle*, Andamans.

Kohassa, Hind.; *Samp-mar*, Hind. in Orissa; *Ala*, Tel. and Tam. (*apud* Jerdon, *l. c.*); *Laug-laut*, Sumatra.

Loko-Rajaliya, Sinhalese; *Kadal-ala*, Tam. in Ceylon.

Adult male. Length to front of core 25·2 to 26·5 inches; culmen from core 1·98 to 2·0; wing 21·2 to 22·5, expanse 71·5 to 78·0; tail 10·0; tarsus 3·4 to 3·8; mid toe 2·3 to 2·4, claw (straight) 1·05 to 1·1; hind toe 1·5, claw (straight) 1·4; height of bill at core 0·71.

Female. Length to front of core 27·0 to 27·75 inches; culmen from core 2·1; wing 22·5 to 24·0, expanse 70·0 to 80·1; tarsus 4·0; mid toe 2·5, claw (straight) 1·2.

A series of Malaccan, Indian, and Cape * birds examined in the British Museum correspond well in size with Ceylonese; and Tasmanian examples, which are very fine, do not, as far as I am aware, exceed the above limits.
Iris hazel-brown; core pale london; bill dark london; legs and feet whitish, or sometimes pale greenish white.
Entire head and neck, with the entire under surface, lesser under wing-coverts, under tail-coverts, and terminal 3¼ inches of the tail pure white; interscapular region, back, and rump dark cinereous grey, becoming darker on the upper tail-coverts; the white feathers at the lower part of the hind neck with dark shafts, and the grey hue appearing lower down on each side of them; wing-coverts, scapulars, and tertials bluish slate-colour, with dark shafts; quills and basal portion of the tail blackish cinereous; under wing-coverts and flank-feathers with black shafts.
I observe no variation in the tints or proportions of the several colours in this bird from all parts of its habitat.

Young. The unfledged nestling † has the iris brown; bill and cere very much as in the adult but more "fleshy," and the base of the under mandible very pale; legs and feet fleshy white. The body is covered with pure white down for about three weeks, when yellowish-brown feathers appear on the nape, and dark brown ones on the scapulars, the primaries coming out blackish at the same time, the whole being tipped with white, which shows most conspicuously on the forehead and crown; the tail-feathers, which are brown, tipped with fulvous, appear simultaneously with the primaries; the chest is clothed with umber-brown feathers, with fine tips of fulvous. The various hues of this first plumage alter somewhat during the first six months until they settle down into the normal hue of the yearling dress.
At a year old the iris is of the same hue as the adult's, the bill has lost its fleshy tint, and the legs are as in old birds. The plumage is as follows :—
Head, neck, and throat pale tawny brown, lightest on the chin, and the tips and margins of the head-feathers paler than the rest; over the eye there is a pale stripe, and above the ears a conspicuous dark patch; the light hue of the hind neck darkens into rich brown on the interscapular region, back, and wings; the lower part of hind neck and the lesser wing-coverts tipped with fulvous; the edge of the wing above the flexure has the feathers broadly margined with buff-white, which with a light patch on the side of the neck, partly concealed by the wing, forms a conspicuous light space in the bird's plumage; the median wing-coverts and the inner feathers of the greater series edged fulvous; the winglet and the quills are deep brown; the upper tail-coverts blackish brown, edged whitish; tail white at the base, with the terminal half blackish, blending with a mottled edge into the white; chest and upper breast chocolate-brown, the lower part edged with fulvous: the breast below this and the belly are whitish, washed at the sides with pale brownish : under tail-coverts whitish; on the under wing the secondary coverts are buff, marked with brownish, the greater row of the primary-coverts brownish, the next fulvous, and the least series brown, edged with buff. At the end of the first year the head, in the example here treated of, got very much paler, and the dark half of the tail faded considerably, while the chest-patch became much lighter.
After the first moult an advance towards the adult dress is made on the head, tail, and under surface; but the back and wings remain in the brown plumage still. The head and hind neck have the bases of the feathers brownish, and the terminal portions fulvous-white, revealing in the centre the dark shaft; the interscapular region and lesser wing-coverts are of a corresponding deep brown, with narrow pale margins; the median wing-coverts have the external portions of the feathers brownish fulvous, and the central brown with darker shafts; the upper tail-coverts are more or less white, mottled and clouded with blackish brown near the tip; the order of coloration in the tail is reversed, the terminal half becoming white and the basal black, but not completely so, some of the inner webs being nearly all white.
The throat and cheeks are fulvous-white, with a dark patch above the ears; the pale fore neck darkens into light earth-brown on the chest, the centres of the feathers being whitish with dark shafts; this part gradually pales, with a mottling of brown, into the whitish of the lower parts, on which, however, the shafts are brown; the lesser under wing-coverts are a rufescent buff, the upper series shaded with brown, and the greater row of the primary-coverts blackish brown.

---

* Specimen " *m,*" p. 308, Sharpe, Cat. of Birds.
† The immature plumages of this Eagle are described from birds reared from the nest in my aviary.

In the third year traces of immature plumage sometimes remain in the form of patches on the chest and some brown feathers in the wing-coverts; but I imagine that, as a rule, the adult dress is then put on.

There was a singular example of this Eagle, exhibiting a phase of plumage bordering on melanism, in the Zoological Gardens last year, and which it may be interesting to notice here. The head and neck were uniform earth-brown, and the back and wings dark brown; the cheeks and throat were pale brownish, and the whole under surface sooty brown; under surface of quills pale greyish, and the lesser under wing-coverts tawny.

Distribution.—The Grey-backed Sea-Eagle is a common bird round the whole of the north and east coasts and down the west side of the island to the lower end of the Puttalam lake. On the Jaffna lake and among the numerous islands off the west coast of the peninsula, as well as on the many back waters and estuaries from Point Pedro to Batticaloa, it is a characteristic ornithological feature of the coast; and further south every river-mouth that debouches on, and every salt lagoon that lines the shore from Kalmunai to Tangalla has its pair of Eagles. On the west coast, from Chilaw to Point de Galle, where the line of coast is less cut up by brackish inlets, it is not so frequent, being there confined to particular localities, such as the Negombo, Panadure, and Amblangoda lakes and the estuaries of Kalatura and Bentota. In the harbour of Galle a pair are often to be seen, and have their head-quarters at the Kogalla Lake or other neighbouring sheet of water, where an abundant supply of fish furnishes them with daily food. It is not, however, confined to the sea-coast; for the large tanks of the Eastern Province, viz. Ambaré, Erakkamum, and others, are frequented by it; and in the northern half of the island it is a permanent resident on all the large inland sheets of water, such as Minery, Kanthelai, and Tissa Wewa tanks. Its presence at the two former of these lakes adds no little to their romantic beauty; there are always one or two pairs there, which breed in the adjacent forest, and probably never leave the vicinity of those fine sheets of water. At Minery there is an eyrie on the great bund of the tank, and at Kanthelai a huge nest existed, until it was cut down, for many years in the fork of a lofty dead tree which towered above the surrounding forest on the north side of the lake.

The Grey-backed Sea-Eagle has a wide geographical range, extending north to south from Northern India to Tasmania, and west to east from the Cape of Good Hope to the Friendly Islands. In India it is chiefly confined to the sea-coast, and on the western side does not extend commonly above Bombay. It is very numerous near Pigeon Island (lat. 14° N.), which forms one of its chief breeding strongholds. At the Laccadive group it is rare, Mr. Hume recording but one specimen, which he saw at the island of Amini. In the other islands of the Indian Ocean, to the south-west, viz. the Seychelles, Mauritius, &c., it disappears, but again appears at the Cape of Good Hope, provided, that is, that M. Verreaux's specimen in the British Museum is correctly labelled. Returning to India we find it more common on the east coast than it is on the west, and at some parts of it it extends far inland, for Mr. Ball records it as by no means rare in Chota Nagpur. In Burmah it is likewise common; Mr. Davison procured it in Tenasserim, and Mr. Armstrong along the coast of the Irrawaddy delta, but found that it did not extend far up the Rangoon river. In the Gulf of Siam, Finlayson says that it frequents the desert islands. In the Malay peninsula and in all the islands of the Bay of Bengal it is numerous, and it inhabits the entire chain of the Malay islands from Sumatra to Timor, and extends thence through Borneo, Celebes, and the Moluccas to the Philippines, in which group it has been found as far north as Luzon. Eastward and to the south of the Philippines it has been found in the Solomon Islands and in New Guinea. Down the entire east coast of Australia to the islands of Bass's Straits and Tasmania it is a common bird, and, according to Gould, extends up the west coast of the insular continent to Swan River; but I have no doubt that it inhabits the entire western seaboard round to Torres Straits. It has not been observed in New Zealand, which, in spite of the peculiar character of the ornis of this country, is somewhat singular in a bird of such wide range.

Habits.—The island of Ceylon being devoid of the larger and more regal members of the genus *Aquila*, the present species must be considered to rank foremost among its Eagles. Though not possessing the courageous nature or the bold aspect of the powerful Mountain Hawk-Eagle, the lofty flight and commanding bearing of this fine bird, together with the associations of foaming shores and flowing tides, amidst which it passes its life, impart to it an interest for the naturalist and sportsman which does not attach to its congeners of the hills and forests. It frequents the open coasts of North and East Ceylon, as well as the estuaries, lagoons, land-locked bays, and salt lakes which form the chief geographical feature of the seaboard. It lives in pairs, and confines itself to one particular spot—a sequestered hill-side, containing its one or two huge banyan trees towering above the surrounding jungle, or the forest-covered bank of some inland sheet of water; here it breeds and passes all its life, sending forth its young to other haunts, which are usually not far from the place of their birth. Sallying out in the early morning, it quietly sails along high above the resounding shore, its wings outstretched and motionless, and its snowy head turning from side to side, as it scans each passing reef for its favourite morsel, the sea-snake; or it sweeps out to sea, with its eye intent on the inhabitants of the blue waters beneath; and keen indeed is its sight. Should any luckless fish venture to the surface beneath those silent wings, his time has come; with half-closed pinions and extended talons the Fish-Eagle descends with a booming rush; a splash, and up he rises with heavy flappings, bearing away his well-caught prey to some favourite rock or tree, beneath which the bones of many a fish and snake testify to the Eagle's feasts. It is partial to the sea-snake (*Hydrophis*), which, basking on the uncovered reef, is an easy prey; it likewise captures crabs, or feeds, if hungry, on any thing dead which it finds on the shore; but its favourite food is fish, for which it will pursue even the Osprey and rob it, as Jerdon remarks, of its well-earned food. During the heat of the day this Eagle often soars to a great height, and as it rises in wide circles, its pinions upturned to the extreme, its flight is grand and majestic*. Its loud cry of *clank, clank, clank*, which is repeatedly uttered in the breeding-season, can be heard at a considerable distance, and often leads to the discovery of its eyrie. It lives admirably in confinement, thriving even when wounded and captured as an adult, and feeds gluttonously on either cooked or raw meat. Mr. Holdsworth mentions one bird in his notes which was reared on the universal rice and curry of the native; and I have no doubt he was rightly informed, for little came amiss to my tame one. He was a cowardly bird in his disposition, standing in wholesome fear of a fierce little Crested Eagle in the same aviary with him, and also allowing himself to be bullied out of many a morsel by a Gannet which at another time kept company with him. A notice of this fine Eagle would be incomplete without quoting from Layard's graphic description of its habits. He says, "The flight of this species is noble and imposing; poised high over the resounding surge, it wheels above on circling pinions, and with extended neck surveys the finny tribes. Here shoals of bonk-nosed fishes swim in their seasonal migrations along the coral reef; there brilliant Chætodons float in the shallows. The tide has partially receded, and the water lies in still crystal pools in the depressions of the reef: soon out dash the Fish-Eagle passes; an abrupt wheel shows his attention arrested; a moment's pause, and down he plunges, his body swaying to and fro. The surface is reached, the legs suddenly thrown out, and with exulting cries he soars aloft, bearing in his talons a writhing snake, eel, or large fish. The efforts of the bird to secure its prey in a proper position are now curious. If a fish is captured the feat is comparatively easy; the talons of the Hawk are gradually shifted until one grasps the prey near the gills and the other near the tail. . . . . With a snake or eel the matter is more difficult, and I have often seen the prey free itself from its captor by its strong writhings; a bite, however, near the head destroys its power, and it is borne away dangling by the neck in the grasp of its destroyer." I observed that my tame birds invariably commenced eating a fish by tearing it at the back of the neck, the head being pointed to the bird's left; small fish, however, up to 5 or 6 inches, they would bolt entire, jerking them down head foremost.

Mr. Hume, in his interesting account of his trip to the Laccadives, contained in 'Stray Feathers,' 1874, has the following observations on the habits of this Eagle at Pigeon Island. At page 423 he writes, "Once that I shot as he swept overhead, high above the stunted trees that concealed me, had in his claws the entire liver and stomach of a goat. . . . . . It is a fine sight to see these Eagles striking one after the other in rapid succession. Soaring far above the island, often, I should judge, from a height of at least 1000 feet they come down with nearly closed wings, and with a rushing roar, like that of a cannon-ball, in a perfectly direct line, making an angle of about $60°$ with the water, which they scarcely seem to reach before they are again mounting with heavy flaps, and with a yard or two of snake hanging *dead* in their talons. One snake I recovered, shooting its captor, less than a minute after it had been seized. It was stone dead (though we all know how tenacious

---

* Gould, in his grand work on the 'Birds of Australia,' remarks that the great breadth and roundness of its pinions and the shortness of its neck and tail give this Eagle, when floating in circles high in the air, the resemblance of a large butterfly.

of life these reptiles are), and had its head and neck pierced through in several places by the Eagle's cruel claws, its whole skull being completely crushed up."

Messrs. Hume and Ball both allude to the shy disposition of this Eagle. The former says that at the Andamans it is exceedingly difficult to procure; and in writing of it as an inhabitant of Chota Nagpur, Mr. Ball remarks that it is extremely wary and difficult to approach. In Ceylon it is certainly when perched a shier bird than *P. ichthyaetus*; but it will frequently fly close overhead, and then affords an easy shot.

Instances have been likewise known in Ceylon of its having carried off wounded birds. Captain Wade writes me that he saw one in the Yala district take away a fallen Duck; and Mr. Bligh relates to me a similar case in which another of these Eagles pounced on a Stone-Plover (*Esacus recurvirostris*) which had been fired at and had fallen in the surf. The robber turned inland with his booty, and flying over the sportsman's head dropped it, on being fired at, almost at his feet.

Nidification.—In Ceylon this species breeds during the months of December, January, and February. It selects for its eyrie an enormous banyan or other tree with stout limbs and taller than the surrounding forest, and there builds a huge nest, sometimes 5 or 6 feet in diameter, and often, to secure a firm foundation, as many deep. The interior is almost flat, and contains a bed of green leaves, in which the eggs are laid, and with which the bird hurriedly but skilfully covers them on leaving or being frightened from the nest. This lining is removed when the young are hatched, and they repose on the twigs beneath.

One of several eyries which I discovered in the neighbourhood of Trincomalie was visited on two successive years; and on the occasion of my second visit I found that the male bird, whose mate I had robbed him of, had procured another, who was quite at home in her new quarters. The nest was at the top of the junction of an enormous aerial root with the parent limb, and up which my coolie progressed at a great speed : the bird sat very close, not stirring from the nest until the man was up to it; while he was ascending the male brought a huge fish to the nest, but on perceiving the intended robbery, flew off with it, leaving the hen still setting. Both birds flew round the nest, swooping down near it and uttering their loud clanking notes; but they did not, as also on all other occasions in my experience, attempt to molest the man. Even when losing their young these Eagles do not exhibit much courage, although they do not fly off, and calmly settle on a distant tree, as I have seen *Polioaetus ichthyaetus* do. The eggs are nearly always two in number, but sometimes only one, as I, on one occasion, took a single incubated egg from a nest. They are dull white and vary in shape, some being very round, while others are long ovals or pointed at one end ; the shell is tolerably rough, and, in general, much stained and soiled. They vary from 3·17 to 2·77 inches in length, by 2·18 to 2·02 in breadth.

On Pigeon Island there is a large nesting-colony of these birds ; they breed in the lofty trees growing on the island, as many as thirty or forty nests being placed in close proximity to each other. Their breeding-time here, according to Jerdon, is during the months of December, January, and February, or the same as it is in Ceylon. Concerning the nesting of the White-bellied Sea-Eagle in the Andamans and Nicobars, Mr. Davison writes to Mr. Hume, " I found this bird nesting on Nancowry Island on the 8th of March ; the nest was a huge mass of sticks, placed between two great branches of a large tree, at an height of about 80 feet from the ground. . . . . . I could not climb the tree myself, and I could get no assistance from the Nicobarese ; they would not go near the nest ; and when I said I would have it taken without their assistance, they earnestly begged me not to touch it, as doing so would be sure to bring fever into the village, and they would all die." This strange idea of the Nicobarese evidently arises from their acquaintance with the incongruous mass of fish-bones, snakes, skeletons, crab-shells, &c. which are always to be found beneath these great eyries, and which are not always of the most odoriferous kind.

Although this species breeds, as a rule, on lofty trees, it alters its habit according to the locality to which it is obliged to confine itself. Mr. Gould, in writing of it as an Australian bird, says, " I could not fail to remark how readily the birds accommodate themselves to the different circumstances in which they are placed ; for while on the mainland they invariably construct their large flat nest on a fork of the most lofty trees, on the islands, where not a tree is to be found, it is placed upon the flat surface of a large stone, the materials of which it is formed being twigs and branches of the Barilla, a low shrub which is there plentiful."

Genus POLIOAETUS.

Bill not so long nor straight as in *Haliaetus*, but deeper and more powerful in proportion to its size; culmen boldly arched from the cere; festoon slightly pronounced; nostrils oval and exposed; cere moderately advanced. Eyelid devoid of lashes. Wings moderate, rounded, the 4th and 5th quills the longest, the 1st subequal with the 9th. Tail moderately long, considerably exceeding the closed wings, rounded at the tip. Tarsus very stout, its upper third feathered in front, the rest covered with broad rectangular scutæ, the posterior with large irregular scales. Toes stout, protected above by rectangular scales nearly to the base. Claws long, much curved and rounded beneath.

POLIOAETUS ICHTHYAETUS.

(THE BAR-TAILED FISH-EAGLE.)

Falco ichthyaetus, Horsf. Tr. Linn. Soc. xiii. p. 136 (1822); id. Zool. Res. Java, pl. 34 (1824).
Pandion ichthyaetus, Vig. Zool. Journ. i. p. 321 (1824); Horsf. & Moore, Cat. B. Mus. East Ind. Comp. p. 53. no. 65 (1854); Schleg. Vog. Nederl. Ind., Valkv. pp. 13, 62, pl. 5. figs. 1, 2 (1866).
Haliaetus ichthyaetus, Cuv. Règne An. i. p. 327 (1824).
Haliaetus unicolor, Gray & Hardw. Ill. Ind. Zool. i. pl. 19 (1830).
Ichthyaetus horsfieldi, Blyth, J. A. S. B. xi. p. 110 (1842).
Pontoaetus ichthyaetus, Gray, Gen. B. i. p. 17 (1845); Blyth, Cat. B. Mus. A. S. B. no. 121, p. 30 (1849); Layard, Ann. & Mag. Nat. Hist. 1853, xii. p. 101; Kelaart, Prodromus, Cat. p. 115.
Polioaetus ichthyaetus, Kaup, Contr. Orn. 1850, p. 73; Jerd. B. of Ind. i. p. 81 (1862); Blyth, Ibis, 1866, p. 243; Hume, Rough Notes, ii. p. 1 (1870); Holdsworth, P. Z. S. 1872, p. 412; Hume, Nests and Eggs, p. 43 (1873); Sharpe, Cat. Birds, i. p. 452 (1874); Legge, Ibis, 1875, p. 278; Hume, Str. Feath. 1875, vol. iii. p. 28; Legge, ibid. p. 362.
White-tailed Sea-Eagle, Fishing-Eagle, Europeans in India; *Tank-Eagle* in Ceylon.
Matchmorol, "Fish-Tyrant," Bengal; *Madhuya*, Hind.; *Jokowuru*, Javanese.
Rajaliya, Sinhalese; *Ala*, Tam. in Ceylon.

Adult male. Length to front of cere 23·0 to 24·0 inches; culmen from cere 1·8; wing 17·0 to 17·5; tail 9·5 to 10·0; tarsus 3·5; mid toe 2·3 to 2·4, claw (straight) 1·2; inner claw (straight) 1·3; height of bill at cere 0·65.

Adult female. Length to front of cere 25·5 to 26·0 inches; culmen from cere 1·8; wing 18·0 to 18·3; tail 10·0; tarsus 3·5; mid toe 2·3 to 2·5. Expanse 58·5.
These measurements are from Ceylon examples.

Iris clear yellow, sometimes tinged with reddish and mottled with brown; bill dark horn, paling to bluish near the cere and bluish fleshy at gape ; cere above leaden, at lower edge bluish ; legs and feet fleshy white, with a bluish tinge ; claws black.

Entire head, upper part of hind neck, and throat cinereous ashy, the crown and nape shaded with brown ; back, rump, scapulars, and wings dark wood-brown, passing on the interscapular region into a paler shade, which blends above into the grey of the neck ; in old birds the latter part is especially pale, and sometimes has the feathers edged light : primaries dull black ; secondaries blackish brown, the inner webs somewhat cinereous ; chest, breast, and upper flanks light wood-brown, blending into the grey of the throat ; abdomen, lower flanks, thighs, under tail-coverts, and tail white, terminal portion (from 1½ to 2½ inches) of tail black ; lesser under wing-coverts umber-brown. When freshly acquired, the hues of the upper surface are much darker than when the bird is in old-feather, in which state the breast fades considerably, becoming a light chocolate-brown. The chin is whitish in some birds, probably those which have for the first time acquired the adult plumage.

Young. In the bird of the year the iris is hazel-brown ; bill and legs much as in the adult.
The nestling is covered with white down.

On becoming fully plumaged at about four months' old, the upper part of the forehead, crown, hind neck, and inter-scapular region are light chocolate-brown, deepening slightly on the scapulars, back, and wing-coverts ; edge of forehead, throat, face, and above the eye, together with the tips and centres of the head and hind-neck feathers, and the tips only of those of the lower part of the neck, buffy white ; tips of the back, scapulars, and wing-covert feathers fulvous-grey, passing with a tawny hue into the brown ; quills blackish brown, all but the longer primaries and the secondaries tipped with the fulvous hue ; the primaries, secondaries, and greater wing-coverts crossed on their inner webs with light bars, paling into whitish at the inner edges ; tail brown, tipped with fulvous, paling beneath the coverts into whitish, and mottled, except on the bars, with fulvous ; a broad, blackish, terminal band, preceded by a narrow undefined bar of the same, on the central feathers only.

Lower part of fore neck, chest, flanks, and breast more or less pale tan-brown, with shaft-stripes and tips of fulvous-grey, which are usually broadest on the chest ; bases of the chest-feathers dark brown ; abdomen and thigh-covert feathers white, mottled at the tips and terminal margins with the pale hue of the lower breast ; under tail-coverts faintly washed with the same : under surface of tail at its base white, mottled towards the terminal band with grey ; lesser under wing-coverts light tawny fulvous ; greater series white, barred with black ; axillaries pale tawny. marked across the centres of the feathers with brown and white.

At the end of the first year the plumage fades, sometimes to an extraordinary degree, the chest and breast becoming whitish, merely washed about the margins of the feathers with very pale tawny grey ; on lifting up the feathers of the chest the brown bases are found in their original state ; the upper surface does not undergo such a change, except that the light tippings, by reason of abrasion, are less conspicuous ; the tail, however, becomes considerably paler than in the freshly-plumaged yearling. The adult plumage, as far as I can ascertain, is not put on until after the second moult.

Obs. In my notes on " Ceylonese Ornithology and Oology " (*loc. cit.*) I pointed out that Ceylon examples of this Fish-Eagle were, as a constant rule, smaller than those from other places. An examination of specimens in the National collection from the Malay region, and a perusal of the dimensions given of late in various articles in 'Stray Feathers,' confirms the opinion that our bird constitutes a small race of *P. ichthyaetus* of Java. This latter is not invariably a larger bird than the Ceylonese, as I have examined a specimen from Sumatra with a wing of 17·7 inches, and that of another collected by Mr. Armstrong in the Rangoon district measures only 18·2 ; on the other hand the type specimen from Java in the British Museum measures 20·0, and an immature bird, presumably a female, so young that it could not be sexed, shot by Mr. Oates in Burmah, had the wing as large as 19·0, both of which latter dimensions I have never known attained to in the Ceylon bird. The Javan bird has a less cinereous brown hue, both above and beneath, than several that I have examined from other places ; but this is a worthless character, as the brown tints are variable in the Ceylon bird, depending entirely on the age of the feather.

As regards the position of this Eagle among its congeners, I have not placed it with the Ospreys, as Mr. Sharpe has done in his 'Catalogue,' but kept it in its hitherto accepted position among the Sea-Eagles. It differs structurally from the Osprey in having two foramina in the sternum, the posterior edge of which is devoid of the tolerably deep notches existing in that of *Pandion*, and in not having the keel, which is also much shallower, prolonged to the edge ; the sternum is likewise weak, narrow, and more angulated than in the Osprey ; the feathers do not want the accessory plumule, and the bony protection or brow above the eye, which does not exist in the Osprey, is present in this genus as in all other Raptors. In the structure of the foot, the outer toe of which is partially reversible, and also in the rounded claws, *Polioaetus* has some affinities with *Pandion* ; but it differs again in its much shorter wings and the habits consequent on this structure.

L

Distribution.—The Fish-Eagle is chiefly an inhabitant of the northern half of the island, frequenting, on the east, the numerous land-locked bays, estuaries, salt lagoons, and large rivers which intersect the coast-line from Elephant Pass to Batticaloa, and on the west, but not so abundantly, similar localities as far south as Chilaw. In such situations as the Jaffna lagoon at its upper end, portions of the Mullaittivu and Kokelai lakes, the Peria Kerretje and other large salt lagoons, and the mouths of the Mahawelliganga it is numerous. It is found throughout the whole of the interior of this part of the island, haunting the large tanks at Kanthelai, Minery, Topare, Anaradjapura, and likewise most of the smaller sheets of water, the village tanks of not more than a few acres having generally each their pair of these noisy birds. In the Eastern Province it is found on the Rugam, Ambaré, and other tanks, and further south occurs, but not so numerously, at the river-mouths as far down as Hambantota. On the western side of the island, to the southward of Chilaw, I have seen it at the head of the Bolgodde Lake, and it is doubtless an inhabitant of the large sheet of water near Amblangoda.

As regards the mainland, the Bar-tailed Fish-Eagle has chiefly an eastern distribution. I find no record of its being found on the western coast ; and though so common in Ceylon, it appears to be almost unknown in Southern India. Dr. Jerdon remarks, in the 'Birds of India,' "I never observed it myself south of Nerbudda. I saw it frequently in the Saugor territories and in Bengal. It extends to Burmah and the Malay countries." Concerning its *locale* in the north of India, Mr. Hume writes, in 'Nests and Eggs,' "I have myself never seen a specimen of the Bar-tailed Fishing-Eagle from any locality westward of Nepal, though I have it from Sikhim and Rangoon ; it is the next (*P. plumbeus*) and not the present species which is so common along the bases of the Himalayas, from Kumaon to Afghanistan." In North-eastern Cachar it is rather rare, but occurs, both there and in the Sikhim Terai, in conjunction with *P. plumbeus*. In the Tenasserim provinces it appears likewise to be uncommon, a single locality for it (Paybouk) being given in the "First List" of the birds of that region ('Stray Feathers,' 1874). It inhabits the coasts of the Malay peninsula, but does not appear to take the Andaman and Nicobar Islands into its range, as since its supposed occurrence there chronicled by Captain Beavan in 'The Ibis' for 1867, on Col. Tytler's authority, no specimens have ever been met with. In Java, the "Jokowuru" is, according to Horsfield, by no means generally distributed, the only two localities at which this naturalist met with it being "on the banks of the river Kediri, in the eastern district, and the other near the middle of the island, on the hills of Prowoto." It has been procured in Sumatra, Borneo, and Celebes, and probably occurs in all the intermediate islands.

Habits.—This fine Eagle frequents the borders of wooded estuaries and salt lagoons, the narrow mouths of rivers which are lined with forest, and the shores of inland lakes and tanks. The open coast it rather shuns, leaving the sway over that to its nobler ally the Sea-Eagle. The wild and secluded tanks of the interior are, however, best suited to the habits of this inveterate fish-eater. At these solitary reservoirs, many of them the persevering work of Lanka's ancient kings, the Fish-Eagle is to be found, sitting motionless on the limbs of the noble trees which line the retaining bunds, every now and then calling to its mate with its singular, far-resounding shout. While the lazy and uncouth crocodile sleeps on the bank beneath, the Eagle overhead eagerly watches its opportunity, with eye intent on some lotus-covered nook, above which hum, in the morning sun, myriads of insects, luring the finny tribes to the surface. On getting sight of a rising fish, the watchful bird launches itself down with a rapid swoop, not pouncing as an Osprey, but raking up the prey with its talons, like the Sea-Eagle. In May and June, when the village tanks of the Vanni are fast drying up under the influence of the parching south-west wind, and one muddy pool, alive with half-dead fish and frogs, is all that is left of the broad December lake, the "Fish-Hawk," in company with a host of Cormorants, Kingfishers, Egrets, and Pond-Herons, spends a prosperous time, and becomes so fat and lazy, that I have seen one fired at with a rifle, from some little distance, refuse to leave his post. In spite of its ample wings, it seldom soars or takes long flights, contouting itself with frequent peregrinations round the tank or lake on which it has taken up its permanent quarters ; but when chased or harassed by the Sea-Eagle, as I have seen it by a pair of these birds which were breeding at Minery, it exhibits considerable adroitness on the wing. It never stoops on its prey with the velocity of either the Osprey or the Sea-Eagle, but glides leisurely over it with outstretched legs. It perches with a very erect pose, and usually, when not watching for fish, seats itself on the top of a tree. Its singular note is one of the characteristic sounds of the forest-begirt tanks in the north of Ceylon. It is a deep, resounding call or "shout," louder than any bird-note in Ceylon, and when heard at intervals during the night, breaks in with startling effect on the stillness of the forest. It is repeated three or four times, and somewhat resembles the monosyllables *koow, koow*. I have not noticed this peculiar cry referred to by Indian writers; and as regards the Javan bird, Horsfield says, in his 'Zoological Researches,' that its cry is like that of the Osprey. The same author confirms my experience of its timid nature, relating that a male bird "on being caught in a snare permitted itself to be seized by the native without making any resistance. When brought to me lying in the arms of the native, apparently conscious of its situation, and without making use of its claws or bill, or exerting any efforts to extricate itself, it suffered itself to be handled and examined very patiently."

In confinement this Eagle thrives well, and is a very docile and quaint bird in its habits. A young one which I reared soon learnt to recognize the person who fed it, and would swallow fish 6 inches in length as fast as they were thrown to it, quickly filling out its crop, and working it, by a muscular effort, to and fro, so as to pass the food into its stomach. It was kept in the same aviary as the Crested Eagle now in the Zoological Gardens, and would, long after it was fully plumaged, sidle up to it, crouch down, chuckling with a low note, as it would do to its parent. As it grew old, it became the noisiest bird I ever had any thing to do with, continually "cawing" for its food, notwithstanding that it was plied with raw meat, lizards, and fish to an alarming extent, and was almost as ravenous a feeder as a Pelican Ibis, if any bird *could* be found to equal the latter in point of appetite. Up to the age of six months, when he, to my great regret, fell a victim to an accident, he very seldom gave vent to the loud call of the wild bird, his note in confinement being a harsh clanking cry.

Nidification.—This Eagle breeds in December and January, building a huge nest of sticks in large trees near the water's edge. If the structure is fixed in a deep fork or awkward situation on the limbs of the tree, the foundation is heaped up until sufficient breadth is attained for the platform, and the result is a fabric of enormous size. I have never obtained the eggs; but a nest which I visited in 1873 contained one nestling, the tame bird above described, and it may therefore be premised that the bird lays two eggs, the other, in this instance, having been addled. The interior of this nest, which was entirely made of good-sized sticks, was flat, and contained no lining or preparation for the repose of the eaglet, which, on my appearing at the edge of its domicile, stood up and placed itself in an attitude of defence, its cowardly parents flying off and seating themselves on distant trees. As to the eggs they are, in all probability, white and unspotted, and about the size of those of *Pol. plumbeus*, its Himalayan congener, which vary, according to Mr. Hume, "from 2·72 to 2·8 inches in length, and from 2·1 to 2·15 in breadth."

I have never seen more than one nest belonging to this Eagle in the same locality; but occasionally it appears they congregate together in the same manner that the Grey-backed Sea-Eagle does—an abnormal habit, arising no doubt from the force of circumstances. Jerdon mentions having found a whole colony of the nests of this Eagle in a single tree on the skirts of a village near the Ganges.

FALCONIDÆ.

AQUILINÆ.

Genus HALIASTUR.

Bill stout, rather lengthened, the direction of the cere straight, the culmen descending from its edge; festoon prominent. Nostrils oval and oblique, exposed. Eyelid furnished with short lashes. Wings long, reaching nearly to the tip of the tail, the 4th quill the longest. Tail short and broad at base, rounded at the tip. Tarsus short, feathered in front slightly below the knee, moderately stout, protected in front by rectangular transverse scales, behind with broad pentagonal scutæ. Toes rather short, the outer considerably longer than the inner, the whole covered with transverse scales. Claws shortish, moderately curved and trenchant beneath.

HALIASTUR INDUS.
(THE BRAHMINY KITE.)

Falco indus, Bodd. Tabl. Pl. Enl. 25 (1783).

Falco pondicerianus, Gm. Syst. Nat. i. p. 265 (1788).

Haliaetus ponticerianus, Cuv. Règne An. i. p. 316 (1817).

Milvus pondicerianus, Jerd. Madr. Journ. x. p. 72 (1839).

Haliastur indus, Gray, Gen. B. i. p. 18 (1845); Blyth, Cat. B. Mus. A. S. B. p. 31; Kelaart, Prodromus, Cat. p. 114 (1852); Layard, Ann. & Mag. Nat. Hist. 1853, xii. p. 101; Horsf. and Moore, Cat. B. Mus. E. I. Co. i. p. 57 (1854); Jerdon, B. of Ind. i. p. 101; Hume, Rough Notes, ii. p. 316 (1870); Holdsworth, P. Z. S. 1872, p. 414; Hume, Nests and Eggs, p. 51 (1873); Sharpe, Cat. Birds, i. p. 313 (1874); Legge, Ibis, 1874, p. 10, et 1875, p. 279.

Haliaetus indus, Schl. Mus. P.-B. *Aquilæ,* p. 19 (1862).

L'Aigle de Pondichery, Briss. Orn. i. p. 450, pl. 35 (1760).

Aigle des Grandes Indes, Buff. Pl. Enl. i. pl. 416 (1770).

Shiva's Kite, Kelaart, Prodromus; *Pondicherry Eagle,* Lath.; *Maroon-backed Kite,* Jerdon, B. of Ind.

Bahmani-chil, Hind.; *Ru-Mubarik,* Mussulmen; *Sunker-chil,* lit. "Shiva's Kite;" *Dhobia-chil,* lit. "Washerman's Kite," Beng.; *Khemankari,* Sanscrit; *Ratta ookal,* Sindh; *Garud-alawa,* Tel.; *Shemberrid,* Yerklees (apud Jerdon). *Lang-bondol,* Sumatra; *Ulung,* Java (Horsf. & Moore, Cat.).

Rajaliya, Sinhalese; *Chem Prându,* Tam.; *Brimalgumoitu,* Portuguese in Ceylon (apud Layard).

Adult male. Length to front of cere 18·0 to 18·5 inches; culmen from cere 1·1; wing 14·5 to 15·0; tail 8·0 to 9·5; tarsus 2·0 to 2·25; mid toe 1·2, claw (straight) 0·7; height of bill at cere 0·5.

Adult female. Slightly larger only. Length to front of cere about 19·0 inches; culmen from cere 1·2; wing 15·0 to 15·3.

Iris brown and mottled in some with yellow; cere yellowish; bill bluish horn, palest at the base beneath; legs and feet greenish yellow.

Plumage above, from the forehead to the lower part of the hind neck, and beneath down to the abdomen, including the flanks, white, each feather with a narrow, blackish, mesial stripe, which includes a fine portion of the web as well as the shaft; in mature birds these are as broad on the head as on the hind neck, but in very old examples are confined to the shaft only; rest of the plumage, with the exception of the longer primaries, greater series of under wing-coverts, and lower surface of tail maroon-red, darkest on the back, lesser and under wing-coverts, and palest on the abdomen, thighs, and under tail-coverts; the shafts, except of the tail, black, these latter fulvous-white; tip of tail the same; longer primaries black, their inner webs rufous from the base to the notch, the under surface rufescent, paling to whitish at the base; primary wing-coverts dark at their outer edges, the inner webs, as well as those of the secondary feathers, crossed with narrow, widely-separated bars of blackish.

Obs. In some examples the black stripes are conspicuously developed on the red feathers of the abdomen, while in others the shafts alone are dark. I have noticed this characteristic chiefly in Bengal examples, which, as a rule, I think, have the stripes on the white plumage bolder than in Ceylonese birds, although they coincide exactly in the hue of the maroon parts. The Ceylonese bird, as regards the white striping (the variation in which has been considered by Mr. Gurney of sufficient value to justify the separation into species of the two Malayan races, *H. intermedius* and *H. girrenera*), comes between the Bengal and the Malaccan bird. The latter (*H. intermedius*), besides having the shaft-stripes reduced to very narrow lines, is of a redder or paler hue than *H. indus*, and appears to be a well-marked race or subspecies.

Young. The nestling has the iris dark brown, the bill and cere brownish, the latter and the loral skin tinged with green; legs and feet greenish.
Body at first covered with white down; when fully plumaged, the forehead, chin, and lower part of cheeks are dull whitish, the ear-coverts brownish, and the head and hind neck fulvous tawny overcome with brown on the lower part of the latter, the centres of the feathers light, and the edges tawny, imparting a streaked appearance: the brown feathers of the hind neck with rufescent central streaks, diminishing to terminal spots on the intenscapular region, which, with the back, scapulars, wing-coverts, and secondaries, is dark brown, paling much on the upper tail-coverts; the scapulars margined terminally with rufous; quills and tail blackish brown; inner edges of primaries white; greater wing-coverts with pale inner margins and conspicuously black shafts; throat and breast isabelline brown, with tawny shaft-stripes, lower parts with the thigh-coverts tawny, with dark shaft-stripes on the thighs. This plumage at the end of the first year, as ascertained by observation of a caged bird, becomes paler throughout.
After the first moult the head and hind neck are rufous, paling at the tips, and with blackish shaft-stripes, the back, scapulars, and wing-coverts a sober brown, with pale terminal margins, the greater wing-coverts with much white on the concealed portions of the inner webs, the upper tail-coverts with broad pale margins, shorter primaries with rufous-brown, outer webs and the inner webs rufescent white at the base; ear-coverts paler than in the first plumage; under surface pale brownish, with light terminal streaks, and the shafts dark in the brown portion of the feather; the abdomen paler than the breast; under tail-coverts and lesser under wing-coverts rufous, median under wing-coverts brown edged pale, greater series pale as in the adult.

After the second moult, the back, wings, and tail assume their rufous or maroon colouring, the head and hind neck are whitish, washed here and there with rufous, and with black shaft-stripes; the face and throat are white, gradually darkening into rufescent fulvous on the chest and upper breast, on which there is again a gradual change to the maroon of the lower parts, the shafts of all the feathers being black. After this stage, the head, hind neck, and breast get whiter by degrees, throwing off all trace of the rufous hue, and the shaft-stripes assume their normal character, covering a portion of the web at the sides of the shaft, which alone is dark in the intermediate stages.

Distribution.—The Brahminy Kite is a well-known and very common bird in Ceylon, being more or less abundant round the whole coast of the island, and occurring about the large tanks and inland waters of the interior. On the seaboard, however, it is local in its choice of habitat, as an instance of which feature I may cite its abundance in Galle harbour, and almost total absence from the equally inviting roadstead at Colombo. It is sometimes seen about the mouth of the Kelani, and in the marshes at the back of Borella, and it occurs sparingly at Negombo and more commonly at Bolgodde; but I never once saw it about the shipping in the Colombo Roads. At Chilaw it commences to be commoner, and continues to increase in numbers at Puttalam and northwards to the Manaar district, where, as well as throughout the whole of the northern maritime region, it is very numerous. At Trincomalie it is abundant, and is a common bird down the coast to the Batticaloa Lake. South of this and in the Hambantota district it is scarcer. I have met with it at Kanthelai, Minery, and other tanks in the northern interior, and I believe it also frequents the Bintenne Lake. In the south-western district it is found about the Sinhalese villages on the Gindurah as far up as the "Haycock" hill; and I have known it to breed as far inland as Oodogamma. I am not aware of its ever having been seen on the upland of Dumbara, or anywhere else in the Kandyan hill-region, although there is no reason why it should not follow as a straggler the course of the Mahawelliganga from the low country to the north of Bintenne, up to the neighbourhood of the highland capital.

The Brahminy Kite is found throughout India on the sea-coast and on all large rivers and jheels, extending eastwards to Burmah, and as far south as the lower parts of the province of Tenasserim, where it is, however, not very common. In the south it is abundant, and at the island of Ramisserum I have always found numbers of this bird. Jerdon remarks that it is rare in the plains of India and in the Deccan, in which latter region Mr. Fairbank, in 'Stray Feathers,' records it as uncommon. In the north-western portion of the empire, Mr. Hume speaks of it as follows ('Stray Feathers,' 1875, p. 448):—"Common enough in Sindh and about the coasts of Cutch and Kattiawar, but almost (if not quite) unknown in the dry riverless regions of Rajpootana. Adam never obtained it about Sambhur, and at Ajmere I only once remember seeing it. Dr. King does not appear to have observed it in any part of Jodhpoor." In Chota Nagpur, Mr. Ball remarks that it is found "near the larger rivers and jheels, but nowhere in abundance." In Lower Bengal it is of course plentiful. Further eastward, Mr. Inglis records it as common throughout the year in Eastern Cachar; and in Burmah Mr. Oates writes that it "occurs in immense numbers in all the tidal creeks of the Pegu plains."

Habits.—In Ceylon the Brahminy Kite is especially a denizen of seaport towns and large villages at the mouths of rivers and salt lakes; it frequents, likewise, land-locked bays, estuaries, and lagoons; but in all of such localities seems to prefer the vicinity of human habitations, probably on account of its garbage-eating habits, to the solitude of the surrounding plains. It collects in great numbers among shipping, flying round the vessels on the look-out for garbage of all sorts, soaring in high circles above their masts, and even settling on the rigging, where it keeps a sharp eye on the galley about the dinner-hour, and is ready to pounce immediately on any thing that may be thrown overboard.

It picks up its food with a graceful swoop, and very frequently devours it while in full flight, proceeding about this operation in the most leisurely manner possible; it may be seen bringing forward its talons with the food it has seized beneath the breast, and with a combined backward and upward pull from the legs and shoulders respectively, fragments are torn off with but little exertion. I have observed it swoop down and pick up a Lizard (*Calotes*) basking on the topmost twig of a low tree, this favourite prey among eastern Hawks no doubt forming a considerable portion of its sustenance. It will capture fish in shallow pools, and is very fond of the small crabs frequenting the foreshores of tidal rivers; it may be often seen devouring its food on the ground or on a large rock or the bank of a paddy-field. It is a tame bird, and exhibits but little fear of man or a gun, sometimes making off with a Snipe which has fallen at some little distance from the sportsman. At Trincomalie it was always a morning attendant at the drawing of the sea-nets; and was just as agile in snapping up any outside fish as its more numerous companions the Crows, and when not particularly successful in its foraging, would pursue a Crow and rob it of its well-earned "sardine." Layard says that he has known it seize a fowl; but, as he remarks, this must be a rare occurrence. Jerdon has seen it "questing over woods and catching insects, especially large *Cicadæ*." It is continually on the wing, and has an easy, buoyant, and powerful flight, being much in the habit of soaring up to a great height, and then launching itself off for a long distance with motionless wings. Its chief characteristic as regards locomotion is its habit and power of sailing steadily up against the wind, with scarcely a movement of its frame, except a twisting of its head from side to side, as it carefully scans the ground beneath and awaits its chance of darting down on some coveted morsel. I have on other occasions witnessed it exhibit considerable skill in catching up a lizard in the air, which it had let fall from its talons while flying off with it. Its favourite note is a weak squealing cry, which it constantly utters on the wing, or while perched on some building or tree-top. In Ceylon it sits much on the fronds of cocoanut-trees in the vicinity of native bazaars, and at night takes itself off in flocks to roost in some favourite spot in the jungle. Numbers of these birds frequented the town of Trincomalie, haunting the harbour and the sea-beaches of Dutch and Back Bays, where they subsisted chiefly on the fish picked up from about the fishermen's nets; about 4 or 5 P.M. they commenced to fly away one after the other to the north, and, passing over the "Salt Lake," retired for the night to the forest between there and Peria Kulam.

Altogether the habits of this bird are as singular as they are interesting, and tend to place it more among the Kites than the true Sea-Eagles. Jerdon very aptly remarks, it may be considered either an aberrant form of *Haliaetus* leading to the Kites, or an aberrant Kite leading to the Sea-Eagles: and this is the position claimed for it in Mr. Bowdler Sharpe's 'Catalogue.'

Nidification.—This species breeds in Ceylon in February and March, nesting in trees on the shores of salt lagoons or paddy-fields. All the nests I have seen have been rather bulky structures, about the size of that of *Herodias alba*, made of tolerably large sticks, and placed in a top branch of moderately-sized trees. The number of eggs is usually two; but mention is made in Mr. Hume's book on Indian oology of four in one instance. The ground-colour is dull white, and the markings, which are scanty, consist of faded reddish or reddish-grey dots, sometimes scattered over the surface, and occasionally confined to the obtuse end; the spottings in some are mixed with small streak-shaped blots; and one egg, taken by Mr. MacVicar in the Western Province, has the appearance of being "dusted" all over with minute pale reddish specks. Five Ceylonese specimens varied in length from 2·04 to 1·88 inch, and in breadth from 1·7 to 1·54.

Layard states that this Kite makes several false nests, and that the male occupies one of them while the female is incubating her eggs near at hand. The chick or nestling has a querulous twittering cry. Concerning its nidification in India, Mr. Hume writes, " It almost invariably makes its nest in the neighbourhood of water, building a rather large loose stick-structure, scarcely if at all distinguishable from those of the Common Kite (*M. govinda*), high up in some large mango, tamarind, or peepul tree. The nest, which is from 18 inches to 2 feet in diameter, and from 3 to 5 inches in depth, with a rather considerable depression internally, is sometimes perfectly unlined, at other times has a few green leaves laid under the eggs, as in an Eagle's nest ; but most commonly is more or less lined, or has the inner part of the nest intermingled with pieces of rag, wool, human hair, and the like."

ACCIPITRES.

FALCONIDÆ.

AQUILINÆ.

Genus MILVUS.

Bill longer and with the tip more hooked than in *Haliastur*; festoon less prominent, the cere more advanced at the sides. Nostrils moderately large, oval, and oblique. Wings very long and pointed; the 3rd and 4th quills the longest, and reaching nearly to the tip of the tail. Tail long and forked. Tarsus short, the front and sides plumed considerably below the knee, the rest covered in front with transverse scutæ and behind with hexagonal scales. Toes longer than in *Haliastur*. Claws similar.

MILVUS GOVINDA.

(THE INDIAN PARIAH-KITE.)

Milvus govinda, Sykes, P. Z. S. 1832, p. 81; Kelaart, Prodromus, Cat. p. 115 (1852); Layard, Ann. & Mag. Nat. Hist. 1853, xii. p. 103; Schlegel, Mus. P.-B. *Milvi*, p. 2 (1862); Gould, B. of Asia, part iv.; Jerd. B. of Ind. i. p. 104 (1862); Blyth, Ibis, 1866, p. 248; Hume, Rough Notes, ii. p. 320 (1870); Holdsworth, P. Z. S. 1872, p. 414; Hume, Nests and Eggs, p. 52 (1873); Sharpe, Cat. Birds, i. p. 325; Legge, Ibis, 1874, p. 10; Hume, Str. Feath. 1875, p. 229, footnote.

Haliaetus lineatus, Gray & Hardw. Ill. Ind. Zool. i. pl. 18 (young).

Milvus cheela, Jerd. Madr. Journ. x. p. 71 (1839).

Milvus ater, Blyth, Cat. B. Mus. A. S. B. p. 31 (1849).

Common Kite, Black Kite, Jaffna Europeans; *The Cheela Kite*, Kelaart's Prodromus.

Chil, Hind.; *Malla Gedda*, Tel.; *Paria Prāndu (apud* Jerdon).

Rajaliya, Sinhalese; *Kalu-Prāndu, Paria Prāndu*, Tam. in Ceylon.

Adult female. Length to front of cere 22·0 to 22·5 inches; culmen from cero 1·2; wing 17·5 to 18·0; tail 10·0 to 11·0; tarsus 2·0 to 2·2; middle toe 1·5, its claw (straight) 0·7; height of bill at cere 0·5; expanse 55·5 (of an example with a wing of 18·0).

Adult male. Length to front of cere 21·0 to 22·0 inches; wing 16·0 to 17·3; tarsus 2·0 to 2·2.

Iris light hazel-brown, sometimes tawny, with brown radii and mottlings between them; cere pale greenish, dusky above; bill blackish, gape and base of under mandible bluish; legs and feet whitish green, greenish yellow, or pale yellowish; claws black.

Head and hind neck brownish tawny, the feathers slightly pale-edged, and each with a fine dark shaft-stripe; on the hind neck the stripes expand slightly, and the ground-colour darkens into the glossy wood-brown of the back, rump, scapulars, and lesser wing-coverts; bill blackish, gape and base of under mandible bluish; legs and feet whitish green, greenish yellow, or of the foregoing parts slightly paler than the rest, those of the least coverts tawny; median wing-coverts lighter than the rest, the webs paling off from the shaft to fulvescent greyish at the edges; primary-coverts, secondaries, and shorter primaries dark brown, the latter somewhat paler on the outer webs; longer primaries blackish brown, the inner webs paling from the notch to the base, and the colour broken up with white interspaces and mottlings: inner secondaries and adjacent tertials crossed with narrow blackish-brown bars, the interspaces being ashen, paling to white on the shorter and innermost tertials; tail ashen brown, with a tawny hue near the shafts of the feathers, and the laterals paling to whitish at the bases of the inner webs, the whole crossed with narrow bars of dark brown, more or less incomplete towards the margins; tips of all but the two outer feathers whitish; a line of blackish above the lorus, and over the ears a dark brown patch; face greyish with the shafts dark; beneath brown, paling to tawny rufous from the lower breast to the under tail-coverts; the centres of the feathers dark brown and the shafts blackish, the web adjacent to the mesial stripes being somewhat paler than the margin; margins of the throat-feathers fulvous, and the basal portions of the webs whitish: on the belly and under tail-coverts the mesial stripes are wanting, the shafts alone being dark; least under wing-coverts deep tawny, the feathers dark-centred; greater series blackish brown with tawny edges: primary under wing-coverts ashen-brown with dark softened bands; basal portion of the 2nd and 3rd quills beneath more or less whitish, the amount of white varying much in individuals, some being quite as dark as *M. affinis*.

Young. In the first, or nestling plumage, the head, back, rump, and wing-coverts are dark brown with a purplish gloss, the feathers of the head and hind neck with terminal whitish-buff "points" or streaks, surrounding a shaft-stripe darker than the rest of the feather; those of the back and rump with terminal margins of a slightly more rufous hue, the wing-coverts and tertials with much deeper tips of fulvous, passing with a rusty tint into the brown, and surrounding a dark shaft-stripe; primaries and their coverts blackish brown, tipped with fulvous, slightly on the longer primaries, and deeply on the rest; the inner webs of the quills mottled with dusky greyish: tail obscure ashen-brown, tipped with fulvous and crossed with indistinct bars (as in the adult) of darker brown. Loral streak and postorbital patch darker, and the latter more extensive than in the adult; throat and lower part of cheeks fulvous, with narrow shaft-stripes of brown; fore neck, chest, breast, and flanks brown, the centres of the feathers rufous, enclosing pointed shaft-stripes of blackish brown; on the lower parts the brown hue pales into brownish fulvous, and the shaft-stripes disappear; tibial plumes and under tail-coverts more rufous still: under wing-coverts dark chocolate-brown tipped with fulvous, the primary-coverts ashen-brown with the outer webs whitish, as is also the edge of the wing; basal portion of primaries beneath scarcely showing any white in some birds, and in others even more than in old birds.

In the following season the terminal margins throughout the upper surface are less conspicuous, and those of the back-and scapular feathers less rufous, the margins of the head- and hindneck-feathers, however, are often more fulvescent, and the dark stripes on the latter part less conspicuous than in the nestling: the tips of the secondaries are likewise less in extent; on the under surface the throat becomes more "lined," the streaks on the chest and upper breast diminish, and their pale borders contrast less forcibly with them, while the ground-colour of these parts is browner than in the youngest stage; the amount of white at the base of the quills beneath varies, but it is usually more extensive during this period.

When not fully adult, the signs of nonage show themselves in the pale tips of the back, scapulars, and tertials, the softened and less intense shaft-lines of the head and hind neck, and the pale borders of the dark chest-strize; the markings of the throat are variable at this stage, the shaft-lines being marked in some and faint in others, while the ground-colour is at times conspicuously rufous; the quills are quite untipped in these birds, and the lower parts more rufous than in adults.

Obs. The difference of opinion among some ornithologists as to what Kites in India should be classed as *M. govinda* and what as *M. affinis* makes it somewhat difficult to define what the Ceylon birds really are, as they present some points of dissimilarity to the types of both these species. If typical *M. affinis* be represented by the small rufous-plumaged Kite inhabiting the east coast of Australia and the Malay Archipelago, and *M. govinda* by the ordinary brown-plumaged bird of the plains of India, having a certain amount of white (which, however, is a variable and uncertain characteristic in Ceylon birds) at the base of the primaries beneath, then the Ceylon Kite has more affinity with the latter than with the former.

From *M. affinis* it differs, as an adult bird, in the less rufous coloration of the head, hind neck, and lesser wing-coverts, and in youth in the less-rufescent character of the upper-surface tippings, a Macassar example being taken for comparison. It is likewise a larger bird, the wings of six examples of *M. affinis* measuring as follows—(Sydney) 15·8, (Australia) 15·0, (Australia) 15·2, (Timor) 16·5, (Macassar) 16·6, (Timor) 16·5. As regards the pale markings of the under wing, adults of *M. affinis* are on the whole darker than Ceylon birds, which, though quite as dark in the young stage, are variable when mature. From the type of *M. govinda* in the India Museum, and similar examples in the British and Norwich Museums, the insular bird differs in the more rufous edgings of the head- and hindneck-feathers, the paler median wing-coverts, more cinereous tail, more conspicuous striation of the upper part of the throat, more ashy hue of the dark chest-stripes, and more fulvous colour of the abdomen and under tail-coverts; but though these differences are numerous, they are less appreciable than are those in the case of the Australasian bird. The Ceylon *Milvus* is also a somewhat smaller bird than the Indian *M. govinda*, Sykes's type, a female measuring 18·5 inches in the wing, and others I have examined 18·0, 17·8, 17·8 and 17·4, while Mr. Hume gives the wing in five females as from 18·25 to 19·10.

M

In several examples of the young of Indian *M. govinda* I have observed that there is more whitish at the base of the primaries than in adults; some juvenile Ceylonese examples have scarcely any, while others have more white than old birds; so that I incline to the belief that this character in the medium-sized Kite is entirely worthless.

In referring to the species *M. govinda*, and speaking of its type in the India Museum, I select the example of the medium-sized Indian Kite, which, I believe, Sykes's description relates to, and which has, on the bottom of the pedestal, the name *govinda* written in pencil by Dr. Horsfield. Sykes's description is too short to identify with certainty the specimen which it refers to; but the smaller bird agrees better with it than with the young example of *M. melanotis* mentioned by Mr. Brooks ('Stray Feathers,' 1876, p. 272). Then there is, in favour of the smaller bird being the type, the indisputable evidence of the habits and locality of the bird referred to by Sykes. He says it is the Common Kite of the Deccan, and is "constantly soaring in the air in circles, watching an opportunity to dart upon a chicken, upon refuse matter thrown from the cook-room, and occasionally even having the hardihood to stoop at a dish of meat carrying from the cook-room to the house." This is not the habit of the larger Kite, which, according to most Indian observers, is a wary bird, and is furthermore not found in the district dealt with by Sykes. Mr. Hume, who has, I conclude, the largest series of Kites of any one in India, says, "I have examined more than 30 specimens of Kites from Bombay, Matteran, Sholapoor, Nattram, and Poona, and I never found one *M. major* among them; nay, when at Bombay and Poona, I specially noticed the Kites, and, while I thought I recognized some *M. affinis*, I can positively affirm there were no *M. major*. . . . . Everywhere in the plains *M. major* is a bird of the jungle, very rarely approaching towns or even villages, and living more on frogs, locusts, &c. than on offal." With regard to the measurements given by Sykes, ornithologists so far back as thirty or forty years ago rarely measured birds in the flesh; and I agree with Mr. Hume that Sykes's bird must have been measured from the skin. The tail, which is 11 inches, is decidedly that of the medium-sized bird, and corresponds in size with that of Ceylonese examples.

Distribution.—The Pariah Kite of Ceylon has a somewhat local habitat, being almost entirely confined to the northern half of the island. Its headquarters are the Jaffna peninsula and the west coast of the Northern Province, as far south as Manaar. It is, singularly enough, notwithstanding its limited range, subject to a seasonal movement from the east coast to the west during the south-west monsoon. Although tolerably common from the peninsula down to Trincomalie, from October until March, scarcely a bird is to be seen in that quarter during the opposite season. I am likewise informed by my friend Mr. W. Murray, of the Ceylon Civil Service, who has made large collections of birds in the Jaffna district, that its numbers are greatly decreased during the same time of the year—a circumstance which may be explained by its retiring into the jungle to breed, and also by its undertaking a partial migration to the southern coasts of India. In the island of Manaar and in the adjoining district of Mantotte it is plentiful, Mr. Holdsworth recording it as very common at Aripo; to the southward of the latter place it occurs in less numbers, taking in the island of Karativoe into its range, down the coast to Puttalam, at which place it is again tolerably numerous in the cool season. South of this it is rare, occurring as far as Madampe and perhaps to Negombo, below which I have never observed it.

In Ceylon it is exclusively a sea-coast bird, except in the very north of the Vanni, where it may now and then be seen about the villages of the interior. I have no record of its occurrence south of Batticaloa, on the east side of the island. Nor does it ascend into the hills as it does in the Nilghiris and Himalayas.

In India this Kite is almost everywhere abundant. It is found alike at seaport and inland towns; and most villages even have their attendant flock, who act the part of scavengers in quickly disposing of everything which it is possible for a bird to digest. In the south it inhabits the Nilghiris, in which hills Mr. Davison says it is very common, ascending to their summits, and often roosting with *Haliastur indus*. In the Travancore hills, likewise, Mr. Bourdillon writes that the Pariah Kite occurs in numbers in the hot weather; and it is to be presumed that the present species is intended, as the larger bird (*M. melanotis*) is not found in the extreme south. Sykes, who first discriminated the species, says it is the common Kite of the Deccan, while at Bombay and up the coast to Sindh, as well as throughout the whole region of Mount Aboo and Northern Guzerat, and in the Kandhala district, it is recorded by various writers in 'Stray Feathers' as very common. It inhabits the southern slopes of the Himalayas, up which it ascends to an elevation of 6000 or 7000 feet. It has been procured by Mr. Ball as far west as in the Suliman Hills, which form the western boundary of the Punjaub. The same writer observes that it is common at Chota Nagpur, and that specimens from the jungle are often intensely dark. In Kashgar Dr. Scully obtained nothing but the large bird, although the late Dr. Stoliczka mentions seeing what appeared to be true *M. govinda* in the hills between Yanjihissar and Sirikul; this, I am inclined to think, was a wrong identification. In the plains of India and at Calcutta it abounds; but from Burmah Mr. Hume has only received what he considers to be the rufous (Australian) species, *M. affinis*, which inhabits the Malay peninsula, the Archipelago, and the eastern coasts of Australia. From the Andamans *M. govinda* appears to be entirely absent; and doubtless, if a Kite is procured from the islands of the Bay of Bengal, it will be the Malayan bird, which, as I have just mentioned, inhabits the peninsula. At the Laccadives, Mr. Hume mentions that a Kite not uncommonly occurs, which must be either *govinda* or *affinis*; and as the former species is represented in Ceylon, it is doubtless the same bird which affects these islands.

Habits.—This Kite, in the north of Ceylon, as it does in India, plays the part of an extremely useful scavenger. There, as in the districts on the mainland frequented by it, it resorts to villages and towns, more particularly those situated on the coast, and, collecting in large flocks, performs the office of devouring all the offal, refuse of human food, thrown out of the doors of native houses, garbage, and decaying organic remains which it can possibly get hold of in the course of the day's peregrinations. At the hauling-in of the morning seine net it is also a constant and regular attendant, disputing with the usual crowd of "Kakas" for the possession of stray fish and crabs rejected by the fishermen. In the town of Jaffna, where it is exceedingly abundant and extremely useful in a sanitary point of view, it resorts in scores, nay, hundreds at times, to the grand old banyan tree upon the fort-ramparts, roosting in it at nights, and perching on its outspreading branches between "meals," sallying out thence to the sea-beach and various parts of the town, as well as to the open fields of the surrounding country. At the beach, attracted by the arrival of fishing-boats and small craft from the adjacent islands, they present a lively scene : scores of birds circle round and fly to and fro with squealing notes and eager glances at the boats beneath them; some glide over the roofs of the houses, and, taking a wider tour than their mates, return again, sailing back through the streets in utter disregard of the busy human throng ; meanwhile their more fortunate companions, alighted here and there on the sand, are discussing dainty (?) morsels of the most various description picked up with a quick and sudden swoop, or robbed from their sable allies the Crows, who stand off at a respectful distance, ruefully "cawing" their disappointment and rage. Layard, who lived for a considerable period in the north of the island, markedly alludes to their daring when pressed by hunger, and says :—"They are bold enough to make frequent depredations on the fish-stalls ; and in one instance I saw a lad of about thirteen years struck to the ground by the sudden pounce of a Kite, who bore off a good-sized fish from a basket the boy was carrying on his head." This statement of its boldness is corroborated by a letter which has lately appeared in 'Stray Feathers,' vol. v. p. 347, in which a correspondent states that a Kite, whose nest had been robbed by the son of a sepoy, persistently watched for the lad, swooping down and attacking him whenever he left the house, ample evidence of which maltreatment was afforded by the appearance of the lad's head and arms.

Jerdon has the following paragraph in his 'Birds of India,' on the habits of the "Chil" in India :—"When a basket of refuse or offal is thrown out in the streets to be carted away, the Kites of the immediate neighbourhood, who appear to be quite cognizant of the usual time at which this is done, are all on the look-out, and dash down on it impetuously, some of them seizing the most tempting morsels by a rapid swoop, others deliberately sitting down on the heaps along with crows and dogs, and selecting their scraps. On such an occasion, too, there is many a struggle to retain a larger fragment than usual; for the possessor no sooner emerges from its swoop than several empty-clawed spectators instantly pursue it eagerly, till the owner finds the chase too hot, and drops the bone of contention, which is generally picked up long before it reaches the ground, again and again to change owners, and perhaps finally revert to its original proprietor. On such occasions a considerable amount of squealing goes on."

The flight of the Pariah Kite is buoyant and easy, the points of its wings being much turned up, and its long tail swayed to and fro as it gracefully curves about and alters its course with motionless pinions. It devours much of its food on the wing ; and what it cannot thus consume it disposes of on the ground. In the north of Ceylon the bare and broken leaves of the Palmyra palm afford it a favourite perch. When not occupied in seeking for garbage it quests about marshes and other open places near the sea-coast for frogs, water-snakes, small crabs, &c. Mr. Holdsworth has observed a large flock at Aripu, feeding on winged termites, which they were taking in the air, with apparently but little exertion, by seizing them in their talons ! The note of this Kite is a tremulous squeal, uttered much when on the wing, or when congregated to feed on any newly-found garbage, when they become very noisy, as observed by Jerdon in the above paragraph.

Nidification.—The Pariah Kite breeds, as I am informed, in the north of Ceylon about May, retiring into the jungle for the purpose, and often building on trees near village tanks or in the vicinity of villages. I have not myself seen their nests; nor have I any description of them as built in Ceylon; I therefore subjoin the following account from Mr. Hume's voluminous notes in his ' Nests and Eggs of Indian Birds :'—"They build, almost without exception, on trees ; but I have found two nests (out of many hundreds that I have examined) placed, *Neophron*-like, on the cornices of ruins. The nest, mostly placed in a fork, but not uncommonly laid on a flat bough, is a large clumsy mass of sticks and twigs, the various thorny acacias appearing to be the favourite material, lined or intermingled with rags, leaves, tow, &c. The birds are perfectly fearless, breeding as freely on stunted trees situated in the densest-populated bazaars or most crowded grain-markets as on the noblest trees in the open fields. Two appears to be the normal number of eggs; but they often lay three." The same author remarks that the variety of types of coloration is countless, and that " the ground-colour is almost invariably a pale greenish or greyish white, more or less blotched, clouded, mottled, streaked, penlined, spotted, or speckled with various shades of brown and red, from a pale buffy brown to purple, and from blood-red to earth-brown. Many of the eggs are excessively handsome, having the boldest hieroglyphics blotched in blood-red on a clear white or pale-green ground. Others, again, are covered with delicate markings, as if etched on them with a crow-quill." The average size of 273 eggs, measured by Mr. Hume, was 2·19 by 1·77 inch.

Bill weak, the tip considerably produced, margin slightly festooned. Nostrils oval, and protected by the long loral bristles. Wings very long, reaching to or beyond the tip of the tail when closed, the 2nd quill the longest, and the 1st and 3rd slightly shorter; the distance between the tips of the secondaries and those of the primaries almost equal to length of tail. Tail slightly sinuated, or even at the tip. Tarsus short and stout, covered throughout with small reticulate scales, its anterior portion feathered for more than half its length. Toes very strong and short, inner toe very slightly longer than the outer one. Claws well curved, acute, and all but the centre one rounded beneath.

ELANUS CÆRULEUS.

(THE BLACK-SHOULDERED KITE.)

Falco cæruleus, Desf. Mém. Acad. R. des Sciences, 1787, p. 503, pl. 15.

Falco vociferans, Lath. Ind. Orn. i. p. 46 (1790).

Falco melanopterus, Daud. Traité, ii. p. 152 (1800).

Elanus cæsius, Savign. Syst. Ois. d'Egypte, p. 274 (1809).

Elanus melanopterus, Leach, Zool. Misc. iii. p. 5, pl. 122 (1817); Gould, B. of Eur. i. pl. 31 (1837); Gray, Gen. B. i. p. 26 (1845); Kelaart, Prodromus, Cat. p. 115; Layard, Ann. & Mag. Nat. Hist. xii. 1853, p. 104; Jerdon, B. of Ind. i. p. 112 (1862); Layard, B. S. Afr. p. 26 (1867); Hume, Rough Notes, p. 338 (1870); Shelley, Ibis, 1871, p. 44; Holdsworth, P. Z. S. 1872, p. 415; Hume, Str. Feath. 1873, p. 21; Jerd. 'Nests and Eggs,' p. 56 (1873); Legge, Ibis, 1874, p. 10; Butler, Str. Feath. 1875, p. 449; Hume, Str. Feath. 1876, p. 462; Inglis, Str. Feath. 1877, p. 16.

Elanoides cæsius, Bonn. et Vieill. Ene. Méth. iii. p. 1206 (1823).

Buteo vociferus, Bon. et Vieill. Ene. Méth. iii. p. 1220.

Elanus minor, Bp. Consp. i. p. 22 (1850).

Elanus cæruleus, Strickl. Orn. Syn. p. 137 (1855); Shelley, B. of Egypt, p. 198 (1872); Sharpe, Cat. B. i. p. 336; Legge, Ibis, 1875, p. 279; Dresser, B. of Eur. pts. xxxv. xxxvi. (1875).

La Petite Buse criarde, Sonn. Voy. Ind. ii. p. 184 (1782).

Criard Falcon, Lath. Gen. Syn. Suppl. i. p. 38 (1787); *Black-winged Kite*, Europeans in India.

Kapasi, Hind.; *Chanwa*, Nepalese; *Adavi Ramadasu*, Tel., lit. "Jungle-Tern" (*apud* Jerdon).

Ukkussa, Sinhalese, West Province.

Adult male. Length to front of cere 11·4 to 12·0: culmen from cere 0·75; wing 10·4 to 10·8; tail 5·2 to 5·6; tarsus 1·3 to 1·4; mid toe 1·0 to 1·1; claw (straight) 0·5 to 0·6; height of bill at cere 0·35. The wings exceed the tail in old birds.

Female. Wing 10·6 to 10·9.

Iris, varying according to age from orange-red to pale scarlet or carmine; cere and base of under mandible yellow: bill black; legs and feet rich yellow, claws black.

Crown, hind neck, back, scapulars, major wing-coverts, and central tail-feathers bluish or ashy grey; forehead, a line above the supercilium, ear-coverts, entire under surface, under wing, upper edge of the same, axillaries, and under surface of tail white; lores, a short supercilium, lesser and median wing-coverts, and the winglet coal-black; quills dark ashen-grey, the shafts black, and the under surface of the primaries blackish, the three lateral tail-feathers whitish, sullied on the outer webs with grey, shafts of all the rectrices black except at the tip.

Young. After leaving the nest, the iris is hazel-brown, and the bill, cere, and legs much as in the adult; in a few months the iris pales to olive-grey.

Crown and nape brownish fulvous, paling into buff over the eyes; upper part of hind neck edged whitish; back, scapulars, and greater wing-coverts slaty brown, broadly edged with fulvous white; quills dark slate, with deep whitish tips; secondary wing-coverts only, black with pale margins; tail with the central feathers brownish slaty, the rest slaty-grey; chin, gorge, and ear-coverts white; throat, chest, and breast richly tinged with buff, paling into the pure white of the lower parts; lores and eye-streak as in the adult.

With age the forehead and chest become whitish, or, in some, pale greyish, while the back and scapulars lose their brown hue and become ashy, but the two latter parts still remain tipped with whitish; the shoulder of the wing becomes blacker before the end of the first year; but the greater coverts, the primaries, and their coverts remain tipped with white until after the next moult. It is not until the bird is fully adult, probably two years old, that the back loses entirely the brown shade, and the lateral tail-feathers their grey hue.

Distribution.—The Black-winged Kite is widely dispersed over the low country, and is a common bird throughout the Kandyan Province, more especially during the cool season (October until April), during which period it breeds in many of the hill-districts. As regards the lowlands, it is not at all uncommon in the south-eastern, eastern, and northern portions of the island, where the characteristic grass-lands, surrounded by forest, or bordering the shores of large tanks or inlets of the sea, and often, too, studded with dead trees, furnish it with a hunting-ground and many a favourite perch. In the extreme north I have seen it in the Jaffna peninsula; and Layard procured at Pt. Pedro.

In the Western Province south of the Chilaw district it is not often seen during the south-west monsoon; but in the dry season it is not uncommon, and has been procured as near Colombo as the cinnamon-gardens.

It occurs in many places in the Galle district, more particularly about citronella-grass estates and young cocoa-nut plantations. I have found it more particularly in the open lands of the delta of the Mahawelliganga and the Batticaloa district, in the low jungles and scattered scrubs between Madampe and Puttalam, and in grassy wastes surrounding the tanks near the south-east coast, than in other parts of the low country.

In the Central Province it confines itself to the open country in Uva, and the patnas and cultivated valleys interspersed with woods which are characteristic of the hills from the neighbourhood of Kandy to the base of the main range, as also to the so-called "plains" surrounded by forest in the latter district, among which I may cite Nuwara Elliya, the Kandapolla, Elk, and Elephant Plains, where it is a well known bird, particularly in the breeding-season.

The Black-winged Kite is a bird of wide geographical range, inhabiting the entire Indian peninsula, South-eastern Europe, and the whole of the continent of Africa. As regards the Indian empire, in which its range has more interest than elsewhere for my readers, it is found in the south of the peninsula, but perhaps not commonly, as it is absent from Mr. Hume's First List of Birds from the Travancore Hills, Mr. Bourdillon not having observed it there. In the Khandalla district it is rare in the vicinity of Ahmednagar; but this is a local peculiarity, for it is fairly plentiful further north. Dr. Stoliczka procured it at the Gulf of Cutch, Captain Butler says it occurs all over the plains of Northern Guzerat; and Mr. Hume records it as plentiful in Nepal, though it is rare in Sindh, which region is probably too barren for its habits. Along the base of the Himalayas it is not uncommon, Mr. Thompson having found it breeding in Lower Gurhwal and the Dehra Doon. About the Sambhur Lake Mr. Adam says it is not uncommon; and Mr. Ball found it tolerably so in the western parts of Chota Nagpur, while in the Satpura hills it was rather abundant. Bearing towards Burmah, we find that in the boundary-district of Cachar it is rare, Mr. Inglis only having seen half-a-dozen specimens in four years; and at Thayetmyo Captain Feilden merely notes its occurrence, while Mr. Oates met with it only in the Arracan hills. In Tenasserim Mr. Hume has reason to think it occurs; and if so, this is its furthest range to the south-east. It has not been met with at the Andamans. In the Laccadives, however, it is a visitant, presumably from the west coast; and Mr. Hume procured specimens at the islands of Amini and Cardamum. Turning towards Western Asia, we find Mr. Danford observing it in Asia Minor *in winter*, and Canon Tristram recording it as a summer visitant to Palestine and haunting the thickets on the Jordan, where it is very shy—the reverse of its nature in Egypt, where it is said to be tame and easy to shoot. In South-eastern Europe it occurs as a straggler; and Lord Lilford mentions having seen a specimen killed in Southern Spain.

The example recorded in 'The Ibis,' 1872, as killed at Harristown Bay on the east coast of Ireland, was probably an escaped bird from some ship.

As regards Africa, Captain Shelley says that it is abundant in Egypt. On the Gold Coast Mr. H. T. Ussher, now Governor of Labuan, observed it in considerable numbers; it frequented there low ground sloping towards the sea, and hawked in the evening towards sunset. Mr. T. E. Buckley found it fairly common in Natal; and Mr. Barratt procured it near Rustenberg; and it is seen in most South-African collections.

Habits.—This handsome bird, frequently called the "White Hawk" in the coffee-districts, affects grass-land surrounded by forest, dry pastures interspersed with low timber, cocoa-nut estates, citronella-grass plantations, and such spots as are open and dotted here and there with large trees. The manna-grass patna teeming with life and here and there broken by strips of jungle is a favourite resort; or, in the upper hills, a tall dead tree by the border of the lonely forest-begirt "plain" forms an equally appreciated look-out. It is usually a solitary bird, and is abroad at early dawn, lazily flapping its way across the silent jungle-glade to some accustomed perch, where it will sit preening its feathers in the rays of the rising sun, and if disturbed will fly off to the nearest prominent tree, of which it invariably selects the topmost branch to rest on. In some places, however, where no doubt it is very plentiful, it forsakes its solitary habit; for Mr. Hume writes in his 'Rough Notes,' that he once saw more than a dozen pairs hunting together over the dry reedy bed of a jheel. I have usually found its diet to consist of lizards and large coleoptera; but it is said to carry off wounded birds in India. It likewise feeds on field-mice and rats; and when quartering over grass-land I have often seen it stop and hover like a Kestrel, but with a slower motion of the wings. Its usual flight is performed with a heavy flapping of the wings; and this action, combined with its short tail and white plumage, imparts to it much the appearance of a Sea-Gull. I have often admired it, showing its handsome plumage off against the dark green forests in the upper hills, as it would leisurely course round the edge of one of the open patnas, now and then stopping when its attention was arrested by something in the grass beneath it, and hovering for a minute, perhaps rapidly to descend with outstretched talons and uplifted wings, or to resume its quiet tour of observation round the forest. Concerning its economy in India Jerdon writes, "It is not very much on the wing, nor does it soar to any height, but either watches for insects from its perch on a tree or any elevated situation, or takes a short circuit over grain-fields, long grass, or thin jungle, often hovering in the air like a Kestrel, and pounces down on its prey, which is chiefly insects, but also mice and rats, and probably young or feeble birds."

In Northern Guzerat, Capt. Butler writes (*loc. cit.*), "it is generally found singly or in pairs. Its *modus volandi* is very varied. Sometimes it flies lazily along like a Gull; at other times it sails round and round in circles, often stopping to hover in the air like a Kestrel, as recorded by Dr. Jerdon. Then, again, when hunting, it flies with quite the swiftness and quite the style of a Falcon. I have seen one of these birds stoop and carry a wounded Quail with quite the rapidity and dash of a Peregrine." Concerning this Kite's note, although it is generally a very silent bird (I have never heard its voice, though I have seen it dozens of times), it is said sometimes to utter loud screams. So far back as 1782, Sonnerat, who met with it in his voyages to India, named it the "Petite Buse criarde," doubtless on account of the loud notes it uttered; and Mr. F. A. Barratt writes, in his "Notes on the Birds of the Lydenburg district," South Africa, of one which he shot:—"It attracted my attention by a harsh cry, high in the air, which I thought to be that of an Eagle; but, to my surprise, I found it proceeded from this bird."

The Black-winged Kite appears to thrive in confinement. Mr. W. Murray, of the Ceylon Civil Service.

kept a young bird, which he took from a nest at Nuwara Elliya, for some time. It partook greedily of meat; and I noticed that it perched with the outer toe reversed. The iris of this bird took two months to change from dark brown to light hazel.

Nidification.—This species breeds from December until March, and, I have reason to believe, resorts in considerable numbers to the hills during its nesting-season. I have known it to build both near and in Nuwara Elliya, in Deltota, and in Kadugunawa, in which latter place I took its nest myself in December 1876. This nest was built in a moderately tall, umbrageous tree, in an exposed situation on one of the patuas of the Kirimettic estate, and within a few hundred yards of the bungalow. It was placed among the topmost leafy branches, supported by a fork so slender that the small boy I sent up had great difficulty in reaching it. It was a very openly constructed fabric, about the size of a common Indian Crow's nest, made of small sticks laid over one another so far apart that daylight could be seen anywhere through it except just in the centre. The interior was flat, and formed of small twigs, on which lay the two eggs. One of these was almost a perfect, and the other a broadish oval, of a dull white ground-colour, in one stippled all over with reddish-brown dots and encircled just beyond the centre with a ring or zone of the same, in the other blotched openly throughout with smeary markings of brownish red, confluent round the smaller end, and mingled in other parts with lighter patches of reddish brown. They measured respectively 1·54 and 1·61 inch in length by 1·23 and 1·17 inch in breadth. The female bird was frightened from the nest by our approach, and flew off with the male, not returning until after we had left with the eggs, and then only to fly heavily round the tree, and make off again to a neighbouring wood.

The nest from which Mr. Murray procured his young bird was situated in the compound of the Agent's house at Nuwara Elliya, and built in the top of an Australian lightwood (*Acacia melanoxylon*). It contained two young. Conflicting descriptions have been given of the eggs of this Kite by various naturalists; and a *résumé* of the information possessed concerning its nidification up to date will be found in 'Stray Feathers,' 1873, as above quoted. So many nests have, however, now been taken and thoroughly identified that the eggs have been satisfactorily proved not to vary more than those of other Hawks. Messrs. Blewitt and Adam in India, and likewise Captain Shelley in Egypt, found the number to vary from three to four; and most of the eggs found by these gentlemen seem to have been more heavily and darkly blotched than mine. From Mr. Adam's account, quoted by Mr. Hume in 'Nests and Eggs,' it appears that the nest is built in less than a week, which is a short time for a hawk to construct its nest. After writing of the discovery of a nest near the Sambhur Lake in July 1872, he says :—"On the 7th of August I sent a man to see if the nest contained eggs; but he found that it had been abandoned and a new nest commenced in one of a group of six Lasora trees (*Cordia myxa*), which stood near to the Khajur tree. He also informed me he had seen the birds together. I inspected the nest on the 10th of August, and found one of the birds sitting on it. The nest was so loosely constructed that with my binocular I could see that it contained no eggs. I again inspected the nest on the 14th August, and found that it contained two eggs. One of the birds sat close on the nest, and could not be frightened off by a man beating on the trunk of the tree with a stick; and this same bird made a swoop at my servant as he was climbing the tree. The nest was situated on the very top of the Lasora tree, and was from 25 to 30 feet from the ground. In shape it was circular; and, with the exception of two or three pieces of Sarpat grass (*Saccharum sara*), there was no attempt at lining. It was about 10 inches in diameter; and the egg-cavity had a depression of about 2 inches." Of the eggs he writes, they "are without gloss; both have a light creamy-white ground, of which, however, little is shown. One had the broad end all blotched over with confluent patches of deep rusty red, while the smaller had numerous spots of a much lighter brownish-red." Captain Shelley, who found these nests at different times in Egypt containing each four eggs, says that in that country the nest is carefully constructed of sticks and reeds, and is smoothly lined with dry leaves of the sugar-cane.

FALCONIDÆ.

AQUILINÆ.

Genus PERNIS.

Bill long, rather weak, curved from the base, the tip not much hooked, wide at the base, the sides slanting from the culmen to the margin, which is not festooned; cere much advanced and bare. Nostrils linear, oblique, overlapped by the membrane of the cere. *Lores feathered like the forehead.* Wings moderately long, pointed; the 3rd and 4th quills subequal and longest. Tail rather long, broad, even at the tip. Tarsus stout, shorter than the middle toe, the upper half plumed in front, and the remainder covered with small reticulate scales. Toes protected above with narrow bony transverse scales; lateral toes rather long and subequal. Claws acute, rather straight, trenchant beneath. Tibial plumes reaching down to the foot. Head usually furnished with an occipital and somewhat scanty crest. Eyes placed in the head posterior to the gape.

PERNIS PTILONORHYNCHUS.

(THE INDIAN HONEY-BUZZARD.)

Falco ptilorhynchus, Temm. Pl. Col. i. pl. 44 (1823).

Pernis ptilonorhynchus, Steph. Gen. Zool. xiii. pl. 35 (1826); Holdsworth, P. Z. S. 1872, p. 414; Sharpe, Cat. Birds, i. p. 347 (1874); Ball, Str. Feath. 1874, p. 381; Hume, Nests and Eggs, p. 56 (1874); Legge, Str. Feath. 1875, p. 364; Butler, ibid. p. 448; Tweeddale, Ibis, 1877, p. 286.

Pernis cristata, Cuv. Règn. An. i. p. 335 (1829); Blyth, Cat. B. Mus. A. S. B. p. 18, no. 82 (1849); Horsf. and Moore, Cat. B. Mus. E.I. Co. no. 74, p. 63 (1854); Jerd. B. of Ind. i. p. 108 (1862); Wall, Ibis, 1868, p. 17; Hume, Rough Notes, ii. p. 330 (1870).

Pernis torquata, Less. Traité, p. 76 (1831).

Pernis ruficollis, Less. *l. c.* p. 77 (1831).

Pernis albigularis, Less. *l. c.* p. 77 (1831).

Pernis apivorus, Temm. and Schl. Faun. Jap. Aves, p. 24 (1850).

The Crested Honey-Buzzard of some authors.

Madhava, Nepalese, from *madhu* (honey); *Shahutela,* Hind., from *shahud* (honey); *Tenugedda,* Tel.; *Ten Prandu,* Tam.; *Jutalu,* Yerklees; *Malsuwari* of the Mharis (Jerdon). *Rajaliya,* Sinhalese.

Adult male. Length to front of cere 23·5 to 24·5 inches; culmen from cere 1·0; wing 15·5 to 15·8; tail 9·0 to 10·3; tarsus 1·9 to 2·0; middle toe 1·9 to 2·1, its claw (straight) 0·95; height of bill at cere 0·38 to 0·4.

Female. Length to front of cere 24·5 to 25·5 inches; culmen from cere 1·0 to 1·1; wing 15·7 to 17·9; tail 10·0 to 11·5; tarsus 1·0 to 2·1; middle toe 1·9 to 2·2, its claw (straight) 1·0 to 1·02; height of bill at cere 0·48. Expanse of an example with a wing of 16·5, 55·0.

The above dimensions of males are taken from four specimens, and those of females from twelve, of Ceylon-killed birds. It is the exception to find a female measuring in the wing more than 17·0 inches. Four in my own collection measure as follows—15·7, 16·5, 16·5, 16·6: eight others, five of which are in the Norwich Museum, and two in Lord Tweeddale's collection, 15·7, 16·8, 16·7, 16·4, 17·5, 16·6, 17·0, 16·4. The last but one is included in a list my friend Mr. Gurney sent me, of two or three birds in Lord Tweeddale's collection, measured by himself, and is most exceptional if the measurement is correct, which I have no doubt it is. The specimen must be an extraordinary and quite abnormal one—a giant among the Ceylonese Honey-Buzzards! I may remark that Mr. Gurney sends me the wing-measurement of a male in the same collection as 17·78. I take it for granted that this specimen has been wrongly sexed by the collector.

Iris golden yellow, yellow mottled with brown, or yellow with a pale outer circle ; cere deep leaden colour ; bill blackish, gape and the base of under mandible bluish ; legs and feet dull yellow, in some citron-yellow. The iris being very variable, I have enumerated the several colours which I have found in dark birds. It is never red as in the North-Bengal race.

Fully adult or very old stage. Crown, hind neck, and upper surface rich dark earth-brown, the tips of the hindneck-feathers often darker than the rest ; back and wing-coverts suffused with a purplish lustre, a short occipital crest of 4 or 5 stiffish ovate feathers attaining a length of 2·3 inches, sometimes black, and at others concolorous with the nape ; the forehead above the eye, entire face, ear-coverts, and throat iron-grey, blending into the surrounding plumage ; quills ashy brown, crossed with three or four widely separated bars of dark purplish brown, and a broad terminal band of the same, the extreme tip pale, the inner webs whitish from the notch inwards, with the interspaces mottled with brownish ; upper tail-coverts, in some examples, tipped with whitish ; tail dark purple brown, crossed by a broad, 2-inch, smoky-grey band about the same distance from the tip, and in some with a narrow bar of the same near the base.

Throat and entire under surface dark chocolate-brown, the feathers dark-shafted ; a dark stripe on each side of the throat, frequently continued across the fore neck in the shape of a gorget ; under wing-coverts at times tipped with fulvous ; under surface of light portion of tail grey. In two very dark specimens I examined in Kandy the feathers of the lower breast and abdomen were pale-tipped.

A younger stage of plumage, but one which represents the generality of apparently adult birds killed in Ceylon, is as follows :—

Above rich sepia brown, the margins of the feathers somewhat paler, and the feathers of the occiput and hind neck, as well as the fore neck and entire hinder surface, a fine chestnut brown, with blackish shafts ; a well-developed crest of black feathers ; the lores and round the eye, in some examples, dark iron-grey mingled with brown, while in others the forehead and above the eye is whitish, the centres of the feathers being concolorous with the crown ; the dark moustachial stripe is present, and, in the darker examples, is black, spreading over the throat and sometimes running up in a point to the chin ; the median wing-coverts are usually light-tipped, and have a considerable amount of white at the base of the feathers ; the quills are not so dark as in the above, and have more white at the tips, the bars being also closer together, and the interspaces more or less crossed with wavy light rays ; in the tail, the lighter or earthy-brown hue is the ground-colour, and contains numerous pale wavy cross rays ; the tip is whitish, and adjacent to it is a brownish deep-brown bar : about 2½ inches above this, across the centre of the feathers, is a narrow bar of the same, and another similar one near the base. The under surface is variable, being in some examples a light fulvous brown, with the stripes very broad ; while in others the strim are almost wanting on the breast ; the colour of the whole breast, however, is more or less uniform and devoid of white spaces in the younger bird ; most of the basal portion of the inner webs of the primaries is white.

In this stage the tail wants the characteristics of the very old bird, viz. the smoky-grey nearly uniform bands ; but the lores and the space beneath the eye have the grey appearance, which is a marked adult sign. The presence of the white forehead in this adult stage, I consider to be quite abnormal, as many younger birds (as will be presently noticed) have it uniform with the head.

Young. In birds of the first year the wing varies from 15·6 to 16·0, the other parts equal those of the adult.

Iris in some yellow, in others brownish yellow, sometimes with a dark inner edge ; cere bluish with greenish patches, in others greenish yellow ; legs and feet greener than in the adult.

Back, scapulars, and wing-coverts darkish hair-brown, the wing-coverts more or less pale-edged, the median series being the lightest, some examples having the lesser rows edged with whitish, and the outer series of primary-coverts broadly margined with the same ; crown and occiput rich tawny brown, the feathers with blackish shaft-stripes ; the hind neck with the larger part of the feather whitish, and the terminal portion pale brown with a dark shaft-stripe ; the crest-feathers blackish brown, broadly margined or tipped with white ; forehead and a broad space above the eye white ; lores and a broad posterior orbital streak dark brown with a slightly greyish shade, inner primaries and secondaries deeply tipped with white, pale brown on both webs, and barred with dark brown, longer primaries with more of the inner webs white than in adults, and with the basal portion of the outer webs light brown, crossed with dark bars alternating with the interspaces of the inner webs ; tail smoky brown, deeply tipped with white, and crossed with four narrow and rather irregular bars of dark sepia-brown, the subterminal one not much broader than the others, and the light portions crossed with wavy light rays ; throat and entire under surface, with the under wing and the edge above the metacarpal joint, pure unmarked white ; ear-coverts pale brownish.

From this stage the first advance towards adult plumage is made (probably after the first moult) by the head, hind neck, and upper surface generally becoming more uniformly dark, although there is usually still a good deal of white about the hind neck ; the dark lores and space behind the eye extend, and the cheeks and face become striated with dark brown, and a series of streaks from the gape down each side of the throat appear as the first signs of the future dark stripe ; the bars on the tail, especially the subterminal one, become broader ; the chest and breast assume blackish-brown stripes, more or less broad, on the white ground, while the lower breast, flanks, and abdomen become, in some examples, barred with brown, and in others washed over the whole feather with the same, the flanks and thigh-coverts generally being the darkest. In this stage, I believe, a considerable advance in the plumage is made by a change in the feather itself ; and hence the great variety in the plumage at this age. The dark grey hue of the lores spreads over the cheeks ; the ear-coverts and forehead become nearly concolorous with the crown ; the broad lateral throat-stripes of black develop and spread across the fore neck, the chin and gorge becoming brownish ; at the same time the bars on the lower parts of those examples having the barred feature spread over the feather, or the brown of the flanks in the other type encroaches gradually on the breast.

Obs. Mr. Gurney has noticed that Ceylonese specimens of this Honey-Buzzard are larger than those from India. As will be seen, the above list contains some very high wing-measurements ; but if an extensive series of Indian birds be examined, I have no doubt some will be found equally large. Mr. Hume gives the largest wing, in six females measured, as 17·25, and Mr. Sharpe, in his Catalogue, the average of a large series as 16·5. Some I have measured in the British Museum are as follows—(Deccan) 16·3, (Nepaul) 16·2, (N. Bengal) 17·4, (Kamptee) 16·3, (Himalayas) 17·1, (Darjiling) 15·0. All our largest specimens have been shot in the hills of Ceylon : and, as I demonstrate below that the species is for the most part migratory to Ceylon, these large birds must be not inferior to their fellows elsewhere, or they must be bred on the hills of the island. Mr. Sharpe has measured an example from Java with the wing 17·8, which favours the idea that Ceylonese birds may migrate from that quarter, although it must be remarked that Javan birds have longer crests than ours. Much has been said about the irregular plumage of the Honey-Buzzards ; but if a series of examples of different ages be examined, a regular gradation in the plumage, from the pale-chested bird up to the one with the grey face (which is an unmistakable sign of age) and the dark under surface, can be noticed. The fact of white-chested birds breeding with dark ones can be easily explained by assuming that there is in the Honey-Buzzard, as in some Eagles, a permanent light phase.

Distribution.—The Honey-Buzzard is to a certain extent a migratory bird to Ceylon, and appears, from what I observed while in the island, to make its appearance first of all on the north and north-east coasts, which leads to the inference that it migrates with the north-east monsoon from Burmah, or perhaps from the southern part of the Indian peninsula, to Ceylon. It used to appear yearly on the coast about Trincomalie during November and December, and then depart into the interior. In 1874 I obtained two newly arrived and very tame examples in the Fort, which is a point of call for many migrants arriving with the north-east wind on that part of the coast. Several other birds haunted the vicinity of the town at the same time ; in the following year, however, scarcely an example was to be seen, although it was comparatively numerous in the Kandy district. It was first recorded as a Ceylonese bird by Mr. Holdsworth (*loc. cit.*), from an adult female shot by Mr. Forbes Laurie in the Madoolkella district, not far from Kandy. It had prior to this been received from Ceylon, but its occurrence omitted to be noticed in print. It locates itself in the northern forests, preferring the vicinity of the tanks which abound in that part ; and many birds remain there yearly, and doubtless breed in those unfrequented haunts. I have seen it in such places during the south-west monsoon, and have likewise received specimens from Avisawela and Kurunegala, in the western part of the island, at the same season of the year. I have shot it in August in the Park country, where it is not uncommon ; and I have no doubt it inhabits the forests between Badulla and Hambantota. In the south-west I have never known it to occur. As regards the mountain-region, it is principally found about Dumbara and other places of intermediate altitude in the direction of Kandy. Occasionally, however, it ascends much above this ; for Mr. Bligh has shot it in Dimbula. It is possible that some of the birds occurring on the hills have been bred there, as they appear to be larger than those which are evidently migrants.

N 2

One of the most interesting points yet to be decided with reference to Ceylon ornithology is that relating to the movements of this fine bird. Whether it comes from Burmah or from South India, or even from Sumatra, remains yet to be seen. If an extensive series could be obtained from South India, a comparison of it with another from Ceylon would easily settle the matter with reference to that quarter.

This species is scattered throughout India, extending into Burmah and a portion of the Asiatic archipelago. It is not unfrequent in the south of India, but appears to be local in its distribution there. Jerdon says of it, in the 'Madras Journal:'—" I have only met with this bird in the jungles of the western coast and Nilghiris. It is by no means common. . . . . I procured a female at the foot of the Conoor pass, and another on the summit of the hills." Mr. Bourdillon appears not to have found it in the Travancore district. Near Khandala and in the western parts of the Deccan it is common; in the region about Mount Aboo and in Northern Guzerat Captain Butler states that it occurs, but not commonly; and at Sambhur it appears now and then as a straggler. Mr. Hume does not record it from Sindh. In the North-west Provinces it occurs; and in Chota Nagpur Mr. Ball has procured it; but it is found in that district sparingly, though this gentleman says that it appears to be common near the Ganges at the north-east corner of the Rajmehal hills; this, however, has reference to the red-eyed race, which is spread through Bengal, and which some think is specifically distinct from the southern bird. The Pegu race, likewise, Mr. Hume considers differerent from the Bengal on account of its smaller size; it appears to be not uncommon there. From Tenasserim I do not find that it has as yet been received; and it has not yet been discovered in the islands of the Bay of Bengal. From Java it is well known; and of late it has been procured by Mr. Buxton in South-east Sumatra, having been also previously known from the island of Banka.

Habits.—Well-wooded districts and large tracts of jungle are the favourite habitat of this handsome bird. It is solitary in its habits, and is partial to the vicinity of water. I have more than once surprised it in shady trees on the borders of forest-rivers or lonely tanks, when it would make off with a straight quick flight to another inviting perch. I have also seen it perched on the tops of high trees in forests, when it much resembles the Serpent-Eagle in the distance. It soars high in the air at times, taking short circles as it ascends, and, according to some observers, has the habit of descending with a rush, much to the terror of the small birds in the neighbourhood. This I have not seen myself, though I have witnessed it soaring at a considerable height. Jerdon observed it attempting to hover, which he said it did in a clumsy manner. Its usual diet consists of honey, which it robs in spite of the attacks of the inmates of the nest, against whose stings, however, its peculiarly-feathered face and lores well protect it. With the honey it also devours the young ones, remains of which I have invariably found in its stomach. It is said also to eat other insects, white ants, and small reptiles; but the latter food, I imagine, is only resorted to when pressed for want of its usual diet. One that was shot in the Fort at Trincomalie was associating with Crows, and flying round the barrack-room at the dinner-hour in company with them, on the look-out for scraps thrown out from the verandahs. Another haunted the fine trees shading the officers' quarters for more than a day, and appeared not to mind the frequent passers-by in the least, finally allowing me to shoot it in the tamest manner.

Its habits do not appear to have been paid much attention to by Indian observers, Jerdon being the only one who has recorded much concerning it. He writes in the 'Madras Journal:'—"I occasionally saw it seated on a tree, alternately raising and depressing its crest, and in the Nilghiris frequently noticed it questing diligently backwards and forwards over the dense woods there. . . . . Their usual flight is rather slow; but I once observed one flying more rapidly than in general, with a continued motion of its wings, and every now and then attempting to hover, with its wings turned very obliquely upwards." He further remarks in the ' Birds of India,' that Burgess mentions his having been told by some natives that, when about to feed on a comb, it spreads its tail and with it drives off the bees before attacking it.

Nidification.—The Honey-Buzzard may possibly breed in the central and northern forests of Ceylon; but I know of no evidence to this effect. In India it breeds from April until July, nesting in the forks of trees. It builds a nest of sticks and small twigs, and lines the interior with green leaves or fresh grass—a common habit with raptorial birds. Captain G. Marshall observes that the female sits very close during the period of incubation, and is not easily driven away from its nest. This is unusual with the Hawk tribe, the majority of which leave their nests when they are approached. The eggs are two in number as a rule; but some nests have been found only to contain one. They are round in shape, of a "whitish pinkish-white or buffy-yellow" ground-colour, and vary much in the character of their markings, although they are usually very highly coloured with blotches and clouds of reddish or purplish brown and dark red, sometimes quite confluent round one end. They average 2·03 inches in length by 1·72 in breadth.

Subfam. FALCONINÆ.

"Outer toe only connected to the middle toe by interdigital membrane; tibia much longer than tarsus, but the latter not contained twice in the former; hinder aspect of tarsus reticulate; bill distinctly toothed." (*Sharpe*, Cat. Birds, i. p. 350.)

ACCIPITRES.

FALCONIDÆ.

FALCONINÆ.

Genus BAZA.

Bill stout, curved rapidly from the base of the cere; tip much hooked, and notched with a double tooth; cere but slightly advanced. Nostrils linear, oblique, covered as in *Pernis* by the superlying membrane. Wings moderate, rounded, with the 4th quill the longest, and the 1st subequal to the secondaries. Tail moderately long, much exceeding the closed wings. Tarsus short, the front and sides plumed for more than half its length; the remainder covered throughout with reticulate scales. Middle toe subequal to the tarsus; lateral toes nearly equal; the whole covered with bony transverse scales. Claws rather straight, the inner less than the middle. Head with an elongated occipital crest.

BAZA CEYLONENSIS.

(THE CEYLONESE CRESTED FALCON.)

(Peculiar to Ceylon?)

Baza ceylonensis, Legge, Str. Feath. 1876, vol. iv. p. 202; Whyte, ibid. 1877, p. 202.

Similis *B. magnirostri* sed cristâ nigrâ latè albo terminatâ, secundariis latè albo terminaliter marginatis: rectricibus 4-fasciatis, plagâ subterminaliter minore quam fascia apicalis brunnea: subtùs fulvescenti-brunneo nec rufescenti transfasciata: gutture fulvescenti nec cinereo lavato.

Adult. The following are the measurements of the two type specimens described by me *loc. cit.*, the larger of which is presumed to be a female :—Length to front of cere (from skin) 16·5 and 16·8; culmen from cere 1·01 and 1·03; wing 11·7 and 12·0; tail 7·6 and 8·0; tarsus 1·5 and 1·5; middle toe 1·35 and 1·45; claw (straight) 0·65 and 0·68; height of bill at cere 0·4 and 0·5; tarsus feathered to 0·5 from the root of the middle toe.

Iris yellow; bill blackish leaden, lower mandible pale at base; cere (judging from the skin) dusky plumbeous; legs and foot yellow; claws plumbeous, pale at base.

Male. Back, scapulars, lesser wing-coverts, and centres of the feathers on the hind neck and head deep brown, paling on the rump slightly, and with a strong purple sheen on the mantle as well as on the under-mentioned caudal bars: the feathers of the back with perceptibly pale edgings, those of the head and hind neck broadly margined with pale tawny, the superciliary region being entirely of this colour, the forehead slightly darker with the shafts of the feathers blackish; crest 1¾ inch in length, black, conspicuously tipped with white; greater secondary-coverts and tertials paler brown than the scapulars, many of the feathers tipped whitish; primaries and secondaries smoky brown, the latter and the inner primaries deeply tipped with white, and the whole crossed with blackish bars, the ulterior one being terminal, inner edges of the primaries white on the lighter portion of the feather; tail drab-brown, pale-tipped, a broad subterminal band of purplish black, and three narrower of the same, the basal one hidden beneath the coverts.

Lores and a stripe behind the eye blackish brown; cheeks and ear-coverts slate-grey, with dark shafts; chin and throat buff, the feathers down the centre with blackish shaft-stripes; chest and sides of the fore-neck almost uniform tawny cinereous, under surface from the chest, with the under wing-coverts and lower surface of the basal portion

2

of quills, white, barred on the breast and flanks with rufescent brown bands equal to the white interspaces, narrower and further apart on the tibial plumes, and almost absent on the under tail-coverts; inner sides of legs buff-white; lesser under coverts crossed with narrow rufous markings, major series with a few transverse brownish patches; lower surface of light portions of tail greyish white.

Presumed female. Has the upper surface generally somewhat paler; but the crown is darker, the blackish central stripes being broader than in the above example; crest consisting of four long feathers 2 inches long; the primaries and secondaries, which are just acquired after moult, very deeply tipped with white; the chest differs in its less uniform hue, having the feathers with broad rufous centres and widely margined with buff-whitish; the under surface is similarly barred, under tail-coverts and wing-lining the same.

Young. The example referred to below as presented by Mr. S. Bligh to the Norwich Museum is a young bird. The posterior tooth is not developed, and the anterior less deep than in the adult.

Its length (from the skin) is 17·0 inches; wing 12·25; tail 8·0; tarsus 1·15.

Above glossy dark brown, the feathers of the back, scapulars, and wing-coverts edged with whitish; centres of the head-and hindneck-feathers brown, with broad margins of fulvous white; crest black, deeply tipped with white, and 1·8 inch in length; primaries and secondaries smoky brown, with blackish bars and white inner edges to the basal portions of the former, similar to the adult; median and greater secondary wing-coverts deeply tipped with white, adjacent to which the brown hue changes into rufous, giving the wing-coverts a rufescent appearance; tail smoky brown, banded with five brown bars narrower than in the adult; under surface white; a very fine chin-stripe of brown, formed by dark shaft-lines on one or two feathers; chest marked with well-defined brown stripes; breast and flanks widely barred with broad pale sienna-brown bars.

Another immature example, in the British Museum, from the collection of Messrs. Whyte and Co., and which is a female, is very similar to the above, but may perhaps be a little older; wing 12·1, tail 7·8, culmen from cere 1·01. The posterior tooth slightly developed, but not so prominent as the anterior.

Iris yellow; feet and tarsi yellowish; head and hind neck fulvous tawny, with dark central stripes increasing in width at the lower part of the neck; the crest black, deeply tipped with white, and 2·0 inches in length; the back and wings are deep brown with a purplish lustre, the feathers margined with rufous brown; greater wing-coverts barred with pale brown; the barring of the quills is the same, and the inner part of the lighter interspaces on the inner web white; tail as above, the tip whitish, and the subterminal dark bar equal in width to the adjacent interspace; lores blackish brown; cheeks and ear-coverts with tawny-brown stria; throat and under surface buff-white; the chin with a pale brown mesial stripe, widening and darkening on the throat; chest marked with broad "drops" of rufous-chestnut, changing on the breast, flanks, abdomen, and shorter under tail-coverts into bars of the same; longer under tail-coverts unmarked; thighs crossed with bar-like spots of rufous.

A third immature specimen has been sent home to the Norwich Museum by Messrs. Whyte and Co., since this article was written. Mr. Gurney writes me that it measured, as he was informed, 18·5 inches in the flesh, and weighed 1 lb. The wing, according to his system of measuring, is 12·5 inches (which would be equal, after my plan, to 12·2 or 12·3), tail 8·5, tarsus 1·5, crest 2·3.

It is older than the specimen presented by Mr. Bligh, "having much less of the white margins to the feathers on the upper surface, and the throat and breast being more fulvous; the tail has 4 bars instead of 5." This latter feature testifies to its age; and I think its plumage may be taken as representing an intermediate stage between the young and the old bird.

Obs. I do not consider this a very good species. It comes very close to *B. magnirostris* from the Philippines; but as this latter has such a remote habitat, I have allowed the slight differences that exist to weigh in favour of keeping the Ceylonese bird distinct for the present. The adult type of *B. magnirostris* is a smaller bird than *B. ceylonensis*: it has the wing 11·1 inches, tail 7·2, tarsus 1·3. The crest is not deeply tipped with white as in the latter, but has the terminal portion of the webs laterally edged with it only; the secondaries and primaries are not deeply tipped with white; and the tail-bands are narrower and five in number; the cheeks are much paler, and the chin-stripe *inconspicuous* and of a light iron-grey colour uniform with the cheeks; the chest is very similar, but the breast- and flank-bands are more rufous than in my bird. This latter characteristic, however, is not to be depended upon. *B. ceylonensis* likewise has a considerable general resemblance to the example in the British Museum, which Mr. Sharpe considers now to be *B. jerdoni*: but this has the head very dark indeed, and is rufous on the cheeks and sides of the head. Mr. Hume's species, *B. incognita* (Str. Feath. 1875, pp. 314-316), from Sikhim and Tenasserim appears to be more closely allied to this species than to the Ceylonese bird, being considerably larger (wing, ♂ 13·12, ♀ 13·75) than the latter: and the specimens described seem, moreover, to be immature.

If identical with any other member of the genus, one would naturally seek to join my bird to *B. sumatrensis*, which has a comparatively adjacent habitat, to it. I have, however, compared this, in company with Messrs. Sharpe and Gurney, with two of the immature examples of the Ceylonese form; and these gentlemen concur with me that the

Sumatran bird, as far as can be proved by the evidence of the single immature example which exists of it, is distinct. The testimony of an immature bird, it must be allowed, is not a very safe one to go upon; but nevertheless, as the specimen exists, it is a larger bird (wing 12·75, tail, very long, 9·6), has a chin-stripe, which is a marked characteristic of *B. ceylonensis*, has the under-surface bars much broader and of a different appearance, and the tippings of the back and scapular feathers fulvous and not white. Unless, therefore, *B. magnirostris* from the Philippines turns out some day to be identical with *sumatrensis* from Sumatra, and both the same as *ceylonensis*, I think the latter species may hold its own, as it can scarcely be one with the Philippine bird, a species not hitherto procured to the westward of those distant islands. As yet every member of the genus (except the curious *Baza lophotes*, totally unlike any other in its plumage) has proved very local in its habitat; and were it not for this fact, it would be difficult to imagine our bird restricted to so small an island as Ceylon *.

Distribution and Discovery.—This interesting Crested Falcon was described by me (*loc. cit.*) from two adult examples which I found in the collection of Messrs. Whyte and Co., naturalists, in Kandy, in August 1876. They were both shot on the same day, the 6th of the same month, by Mr. F. H. Davidson, of Matale, on the Kudapolella estate. In May of the same year, however, I had met with an immature specimen (the one now in the Norwich Museum) at Mr. Bligh's bungalow, and identified it from Mr. Sharpe's plate in the 'Catalogue of Birds,' vol. i., as *B. sumatrensis*. This example was therefore the first that came under my notice; it was shot in the early part of 1875 by a Mr. Colville, near Nilambe, in the Kandy district, and preserved in Messrs. Whyte and Co.'s establishment. In the beginning of last year the immature bird referred to above as now in the National collection, was procured near Kandy by Messrs. Whyte and Co.'s collectors; and a third example has been lately sent by this firm to the Norwich Museum, a female, and shot in the Central Province on the 3rd of January last. Since the publication of my account of the species, Mr. A. Whyte has stated, in a paper which appeared in 'Stray Feathers,' August 1877, that the " bird was *discovered* † by us *eight years ago*, a pair having been shot by one of our collectors not far from Kandy." With regard to this pair Mr. Whyte writes to me lately as follows :—" They were shot on the same day, from the top of Oodoowella crag, about four miles from Kandy, by a Singalese collector, Carolis, in the fall of 1870; since then at least ten specimens of the bird have passed through our hands; and I can quote Kandy, Matale, Rattota, and Deltota as among the situations in which it has been found." It would appear, therefore, that it has only been procured within the very limited district stretching from Matale 10 miles north of Kandy, to Deltota, about 12 miles, in a direct line, to the south of the town. This part of the hill-region of Ceylon, it should be remarked, is that in which most of the birds are shot that are sent to Messrs. Whyte and Co.'s for preservation, inasmuch as they can be forwarded by Coolie runners, and skinned before suffering from the decomposing effects of tropical heat; it is not, therefore, to be inferred that the habitat of the Ceylon *Baza* is restricted to such a very small tract of country as this, but rather that it is a hill-bird scattered throughout the

---

* I have just heard, since correcting the proof of this article, from Mr. Hume, that he has lately received a young specimen of a *Baza* from the Wynaad, which he considers must be identical with this species. Mr. Hume has not, as far as I am aware, seen examples of *B. ceylonensis*; but his surmise may be correct. I accordingly put it doubtfully " peculiar to Ceylon."

† In the interests of Ceylon ornithology I am constrained to make some remarks on Mr. Whyte's note on this species. Were it not my aim to give a faithful history of all the peculiar Ceylonese forms, I should not have referred to the subject. It is difficult to see in what sense the writer uses the word "discovered." The species was in reality discovered by the collector who shot it ; for the specimens were afterwards skinned, sold unidentified, and lost for ever to science ! In continuation of the above paragraph, follows :—" Three more specimens have been *collected by us*, one of which Captain Legge obtained from us." Two of these I will remark are comprised in the pair *shot by Mr. Davidson* and sent to Messrs. Whyte and Co.'s for preservation, one of which Mr. Whyte sold me under the impression that it was a Crested Goshawk (a not unlikely mistake for one who had formed no acquaintance with the genus *Baza*) : and the other he sent me on the order of Mr. Fraser, of Colombo, a friend of Mr. Davidson, and who kindly presented it to me. The words *collected by us*, in reference to this pair are therefore misapplied. When I wrote to Mr. Whyte, shortly after the purchase of the type specimen, that it was a new *Baza*, I much wish that he had informed me of his having previously received a pair. I could then have made inquiries concerning the birds, and should perhaps have succeeded in tracing them to their destination; in which case I could have verified Mr. Whyte's identification.

Central-Province subranges, although it has not yet been recorded beyond the vicinity of the Kandyan capital.

Habits.—I am unable to furnish any information concerning the habits of this species, beyond that I learn it frequents the borders of forests, the vicinity of steep-wooded hill-faces and patnas interspersed with jungle. When killed it has doubtless been met with in such localities; but as a rule it will be found, like its congeners, to be a forest-loving species, like *Baza lophotes* and *B. reinwardti*.

The front figure in the Plate accompanying this article is the adult male bird killed at Matale, and that in the background the young bird sent home by Messrs. Whyte and Co. to the British Museum.

BÁZA LOPHOTES.

(THE INDIAN CRESTED FALCON.)

Falco lophotes, Temm. Pl. Col. i. pl. 10 (1823).

Buteo cristatus, Bonn. et Vieill. Enc. Méth. iii. p. 1220 (1823).

Baza syama, Hodgs. J. A. S. B. v. p. 777 (1836).

Baza lophotes, Gray, List Gen. B. p. 4 (1840); Blyth, Cat. B. Mus. A. S. B. p. 17 (1849); id. J. A. S. B. xix. p. 325 (1850); Kelaart, Prodromus, p. 115 (1852); Layard, Ann. & Mag. N. H. 1853, xii. p. 102; Horsf. & Moore, Cat. B. Mus. E. I. Co. i. p. 62. no. 72 (1854); Jerd. B. of Ind. i. p. 111 (1862); Hume, Rough Notes, ii. p. 337 (1870); Holdsworth, P. Z. S. 1872, p. 415; Sharpe, Cat. B. i. p. 352 (1874); Walden, Ibis, 1876, p. 341.

Hytiopus syama, Hodgs. J. A. S. B. x. p. 27 (1841).

Hytiopus lophotes, Blyth, J. A. S. B. xii. p. 312 (1843).

Pernis lophotes, Kaup, Contr. Orn. 1850, p. 77.

Baza indicus, Bp. Rev. et Mag. de Zool. 1854, p. 535.

Cohy Falcon, Lath. Gen. Hist. i. p. 165, pl. x. (1821).

Black-crested Kite, "*Baza*," *Cohy Falcon, Cohy Pern,* in India.

Cohy of the Parbuttics; *Syama,* lit. "Black," Nepal.

Adult male*. Length to front of cere 12·5 inches; culmen from cere 0·8; wing 9·2 to 9·4, expanse 30·5; tail 5·0 to 5·5; tarsus 1·05 to 1·1; middle toe 1·0 to 1·1, claw (straight) 0·47; height of bill at cere 0·35.

No difference in size exists between examples from Nepaul, Ceylon, and Pinang.

Iris brownish red: cere bluish leaden; bill pale bluish leaden, darker at the sides above the tooth; legs and feet pale bluish, claws black.

Entire head, throat, body above, wing-coverts, longer scapulars, quills, tail, and body beneath from the upper breast black, with a dark green gloss above and on the under tail-coverts. A long occipital crest of 3 or 4 narrow feathers 2½ inches in length; tertials and some of the concealed scapulars rufous towards the tips; a broad edging of the same near the extremities of the secondaries; tertials and scapulars white across the middle, showing conspicuously on the longer feathers, the terminal portions of which are black.

Chest pure white, succeeded by a band of deep vinous chestnut, many of the feathers of which are edged with black; below this the black sides of the breast are overlaid with long ochraceous white plumes, meeting across the body below the band, and barred down the sides with the chestnut; lower surface of quills and tail stone-grey, with a dark patch near the tips on the outer portion of the latter.

The black plumage underlying the stiff breast-plumes is a singular character in this bird's attire.

Young. In the bird of the year the anterior tooth is less developed than in the adult, and the second or posterior notch is *not developed*; the crest is of much the same length as in the old bird.

The chief characteristic is the great amount of white and rufous, handsomely intermingled, on the wings and scapulars. Head and upper surface dusky black, with a rufescent tinge on the back-feathers everywhere but at the tips: the scapulars and tertials are vinaceous rufous, with their centre portions white, and a bar of the same extends across the outer webs of the secondaries in the same position as the rufous edgings in the adult; lateral tail-feathers paler than the rest and tipped with white; throat a brownish or paler black than the head; the white of the chest narrower than in the adult; the pectoral band a paler and handsomer rufous, variable in width, and only continued in bars on the breast-plumes to a very limited extent; the abdomen and underlying breast-feathers with pale edgings; under surface of tail wanting the black patch.

---

* An example in the British Museum from Nepaul, which has a wing of 9·4 and is not sexed, may be a female; a Ceylonese male, however, measures 9·3.

With age the back becomes blacker and more glossy, and the rufous colouring of the scapulars and tertials gradually gives place to the nigrescent adult hue: the white patch on the outer webs of the secondaries becomes rufous at the margins, and then black near the shafts, till in the old bird it finally disappears altogether.

Obs. The immature plumage of this bird appears not to have been hitherto described. In looking over the specimens in the National Collection, I came upon the example treated of above, which is undoubtedly in yearling plumage. The absence of the posterior tooth, the undeveloped crest, the pale edgings of the abdominal feathers, and the appearance of the under tail-coverts unmistakably indicate its immaturity, and have furnished a key by which at last the gradations in the plumage of this interesting species may be understood. The existence of this specimen precludes the possibility of the bird shot by Col. Tickell (J. A. S. B. 1833, p. 569) being the young of this *Baza*. This example was 15 inches in length, had a "fine long occipital crest black with white tips ; the head, nape, and wing-coverts clouded with ashy and rusty ; back clouded with brown ; lower parts white, with a streak of black down the centre of the throat, and with rusty bars on the breast and belly." This bird cannot be referable to *B. lophotes*; but it may be *Sp. albonigra* or another species of the genus *Baza* (*B. jerdoni* ?).

Distribution.—This beautiful Falcon is one of our rarest raptorial birds, and is, as far as observation has hitherto tended to prove, a cool-season migrant to Ceylon ; and the fact of its having been observed to be migratory to Burmah and the east coast of India is, I think, for the most part, confirmatory of this belief. During its visits to the island it appears to confine itself mostly to the low country, and to be most partial to the northern half of the island. It was first recorded from Ceylon by Edgar Layard, who obtained a specimen near Jaffna, and who speaks in his "Notes" of another having been procured by Mr. Mitford, of Ratnapura. Subsequent observers do not seem to have met with it until Mr. Bligh obtained another, which was caught near Lemastota. In January 1876 I came suddenly upon a little troop of five in close company, and out of them secured an immature male. In the following October I saw another example near Ambepussa ; and in January last year (1877), through the kindness of Mr. Chas. Byrde, of the Ceylon Civil Service, I received a second specimen, shot at Pasyala, in the Western Province. Mr. Simpson, of the Indian Telegraph Department, who has spent much of his time in the northern forests, and who is an accurate observer of birds, informed me that he had seen this Falcon at Kanthelai tank. Mr. Holdsworth mentions having seen specimens from the Kandy district which, with the exception of the evidence afforded by the Lemastota specimen, is the only record we have of its occurrence in the hill-region.

This species has a limited geographical distribution. As far as can be judged, it has its head-quarters in Assam and Burmah, and migrates thence down the east coast of India to Ceylon. Jerdon procured one specimen on the east coast near Nellore ; and he remarks that it is occasionally killed at Calcutta, and is spread very sparingly throughout India. Of late years, however, it has not been recorded from the Deccan, North-west Provinces, Chota Nagpur, nor any of the western districts, the ornithology of all which regions has been so fully worked out in 'Stray Feathers'; neither has it been recorded from the Travancore, Palani, nor Nilghiri forests. It can only therefore locate itself in few places (and those far between) when it makes its annual visits to the Peninsula. The strangest feature in its distribution is, that it is likewise nothing more than a migrant to Burmah and Tenasserim. In the latter district Mr. Davison found it not uncommon in December and January throughout the southern parts of it ; but no mention is made of its occurrence at other seasons, so that it is undoubtedly non-resident on the eastern side of the Bay of Bengal. There are specimens from Malacca and Pinang in the British Museum ; but it has not been met with in the Andamans or Nicobars. Neither Mr. Oates nor Capt. Feilden appear to have found it in Upper Pegu ; but in North-eastern Cachar, which lies to the north of it, Mr. Inglis found it consorting together, in November, in the same sociable manner that I did in the northern forests of Ceylon. Where, then, is its home throughout the greater part of the year ? Where are those birds bred which mysteriously visit the above-mentioned regions for so short a time and again vanish as suddenly as they appeared ? The northern portion of Burmah, together with the immense Chinese provinces of Yunan, Sechuen, and Quei Chou, which lie to the north and north-east of the Burmese kingdom, are traversed here and there by extensive mountain-systems, such as the Palkoi, "Snowy," and other ranges—a vast and little-known ornithological district extending over 12° of latitude, all of which forms a territory sufficiently large to furnish a home for a bird of far less local disposition than a *Baza*. It is pretty certain that this species does not inhabit the more eastern parts of the Celestial Empire, for Père David makes no mention of its occurrence there or in the Moupin mountains in his new work on the Birds of China.

Habits.—This "Baza" frequents forest or large tracts of jungle, and usually keeps to districts of no considerable altitude. It appears to be more gregarious than most Hawks; for with the exception of the Kestrels and Kites, none seem to be so fond of each other's company. The little troop that I met with more resembled Pigeons in their actions than birds of the hawk-tribe; three were seated among the branches of one tree, and two others flew from branch to branch close by; when I approached the whole made off with short flight, from tree to tree, during which movement I dropped my bird. They had a quick irregular mode of flying, and with their white chests and handsome wings, contrasted against the green foliage, had a very unhawk-like appearance. I notice, with regard to their sociability, that Mr. Inglis, in the "First List of Birds from Cachar" ('Stray Feathers,' vol. v.), speaks of finding three in company with Bulbuls and King-Crows. Jerdon remarks that it is entirely insectivorous in its diet; and a pair that Mr. Mitford met with near Ratnapura, referred to by Layard in his notes, were catching bees on the wing, and also by darting at them as they issued from their hive; they sat on the dead branches of a tree, and raised and depressed their crests, and this they have the power of doing vertically, like the Crested Swift (*Dendrochelidon coronata*). Layard's specimen had a Lizard (*Calotes viridis*) in its stomach; and one of my birds, which was shot by Mr. Chas. Byrde, sitting in a jack-tree near the Rest House at Pasyala, had been feeding on Coleoptera. I know nothing of its note, nor can I find any thing recorded concerning it.

Jerdon writes of it in the 'Birds of India':—"It is almost entirely insectivorous in its habits, and keeps to the forests or well-wooded districts. It takes only short flights, and certainly is not usually seen soaring high in the air, as Mr. Gray says in his 'Genera of Birds.'"

Comparatively little is known concerning any of the Malayan members of this interesting genus, conspicuous in which, for its singular and beautiful plumage, is the present species. It is therefore to be hoped that naturalists in India and Ceylon will, when they have the good fortune to come upon it in their wanderings, pay particular attention to its actions and habits, as far as their opportunity will permit of.

FALCONIDÆ.

FALCONINÆ.

Genus FALCO.

Bill very stout and strong, short, the tip well hooked, and its margin indented with a deep notch or tooth; culmen curved gently from the base of the cere; cere well advanced. Nostrils circular, exposed, and with a tubercle. Wings long, much pointed, reaching in some to the tip of the tail; the 2nd quill the longest, the 1st subequal with the 3rd, and notched near the tip on the inner web; secondaries falling short of the primaries by more than half the length of the tail. Tail moderately short, stiff, and somewhat cuneate at the tip. Tarsus shorter than the middle toe, plumed somewhat below the knee, covered in front with small hexagonal scales. Toes very strong; middle toe much longer than the outer, which exceeds the inner; the whole shielded with narrow transverse scales nearly to the base. Claws much curved and acute.

FALCO PEREGRINUS.

(THE COMMON PEREGRINE.)

Falco peregrinus, Tunstall, Ornith. Brit. p. 1 (1771); Gm. S. N. i. p. 272 (1788); Gould, B. of Eur. pl. 21 (1837); Blyth, Cat. B. Mus. A. S. B. p. 13. no. 63 (1849); Kelaart, Prodromus, Cat. p. 115 (1852); Layard, Ann. & Mag. Nat. Hist. 1853, xii. p. 101; Horsf. & Moore, Cat. B. Mus. E. I. Co. p. 16. no. 18 (1854); Jerd. B. of Ind. i. p. 21 (1862); Gould, B. of Gt. Britain, pt. 1 (1862); Blyth, Ibis, 1866, p. 234; Hume, Rough Notes, i. p. 49 (1869); Jerd. Ibis, 1871, p. 237; Delmé Radcliffe, ibid. p. 363; Swinhoe, P. Z. S. 1871, p. 340; Holdsworth, P. Z. S. 1872, p. 410; Hume, Str. Feath. 1873, p. 367, et 1874, p. 140; Swinhoe, Ibis, 1874, p. 427; Legge, Str. Feath. 1875, p. 360; Hume, ibid. p. 443; Scully, Str. Feath. 1876, p. 117; Hume, ibid. p. 461; Dresser, B. Eur. pts. 47, 48 (1876).

Falco communis, Gm. S. N. i. p. 270 (1788, *ex* Buff.); Sch. Vog. Nederl. p. 6, pls. 1–3 (1854); Sundev. Sv. Fogl. p. 206, pl. 26. fig. 2 (1867); Sharpe, Ann. & Mag. Nat. H. 1873, xi. p. 222, et Cat. B. i. p. 376 (1874); David and Oustalet, Ois. de la Chine, p. 32 (1877).

Falco calidus, Lath. Ind. Orn. i. p. 41 (1790).

Falco lunulatus, Daud. Traité, ii. p. 127 (1800, *ex* Lath.).

Falco anatum, Bp. Comp. List B. Eur. & N. Am. p. 4 (1838, *ex* Audubon); Scl. et Salv. Ibis, 1869, p. 219.

Falco micrurus, Hodgs. in Gray's Zool. Misc. p. 81 (1844).

Le Faucon, Briss. Orn. i. p. 321 (1760).

Le Faucon pèlerin, Briss. Orn. i. p. 341 (1760).

Le Faucon sors, Buff. Pl. Enl. i. pl. 470 (1770).

Oriental Hawk, Behree Falcon, Latham, Gen. Syn. Suppl. p. 34* (1787).

"*Falcon*" (female), "*Tiercel*" (male), in Falconry; "*Duck-Hawk*" in America.
Bhyri (female), *Bhyri bacha* (male), Hind.; *Bhyri Dega*, Tel.; *Dega*, Yerklees (*apud* Jerdon); *Bahri* or *Water-haunting Bird*, Turkestan (*apud* Scully); *Basi*, Persia (*apud* Pallas); *Raja wali*, Malay; *Sikap lang*, Sumatra (*apud* Raffles); *Laki Angin* of the Passmumahs; *Halcón*, Spain.
Ukussa, Sinhalese.

Adult male. Length to front of cere 15·2 to 16·0 inches; culmen from cere 0·8; wing 12·6 to 12·8; tail 6·5; tarsus 1·9 to 2·05; middle toe 1·85 to 1·9, claw (straight) 0·65 to 0·7.

Adult female. Length to front of cere 17·5 to 18·5 inches; culmen from cere 1·05 to 1·2; wing 14·0 to 14·6; tail 7·3 to 8·5; tarsus 2·1 to 2·2; middle toe 2·1 to 2·3, claw (straight) 0·75; height of bill at cere 0·45 to 0·48. Weight of a female (wing 14·5) killed at Trincomalie 2 lb. 4 oz.

Iris dark hazel-brown; eyelid and cere above nostril rich yellow, greenish near the gape; bill pale blue at the cere and yellowish at the base beneath, darkening to blackish at the tips; legs and feet yellow.

Above bluish ashen, darkening into blackish or blackish brown on the head and hind neck, and paling into bluish grey on the rump and upper tail-coverts, all the feathers with dark shafts conspicuous on the back and scapulars, and banded with narrow, softened, wavy bars of cinereous blackish from the hind neck downwards; on the rump and upper tail-coverts these markings take a spear-shaped form; bases and sides of the feathers, in many examples, on the hind neck rufescent; lesser wing-coverts edged pale; quills dark brown, pervaded with ashy on the outer webs; the tips finely edged with greyish, the inner webs barred with rufescent grey or greyish white; tail dusky ashen, palest at the base, and crossed with narrow wavy bands of blackish, and tipped deeply with buff-white.

Forehead usually whitish close to the cere; lores, cheeks, and a short broad moustachial streak black; chin, throat, fore neck, and all beneath with the under wing-coverts white, tinged on the upper breast with faint isabelline grey, and often on the lower parts with bluish grey; the throat and fore neck unmarked, the chest streaked with narrow shaft-stripes of brown, which change gradually on the upper breast into the narrow wavy bars of blackish brown of the whole under surface and thighs; under wing-coverts with broader bars of the same.

Obs. It is the opinion of many naturalists, and among them Mr. J. Hancock, who has made the Falcons a life-long study, that the Peregrine, as well as other members of the genus, acquires its adult plumage at the first moult [*]. From observations I have made of a number of specimens in the barred plumage, but showing here and there a thorough immature feather, it seems evident to me that the change does take place in the second year. Notwithstanding this, however, it is equally evident that modifications take place in the adult plumage as the bird grows older; the streaks on the chest become finer and less numerous, and the change to the bars just beneath is more sudden than in the two-year old. Considerable variety exists in the depth of hue of the upper surface in birds from different parts of the world, and some examples are very rufous beneath; an instance of this coloration is afforded in the bird now in the Zoological Gardens, captured off Yucatan, which is almost as rufous as the Indian Peregrine. Asiatic-bred birds shot in India seem to be, as a rule, very heavily streaked on the chest.

"*Young male on leaving the nest* (Sharpe, Cat. B. i. p. 378). Brown, all the feathers edged with rufous, a clear greyish shade pervading the upper surface, and particularly distinct on the secondaries; head and neck rusty buff, the sides of the crown and occiput, the nape and hind neck, the feathers behind the eye, and the moustachial line mottled with blackish; under surface of the body rusty buff, with longitudinal median spots of dark brown, fewer on the thighs, and changing into bars on the under wing- and tail-coverts; throat paler and unspotted."

The bird of the year attains almost its full size before the first moult, and has the cere and bill much as in the adult; legs and feet greenish yellow.

When fully acquired the plumage is as follows:—Head, back, and wings dark brown, paling, in some, on the rump into light umber-brown, in others into cinereous brown, the feathers more or less edged with rufescent brown,

---

[*] Mr. Hancock argues from the testimony of caged specimens in his possession, which have invariably acquired the barred plumage at the first moult. Now all who have kept Raptors in confinement know that they are *slower* in acquiring their adult plumage, owing to loss of vigour, than when in a wild state; if, therefore, the Peregrine makes the sudden change in captivity, how much more must it do so in a state of nature.

paling on the scapulars into fulvous ; front of the crown and the forehead whitish or fulvous, with the centres of the feathers blackish ; sides of the hind-neck feathers marked with the same ; shafts of the scapulars and upper tail-coverts black, and the tips of the latter part deeper than elsewhere ; quills brownish black, barred on the inner webs with rufous-grey ; tail cinereous brown, crossed with incomplete bars of rufous- or fulvous-grey and tipped deeply with whitish.

Cheeks and moustachial streak blackish brown, the white portion of the ear-coverts streaked with the same ; chin, throat, and entire under surface white, in some slightly tinged with rufescent on the lower parts, and boldly streaked from the chest downwards with umber-brown ; the markings are usually broader on the flanks, and in very many examples, even at this age, have a bar-like form ; on the under wing-coverts the brown predominates, the white markings being confined to the tips.

Distribution.—The Peregrine was first recorded from Ceylon by Layard, who gives an account (*loc. cit.*) of shooting three specimens at Pt. Pedro in the month of January. Doubt has been thrown by the late Dr. Jerdon and others on Layard's identification, chiefly on account of the latter's statement that he found them nesting ; but I have carefully examined the two specimens that still exist in the Poole collection—an adult and an immature bird ; and there they are, veritable Peregrines, in spite of their having been found breeding in so strange a latitude as Ceylon. It appears to confine itself principally to the sea-coast during its visit to Ceylon, which is of course during the north-east monsoon. During the latter part of 1872 a pair frequented the Fort-Frederick cliffs at Trincomalie ; but, fortunately for themselves, eluded several attempts I made to procure them ; they tenaciously kept to one place on the face of the great "Sämi" rock, where they commanded any approach to their haunt either by land or sea. In February 1874 Mr. R. Pole, of the Ceylon Civil Service, shot a fine female at Puttalam, which is now in the British Museum, and was the first procured since Layard's time, as far as I am aware. In October of the following year I failed in killing one which frequented the dead trees in the bed of the newly-restored tank at Devilane ; but on the 28th of the same month I succeeded in shooting a female on the cliffs at Fort Frederick. During the cool season of 1876-77 another example, also a female, judging by its size, was observed by myself on two occasions in the cinnamon-gardens near Colombo ; and in December of the same season I met with and wounded a second at the top of Allegalla Peak.

Beyond this latter locality, I do not know of any place in the mountain-zone in which it has been observed. This fine hill, which is one of the bulwarks of the mountain-range of Ceylon, rises 3400 feet sheer out of the low country, and consequently furnishes the present species with a seasonal shelter and the next with a permanent home.

The Peregrine is a cold-weather visitant to the peninsula of India, the Laccadive and the Andaman Islands ; but a good many birds, probably young, remain behind in India, and take up their quarters on the borders of extensive jheels and tanks, attracted by the quantity of wildfowl and waders, which form their chief sustenance. It arrives, says Jerdon, in India, about the first week in October, and departs again in April, and during its visit is less abundant on the west coast than on the east. It is common in Burmah, and finds its way, according to Mr. Hume's observations, to the Andamans *viâ* Cape Negrais. Professor Schlegel records it from Sumatra ; and on the east coast of China Mr. Swinhoe says that it is a permanent resident. Père David, however, remarks that it is driven by the Saker out of the south of China. It is not uncommon in Japan. It is spread throughout Central Asia, extending northwards into Siberia, and, according to Dr. Scully, remains about Yarkand even in the winter. Canon Tristram found it all times of the year in suitable localities on the coast, but to the eastward of the watershed of Central Palestine he never observed it.

It is distributed throughout the continent of Europe to the extreme north, and it occurs likewise in the islands of the Mediterranean. It is found chiefly on the coast-line of Northern Africa, being, however, not very abundant in Egypt, though it is, according to Mr. T. Drake, numerous in Tangiers and Eastern Morocco ; southward it extends its range to Natal and the Cape. From the Canary Islands MM. Berthelot and Bolle record it ; but it does not seem to have been noticed in Madeira. In the New World it enjoys a very wide range ; commencing in Greenland it extends down the east coast to South America, and spreads across the continent to Vancouver Island, and thence along the entire Pacific coast of the continent to Peru, being replaced in Chili and to the south of that country still by *Falco cassini*, a species somewhat akin to the Australian Peregrine, *F. melanogenys*. It is not my province to go so minutely into its distribution as to record those localities from which it is absent ; but from the above sketch of its habitat it will be seen that the Peregrine has one of the widest ranges of the birds of prey, rivalling even the Osprey in its wandering

Habits.—This noble Falcon is perhaps too well known to need much comment on its habits. Bold, swift on the wing, and keen-sighted to a degree, as well as extremely docile in confinement, the female has long been celebrated for its employment in the ancient and royal pastime of Falconry; and although this sport has declined much in Europe during the last century, it is still practised to a certain extent both on the continent and in England, the birds used with us being brought over principally from Holland, where they are netted. In India it has always, in common with the next species, been prized by the natives for Falconry, and is still trained there for that purpose; but used to be so, according to Jerdon, much more than now. He writes, in the 'Birds of India,' "It is trained to catch Egrets, Herons, Storks, Cranes, the *Anastomus, Ibis papillosa, Tantalus leucocephalus*, &c. It has been known, though very rarely, to strike the Bustard. Native falconers do not train it to hunt in couples, as is done in Europe sometimes. I may here mention that the idea of the Heron ever transfixing the Hawk with its bill is scouted by all native falconers, many of whom have had much greater experience than any Europeans. After the prey is brought to the ground, indeed, the Falcon is sometimes in danger of a blow from the powerful bill of the Heron, unless she lays hold of its neck with one foot, which an old bird always does. When the Kulung (*Grus virgo*) is the quarry, the *Bhyri* keeps well on its back to avoid a blow from the sharp, curved inner claw of the Crane, which can, and sometimes does, inflict a severe wound."

Jerdon comments on the curious mistake that artists, even Landseer included, have made in depicting the Peregrine as striking with its bill! This erroneous idea, however, is not confined to artists, for I have more than once seen it in the writings of naturalists. No raptorial bird that I have ever heard of uses its bill either for defence or offence; this organ is constructed for, and only used in, tearing the food on which the bird subsists. The talons alone are used in striking the quarry and in fighting or defending itself against attacks from any source whatever. I have kept half a dozen species of diurnal birds of prey, and have often had occasion to catch them by hand; but have never known one to use its bill when caught further than in giving a very incipient sort of peck. It is well known what a tremendous wound the Peregrine inflicts with the hind claw when striking its quarry; and in America, where it is called the "Duck-Hawk," on account of its partiality for ducks, these birds have been found with the whole back ripped up by the stroke of the Hawk's sharp talon, combined with the great momentum of its downward swoop. Peregrines have their favourite localities in India and Ceylon, which they tenaciously keep to throughout the season; they usually take up their quarters near water, and are very partial to sea-coast cliffs, which afford them a tolerably secure refuge. The birds that almost annually frequent the rocks at Trincomalie feed on the Pigeons frequenting the islets lying off the coast some 12 miles to the north. I observed them flying home at usually about nine or ten o'clock, when they would shelter themselves during the heat of the day, and sally out again in the afternoon. The favourite food of the Peregrine in India consists of waterfowl and waders, the latter being chiefly preyed upon by those birds which frequent the sea-coast. Mr. Adam writes that at the Sambhur Lake "they sit on stakes which are required to form a low retaining wall to separate a portion of the lake-water for the formation of salt, and from these perches they pounce on the numerous waders which feed along this wall." It is well known to what an extent Coleoptera are preyed upon in the East; and Mr. Pole assures me that the specimen he shot at Puttalam was flying round his compound at dusk, and appeared to be darting at the large beetles which were swarming in the air at that hour.

The ordinary flight of the Peregrine is regular and straight on end, being performed, as in other Falcons, with a quick wing-stroke; it is moderately swift, but nothing out of the common; when, however, it is in pursuit of a quick-flying quarry, such as a pigeon, duck, or limicoline bird, its wonderful powers of progression are fully brought out, and in making its final dash on the doomed victim its speed for the moment is estimated at 160 miles an hour.

Nidification.—As the Bhyri is not known to breed in India, the fact of its having been found nesting at Pt. Pedro by Layard has been a matter of dispute. As mentioned above, I have identified Layard's birds, and they are not the Jugger (*F. jugger*), as has been suggested; and consequently the interesting fact remains that the species (probably quite an abnormal occurrence in tropical latitudes) has bred in Ceylon. He writes as follows :—"I found them breeding in a palmyra tope on the left-hand side of the road from Jaffna to Pt. Pedro; the nest a rough structure of sticks laid on the dead '*matties*' or fronds of the palmyra, from which the leafy parts had been cut away. . . . . I shot the first specimen early in the month (January); but the female was so shy that, though I long remained concealed near the nest, she never afforded me a shot, and I was obliged to return home without her. I was surprised to find another male at the same nest when I revisited the spot at the end of the month, and procured both him and his mate with a double shot."

Schlegel affirms that the Peregrine has bred in Sumatra; and Swinhoe found it nesting on the cliffs of North rock, in the province of Shantung, North China, and remarks that it appears to be a resident species down the whole length of the Chinese coast, young birds in their down having been brought to him at Amoy. No further testimony beyond that of these three writers is forthcoming of its breeding in the south-east of Asia or in the Indian empire southward of the Himalayas. Dr. Adams is supposed to have found its nest on the banks of the Indus; but the occurrence is mentioned with doubt, as to the correct identification of the bird, by both Jerdon and Hume; and the latter does not include it in his list in ' Nests and Eggs.' In more northerly latitudes it usually chooses an inaccessible cliff on which to build and rear its young. There, on some ledge which it deems secure from the attack of man, it constructs a nest of sticks, often mingled with the bones of its quarry, which, collecting year after year, have at last become part and parcel of the structure. The eggs are either three or four in number, and vary both in size and markings, these characters depending on the age of the bird. In Mr. Hewitson's plate (vol. i. of his ' British Birds' Eggs ') are two examples: the first laid by an old bird, and measuring 2·13 by 1·7 inch; the second by a younger bird, not exceeding 1·92 by 1·55. In the larger of the two the general colour is reddish white, closely freckled, except at the small end, with brick-red, and blotched openly over that with reddish brown, the markings on the smaller half being the largest. The second egg is not so decided in its markings, is of a paler ground, covered with a stippled wash of pale reddish, in which there are a few darker clouds and several openly distributed large blotches round the centre.

FALCO PEREGRINATOR.

(THE INDIAN PEREGRINE.)

Falco peregrinator, Sund. Phys. Tidssk. Lund, 1837, p. 177, pl. 4; Gray, Gen. Birds, i. p. 19 (1844); Blyth, Cat. B. Mus. A. S. B. p. 14 (1849); Layard, Ann. & Mag. Nat. Hist. 1853, xii. p. 102; Gould, B. of Asia, pt. 3 (1851); Blyth, J. A. S. B. xix. p. 321 (1851); Horsf. & Moore, Cat. B. Mus. E. I. Co. p. 18. no. 20 (1854); Jerd. B. of Ind. i. p. 25 (1862); Hume, Rough Notes, i. p. 55 (1869); Jerd. Ibis, 1870, p. 237; Holdsworth, P. Z. S. 1872, p. 410; Sharpe, Ann. & Mag. Nat. Hist. ser. 4, xi. p. 223 (1873); id. Cat. B. i. p. 382 (1874); Legge, Str. Feath. 1875, p. 195; Hume, Nests and Eggs, p. 23 (1874); Walden, on Col. Tickell's MS. Ill. Ind. Orn., Ibis (1876), p. 342.

Falco shaheen, Jerd. Madr. Journ. x. p. 81 (1839); id. Ill. Ind. Orn. pls. 12 & 28 (1847).

Falco sultaneus, Hodgs. in Gray's Zool. Misc. p. 81 (1844).

Falco ruber, Schl. Mus. P.-B. Falc. p. 5 (1862).

The Shahin Falcon, Jerdon, B. of India; *Royal Falcon* of some.

Shahin, "Royal bird" (female), *Kohee Koela* (male), Hind.; *Jawolum*, Tel.; *Wallur*, Tam. *Ukussa*, Sinhalese.

Adult male (from Ceylonese and Indian examples). Length to front of cere 13·0 to 14·2 inches; culmen from cere 0·9 to 1·0; wing 11·4 to 11·6; expanse about 34·0; tail 6·0 to 6·4; tarsus 2·0; middle toe 2·1; claw (straight) 0·7; hind toe 0·85, claw (straight) 0·9; height of bill at cere 0·45.

Female. Length to front of cere 15·0 inches; culmen from cere 1·1; wing 12·0 to 13·3, expanse of the latter 38·2.

A male from Ceylon measured 11·6, and a female 12·8 in the wing.

Iris dark umber-brown; cere, eyelid, and gape ochre-yellow; bill dark plumbeous, changing to greenish near the cere; legs and feet chrome-yellow, claws black.

Head, hind neck, and upper back ashy blackish, deepest on the sides of the neck and paling gradually into bluish ashy on the rump and upper tail-coverts, the latter part being the lightest; all the feathers with dark shafts, the scapulars and wing-coverts edged with pale ashy and the lower back and tail-coverts crossed on the centre of the feathers with dark wavy bars, often concealed by the tips of the overlying feathers; lesser coverts darker than the median; quills blackish brown, the shorter primaries slightly pervaded with grey, and the whole narrowly barred on the inner webs with fulvous or light rufous-grey, according to the age of the bird; the secondaries paler than the primaries, and tipped with dull whitish; tail ashy blackish, tipped with rufescent and barred chiefly at the base with softened slaty markings; edge of the forehead buff with dark shafts.

Cheeks and moustachial stripe black, blending into the paler hue of the head; chin and throat rufescent white, passing on the chest into pale rufous, and from that into the rich rufous of the breast, flanks, and lower parts; shafts of the chest-feathers darker rufous than the web; flanks and under tail-coverts crossed on the centre of the feather with narrow lines of blackish; under wing-coverts dark rufescent, with darker shafts and cinereous black barrings; greater row brownish, barred with rufescent.

Obs. The rufous of the under surface is variable in depth, notwithstanding that the bird may be fully adult. Ceylonese examples in my collection correspond well with Indian, old birds, devoid of any barring on the breast, being scarcely less dark on the head and hind neck than the blackish-headed Nepaul birds (*Falco atriceps*, Hume).

Some examples in the British Museum from Northern India present puzzling characteristics. There is one from Simla, presented by Capt. Pinwill, which has the appearance of a rather small Common Peregrine with a very rufous under surface. The feathers of the back and rump and the scapulars are as much barred as in *F. peregrinus*; the chest is marked with fine mesial points like that species; the breast and lower parts are rufous-grey, and barred with narrow cross rays of blackish brown as in an old Peregrine, with the exception that the markings are closer together; the flanks and under tail-coverts are likewise tinged with bluish grey.

Young. Wing of a male 10·6 inches. Core yellowish, tinged with green, in some entirely bluish; legs and feet greenish yellow.

Above brownish black, the feathers of the back and wing-coverts with fine pale margins, the scapulars tipped with rufous and some of the concealed portions of the feathers barred with the same; rump edged with rufous, upper tail-coverts tipped and barred with a paler hue; quills deep brown, the bars of the inner webs more rufous than in the adult; tail barred obscurely with rufous, which on the central feathers is of a dusky hue.

Cheeks and moustachial stripe blackish brown; throat and chest white, passing into rufescent buff on the breast and flanks; the chest and the white space above the moustache streaked with shaft-lines of brown, expanding at the tip; breast streaked broadly with brown, the lower flank-feathers deeply tipped and marked with bar-like spots of the same: the abdomen, under tail-coverts, and thighs are paler than the breast, the former streaked similarly to the chest and the thighs more boldly marked, some of the longer feathers having bar-like spots; under tail-coverts barred with brown; under wing-coverts whitish, with irregular cross-markings of brown.

At the first moult the following change takes place:—the rump and the base of the tail assume a cinereous hue, the edgings of the scapulars are less conspicuous, the bars of the primary inner webs become paler and the shaft-stripes on the chest narrower, the breast and flanks darker rufous, this hue extending to the belly and thighs, and the stripes on the flanks turn into bars.

The back and rump from this stage onwards begin to turn grey, the shafts of these parts and of the scapulars standing out darkly; the stripes on the centre of the breast disappear altogether in some examples, leaving the flanks barred to a greater or less extent.

Distribution.—This bold and handsome Falcon was recorded by Layard (*loc. cit.*) as having been shot by his collector and servant near the beautiful upland plain of Gillymally. The account of the specimen in question referred chiefly to its long wings causing the native "Muttoo" to think that it was a "large Swift," deceiving Layard also, who says of the bird, "which I also mistook for a Swift, so much did its wings overlap its tail." I have carefully examined the whole collection at Poole, and there is not in it any example of *F. peregrinator*; but there is one of a female *Falco severus*, a bird not recorded by Layard in his list. I am therefore of opinion that he did not correctly identify the bird shot on the occasion in question, but that it was in reality a specimen of the Indian Hobby, to which his remarks as to length of wing &c. would relate with correctness. I have written to him on the subject; and in his last letter to me from New Caledonia he says that he has no doubt the bird was the latter species. Should this seem be correct it is difficult to say when the bird was first discovered in Ceylon; but I imagine that my reference in 'Stray Feathers,' 1875, to the Pigeon-Island specimens is the first actual record of the bird's occurrence in the island. It is resident in Ceylon, but by no means common, and frequents such very retired spots or inaccessible cliffs that it is rarely met with by the ordinary sportsman. A pair usually affected the cliffs at Fort Frederick during the cool season, dividing their time between foraging on the mainland and making inroads upon the Rock-Pigeons which swarmed at the island beyond Nilävele. At this spot a pair out of three or four birds which had taken up their abode on the northern face were killed by myself and a brother officer in October 1874. This island is an out-of-the-way locality, which, stocked as it is with fine pigeons, forms a welcome refuge for the Shahin. As it has so seldom been shot in Ceylon, I quote here the following passage from my notes in 'Stray Feathers':— "The islet is situated 1½ miles north of Trincomalie at about 1½ mile from the mainland. Near this place, about ½ a mile nearer the shore, is another rocky islet frequented by flocks of *Columba intermedia*, which furnish many a dainty meal for the Royal Falcon. Pigeon Island itself is rarely visited except by fishermen, who can only land at the south side, where there is a little beach backed by a tangled thicket, which rises gradually to the pinnacle in the centre, whence the northern side descends in the form of a perpendicular face right into the sea. This cliff, under which it is very difficult to pass on foot, forms a splendid shelter for the Shahin; for he can perch and roost on the shelves which jut out into the numerous crevices in the face of the rock without being disturbed by any one in the island who does not choose to scramble along the almost inaccessible rocks at its foot. I visited the spot on the 6th October 1874, in search of pigeons, and finding none, was clambering over the rocks on an adjoining islet, separated at high water from the main portion, when I espied a large Falcon coming along over the water and making for the cliff. I quickly turned back, reached the cliff, and got out on to an enormous boulder which enfiladed the precipice, affording a good view of the whole of it, but not a vestige of the Falcon was to be seen. I then determined to get right underneath, and jumped across a chasm to a lower boulder, from which I could see almost every spot in the precipice ; but still no falcon. I then shouted, and out shot three splendid fellows, which I missed with my first barrel ; but back they came, dashing up to the rock, and not caring the least for my shot, when bang went the weapon, and down came a fine fellow between two large rocks, where I judged him to be safe, and then fired several shots at impossible distances at the other two, which wheeled and dashed round the summit of the hill in such a manner that I thought they must be breeding. After a while the third bird made off, the second disappearing suddenly from the battle-field. Thinking it was about time to pick up my dead bird, I made my way across and through the water to the spot where I had dropped him, and to my extreme disgust found that he had fallen into a sluice, out of which the first receding wave must have carried him. Not a sign of my prize anywhere ; high and low I searched, and at last gave up in despair, convinced that a monstrous blue rock-fish, with which the water beneath the cliff swarmed, had long since polished him off ! On returning to the other side of the island, where my companion was hungrily waiting breakfast, the first sight that greeted me was a magnificent winged *Shahin* hanging by his knotted primaries to the branch of a tree. My companion (Major Sir John Campbell) had dropped him as he shot past ; and hence his sudden disappearance from my side of the island." Elsewhere in the lower country I have met with the Shahin in the Friars-Hood district, and at Yakkahatua mountain near Avisawella ; and Captain Wade, 57th Regt., shot a fine adult specimen at Tissa-Wewa Tank, near Anarad-japura, in December 1875. In the hill-zone it is more often seen, and no doubt breeds in the mountains. I killed an old male at the top of the celebrated Yakka rock, Hewahette, in May 1876, and in the following month Mr. Bligh procured another in Haputale. During the same season a young bird, which I saw after-wards alive in the possession of Messrs. Whyte and Co., was caught in the act of dashing at some pigeons near Kandy. I have seen it on the Alagalla Peak, in the precipices of which I have reason to believe it nests.

This Falcon was first described by Sundevall from a specimen which settled on the vessel he was sailing in, " in lat. 6° 20′ N., between Ceylon and Sumatra, rather nearer the last-named island, and at least 70 Swedish miles from the nearest land, viz. the Nicobar Islands." From what follows in the Professor's remarks on this occurrence, he was of opinion that it was either flying to or from Sumatra. It has not, however, been discri-minated from that island ; and it is more probable that the specimen in question was on its way to or from the Nicobar Islands, but where also it has not been found up to the present time. It is said by Jerdon to be found " throughout the whole of India, from the Himalayas to the extreme south, extending into Affghanistan and Western Asia." As regards the two latter regions I imagine that it has been here confounded with another species, as the bird does not appear to extend beyond the confines of the Indian empire, and the northern race, inhabiting even the Himalayas, is separated as *F. atriceps* by Mr. Hume. I have, however, examined individuals in the National Collection from Nepaul, and they are not separable from Ceylonese specimens. It is more often found in Central than Upper India, and is more frequent still in the South, inhabiting the Nilghiris and breeding there. In the Carnatic it is seldom met with ; but in the Eastern Ghauts it is tolerably common, according to Jerdon, breeding there, and migrating in the young stage to the former locality. As this writer has stated, it is no doubt far from being a common bird, confining itself to forest-clad districts. I observe that it is not mentioned in Mr. Fairbank's list of birds from the Palani Hills, nor in Mr. Bourdillon's from Travancore, although Jerdon shot it in the latter district. Col. Tickell states that it is a commoner species in Burmah than in India, and that he frequently observed it on the sea-side at Amherst. It must be local, however, in Burmah, as I do not find it recorded thence by any of the naturalists whose work has been described of late years in ' Stray Feathers.' With regard to the specimens of this Falcon said to have been procured at sea in the Indian Ocean, I have to remark that the bird mentioned by Mr. Whyte (Ibis, 1877, p. 149) as being captured in the Gulf of Socotra, and belonging to the present species, has eventually proved to be a Common Peregrine ; and I am strongly of opinion that the source to which the presence of another (mentioned in a footnote, ' Stray Feathers,' 1877, p. 502, as being procured in 1833 on board ship between Mauritius and Madagascar) might be traced is that which has led to many mistakes in "distribu-tion," viz. an escape from a state of confinement.

Habits.—In Ceylon the Indian Peregrine frequents lofty mountain-precipices or inaccessible cliffs on the sea-coast. It is an excessively shy bird, retiring when not engaged in the pursuit of its quarry to sequestered ledges, and easily escapes all notice, unless observed to fly towards its retreat. It is as bold and courageous in the hunt as its larger and more esteemed congener; but of course is not so powerful in its attack on large birds. It is taught to catch partridges, florikin, and jungle-fowl by native falconers in India, and is usually caught by the ordinary contrivance of bird-lime, with which it comes in contact on stooping at a decoy-bird. Jerdon, who narrates, in his work on the 'Birds of India,' that it is trained for what is called "a standing gait," or the art of hovering or circling in the air over the falconer and his party, says that "it is indeed a beautiful sight to see this fine bird stoop on a partridge or florikin which has been flushed at some considerable distance from it, as it often makes a wide circuit round the party. As soon as the Falcon observes the game which has been flushed, it makes two or three onward plunges in its direction, and then darts down obliquely with half-closed wings on the devoted quarry with more than the velocity of an arrow." I can testify to the accuracy of this account of the Shahin's powers of flight, as I was once myself an eye-witness to its capturing a Palm-Swift at Trincomalie. A little colony of these birds had their nests in a solitary palmyra-palm which grew near the sea-beach; and one evening I observed one of these Falcons, which had been haunting the cliffs of the Fort, dash past me, and, mounting higher and higher, go away at a tremendous pace, and with a twisting flight, for about 300 yards. I could not see at the moment what he was pursuing, as it was getting dusk; but he suddenly checked himself and shot down with meteoric swiftness almost into the sea. I then perceived a poor little Swift just in front of him; close to the surface of the water it dashed along in a horizontal direction for about 100 yards, closely pursued by the Falcon, and then twisted hither and thither for the space of a few seconds, the Shahin following its every movement, until he struck it with his talons, and, seizing it in his bill, flew past me to the cliff. These Falcons frequently sally out thus from their perch about sunset, and make a meal off the first unlucky bird that crosses their path; and they would seem to have rather a partiality for Swifts and Swallows, for I noticed the bird I shot at the Yakka rock dart at a Swallow that was flying about the cliff. They may be always distinguished from the Peregrine on the wing, even at some little distance, by their smaller size and by the conspicuous blue-grey of the rump. I have now and then observed them perch on trees; but I think it is the exception for them to do so, as they prefer the rocks of the precipices about which they almost entirely live. This species lives exclusively on birds; and Jerdon remarks that in India it kills large quantities of game, partridges, quails, &c., and that it is very partial to parrakeets. He observes, further, that its habits vary according to the locality in which it lives, birds from open districts, where they require to be more on the wing in pursuit of their prey than in forest districts, being by far the best fliers and the most useful in falconry. It is more highly prized by the natives than any Falcon in the East, the Peregrine being considered even second to it.

Nidification.—But little is known concerning the nidification of this Falcon. I have no doubt whatever but that it breeds in such localities as the Yakka rock, Alagalla Peak, and perhaps in the low country in hills like Yakdessagalla, Rittagalla, Friars Hood, &c. It nests usually on inaccessible cliffs. Jerdon mentions three eyries in India—one at Rutoor, another in the Nilghiris, and a third near Mhow. It builds a nest of sticks on a projecting or receding ledge of rock, and sometimes takes possession of the old nest of another Raptor. Mr. Hume speaks of an egg taken by Mr. Blewitt in the Raipoor district as being narrow and oval, of a pale pink ground-colour, clouded with pale purplish, and finely speckled and spotted with deep reddish brown. It measured 2·0 by 1·43 inch. This egg was taken in January; but Jerdon says it lays also in March and April.

FALCO SEVERUS.

(THE INDIAN HOBBY.)

Falco severus, Horsf. Trans. Linn. Soc. xiii. p. 135 (1822); Blyth, Ibis, 1863, p. 8; Schl. Vog. Ncderl. Ind., Valkv. pp. 4, 45, Taf. 2. figs. 2, 3 (1866); Radcliffe, Ibis, 1871, p. 366; Sharpe, Cat. B. i. p. 397 (1874); Bourdillon, Str. Feath. 1876, p. 354.

Falco aldrovandii, Temm. Pl. Col. i. pl. 128 (1823).

Falco rufipedoides, Hodgs. Calc. Journ. N. H. iv. p. 283 (1844).

Falco guttata, Gray, Cat. Accipitr. Brit. Mus. p. 26 (1844).

Hypotriorchis severus, Gray, Gen. of B. i. p. 20 (1844); Blyth, Cat. B. Mus. A. S. B. p. 14 (1849); Horsf. & Moore, Cat. B. Mus. E. I. Co. i. p. 22 (1854); Jerd. B. of Ind. i. p. 34 (1862); Wallace, Ibis, 1868, p. 5; Hume, Rough Notes, i. p. 87 (1869); Holdsworth, P. Z. S. 1872, p. 410; Walden, Trans. Zool. Soc. viii. p. 33 (1872).

The Severe Falcon, apud Horsf. & Moore.

Dhuti (female), *Dhuter* (male), Hind.

Allap-Allap-Ginjeng, Java.

Adult male. Length to front of cere 10·3 to 10·5 inches; culmen from cere 0·65; wing 8·0 to 9·0; tail 4·5; tarsus 1·1 to 1·2; middle toe 1·2, claw (straight) 0·5; height of bill at cere 0·27.

In 'Stray Feathers,' vol. iv. p. 355, the wing of a male shot in Travancore is given at 9·25. This is *most exceptional*, or it is a misprint for female. Two Ceylonese examples, one of which, on account of its small size, must be a male, measure 8·6 and 9·0. Three males in the Norwich Museum, from the Philippines and Java, measure 8·3, 8·6, and 8·7; two others in the British Museum do not exceed 8·3; Jerdon, however, gives the wing of a male as 9·0, from which I have taken the above limit.

FALCO CHICQUERA.

(THE RED-HEADED MERLIN.)

Falco chicquera, Daud. Traité, ii. p. 121 (1800, *ex* Levaill.); Less. Traité, p. 90 (1831); Gould, Cent. B. Him. Mts. pl. 2 (1832); Blyth, Cat. B. Mus. A. S. B. p. 14 (1849); Sharpe, Cat. Birds, i. p. 403 (1874).

Hypotriorchis chicquera, Gray, Gen. B. i. p. 20 (1844); Kelaart, Prodromus, Cat. p. 115; Layard, Ann. & Mag. Nat. Hist. 1853, xii. p. 102; Horsf. & Moore, Cat. B. Mus. E. I. Co. i. p. 23 (1854); Jerd. B. of Ind. i. p. 36 (1862); Holdsworth, P. Z. S. 1872, p. 410; Butler, Str. Feath. 1875, p. 444.

Æsalon chicquera, Kaup, Class. Säug. n. Vög. p. 111 (1844).

Chicquera typus, Bp. Rev. et Mag. de Zool. 1854, p. 536; Hume, Nests and Eggs, p. 10 (1873).

Turumtia chicquera, Blyth, Ibis, 1863, p. 9.

Lithofalco chicquera, Hume, Rough Notes, i. p. 91 (1870); Anderson, P. Z. S. 1871, p. 681.

Toorumtee, Europeans in India.

Turumti, Turumtari, Putri mutri (female), *Chetwa* (male), Hind.; *Jellaganta, Jelgudda*, Telegu; *Jelkut*, Yerklees (*apud* Jerdon).

Adult male. Length (from skin) to front of cere 10·5 inches; culmen from cere 0·7; wing 7·0 to 8·1; tail 5·0; tarsus 1·3 to 1·5; middle toe 1·3, claw (straight) 0·48; height of bill at cere 0·35.

Adult female. Length to front of cere 11·0 to 12·0 inches; culmen from cere 0·65; wing 9·2 to 9·7; tail 4·9; tarsus 1·3 to 1·4; middle toe 1·3, claw (straight) 0·5. The wing sometimes reaches 0·5 beyond the tail.

Iris deep brown; cere and bill at base yellow, the upper mandible and tip of the lower blackish; legs and feet yellow, claws black.

Entire face, head, hind neck, and interscapular region glossy black, paling into blackish slaty on the back, wings, rump, and upper tail-coverts; the feathers on these parts have the shafts black and the bases blackish brown, the slaty hue being confined to the tips of the feathers; on the head and hind neck there is an ashen hue; quills blackish brown, the inner webs more or less barred with rufous (in some very old birds these are almost absent or reduced to pale transverse dashes); tail slaty black, tipped finely with rufous, and in some with a subterminal band, such examples having the outer feathers with rufous or greyish bars on the inner webs. Some examples have undefined slaty bars across the whole tail.

Throat and fore neck buff, tinged with rufous, the colour running up into the sides of the neck, all beneath from the fore neck, with the thighs, under tail- and under wing-coverts, deep chestnut or ferruginous; under primary-coverts paler rufous, barred with black; the remainder of the wing-lining with black shaft-lines; sides of the chest with a few black patches, running into the black hind neck; middle of the chest usually with a few black shaft-lines.

Young. The immature bird is almost as dark above as the adult; but the exposed portions of the sides of the hind-neck feathers are more or less rufous, the central tail-feathers are crossed with greyish markings, and the inner webs of the remaining feathers barred with rufous; extreme tips of the secondaries whitish; chin and throat as in the adult, the rufous of the under surface not quite so deep; the chest streaked with drop-shaped strias of black, and the breast and flanks marked with oval central drops, the thighs and under tail-coverts with central streaks, longer and narrower than the breast-markings.

Distribution.—The handsome Indian Hobby can only be classed in our lists as a straggler, having been but twice procured in the island. The first record of it as a Ceylonese bird is contained in Mr. Holdsworth's Catalogue (*loc. cit.*), from a specimen shot by Mr. Bligh, at Catton Estate, Haputale; but from recent investigation, as noticed in the preceding article, I find that Layard killed another example, which is, in all probability, referable to his *Falco peregrinator* shot at Gillymally; and he therefore must be looked upon as the discoverer of the species in Ceylon. I imagine that both these specimens were killed during the cool season, and that without doubt the species is migratory to Ceylon, as it is to South India.

This Hobby is a bird of fairly wide distribution, being found throughout the whole of the Indian peninsula

---

Adult female. Length to front of cere (from skin) 13·0; wing 8·5 to 9·1; tail 6·5 to 6·8; tarsus 1·6 to 1·7. Weight 8·5 oz. (*Hume.*)

The above measurements are from N. Bengal and Nepaul specimens.

"Iris rather light brown; orbits yellow; bill greenish yellow at base, bluish black at tip; legs and foot pure (slightly orange) yellow." (*Hume.*)

Head, back, and sides of neck cinnamon-rufous; a moustachial streak of a paler hue than the head, between which and the eye is a blackish streak; a dark superciliary line; back, rump, scapulars, and wing-coverts bluish slate, paling gradually towards the tail, and blending somewhat into the hue of the neck; the feathers of these parts with dark shafts; wing-coverts at the point of the wing barred with blackish grey; feathers along the ulna edged with rufous, and beyond this the edge of the wing is buff-white; primaries deep brown, the inner webs barred narrowly with white, not reaching on the terminal half to the edge of the feather; primary-coverts and secondaries slate-grey, the inner webs albescent and barred with blackish grey; tail pale bluish grey, lighter than the coverts, deeply tipped with greyish white, and crossed with a broad subterminal black band, the remainder crossed with narrow widely separated rays of blackish grey.

Chin, throat, sides of the face, and under surface white, barred from the breast downwards with blackish grey or dark slate-colour, and the markings on the centre of the breast somewhat pointed at the middle of the feather; flanks more heavily barred than the breast; under wing-coverts white, the external feathers with dark mesial lines, the inner ones barred like the chest.

Females that I have examined in the British Museum have the under wing-coverts more darkly barred than males. The tail-band appears to fade very much in this species, turning brown when the feathers become old.

from the Himalayas to Travancore, and likewise in the Malayan peninsula, whence it extends through the whole Asiatic archipelago by way of Celebes and New Guinea to the Philippines. I have seen specimens of it from Java, Salwati, Borneo, and Makassar; and it in all probability inhabits many of the smaller islands in the Malayan region. In India it is chiefly confined to the Himalayas; but it is not very numerous even there, and does not extend to the north of this range. A few visit the plains in the cool season, and it is often killed in the neighbourhood of Calcutta. Colonel Radcliffe procured it near Futteghur in 1866; and Mr. Bourdillon in Travancore, where it is a winter visitor only. It does not appear to have been detected often, if at all, in Burmah, as I can find no record of its occurrence there in 'Stray Feathers;' it has, however, been found further south in the peninsula, and it will very likely be met with some day in Tenasserim or Pegu.

Habits.—This stout little Falcon frequents mountainous country, dwelling chiefly about heavy jungle and

---

Distribution.—I assign my notice of this bird as belonging to the Ceylonese ornis to a footnote for the same reason as in the case of the Sparrow-Hawk, viz. that its occurrence on the island is not a matter of absolute certainty. Layard writes (*loc. cit.*), " I saw this pretty Hawk in the flat country near Pt. Pedro, but could not got a shot at it: I cannot, however, be mistaken in the bird, as I long watched it with my telescope." He writes me from New Caledonia, "You may safely include *Hypotriorchis chiquera*;" and I therefore do so in the way I have adopted for the treatment of those species which have not been actually procured. There is no reason whatever against inferring that this little Falcon now and then visits the northern shores of Ceylon, as it is found in the extreme south of the peninsula. Jerdon says that it is spread throughout India from north to south, but is rare in the forest-districts, as it chiefly affects open country in the vicinity of cultivation. It does not appear to be procured so often in the south as in the northern parts of India and on the outskirts of the Himalayas—the province of Nepaul, to wit. Captain Hayes Lloyd found it common in the Kattiwar district, Western India; and further north in the northern Guzerat region, Captain Butler writes, in 'Stray Feathers,' "it is not very common, but appears to be distributed pretty evenly throughout the plains." In the eastern parts of the peninsula it is not so common. Mr. Ball says that it is of very rare occurrence in Chota Nagpur, and that he only once observed it in the Satpura hills."

Habits.—This pretty little Merlin is a most courageous bird, and appears to be a general favourite with sportsmen in India on account of its boldness, powers of flight, and interesting habits. It frequents compounds, groves of trees, the edges of isolated woods, or even single trees in open country, whence, Jerdon remarks, it "sallies forth, sometimes circling aloft, but more generally, especially in the heat of the day, gliding with inconceivable rapidity along some hedge-row, bund of a tank, or across some fields, and pouncing suddenly on a Lark, Sparrow, or Wagtail." It often haunts in pairs, and sometimes hovers for a few seconds like a Kestrel. It feeds on small birds almost entirely, but will occasionally kill the smaller mammals, Mr. Hume recording that he has found the remains of squirrels in their stomachs; they have also been known to fly at Bats in the dusk of the evening. It is occasionally used by falconers, and flown at small game and also at the Roller and at Pigeons. Jerdon writes, "In pursuit of the Roller it follows most closely and most perseveringly; but is often baulked by the extraordinary evolutions of this bird, who now darts off obliquely, then tumbles down perpendicularly, screaming all the time and endeavouring to gain the shelter of the nearest tree or grove." Captain Butler gives a most interesting account (*loc. cit.*) of the performances of one of these brave little birds, which I subjoin here:—"Upon one occasion I remember shooting into a small flock of *Cursorius gallicus*, wounding two and killing a third. One of the wounded birds, before falling, flew 'pump-handling' for some distance close to the ground, and the other one towered. One of those beautiful little Merlins at once appeared on the scene, and followed in pursuit of the towering bird to a height of 300 or 400 feet. As soon as the Courier became aware of his presence he closed his wings and dropped to the ground like a stone, followed of course by the Turumti, who stood erect by his side on my arrival, staring at him as if it was the first bird he had ever seen. On my approaching the spot the Courier again took wing followed by the Merlin; and thinking he might fly some distance, I shot him. The Merlin took no notice whatever of the report of the gun, but made a stoop at the falling bird and accompanied it to the ground. I then walked up to the spot and drove him away.

"After picking up the Plover I turned round and, to my unutterable surprise, I saw the Falcon on the top of the other wounded bird. I ran up to them, and found a desperate struggle going on: and it was not until I nearly knocked the plucky little fellow over with a stone that I induced him to leave his intended meal."

The cry of this species is a shrill angry scream.

forest, which furnish it with a supply of small birds, on which it is said chiefly to feed. Mr. Bligh's specimen was shot hawking after dragonflies; and no doubt the bird feeds as well on lizards, which form a large proportion of the food of most Indian Raptors, from the Hawk-Eagle downwards. It is said to be crepuscular in its habits. Mr. F. Bourdillon, in 'Stray Feathers' (vol. iv. p. 354), says that its cry is shriller and weaker than that of the Kestrel; he is also of opinion that it breeds in Travancore. Wherever it does it must be in remote or inaccessible forests, for nothing appears yet to be known concerning its nest and eggs. Jerdon remarks that it nidificates on trees; and Mr. Thompson, as quoted in Mr. Hume's 'Rough Notes,' says that it breeds in Kumaon. He writes, "These birds regularly resort to the dense forests on the lower ranges of Kumaon and Gurhwal about April. In June I watched a female bearing a small bird away, but could not find where she took it to. I infer from this that she must have had a nest of fledged young ones, as there were lots of fine trees standing close to where she passed me, and where she might have stopped to pluck her quarry. Later observations confirm me that the bird breeds about April in our lofty and dense forests."

Nidification.—The Toorumtee nests exclusively on trees, making its own nest, and building a fresh one every year. It is neatly built of both stout and fine twigs closely put together, and lined with fine roots and vegetable fibres, mixed sometimes with straw, feathers, or pieces of rag, which are firmly interwoven with the body of the nest. It is generally fixed in the fork of the top branch of a large tree, such as a mango, peepul, or tamarind, where these are to be found; but where they do not exist, it is placed in small trees, sometimes not more than 10 feet above the ground. The eggs are usually four in number; but sometimes three and five are laid. They vary from "a pale yellowish brown, with just a few reddish-brown specks, to a nearly uniform dark brownish red, obscurely mottled and blotched with a somewhat darker red." Sometimes there is a ring of feeble blotches round the large end, and at others a zone of darker markings round the middle. They average 1·06 inch in length by 1·27 in breadth. The breeding-season lasts from January till May.

ACCIPITRES.

FALCONIDÆ.

FALCONINÆ.

Genus CERCHNEIS.

Bill shorter and more suddenly curved than in *Falco*; wings as in that genus, but the 1st quill shorter, and the 1st and 2nd notched on the inner web. Tail longer than in the last; tarsi longer and more feeble. Lateral toes nearly equal, the scales rectangular up to the base of the toes.

Of small size. Sexes generally differing in coloration. Sternum weaker than in *Falco*.

CERCHNEIS TINNUNCULUS.

(THE COMMON KESTREL.)

Falco tinnunculus, Linn. Syst. Nat. i. p. 127 (1766); Gould, B. of Europe, i. pl. 26 (1837); Yarr. Brit. B. i. p. 52 (1843); Schl. Vog. Nederl. pls. 9, 10 (1854); Sharpe & Dresser, B. of Eur. pt. 2 (1871); Newt. ed. Yarr. Brit. B. p. 79 (1872).

Falco alaudarius, Gm. S. N. i. p. 279 (1788).

Cerchneis tinnuncula, Boie, Isis, 1828, p. 314; Sharpe, Cat. Birds, i. p. 425 (1874): David & Oust. Ois. de la Chine, p. 36 (1877).

Tinnunculus alaudarius, Gray, Gen. Birds, i. p. 21 (1844); Blyth, Cat. B. Mus. A. S. B. no. 69, p. 15 (1849); Kelaart, Prodromus, Cat. p. 115 (1852); Layard, Ann. & Mag. Nat. Hist. 1853, xii. p. 102; Horsf. & Moore, Cat. B. Mus. E. I. Co. ii. p. 13 (1854); Gould, B. Gt. Brit. pt. ii. (1862); Jerdon, B. of Ind. i. p. 38 (1862); Tristram, Ibis, 1865, p. 259; Hume, Rough Notes, i. p. 96 (1869); Holdsworth, P. Z. S. 1872, p. 410; Du Cane Godman, Ibis, 1872, p. 165; Hume, Nests and Eggs, p. 21 (1873); Legge, Ibis, 1874, p. 10; Scully, Str. Feath. 1876, p. 120; Brooks, Str. Feath. 1876, p. 228.

Cerchneis alaudarius, Hume, Str. Feath. 1876, p. 460.

L'Epervier des Alouettes, Briss. Orn. i. p. 279 (1760).

La Cresserelle, Briss. Orn. i. p. 393 (1760).

"*Windhover*," "*Stonegall*," popularly in England; *Cernicalo, Primilla*, Spanish (Saunders); *Francilho*, Portuguese; *Sweef*, Moorish (Drake); *Narzi* (female), *Narzanak* (male), Hind. (Jerdon); *Khurmutia, Kurumtia, Karontia*, Hind. (Blyth); *Nardunak*, Scinde; *Gytthin, Tondala-muchi gedda*, lit. "Lizard-killing Kite,"Tel. (Jerd.); *Kurganak*, Turkestan (Scully). *Ukussa, Kurullagoya*, Sinhalese; *Walluru*, Ceylonese Tamils.

Adult male. Length to front of cere 12·5 to 13·5 inches; culmen from cere 0·6; wing 0·6 to 10·2 (0·7 to 9·9 being the average); tail 6·5 to 7·0; tarsus 1·4 to 1·6; mid toe 1·05 to 1·15, claw (straight) 0·45; height of bill at cere 0·35. In a large series examined in the British Museum I find that Asiatic examples measure as much as, if not more than, European, the largest specimen, having a wing of 10·2, being from Behar.

Iris dark brown ; cere and eyelid chrome-yellow ; bill adjoining cere and at the base beneath paler yellow, darkening to bluish at the tips ; legs and feet chrome-yellow, claws black.

Head, back, and sides of neck, together with the moustachial stripe, ashy bluish, the feathers with dark shafts ; lower part of hind neck, back, and wing-coverts cinnamon-rufous, with an arrow-headed blackish-brown spot, more or less broad, at the tip of each feather, developing on the tertials into a subterminal bar ; quills and their coverts ashy brown, the inner webs with bar-like indentations of rufescent white, and all with pale tips ; rump, upper tail-coverts, and tail clear ashy blue, with a broad subterminal blackish bar across the latter ; edge of forehead, lores, throat, and the space between the cheek-stripe and the ear-coverts buff-white, the latter more or less shaded with ashy ; beneath, from the throat, rufescent white, in some specimens isabelline grey, the feathers of the chest and upper breast with dark brown stripe, and those of the breast and flanks with terminal drop-shaped spots : the lower parts and under tail-coverts unspotted, the thighs being, as a rule, more rufous than the abdomen : under wing-coverts marked with pointed central spots.

The above is a description of a fully-aged bird, in which a bluish cast often pervades the entire upper surface.

In a slightly younger stage of adult plumage, and one in which most birds are procured in Ceylon, the head is more or less washed with rufous, the markings of the back and wing-coverts are larger, and the rectrices, though blue, present in various degrees a certain amount of barring ; in some this appears on the central feathers, either at the base or down most of the web, in others these are devoid of any markings, while the inner webs of the outer feathers are crossed with narrow transverse lines. In this stage, however, the head is very variable, being not always tinged with rufous as above stated, but, at times, as blue as in the oldest birds.

Young. In the bird of the year the wing averages from 9·5 to 9·8 inches in the male, and slightly more in the female. Soft parts much the same as in adult ; cere slightly greenish in some ; legs and feet not so bright in hue.

Head and upper surface dusky rufous, usually paler on the hind neck and rump than elsewhere ; feathers of the head and hind neck with broad brown stripe, and the back, rump, and wing-coverts crossed with broad bars of brown, the shafts being of the same colour ; quills brown, tipped pale, most deeply on the secondaries, the inner webs partially crossed from the edge with rufous or rufescent yellow ; tail dark rufous in some, yellowish rufous in others, with continuous or interrupted bars of blackish brown, and a broad terminal band of the same, the outer feathers paler than the rest ; forehead and round the cere fulvous, with dark shafts to the feathers, a broad, dark, moustachial stripe crossing the gape from the lores ; ear-coverts fulvous-grey, shading off into brownish ; throat and under surface rufescent white, palest on the abdomen and under tail-coverts, which, with the chin and gorge, are unstreaked ; chest, breast, and flanks streaked with brown, the markings on the chest being broader than on the breast, on the lower part of which they diminish to mesial lines ; under wing-coverts buff-white, lined and spotted (on the longer feathers) with dark brown. In a large series no constant variation can be found between male and female, notwithstanding that females appear sometimes to have the thigh-coverts and lower parts more striated than the other sex.

In the next stage towards the mature dress the tail, at moult, becomes ashy blue, completely barred with brown, the upper tail-coverts changing at the same time to bluish, and the bars on the back and wings fading out. Birds are occasionally found with the tail composed, in the same moult, of the adult blue and immature red feathers.

Adult female. Length, including culmen, 14·0 to 15·0 ; wing *averaging* more than the male's, but seldom exceeding the highest of the above dimensions. In an immature example shot at Colombo it measures 10·5.

Upper plumage of a browner rufous than the male ; in very old birds the head with a bluish cast, the stripes and bars respectively of the head and back narrower and darker than in the young, and the latter slightly spear-shaped ; in some birds, as in the male, a faint ashy hue is perceptible on the upper surface ; upper tail-coverts bluish ashy, with either spear-shaped streaks or narrow mesial stripes on the longer feathers, or the whole crossed with narrow bars ; tail bluish ashy at the base and down the centre of all the feathers, the edges shading into rufous and crossed on both webs with narrow bars of blackish brown, incomplete at the base of the central feathers ; tip as in the male ; under surface a paler, but generally finer rufescent than in the male, and more boldly streaked on the chest and spotted on the breast ; in examples which are heavily barred above the flanks have transverse markings. Birds not fully adult betray their youth in the greater amount of rufous on the tail and its coverts.

Obs. It has lately been ascertained that the female Kestrel is capable of acquiring a somewhat masculine plumage, a pair having been shot at the nest at Aldenham, Hertfordshire, in 1874, in which the female had the tail bluish, barred with black. An account of this remarkable occurrence is given by Mr. Sharpe in the 'Proceedings of the Zoological Society,' 1874, p. 580, pl. 68.

Distribution.—This well-known bird, the "Windhover" of the English farmer, migrates, in the cool season, freely to Ceylon, remaining in the island until the usual time of departure, the coming-in of the following south-west monsoon, when it takes wing for its breeding-haunts in more northern climes. It spreads over the whole island, without respect to locality or elevation, frequenting the entire seaboard, the low country of the interior, and the elevated plateau of the main range, while the intermediate coffee-districts come in for an equal share of its patronage. It is commonly met with about Colombo, frequenting the cinnamon-gardens and cocoanut-groves along the Galle road ; at the southern port it always takes up its quarters in the huge ramparts fronting the esplanade ; and at Trincomalie it is numerous in the season, dwelling in the lofty precipices and mural rocks encircling the Fort, and sallying out to the extensive esplanade in search of food. In the Jaffna peninsula and round the north coast of the island generally it is plentiful, and it is likewise common on all the adjacent islands of Palk's Straits. Although abundant in Ceylon, it never occurs there in flocks similar to those that have been seen by Blyth near Calcutta, or by Captain Shelley in Egypt.

The permanent habitat of the Common Kestrel is the entire continent of Europe and Northern Asia, whence it migrates in the winter into Northern Africa, the Indian peninsula, and North China, and it occasionally wanders into South Africa and even into the Seychelles. Although it leaves India for the most part in the breeding-season, it remains in the Himalayas in considerable numbers, and on the other side of the chain it is, according to Dr. Scully, a permanent resident in Turkestan. It appears to be only a winter visitor to Burmah, as neither Capt. Feilden nor Mr. Oates record it as remaining there in the hot season. The latter gentleman says that it is common in the Pegu plains ; but it does not continue its migration as far south as Tenasserim, for I do not find any mention of it in either the first or second list of birds from that province contained in ' Stray Feathers.' Blyth found it very common in Lower Bengal, where it was seen by him in parties of twenty or thirty together. In Chota Nagpur Mr. Ball says that it is tolerably abundant in most parts ; the same remark applies to nearly all parts of the Indian peninsula, for this little Hawk is dispersed throughout the whole of it, irrespective of elevation ; there are, however, some districts in which it is not so numerous, for Mr. Hume found it numerically scarce in the plains of Sindh. Mr. Bourdillon remarks that it is a winter visitor to the Travancore hills, and that it breeds there ; on Mount Neboo, 7000 feet high, in the Palani Hills, which form an eastern offshoot of the same range, Mr. Fairbank observed it until June, and remarks that he thinks it resides permanently there. It has also been found to be a permanent resident in some parts of the Nilghiris ; but Mr. Hume says these southern birds belong to a "smaller and markedly deeper-coloured race," which is perhaps peculiar to the south of the peninsula, and may merit entire specific separation from its ally, which is migratory to the whole country as well as to Ceylon, the latter place forming the southernmost limit of its wanderings. It is certainly remarkable that while a vast stream of Kestrels overruns annually the whole of the region in question, there should exist a certain quantum which, in addition to a different character of plumage, should possess the peculiar habit of remaining stationary and breeding in the hills of the extreme south of the peninsula. This peculiarity in the Kestrel's economy is not, however, confined to South India ; the same occurs in Madeira and in Abyssinia ; and Mr. Sharpe solves the difficulty by pointing out that there is undoubtedly a dark resident form of this species to be found in certain localities along the southern limit of its habitat. As regards South India, I imagine that the Kestrels found in the low country of this region belong to the migratory class, as they certainly do in the island of Ramisseruin, where they are very numerous during the north-east monsoon. Mr. Hume remarks that the Kestrel is the commonest Raptor in the cold season at the Laccadive group, and that the specimens he obtained "were all of the European type," which is, of course, the case with those in Ceylon.

In the northern part of the sister continent of Africa, the Kestrel locates itself in great numbers during the winter. Captain Shelley found it swarming in Egypt, and once saw as many as one hundred together in a clump of palm trees, attracted there by the clouds of locusts which were passing them. In Tangier and Eastern Morocco, Mr. T. Drake found it common ; and beyond this, towards the Atlantic, it wanders sometimes into Western Africa. Mr. Godman found it, however, common in the Canaries and in Madeira, the birds in the latter place being resident and belonging to the dark race. With regard to Europe and Northern Asia, the permanent habitat of the species, my limited space compels me to pass over its distribution there ; and I would merely remark that Canon Tristram found it especially abundant in Palestine, inhabiting every variety of locality, and breeding gregariously in the ruins characteristic of that country. Mr. Sharpe remarks that the Japan Kestrels are the largest and darkest of any of the races of this species.

Habits.—As in England so in Ceylon, the Kestrel prefers open to wooded country, taking up its abode near commons, pasture-land, brush-covered plains, large tracts of dried-up "paddy-field," and any locality on which its prey (lizards, mice, and small mammals) is exposed to its view. It takes up its quarters on high rocks and precipices, always returning to the same roost, and is very regular in its hours for coursing over the surrounding country in quest of food. Two or three pairs lived annually in the ramparts and cliffs at Trincomalie; in the morning they sallied out, returning for their midday rest about 10 o'clock, and passed the heat of the day under some projecting points in the lofty mural sea-face, now and then flying round the Fort, or alighting on the pretty parade-ground, surrounded with fine old trees, in which they often engaged in fierce and quarrelsome harangues with one or two Goshawks (*Astur badius*), who objected strongly to the annual invasion of their territory by the smart little Kestrels. In the afternoon, about 3 o'clock, they departed again on their rounds, and were to be seen until evening about the esplanade or among the Suriah trees (*Ibiscus*) lining the public roads. The Kestrel has a rapid flight, sustained with quick beatings of the wing, and is capable of making sudden and very swift stoops on its prey. It, however, usually hovers over such animals, reptiles, and insects as it feeds on in the remarkable manner for which it is so well known, and drops suddenly down, extending its talons as it reaches the ground, and then usually devours what it has captured on the spot. The skill with which it poises itself, after hovering for some seconds, its wings perfectly motionless and its body suspended, as it were, from the heavens by an invisible thread, is marvellous. I have seen it in such a position in a strong wind, not precisely facing the direction of the current of air, but with one wing pointed up to it—the primaries of which, yielding to the force of the wind, every now and then would give back, but as quickly spring forward into their normal position, while the rest of the body remained unmovable!

Its principal food consists of lizards and large beetles in Ceylon; but in Europe it is an inveterate destroyer of field-mice, although it is not generally accredited with such useful habits, but rather believed to be an enemy of the game-preservers, and frequently pays the usual barn-door penalty at the hands of ignorant keepers. Professor Newton remarks, however, in his late edition of 'Yarrell's British Birds,' that "it does occasionally kill and devour small birds, and at times the young of larger ones." Concerning this assertion I have only to remark that I feel convinced such occurrences are exceedingly rare, and that in Ceylon I am sure it is entirely an insect- and reptile-feeder. It has been known to catch cockchafers on the wing, seizing them in its claws and devouring them while flying. It doubtless kills the locusts, on which it is said to feed in Egypt, in the same manner. Of its habits in Yarkand and Turkestan, generally, Dr. Scully writes :—"It feeds chiefly on mice, lizards, and *grasshoppers* ; the Yarkanders add frogs and, in winter, sparrows. In the stomach of a Kestrel killed at Yepchan, I found, among other things, a rat's tail 6 inches long." Messrs. Sharpe and Dresser, in their admirable work on the Birds of Europe,' quote from the remarks of Dr. E. Hamilton, on the habits of this species, the following passage concerning its vermin-killing propensities :—" I have trained the Kestrel myself to come to the lure, but never could get it to swoop at birds, although I have starved it almost to death ; but put a mouse before it, and it would immediately take it. . . . . Birds, when given, were always left half plucked or half uneaten, as if distasteful."

Notwithstanding the evidence I have adduced to show my readers that this favourite little Falcon is in the main harmless as regards bird-life, it cannot be entirely absolved from all such offence, for it has been known to suck the eggs of the Missel-Thrush, and also to carry off very young Partridges in England ; but it must be said that the latter crime seems very unnatural when taken in conjunction with Dr. Hamilton's evidence as to its distaste for bird-flesh. As has been shown above, this Kestrel is somewhat gregarious when it becomes very numerous in a locality ; but it is not so much so as the next species. In some parts of Asia its tameness and sociability are remarkable. Dr. Scully remarks, in his interesting journal of his trip to Turkestan, " a couple of Kestrels (*Tinnunculus alaudarius*) seem to have taken up their abode here ; they fly about from the rafters of the verandah to the poplar trees just outside my room."

Nidification.—The Kestrel was formerly supposed not to breed within Indian limits ; of late years, however, since so much attention has been paid to the subject of ornithology, it has been found nesting in the Himalayas and outlying districts in Cashmere, in the Central Provinces, and in the Nilghiris. This latter locality, however, is only resorted to by the small dark-coloured resident species, which appears to be peculiar to those hills. Its favourite situations in India for building its nest are crevices and overhung ledges in the face of high cliffs; but it has also been found nesting in trees, and in England, as is well known, frequently takes possession of a crow's nest. The structure is sometimes bulky, and at others the reverse, the requirements of the situation no doubt determining the design. It is made of sticks and lined with small twigs and grass-roots, sometimes intermingled with pieces of rag. The eggs are usually four in number, of a brick- or blood-red ground-colour, freckled or spotted with deep red, with occasionally a few blotches or clouds of the same. The average of 19 eggs, according to Mr. Hume, was 1·57 by 1·21 inch.

In some countries the Kestrel breeds together in colonies, and even in company with other birds. Canon Tristram remarks, in his "Notes on the Ornithology of Palestine":—"It is generally gregarious, ten or twenty pairs breeding in the same ruins, and rearing their young about the end of March. It often builds its nest in the recesses of the caves which are occupied by Griffons; and is the only bird which the Eagles appear to permit to live in close proximity to them. At Amman, too, it builds in the ruins in company with the Jackdaws; and in several places, as at Lydda and Nazareth, large colonies are mixed indiscriminately with those of the following species (*Tinnunculus cenchris*)."

CERCHNEIS AMURENSIS.
(THE AMURIAN KESTREL.)

Falco vespertinus, Schrenk, Reis. Amurl., Vög. p. 230 (1860).
Erythropus vespertinus, Horsf. & Moore, Cat. B. E. I. Co. Mus. no. 13, p. 14 (1854); Swinh.
    Ibis, 1861, pp. 253, 327, et P. Z. S. 1862, p. 315; Jerd. B. of Ind. i. p. 40 (1862);
    Hume, Rough Notes, i. p. 106 (1869); Jerd. Ibis, 1871, p. 243; Legge, Str. Feath.
    1873, p. 487; Hume, Str. Feath. 1875, p. 22.
Falco vespertinus, var. *amurensis*, Radde, Reis. Sibir. ii. p. 102, Taf. 1. figs. 1–3 (1863).
Erythropus amurensis, Gurney, Ibis, 1868, p. 41, pl. 2; Swinhoe, P. Z. S. 1870, pp. 436,
    448, 1871, p. 340, et Ibis, 1873, p. 96, et 1874, p. 425; Hume, Str. Feath. 1875,
    p. 22.
Tinnunculus amurensis, Gray, Hand-l. B. i. p. 23 (1869).
Falco raddei, Finsch u. Hartl. Vög. Ostafr. p. 74 (1870).
Cerchneis amurensis, Sharpe, Cat. B. i. p. 445, et Str. Feath. 1875, p. 303; Legge, ibid.
    p. 362; Inglis, ibid. 1877, p. 6; Hume, ibid. p. 7.
Falco amurensis, Dav. & Oust. Ois. de la Chine, p. 34 (1877).
Le Kobez de l'Amour, David & Oustalet.
The Eastern Red-footed Kestrel, Radde's Kestrel, The Orange-legged Hobby, The Red-legged
    Hobby, The Red-legged Falcon of Indian writers; *White-clawed Kestrel*, Blyth; *Ingrian*
    Falcon, Lath. (*apud* Horsf.).
Karjanna, Karjoona, Hind. (Jerdon).

Adult male. Length to front of cere 9·8 to 10·7 inches; culmen from cere 0·5; wing 8·8 to 9·2 (in a South-African
    example in the British Museum, 9·5); tail 4·75 to 5·0; tarsus 1·1 to 1·2; mid toe 0·95 to 1·0, claw (straight)
    0·38.
Iris brown; cere (all but the tip of bill), orbital skin, and eyelid gamboge-yellow; tip of bill dark leaden; legs and feet
    gamboge-yellow, claws whitish horn-colour.
Head, hind neck, back, wings, and tail dark slate, the forehead darker than the crown, and the rump paler than the
    back; the interscapular region, scapulars, and shoulder of the wing suffused with blackish; quills slate-grey, the
    shafts black and the inner part and under surface of web brownish; shafts of the tail-feathers brownish black;
    cheeks, cheek-stripe, and ear-coverts concolorous with the head; chin, throat, sides of neck, breast, and flanks
    bluish ash-colour, palest on the throat, the shafts of the breast-feathers more or less conspicuously dark; belly,
    thighs, and under tail-coverts dark rufous: *under wing-coverts and axillary plume white*, with dark shafts in
    examples which have them conspicuously so on the breast.
A younger stage of the adult phase has the throat and round the cheek-stripe whitish, and the under parts paler ashy
    than in the above-described old dress; the abdomen and under tail-coverts are also paler or yellowish rufous.

Adult female. Wing 9·4 to 9·6 inches.
Soft parts as in the male.
Above leaden-grey, paling to bluish ashy on the rump and tail; the head and nape suffused with brownish, and the
    feathers with dark shaft-stripes; bases of the hind-neck feathers rufous-yellow, showing across that part; back
    and wing-coverts barred with brownish black; quills blackish, the inner webs crossed with bar-like spots of whitish;
    tail crossed with seven or eight blackish-brown bars and a broad subterminal band of the same.
Forehead, throat, cheeks, and chest whitish buff: lores and a narrow moustachial streak blackish; the throat unstreaked,
    but the chest with mesial stripes, and the breast and flanks with arrow-headed bars of blackish brown; abdomen,
    thighs, and under tail-coverts pale rufescent yellowish, this hue extending further up the breast in some than in
    others; under wing-coverts white, barred and " lined " with blackish brown.

Young. In the British Museum are a pair of nestlings taken in China from the same nest—one with the wing 7·5, the other 7·6 inches. They are not sexed; but the larger of the two may be presumed to be the female.

The presumed male is the darker in colour, the ground-colour of the under surface being rufous-buff; the head and nape brownish slate, the feathers with dark shafts; the back, scapulars, and wing-coverts, as also the rump and tail, paler slate-colour, the whole crossed with blackish-brown marks, and each feather with a deep cinnamon-rufous tip; on the hind neck there is a rufescent-buff collar, formed by the lateral edges of the feathers; tail barred with interrupted bands of dark brown, and a broader one at the tip; quills slaty black, tipped with fulvous-white, and the inner webs crossed with transverse spots of white; thighs unmarked.

Cheeks and a small moustachial stripe and ear-coverts blackish; chin and throat buff-white, deepening into the more rufescent hue of the under surface; the breast with broad central drops of blackish to the feathers, which change into arrow-headed bars on the lower flanks.

The female has the tail less barred than the male, and the under surface buff, with broad spear-headed mesial marks on the chest-feathers, and the markings on the breast have the same character, instead of being plain stripes as in the male; the thigh-plumes marked with dark mesial lines.

As the bird grows older the rufous margins fade, and the barrings become more subdued, the bird of the year presenting a cinereous-brown hue on the head, hind neck, upper back, and scapulars; the feathers of the upper parts with pale margins and dark shafts, and those of the wing-coverts and scapulars fulvous-grey, while the tips of the rump and upper tail-coverts are greyish white and tolerably deep; the primaries and secondaries are tipped with white, the latter more deeply, the quills are dark cinereous brown, the inner webs crossed with white, similarly to the adult female; tail bluish ashy, with 10 or 12 narrow brown bars.

The cheeks, ear-coverts, a space behind the eye, and a more or less developed moustachial streak blackish; throat, sides of the neck, and under surface to the lower breast in some birds quite white, while others have these parts buff: the abdomen, thighs, and under tail-coverts are likewise variable from buff to almost white; forehead whitish: the throat and sides of the neck are unmarked, but the breast is striped with bold dashes of blackish brown, almost covering the feathers on the chest and upper breast; flanks often barred with the same.

The Ceylonese specimen mentioned below was a male in this stage of plumage; the lower parts were scarcely tinged with buff.

The young of both sexes at this stage much resemble the adult female, except that the under tail-coverts and abdomen are often nearly white; these latter parts vary, however, considerably in birds of the first year.

Obs. The transition in the male bird from the greyish barred upper plumage to the adult dark slate is complete in one moult; but the throat in several examples that I have examined has remained white, probably not acquiring the bluish ash-colour until the next season. The distinguishing character in this species consists in the white under wing-coverts, which suffice to separate it from the western Red-footed Kestrel, which has those parts dusky bluish grey: this character was first pointed out in 1863 by Radde, previous to which time the two species were confounded: and the bird spoken of by Jerdon, Horsfield, and others must be referred to the eastern race, and not to *E. vespertinus*, the western form. This latter has not been found nearer India than the Tumlionahian Steppe, Western Siberia. There Radde procured it, and, travelling eastwards, met with no species of Red-footed Kestrel until *C. amurensis* appeared in Amoorland.

Distribution.—The occurrence of this pretty little Kestrel in Ceylon is perhaps one of the most interesting in the history of Ceylon ornithology. The solitary example procured by myself at Trincomalie in December 1872, and recorded as *Erythropus vespertinus* (*loc. cit.*), is the only one that has yet been found in the island or in any part of Asia so far to the south. It was a straggler which found its way under the influence of the north-east wind to the shores of Ceylon; and, judging by the thin state of its body, had only just terminated its flight across the briny deep or down the east coast of India.

The eastern Red-footed Kestrel has its head-quarters in Amoorland and North China, and in the cold season performs, perhaps, the most singular migration of any known bird, encroaching on the path of its near ally, *Cerch. vespertina* (with which it was long confounded, until Radde discovered the difference between the two species), and actually reaching the southernmost regions of Africa. It passes from its home in North China into Burmah, Nepaul, and other sub-Himalayan provinces, Lower Bengal, Central India, and terminates, as a rule, its Indian migration in the Nilghiris, in which hills Jerdon killed it, but where, I imagine, it is very rarely seen. Mr. Hume has received it from Madras. From the north of India the migratory stream sets westward through Asia to the east coast of Africa, along which it flows to the Zambesi district, and thence southward to Natal, Damara Land, and Cape Colony. It was first known from Natal, whence Mr. Ayres sent specimens to Mr. Gurney in 1867, having found it numerous in that province; subsequently he observed numbers in the Transvaal, in December 1870, but did not meet with them in that district on any other occasion. It was procured in the Cape Colony by Mr. Andersson, and there are specimens in the national collection from Zambesi, presented by Dr. Livingstone and Dr. Kirk. It does not appear to have been yet observed in Egypt, which leads to the inference that it passes into Africa by way of Arabia, probably entering the continent in Abyssinia, and thence passing along the east coast to the south. It is at times numerous in Cachar, where it arrives, according to Mr. Inglis, in October, and disappears after February; during the former month, in 1875, he met with a flock of some hundreds of them hawking on a new tea-plantation. In Upper Pegu, Capt. Feilden has procured it, or met with it, in January and February; but it does not appear to wander down the Malay peninsula; for it has not yet been recorded from Tenasserim. About April it returns from its migrations to China, Mr. Swinhoe recording its arrival in Chefoo in that month, and at the same time Père David says it makes its appearance in the plains of China and Mongolia.

Habits.—This Red-legged Hobby has most of the habits of its near ally *C. vespertina*, resembling it in its gregarious tendencies, its crepuscular manner of hawking, and its insectivorous diet, while it also has somewhat in common with the ordinary Kestrel of Europe, hovering like it over its prey, though not in the same motionless manner peculiar to the "Windhover." During its migrations from Northern China to other countries, it associates in large flocks to an extent unusual in the Falcon tribe. As mentioned above, Mr. Inglis observed this habit in Cachar; and in South Africa, Mr. Ayres "found a lot of these pretty Falcons hunting with much assiduity; they were crossing backwards and forwards over the driest end of the swamp with an exceedingly rapid flight, and were taking insects on the wing." The favourite food of this bird appears to be grasshoppers, cockroaches, beetles, &c., to the first of which it is most partial. Its love of insect diet leads it to frequent open commons, plains, downs, dry marshes, and such like. It likewise feeds on white ants, which, indeed, do not seem to come amiss to any Indian bird, from the long-winged Kite down to the fruit-eating Barbet. The example I shot in Ceylon was frequenting the dried-up esplanade at Trincomalie; it constantly hovered near the ground, and then descended, proceeding, on alighting on the grass, to jump on the grasshoppers which attracted it, seizing them in its talons and devouring them on the spot. Contrary to its usual habit, it was very tame, the cause of which doubtless lay in its meagre frame; for though its stomach was distended with grasshoppers, it had but little flesh to boast of. The late Mr. Swinhoe is the only writer who speaks of this species killing birds, which it appears to do in China; and it is even trained at Chefoo to hawk small birds, which, it must be remarked, is somewhat unusual for a Kestrel.

Nidification.—The Amurian Kestrel breeds in China in the nests of Magpies, which build in tall trees of avenues or gardens near dwelling-houses. Swinhoe found it laying in the nests of the two species which inhabited the vicinity of Chefoo, viz. *Pica media* and *Cyanopolius cyanus*. The former builds a domed nest, and the latter, the Blue Magpie, an open one; but both come in for the patronage of this little Kestrel. The time of breeding is in May; but no information is given concerning the eggs, as the nests were in inaccessible trees, or, at any rate, such as baffled the attempts of the Chinese coolies to climb them. I am unable to discover any additional reference to these birds breeding beyond that contained in Swinhoe's notes from Chefoo to which I have alluded, and which are contained in 'The Ibis,' 1874, p. 428.

Suborder PANDIONES.

Differs structurally from Falcones in having the outer toe reversible. Plumage very close and compact, otherwise as in Falcones.

Genus PANDION.

Tip of upper mandible much lengthened, curved at right angles to the commissure and very acute; lobe variable in development. Wings long and pointed, exceeding the tail; the 3rd quill generally the longest. Tail of 12 feathers and even. Tarsi short, stout, reticulated, as are also the toes to the last joint. Soles of the feet prickly, and the claws *rounded beneath*, much curved; outer toe reversible. Feathers wanting the accessory plumule.

PANDION HALIAETUS.

(THE OSPREY.)

Falco haliaetus, Linn. S. N. p. 129 (1766).
Pandion fluvialis, Sav. Descr. Egypte, Ois. p. 272 (1809).
Aquila haliaetus, Meyer in Mey. u. Wolf, Tasch. i. p. 23 (1810).
Accipiter haliaetus, Pall. Zoogr. Rosso-As. i. p. 355 (1811).
Balbusardus haliaetus, Flem. Brit. An. p. 51 (1828).
Pandion haliaetus, Less. Man. d'Orn. i. p. 86 (1828); Gould, B. of Eur. pl. 12 (1837); Gray, Gen. B. i. p. 17, pl. 7. fig. 5 (1845); Blyth, Cat. B. Mus. A. S. B. p. 29 (1849); Bp. Consp. i. p. 16 (1850); Schleg. Vog. Nederl. pl. 30 (1854); Horsf. & Moore, Cat. B. Mus. E. I. Co. p. 52 (1854); Jerd. B. of Ind. i. p. 80 (1862); More, Ibis, 1865, p. 9; Tristram, Ibis, 1865, p. 253; Hume, Rough Notes, i. p. 231 (1869); Gould, B. of Gt. Brit. pt. xvii.(1870); Newton, ed. Yarr. Brit. B. i. p. 30 (1871); Holdsworth, P. Z. S. 1872, p. 412; Shelley, B. of Egypt, p. 203 (1872); Sharpe, Cat. Birds, i. p. 449 (1874); Dresser, B. of Eur. pt. 49 (1876).
Pandion alticeps, Brehm, Vög. Deutschl. p. 33 (1831).
Pandion carolinensis, Audub. B. N. Am. pl. 81, et Orn. Biogr. i. p. 415 (1831).
Pandion indicus, Hodgs. in Gray's Zool. Miscel. p. 81 (1844).
The Fishing-Hawk, Catesby, N. H. Carol. i. pl. 2 (1731).
Le Faucon pescheur de la Caroline, Briss. Orn. i. p. 362 (1790), also *Aigle de Mer*, Briss. ibid. p. 440.
Le Balbuzard, Buff. Pl. Enl. i. pl. 414.
"*Fish-Hawk*," popularly, America and England; *Fischaar, Fischhabicht*, German; *Fisch-*

Arend, Dutch; *Aguila pescadora,* Spanish; *Aguia presqueira,* Portuguese; *Halász-Sas,* Transylvania. *Mesago,* Japanese (Blakiston).

Machariya, Hind.; *Verali addi pong,* Tam. (Jerdon); *Matchmoról,* Beng.; *Macharang,* Nepaul; *Wonlet,* Arracan; *Ile-pew,* "Fish-Tiger," Chinese (Swinhoe).

Adult female. Length to front of cere 20·0 to 23·0 inches; culmen from cere 1·5 to 1·65; wing 19·5 to 20·5; tail 8·0 to 9·0; tarsus 2·2 to 2·3; longer anterior or middle toe 1·7 to 2·0, its claw (straight) 1·15; height of bill at cere 0·55. Weight 3¼ lb. (*Jerdon*).

Male. Wing 18·0 to 19·0 inches; tarsus 2·0 to 2·2; middle toe 1·6 to 1·8.

The above measurements are taken from a series of Asiatic examples, one from Beloochistan being the largest.

Iris yellow; cere plumbeous; bill black, paling to bluish at the gape and base of under mandible; legs and feet greenish in some, yellowish in others; claws black.

The colours of the legs and feet are variously given as greenish and yellowish. An example shot in Ireland, May 14, 1873, and examined in the flesh by myself, had the soft parts as follows:—Iris reddish yellow; cere dark plumbeous; bill blackish horn-colour, paler at gape; legs and feet pale bluish, slightly tinged with green.

Head and hind neck white, the feathers on the centre of the crown, above the eye, a postorbital band running over the ears and down the side of the neck, as also the terminal half of the occipital crest blackish brown, but less in extent in very old birds; some of the feathers on the side of the nape with dark shaft-stripes; lower part of hind neck, back, scapulars, and wing-coverts glossy pale brown, with a purplish lustre in newly-plumaged birds; longer primaries (from the first to the fifth) black-brown; the remainder and the secondaries paler brown, tipped with dull white; the inner webs white towards the base, with the brown hue partially divided into bars; upper tail-coverts tipped with white; tail sandy brown, tipped with whitish, and crossed, except in very old birds, with subdued bars of a darker hue; inner webs of the lateral feathers white; the shafts of the rectrices white; beneath white.

The cheeks striped with brown, and the chest washed with fulvous, with streaks of brown in many examples; flanks streaked partially with brown; under wing-coverts barred with brown and tipped with fulvous, those beneath the edge of the wing browner than the rest.

Obs. The brown colour of the chest seems to be an individual variation, independent, in some cases, of age, although it appears to be, as Mr. Sharpe remarks, generally more marked in the old birds, which are plainly distinguishable by their unbarred tails. In these latter, however, it varies in extent and character, being accompanied, when very marked, by an encroachment on the throat of the side-neck stripe. In some examples the underlying crest-feathers are often rufous, this being a remnant of the immature plumage, which appears to remain in such birds; in other specimens the wing-coverts retain a certain amount of pale edging.

Young. Mr. Sharpe (*loc. cit.*) describes the nestling as being "covered with down of a sooty-brown colour, except along the centre of the back, along the carpal bend of the wing, on the breast and flanks, where it is dusky white; all the feathers of the back are dark brown, with a broad tip of ochraceous buff; crown and ear-coverts blackish; eyebrow and throat white."

Bird of the year *. Above chocolate-brown, the back, scapulars, and wing-coverts with sharply defined white tips to the feathers, separated from the brown by a buff margin; the wing-coverts more conspicuously edged than the back; the postorbital stripe broader than in the adult, and terminally edged with fulvous; the white sides of the nape and the back of the neck not striated as in the adult; primaries and secondaries deeply tipped with fulvous-white; upper tail-coverts margined and tipped with fulvous; tail barred with six or seven narrow bands of brown, conspicuous on the central rectrices; fore part of crown dark brown; crest-feathers often edged with fulvous; beneath white, chest sometimes unmarked, at others washed with fulvous and streaked with brown as in old birds.

Distribution.—This cosmopolitan bird of prey, as a matter of course, takes the island of Ceylon into its range, visiting its northern parts in fair numbers during the cool season, a few birds continuing their course to the extreme south. Although previously received by Lord Tweeddale from Ceylon, Mr. Holdsworth, in his catalogue (*loc. cit.*), was the first to include the Osprey among the birds of the island, having observed it on a

---

* A Tangier example and one from Nootka Sound, North America, are identical in plumage.

R 2

solitary occasion in Galle harbour, in which locality I myself saw it in March 1872. During the period of its stay it is tolerably common on the Jaffua lake and about the shoal water surrounding the adjacent islands and skirting the coast as far south as Manaar. In this latter island, Mr. Simpson, of the Indian Telegraph Department, who resides there, informs me it is to be met with all the year round. Should this gentleman be correct in his observations, it is in all probability a resident likewise in the neighbouring island of Ramisserum, where I have seen it in January and March in greater numbers than anywhere else. Lord Tweeddale's specimens were procured in the north of the island. At Trincomalie I observed it on several occasions during two successive seasons, and in February 1877 I met with an example near Morotuwa, at the head of the Panadura lake, and not ten miles from Colombo. I have no record of the Osprey having been seen on any of the large inland tanks in the north of the island, and I believe it confines itself exclusively to the sea-coast in Ceylon.

The Fish-Hawk, although nowhere very numerically abundant, inhabits suitable localities throughout the entire globe, with the exception of the island continent of Australia, its adjacent islands, including New Zealand, parts of the Malay archipelago, and all but the northern parts of South America. There are certain places in which it is not found, and some from which it is unaccountably absent, such as the Black Sea, from which, according to Mr. Dresser's remarks of Prof. Nordmann's experience, it has not yet been recorded. In Palestine its absence from one spot is noteworthy; Canon Tristram remarks (*loc. cit.*), "In spite of the amazing abundance of fish in the Lake of Galilee, we never noticed this bird there, probably because of the absence of suitable cover." In other places it was always common in winter and in spring, and on the lagoons near the mouth of the Kishon it was to be seen perched on the naked stumps projecting from the water, a similar habit to that which I have observed in the north of Ceylon. It does not appear to be very common in Spain. Lord Lilford says that it is found about Valencia; and Mr. H. Saunders discovered it nesting in May on a crag, 700 feet above the sea, on the island of Dragonera. In Corsica and Sardinia it is more often met with in winter than in summer. It inhabits the extreme north of Europe, and breeds as high as Archangel, near which Messrs. H. Brown and Alston observed one of its eyries on the top of a gigantic blasted pine. It used to be a common bird in the British isles; but the constant robbery of its eyries has in the end tended effectually to drive it from its accustomed haunts. It still breeds in a secluded spot in Invernesshire, is occasionally met with in different places round our coasts, and now and then pays unexpected visits to some of our inland waters—an account of one of which to the large reservoirs of the Paddington Canal, in Bucks, is given in 'The Ibis' for 1865, by the Rev. H. Crewe.

Turning to the New World, we find it recorded from many parts of the northern portion of the continent, both on the seaboard and far inland, from many parts of the States, from Honduras, from the Antilles, and from the northern parts of S. America. In Africa it is not so plentiful. Captain Shelley found it frequenting the rocks on the banks of the Nile. Mr. Taylor met with it commonly in Egypt, and Mr. Drake did the same in Tangiers. Finsch and Hartlaub record it from the eastern parts of the continent, and Layard from South Africa. Governor Ussher, however, did not procure it on the Gold Coast, and it is, no doubt, less common on the western side of the continent than on the eastern. It is found throughout the northern parts of Asia, down the eastern coasts to Formosa, and occurs in Central Asia from Siberia to the Himalayas, which brings us to the closer consideration of its distribution in India, in which empire it is, during the cold season, as common as anywhere in the world. Jerdon remarks, "It is spread over all India, most abundant, of course, along the coast, where there are numerous backwaters and lagoons, but common along all the large rivers, and generally found at most of the larger lakes and tanks, even far inland." Mr. Hume, in his 'Rough Notes,' says, "The Osprey is found throughout the lower ranges of the Himalayas in the rocky gorges of all the larger streams, and along the course of the Ganges and the Jumna from their mouths almost to their sources. I have, from time to time, observed it in Cawnpoor, Etawah, Agra, and Allyghur districts. I met with it also on the Sutledge, at the Sambhur Lake, and the Nugjufgurh jheel; and I recently shot a very fine one close to Saharunpoor, on the western Jumna canal." He likewise found it, though not in any considerable numbers, on the Indus and larger sheets of water in Sindh; but on the coast, particularly in Kurrachee harbour, it was much more common. From these remarks it will appear evident that the Osprey is well distributed throughout northern India; but in the southern part it is apparently chiefly confined to the sea-coast; it is common enough at Ramisserum Island, and I have no doubt is equally so on both coasts of the peninsula; but I find no record of its having been noticed of late years in the inland districts. On the opposite side of the Bay of Bengal it is rare. It is not mentioned in Mr. Armstrong's list of the birds of the Irrawaddy delta; it is a scarce winter visitor in Tenasserim, and has not been noticed at all in the Andaman Islands. It has not been observed in either Sumatra, Java, or Borneo, and to the south-east is replaced by the closely allied but smaller Australian species, *P. leucocephalus*.

Habits.—In Ceylon the Osprey frequents the sea-coast, salt lagoons, the estuaries of rivers, and land-locked brackish lakes; but in the several continents of the globe is also found on great rivers throughout their course and on large freshwater lakes. It feeds entirely on fish, and is a most persevering fisher and skilful captor of its finny prey. Its mode of precipitating itself headlong from a great height, with an almost unerring aim, sometimes even disappearing under water in the force of its downward plunge, and emerging with its well-caught prey, has been the admiration of all lovers of nature acquainted with this fine bird. It seizes with its talons, suddenly darting out its legs as it reaches the surface, and doubtless, when striking at a fish swimming somewhat beneath the surface, having thrown all its strength into the effort and acquired its maximum amount of momentum, it is unable to check its progress, and consequently disappears for a moment. The fact of its using its talons in taking its prey militates against all capability of pursuing it under water; and although an instance has been known, as cited by Professor Newton at page 31 of his edition of 'Yarrell's British Birds,' of an Osprey having been caught in a fishing-net, it is evident that this must have been spread just beneath or, more probably, on the surface of the water. When flying about not in search of food, the Osprey proceeds at a moderate speed with tolerably quick and regular beatings of its long wings, and does not exhibit any great powers of flight. When in pursuit of fish, however, its actions are very different. On one occasion, near Trincomalie, I was startled, while intent on getting within shot of some Turnstones, by a booming noise above my head, and on looking up perceived an Osprey on its headlong course into the lagoon; launching out its legs, it dashed into the water, throwing up a quantity of spray, and immediately rose again, appearing to have missed its aim. Sir John Richardson remarks that should the fish have moved to too great a depth, it "not unfrequently stops suddenly in its descent, and hovers for a few seconds in the air like a Kite or Kestrel, suspending itself in the same spot by a quick flapping of its wings." I have seen it soaring at a considerable height, and could always recognize it at a distance from the Sea-Eagle in Ceylon by its long wings and quickly-performed circling; on descending again, it would often perch on the top of a dead tree, or make off to its favourite perch, the "guide-posts" of the channels in the shoal waters of the Jaffna peninsula. On such posts in the Paumben channel it may also be seen perched on any day in the cool season. At Manaar it roosts on the tops of dead and denuded palmyra trunks, coming to the same spot every evening some time before sundown, and flying about the palm-grove until dark. The reversible outer toe, which is so peculiar in the Osprey, and which exists to a limited extent in the Fish-Eagle, is no doubt a provision of nature to enable these birds, by an onward stroke, to strike or "rake" into the flesh of their quarry with their powerful hind claws, while with the anterior claws they clutch the fish which has been struck. It is stated that the Sea-Eagle (*Haliaetus leucogaster*) frequently robs it of its prey, pursuing it until it lets fall its well-earned prize, when the robber Eagle swoops down and bears it away.

Nidification.—In the opinion of many ornithologists, the Osprey breeds within the Indian limits; but I am not aware that its eggs have been taken south of the Himalayas. Mr. Hume has seen its nest in Kumaon, and Mr. Thompson believes it to breed on the Ganges above Hurdwar. Dr. Jerdon says, in his 'Birds of India,' that it breeds "in this country on trees;" but he does not seem to have procured its eggs. Elsewhere, in Europe, America, and Africa, its nest has frequently been found, examined, and described by naturalists. Mr. Wilson, writing of it in America, says that it is "externally made of large sticks, from half an inch to an inch and a half in diameter, and two or three in length, piled to the height of four or five feet, and from two to three feet in breadth; these were intermixed with corn-stalks, seaweed, pieces of turf in large quantities, and lined with dry sea-grass, the whole forming a mass observable at half a mile distance, and large enough to fill a cart, and formed no inconsiderable load for a horse." These materials are so well put together as often to adhere in large fragments after being blown down by the wind. During the time the female is sitting, the male frequently supplies her with fish, though she occasionally takes a short circuit to the sea herself, but quickly returns again. The Fish-Hawk lays in April, May, and June in northern climes. The number of eggs is nearly always three. Mr. Dresser remarks that in a large series he has found most of the eggs to be white, richly spotted and blotched with deep chestnut-red, and sparingly marked with a few purplish-grey under-lying blurs or markings; in one or two the deep red blotches were so close as almost to hide the ground-colour.

The specimen figured in Hewitson's 'British Birds' Eggs,' pl. iii., has the larger markings in the form of softened clouds of brownish red in a zone round the large end, while the rest of the surface of the shell is tolerably profusely covered with linear blotches of a paler hue. The average dimensions are 2·5 inches by 1·9. The singular fact is recorded by Wilson, that "the most thriving tree will die in a few years after being taken possession of by the Fish-Hawk;" and he remarks that this is attributed to the fish-oil and to the excrement of birds, but is more probably occasioned by the large heap of wet salt materials of which the nest is com-posed. The Osprey, if it really does remain all the year round in the island of Manaar, may perhaps breed on some of the gigantic Baobab-trees (*Adansonia digitata*) which are common there, and supposed to have been introduced by Arabs centuries ago.

Suborder STRIGES.

Outer toe reversible. Eyes directed forwards and encircled by a facial disk. Nostrils generally hidden by stiff bristles. Plumage soft and fluffy. Tibia more than double the length of tarsus. Tail with twelve feathers. (*Sharpe*, Cat. B. ii. p. 1.)

Fam. BUBONIDÆ.

Hinder margin of sternum always deeply cleft, two or more notches being present; furcula free; inner margin of middle claw not serrated. (*Sharpe*, Cat. Birds, ii. p. 1.)

Subfam. BUBONINÆ.

Head usually furnished with two large tufts at the sides of the forehead; facial disk imper-fectly developed above the eye; ear-conch small and without an operculum. Tarsus stout, nearly always feathered.

Genus KETUPA.

Disk undeveloped above the eye. Bill very powerful. Nostrils oval, situated at the anterior margin of the cere, and covered by the long loral bristles. Head very large, and furnished
with two long ear-tufts. Wings ample, falling considerably short of the tail; the 5th quill the
longest, the 3rd and 4th slightly shorter. Tail moderate. Legs and feet very powerful. Tarsus
longer than the anterior toes, feathered in front slightly below the knee, the rest bare and the
toes finely reticulate. Claws stout, rather straight, the inner anterior one only trenchant beneath;
soles furnished with fine spicules.

KETUPA CEYLONENSIS.
(THE BROWN FISH-OWL.)

Strix ceylonensis, Gm. S. N. i. p. 287 (1788).
Strix leschenaulti, Temm. Pl. Col. ii. pl. 20 (1824).
Scops leschenaulti, Steph. Gen. Zool. xiii. pt. 2, p. 571 (1825).
Scops ceylonensis, Steph. *l. c.* p. 54.
Strix hardwickii, J. E. Gray, Ill. Ind. Zool. ii. pl. 31.
Cultrunguis nigripes, Hodgs. J. A. S. B. v. p. 364.
Cultrunguis leschenaulti, Jerd. Madr. Journ. x. p. 90.
Ketupa ceylonensis, Gray, Gen. B. i. p. 38 (1845); Blyth, Cat. B. Mus. A. S. B. p. 37 (1849);
    Kelaart, Prodromus, Cat. p. 116 (1852); Layard, Ann. & Mag. Nat. Hist. 1853, xii.
    p. 107; Horsf. & Moore, Cat. B. Mus. E. I. Co. i. p. 77 (1854); Jerdon, B. of Ind. i.
    p. 133 (1862); Tristram, Ibis, 1865, p. 261; Hume, Rough Notes, p. 379 (1870);
    Swinhoe, P. Z. S. 1871, p. 343; Holdsworth, P. Z. S. 1872, p. 417; Hume, Nests and
    Eggs, p. 64 (1873); id. Stray Feath. 1873, p. 431; Legge, Ibis, 1874, p. 11; Ball,
    Str. Feath. 1874, p. 382; Hume, ibid. 1875, p. 38; Legge, ibid. 1875, p. 198, id. Ibis,
    1875, p. 279; Sharpe, Cat. B. ii. p. 4 (1875); Armstrong, Str. Feath. 1876, p. 300:
    Inglis, ibid. 1877, p. 16.
Great Ceylonese Owl, Brown, Ill. Zool. p. 4; *Ceylon Eared Owl*, Latham, Gen. Syn. i. p. 120;
    Great Brown Owl, *Great Horned Owl*, Europeans in Ceylon.
Amrai-Ka-ghughu, *Ulu*, Hind.; *Utum*, Beng.; *Teedook*, Arracan.
Bakamûna, Sinhalese, lit. " Fish-Owl"; *Anda*, Ceylon Tamils; *Oomuttanloovey*, Tamil (*apud*
    Layard).

Adult male and female. Length to front of cere 10·5 to 20·0 inches; culmen from cere 1·4 to 1·65; wing 14·5 to 15·8, expanse 51·0 to 54·3; tail 7·0 to 8·0; tarsus 2·5 to 2·8; mid or outer anterior toe 1·5 to 1·85, claw (straight) 0·6 to 0·95; height of bill at cere 0·5 to 0·67.

The above measurements are taken from a series of seven Ceylonese examples. Specimens of both sexes vary in size *inter se*; but males appear to be as a rule the larger of the two. Three males measure in the wing 15·3, 15·0, 14·5; four females 14·4, 14·5, 15·0, 14·5. In the size of tarsus, foot, and bill there is not the same preponderance in favour of the males. .

Iris fine clear golden yellow; cere dark olivaceous green; bill olivaceous green, in some greyish green; a dark brownish-green patch on the sides of the mandible; under mandible usually paler than the upper; tarsi and feet olivaceous greenish, the skin between the reticulations yellowish; in some examples the entire tarsus is yellowish, with the joints and toes greenish.

Above light vinaceous brown, darkest on the interscapular region and upper scapulars, and palest on the rump and upper tail-coverts; the feathers of the head, sides and back of neck with broad mesial streaks, increasing on the back and scapulars to drop-shaped patches of dark brown; the light portions of the hind-neck feathers cross-rayed with brown : shorter lateral scapulars with white outer webs and dark shaft-stripes; wing-coverts dark brown, the lesser series edged with light tawny : the greater zone of secondary coverts crossed with interrupted bars of fulvous whitish ; the inner feathers of this and the series above it with large white patches at the tips of the outer webs : the inner feathers, as well as the tertials and longer scapulars, pale at the outer edges and tipped and mottled in a bar-like form with buff : primary-coverts, primaries, and secondaries dark brown, barred and tipped with dusky buff, paling to white on the outer webs of the longer primaries, and subdued with a brownish hue on the inner webs of all the quill-feathers : basal portion of the inner webs white : rump and upper tail-coverts with narrow mesial brown lines, the feathers faintly edged with fulvous : tail concolorous with the secondaries, and tipped and crossed with four bars of dusky buff, paling to whitish at the base of the inner webs of the lateral feathers; face tawny brownish, the base of the loral plumes white and the terminal portion of the shafts black; an orbital fringe of bristly blackish feathers ; ear-tufts concolorous with the crown.

The feathers of the lower cheeks and at the side of the throat with narrow dark shaft-lines; throat more or less white (in some more than in others), with narrow shaft-stripes of brown ; breast, flanks, and under surface delicate fawn-colour, darkest on the upper breast, and paling considerably in some on the belly and longer thigh-plumes ; each feather with a deep-brown lanceolate shaft-stripe, and crossed with fine, wavy, brown rays, which are more conspicuous on the lower parts than on the breast : thighs pale tawny, unmarked; under tail-coverts lightly striped ; lesser under wing-coverts fulvous tawny, with dark brown shaft-stripes widening at the tips ; greater series white at the base, with the terminal portion blackish brown : white basal portions of the primaries tinged with yellow, as also the under surface of the adjacent bars.

Obs. The white throat-patch varies in size independently of the general light or dark hue of the under surface, specimens which have markedly fulvous breasts having the white gorge as large as any other. This, then, being a variable character in (although such has been maintained) of no value whatever as a distinguishing feature in the Ceylonese birds from any others. The latter are, however, as a rule smaller than some North-Indian and Burmese specimens; but this is a *constant character* of Ceylonese representatives of Indian species. That the Brown Fish-Owl varies in size in India will be seen from the following measurements :—Three males, wing 15·0 to 15·7 inches, three females 16·5 to 18·0 (*Hume*, 'Rough Notes'): two males, wing 16·2 and 16·0 (Chota Nagpur, *Ball*); several females 14·9 to 15·7 (Irrawaddy delta, *Armstrong*); one female 15·6 (Burmah), and two examples unsexed, 15·3 and 14·8 (Burmah and N. Bengal; measured by myself in British Museum). Two of these last three coincide exactly with Ceylonese specimens in general hue, white throat-patch, and transverse breast-striations: the third presents an individual peculiarity in its very rufescent character throughout, and almost total absence of chest-striations like *K. flavipes.* Lastly, a Cochin-China example in the national collection is quite similar to those from Ceylon.

Distribution.—This Fish-Owl is very generally diffused throughout the low country, where localities suitable to its habits occur ; but it does not appear to ascend much above the level of the deep valleys of the Kandyan Province. In the western and southern portions of the island it is found near the banks of all the large rivers, and very often about wet paddy-fields; but it is not so numerous in that division as on the eastern side or in the forests of the northern half, where its chief haunts are the borders of all the inland tanks, both large and small, and the forest-lining of the large rivers. It is found on the sea-coast where the jungle clothes the actual shore, as it does round the magnificent harbour of Trincomalie and at many other localities. In the Seven Korales it is very abundant; I have met with it there, in every sort of locality, from the insignificant water-hole ornamented with a solitary banyan, to its favourite haunt the huge Koombook-tree spreading its massive arms over the dried-up, sandy river-bed.

In the Kandyan Province it follows the banks of the Mahawelliganga from the low country into the valley of Dumbara, being a well-known bird about Peradeniya, and occurs in the valley of this river, as well as in those of its affluents up to about 3000 feet. Mr. Laurie has procured it in Kalebokka, about the same elevation, and it is likewise found in the Badulla district. In the southern ranges I have met with it up to 2000 feet, and I have no doubt it occurs generally, though not in any numbers, throughout that hill-district.

Elsewhere the Brown Fish-Owl is found throughout India in suitable spots, from the Himalayas to the extreme south, ranging into Assam, Arakan, and Burmah, as far south as the province of Tenasserim, beyond which it ceases to extend, being replaced in the Malayan peninsula by the smaller species *K. javanensis*, which likewise inhabits the Burmese kingdom. Concerning the Irrawaddy-delta district, Mr. Armstrong writes that this Owl is tolerably abundant in the thin forest-jungle surrounding the jheels between Elephant Point and China-Baker. In Cachar, Mr. Inglis records it as rather common; and about Thayetmyo and Tonghoo, Messrs. Feilden and Oates remark that it is also common. From Chota Nagpur Mr. Ball records it, and Mr. Fairbank from the Sahyadri mountains. Jerdon says that it is found in the Nilghiris to a considerable elevation, being not rare in Ootacamund. Mr. Fairbank likewise has it from the Pulani hills, although it has not yet been procured in the Travancore range. It does not appear to be found either in the Deccan or in Sindh. Beyond the limits of the Indian region Mr. Swinhoe has procured it near Hongkong; and in an equally remote locality to the west, viz. Palestine, Canon Tristram has found it. Its occurrence in these widely distant places is very remarkable. As regards the Holy Land, Canon Tristram writes, in 'The Ibis' for 1865, " We can only point to one locality as the certain residence of this bird in Palestine." It " was found by us in the wild wooded glen of Wady el Kurn, running up from the Plain of Acre. We discovered it accidentally, and at first took it for the *Bubo asculaphus*, when it bolted out of the dense foliage of a great Carob-tree under which we were standing; we thus put up no less than four individuals in two days."

Habits.—This large Owl loves the vicinity of water, haunting the banks of rivers, tanks, inland salt lagoons, the borders of sea-bays, and woods surrounding rice-fields. All who have visited the tanks in the north and east of Ceylon must be familiar with this fine bird, which is so often surprised napping in the lofty trees growing on the embankments or so-called " bunds." Its powers of vision in the day are not quick, but they are tolerably clear; on hearing the footsteps of man, it raises its large ear-tufts, and, bending down its head, stares stolidly down from its lofty perch among the green boughs, and as soon as it becomes aware of the nature of the intruder on its retreat, hastily launches itself out of the tree, and is not easily approached a second time. It is much more common in wild forest country combined with water than in cultivated districts. It sallies out in the evening with great regularity. As soon as the sun begins to sink behind the surrounding forest, it may be noticed, flapping noiselessly round some secluded cheena, or leisurely crossing the lonely tank, resounding at the hour of sunset with the booming of innumerable frogs, to the nearest conspicuous tree, and there gives out its sepulchral groan. This gloomy salutation is usually responded to by its mate, who perches close at hand, and answers by a double note, the two lonesome sounds resembling the words *gloom, oh-gloom.* At night I have often heard these notes repeated by a pair without intermission for many minutes. Layard remarks that when alarmed during the day they utter a loud hiss, subsiding into a growl. They appear to have an accustomed place of roosting, for Mr. Holdsworth notices that they " perched day after day on the same branch." This is very often in an exposed situation, and it frequently falls to their lot to be mobbed by a flock of garrulous Bulbuls, King-Crows, and other Owl-hating small birds. Fish is the favourite food, and, in fact, the usual diet of this species; but when this is not procurable, small mammals, reptiles, and even insects are devoured by them. In the stomach of one example I found a snake (*Haplocercus ceylonensis*) and some large beetles. As a proof of their miscellaneous diet, and also of their voracity, I may mention that a pair of Fish-Owls which were kept by Sir Charles Layard in the same aviary with a Brahminy Kite, fell one night upon their luckless companion, and, after slaughtering him, forthwith proceeded to devour him completely. Further, Mr. Hume records, in 'Nests and Eggs,' finding the remains of quails, doves, and mynahs in the nest of a pair on the Jumna; and in 'Stray Feathers,' vol. v. p. 16, Mr. J. Inglis writes as

s

follows concerning the food of this species :—" I once surprised a pair of them feeding on the carcase of an alligator which I had shot a few days previously."

Nidification.—In Ceylon the Fish-Owl breeds from February until May. It nests in hollow trees or in crevices in rocks near water. The nest is scanty, consisting of a few sticks, and when placed in holes of trees of nothing but the bare wood or, perhaps, a few leaves. The eggs are usually two in number, broad ovals in shape, tolerably glossy in texture, and pure white. Two that I examined from the Kurunegala district measured 2·29 by 1·72 and 2·3 by 1·78 inch.

In India this Owl sometimes tenants the deserted nest of a Fishing-Eagle, carefully lining it with grass and feathers, and occasionally constructs its own nest in the recess of a large upright fork.

Head and disk much as in the last; bill slightly longer; nostrils more oval; ear-conch rather small. Wings moderately short; 4th and 5th quills subequal and longest, and falling short of the tail by more than the length of the middle toe. Tail moderate, even. Tarsus rather short, very stout, and feathered down to the foot. Feet very large; inner toe subequal with the middle one, the outer very short; inner claw very large and long.

BUBO NIPALENSIS.

(THE FOREST EAGLE-OWL.)

Bubo nipalensis, Hodgs. As. Res. xix. p. 172; Sharpe, Cat. B. ii. p. 37 (1875).

Huhua nipalensis, Hodgs. J. A. S. B. vi. p. 362; Jerd. B. of Ind. i. p. 131; Blyth, Ibis, 1866, p. 254; Jerd. Ibis, 1871, p. 346; Blyth, Ibis, 1872, p. 89; Hume, Rough Notes, p. 378 (1873); id. Str. Feath. 1873, p. 431, and 1874, p. 468.

Etoglaux nipalensis, Hodgs. J. A. S. B. x. p. 28.

Bubo orientalis, Blyth, Cat. B. Mus. A. S. B. p. 34 (1849); Horsf. & Moore, Cat. B. Mus. E. I. Co. no. 80 (in part), p. 72 (1854).

Ptiloskelos amhersti, Tickell, J. A. S. B. xxviii. (1859) p. 448; id. MS. Ill. Ind. Orn. vol. ii.

Huhua pectoralis, Holdsw. P. Z. S. 1872, p. 416.

Huhu, *Huhu chil*, Nepalese (*apud* Hodgson).

Loko Bakamūna, Sinhalese; *Peria-anda*, Tamils in Ceylon.

Adult female. Length to front of cere 23·0 to 24·0 inches; culmen from cere 1·7 to 1·9; tail 8·3 to 9·5; tarsus 2·75; mid toe 2·0, claw (straight) 1·75; height of bill at cere 0·8. Expanse of one with a wing of 17·2, 56 inches.

Male. Wing 15·3 to 17·0 inches.

The following are individual measurements of a series of seven examples:—

	Wing. in.	Tail. in.	Tarsus. in.	Weight.	Museum or Collection.
♀ . Kalebokka	18·0	8·3	2·75	....	T. Butler, Esq., Clapton.
♀ . Haputale	17·2	9·5	2·75	3·75 lb.	Norwich.
♀ ?. Kandy district	17·2	9·2	2·50	....	Norwich.
Juv. ♂ . Kandy district	17·0	8·5	2·50	....	Colombo.
♂ . Maskeliya	16·0	7·4	2·50	....	R. Cobbold, Esq., Ardleigh.
♂ ?. Kandy district	16·2	8·1	2·50	....	British Museum.
♂ . Kandy district	15·3	....	....	....	E. Holdsworth, Esq.

Iris "yellowish brown " (*Blyth*, in opist.); bill olivaceous brown; cere olive; feet brownish, claws dark brownish horn.

Forehead, crown, and all above glossy sepia-brown, barred on the head and hind neck with narrow cross rays of fulvescent white, and on the back, scapulars, and wing-coverts with broader bars of buffy, including a terminal band of the same, which, on the longer scapulars and greater wing-coverts, is mottled with brown; inner webs of ear-tufts (which vary from 2·5 to 3·0 inches in length) barred with buff-white; outer webs of the lateral scapulars buff, the inner barred with mottled bands of a paler hue; outer webs of the anterior wing-coverts conspicuously banded with buff; primary-coverts crossed with pale mottled bars, duskier on the inner webs; primaries and

s 2

secondaries banded with smoky grey across both webs, the inner paling to buff at the edges: tail deeply tipped and crossed with fine narrow mottled bars of dusky buff; these widen and are paler towards the base.
Lores and facial disk greyish: chin, throat, and under surface whitish, washed here and there on the breast with buff: chin and ruff-feathers barred narrowly with brown: fore neck and chest banded with regular bars of the same, the distance between which increases on the upper breast; on the breast, pectoral plumes, flanks, and under tail-coverts the bars increase in width, take a pointed or slightly spear-shaped form, and are very far apart, but are three in number, as on the chest: bars of the under tail-coverts paler and narrower than those of the breast; tarsi narrowly barred with brown; under wing-coverts buffy white, marked with bar-like spots and pointed bars of brown.

Obs.   The above is a description of the largest and most mature bird* I have met with. One, probably in the plumage of the second year, has the barrings of the upper surface more buff and generally broader, the markings of the head and hind neck, especially, showing a more yellow hue than in the old bird; the scapulars have more of the buff hue on the inner webs, the markings of the wing-coverts and barring of the tail show the same characteristic; the bands on the throat and round the edge of the disk are more coalescent, those on the chest closer together, and there is a more sudden increase in the width of the interspaces on the breast than in the above example; tarsi not so strongly barred.

Young.   The nestling has the iris brown: bill fleshy white: feet dull yellowish, claws dusky.
Above and beneath white; the head and hind neck narrowly barred with brown; the back, scapulars, and wing-coverts openly banded with the same and tinged with rufescent buff, the edges of the bars whitish, contrasted with the buff ground-colour; quills dark brown, with handsomely mottled bars of smoky grey; tertials whitish, barred similarly to the scapulars; forehead and disk white: orbital fringe dark; tail smoky grey, banded with blackish brown; beneath, the under surface is tinged with greyish, and marked throughout with narrow, wavy, blackish-grey cross bars; legs white, unbarred.

Bird of the year.   After putting off the nestling dress, the bill becomes more olivaceous; the upper surface is light glossy sepia-brown, with all the pale markings bolder and yellower than in the adult; the bars on the head, hind neck, shoulder of the wing, and least wing-coverts are greyish buff; on the scapulars and greater secondary wing-coverts they are rich buff, broad and mottled conspicuously with brown: primaries and secondaries tipped and barred with pale brownish, paling on the inner webs into brownish buff: basal portion of primaries buff; tail brown, tipped deeply and banded with four bars of buff, mottled with the ground-colour.
Lores, face, and ear-coverts greyish, the former with blackish shaft-lines, and the latter with indistinct cross lines of brown; fore neck and sides of throat whitish, changing on the breast and under surface into buffy white; ruff and neck as far as the centre of the chest barred with brown more closely than in the adult, on the breast the space between the bars increases gradually to the lower parts, and on the flanks and pectoral plumes the markings are pointed; legs barred narrowly with undefined marks of brownish; under wing-coverts buff, barred like the under surface.

Obs.   The distinctive characteristic of the immature bird is the difference in width of the chest and breast interspaces, giving the appearance to a casual observer of a coalescence on the former region. Whether this character or not led to the distinctive name *pectoralis* of Jerdon for the South-Indian bird I am unable to say; it is common to nearly all Ceylonese young birds, the only exception to the rule that I know of being that of the young male (?) in the collection of Mr. Holdsworth. The ground-colour of the under surface in this is more fulvescent than in other birds which have come under my notice; the bars are not spear-shaped on the lower parts, and approach gradually from there up to the chest, where they are very close together. Mr. Hume observes ('Stray Feathers,' vol. i. p. 431) that the markings on the chest are variable in Himalayan birds also.
As regards the supposed distinctness of the Nepaul form from the Ceylonese, after giving considerable time and attention to the subject, and examining the specimens of the former in the British and Norwich Museums, I must support Mr. Hume in considering them identical. Ceylonese birds are, doubtless, as a rule smaller than northern; and guided by this, together with the peculiar feature exhibited in the widely-separated pointed bars of the lower parts, I was disposed for some time to follow Mr. Holdsworth in diagnosing them as *H. pectoralis*: but the fact of the old birds, such as the fine example shot by Mr. Laurie, and two others which I have seen in Messrs. Whyte and Co.'s establishment, coinciding exactly with Himalayan examples, settles the matter, I think, beyond dispute. In such the character of the under-surface barring, the coloration of the scapulars, and even the diminishing of

* In the possession of Mr. T. Butler, Knighton House, Clapton; shot by Mr. Forbes Laurie in Kalebokka.

the transverse markings of the head to marginal indentations are precisely similar to the like conditions in the Himalayan bird. Should further investigation, aided by the examination of a larger number of *fully* adult birds than I have been able to get together, lead to the discrimination of the insular race as altogether a smaller one than the North-Indian, I would propose the specific name of *blighi* for the former, as Mr. Bligh was, I believe, the first to procure, or, at any rate, to bring to the notice of ornithologists, the species in Ceylon. An inspection of Jerdon's figure of *H. pectoralis* in the 'Madras Journal,' 1839, vol. x., and a perusal of the description in the text of the lower plumage, does not strengthen the conviction of its identity with the Ceylon bird. The drawing shows a band across the chest, formed by a brownish *ground*-colour, and not by a coalescing of the bars, such as is never seen in the youngest of Ceylonese specimens. The description (p. 80) is in part as follows:—" Beneath white, feathers barred with brown, numerously on the throat, less so on the belly and vent, and the bars are larger and take an arrow-headed form; *a narrow pectoral band of brown with a golden tinge, and edged buff as above.*" The latter characteristic is not represented in Ceylon specimens, and reads as if it had been an abnormal one in Jerdon's bird. With regard to the superior size of Himalayan *nipalensis*, 3 adult examples in the British and Norwich Museums, irrespective of sex (which is not recorded on the labels), measure in the wing 17·5, 18·0, and 18·2 inches. In the Norwich example there is an extra bar on the feathers of the lower surface, which peculiarity likewise exists in Mr. Laurie's bird, described above.

Distribution.—This splendid Owl, the largest and most powerful of its tribe in Ceylon, is a pretty general inhabitant of the mountain-region of the island from about the level of the Dumbara valley to the upper ranges. I have never met with any examples of it from the southern coffee-districts, but have no doubt that it occurs there, and that it may have not unfrequently been killed on the estates in that part of the island. In the Kandyan Province it has been procured in the districts of Matale, Kalebokka, Dumbara, Pusselawa, Maskeliya, and Haputale. It is, however, a comparatively recent addition to the avifauna of Ceylon, having been added to the list of birds by Mr. Holdsworth in his catalogue dated 1872. The specimens brought under his notice were procured by Mr. S. Bligh in 1867 in the Kandyan district; and this gentleman has therefore the credit of discovering this fine addition to the Ceylon Strigidæ. Among the several fine examples which have been procured since Mr. Bligh's first specimens are a female in magnificent plumage shot by Mr. Forbes Laurie in Kalebokka, an equally fine bird killed by Mr. C. Cobbold in Maskeliya, an adult female procured by Mr. Bligh at Lemastota, and a male killed by coolies on his estate at Catton, in addition to all which not a few specimens have found their way to the establishment of Messrs. Whyte and Co. in Kandy. An immature example from this source is now in the Colombo Museum, and another in the British Museum. I have no information of this species having ever been shot in the low country; but doubtless on more extended research it will be found tolerably low down in the Peak forests, and I should not be at all surprised to see it occur in the ranges just above Gillymally. Elsewhere the Eagle-Owl is found in the Himalayas and the Nilghiris (if Jerdon's *pectoralis* should prove not to be a good species). Eastward of India proper it ranges into Tenasserim.

Habits.—This fine bird, as its English name implies, is a denizen of woods and forests; in Ceylon, however, it is, on the whole, more partial to isolated patna-woods than to the gloomy interior of the large jungles, and is doubtless attracted thither by the abundance of bird-life in these cool and retired ravines. It is usually found roosting in shady trees in the most confined portions of patna-dells, down which sparkling streams tumble, shut in by steep wooded banks. Should its retreat, as is often the case, border a coffee-estate, the Eagle-Owl levies contributions on the pigeons and poultry of the neighbouring bungalow, and falls a victim to the gun of the *Doré*. In such manner one of the above-mentioned examples was killed from the roof of his house by Mr. Cobbold in Maskeliya, after it had, as I am informed, decreased the population of the adjoining dove-cot. It is said to feed principally on birds, and very likely also preys on the large squirrels (*Sciurus tennantii*) common in the hill-jungles, occasionally perhaps killing hares, which are plentiful on most of the patnas in the Central Province.

In the Himalayas Hodgson asserts that it kills pheasants, and sometimes fawns of the smaller species of deer. Its exceedingly powerful talons and massive legs would certainly enable it to capture as large animals as most Eagles. I believe it to be strictly nocturnal in its habits; and Doctor L. Holden, formerly of Deltota, who observed something of its habits, informed me that it was very shy, quickly taking flight in the day when its haunts were invaded. The note of *H. pectoralis* is said by Jerdon to be a "low, deep, and far-sounding moaning hoot," and most probably resembles that of the present species. The vocal powers of the latter are, however, not restricted to a hoot; for Major Fitzgerald, R.A., in writing to Mr. Gurney in November last, and as quoted in 'The Ibis' (January 1878) by the latter gentleman, remarks of a caged bird that he had kept for years :—" In confinement the bird became quite tame, and would utter cries of pleasure at recognizing the hand that fed it. . . . . . . It was, I think, a female ; and during the period which might probably be its nesting-season, was in the habit of uttering a peculiar and incessant cry."

Nothing is known of the nidification of this species. It probably builds a stick-nest in the hollow of some large trunk, or on a deep and capacious fork between two limbs overshadowed by thick foliage, or perhaps it may deposit its two eggs in holes in large trees, merely on the rotten wood generally found at the bottom of the cavity. I commend the subject to my ornithological friends in Maskeliya, Haputale, and other likely districts in the Ceylon hills. A knowledge of this bird's breeding-habits would be a grand acquisition to the ornithology of the island.

Genus SCOPS.

Of small size. Cere prominent, the nostrils oval and pierced in the anterior margin; margin of the bill curved throughout. Ear-tufts large. Wings long and ample, reaching to the tip of the tail in some; the 4th quill the longest, 3rd and 5th slightly shorter. Tail moderately short and rounded at the tip. Tarsus long, nearly always feathered to the foot. Toes generally naked, the anterior toes subequal; the inner posterior toe rather short, finely reticulate except at the tips, which are covered with two or three transverse scutes; claws well curved and acute.

SCOPS BAKKAMUNA.

(FORSTER'S SCOPS OWL.)

Strix bakkamuna, Forster, Indische Zoologie, 1781, p. 18, pl. iii.; Lath. Ind. Orn. i. p. 56 (1790).

Otus indica, Gm. S. N. i. p. 289 (1788).

Scops griseus, Jerd. Madr. Journ. xiii. pt. 2, p. 119.

Ephialtes lempigi, Kelaart, Prodromus, Cat. p. 116 (1852); Layard, Ann. & Mag. Nat. Hist. 1853, xii. p. 106; Jerdon, B. of India, i. p. 138 (1862); Legge, Ibis, 1874, p. 11.

Ephialtes bakkamuna, Holdsworth, P. Z. S. 1872, p. 417; Hume, Str. Feath. 1873, p. 433; Legge, J. A. S. (C. Branch), 1874, p. 17; Ibis, 1875, p. 279.

Scops bakkamuna, Hume, Nests and Eggs, p. 69 (1873).

Scops malabaricus, Sharpe, Cat. Birds, vol. ii. p. 94 (1875).

Die Horn-Eule Bakkamuna, Forster.

The Ceylon Hawk-Owl, also *The Little Horn Owl*, of Pennant; *The Little Eared Owl* of some; *The Lempigi Owl*, Kelaart; *Koorooi*, Portuguese in Ceylon (Layard).

Punchi-Bassa, Sinhalese, lit. "Small Owl;" *Sin-anda*, Tamil; *Motu (apud* Layard).

Adult male and female. Length 7·8 to 8·1 inches; culmen 0·6; wing (usually) 5·7 to 6·0, expanse 19·5 to 21·0; tail 2·5 to 2·7; tarsus 1·4 to 1·5; mid toe 0·65 to 0·8, claw (straight) 0·4 to 0·43; height of bill at cere 0·3 to 0·32.

A specimen in Layard's collection, now at Poole, measures 6·4 in the wing, and another in the British Museum 6·2: these are very exceptional dimensions. Between the sexes (that is, in Ceylonese examples, which alone are here-treated of) there is no constant difference in size; two males in my collection measure 5·9 and 6·0 inches; three females 5·7, 5·8, and 6·0 inches. Examples from the Kandy district are, I think, as a rule, larger than low-country ones.

Iris reddish yellow, sometimes mottled externally with chestnut-brown, in others chestnut of various depths; cere olivo-brown; bill greenish horn-colour, pale at the base and dark brown at the tip; feet brownish olive, in some greenish, soles sickly yellow, claws dusky pale at the base.

General hue of back, tail, wing-coverts, tertials, and secondaries earth-brown, with blackish-brown mesial stripes to the feathers, particularly on the back and scapulars, and both webs marked with transverse spots of dusky fulvous and also mottled with brownish grey, on the secondary wing-coverts the fulvous or buffy markings are chiefly conspicuous on the outer webs; outer webs of external scapulars and a more or less defined collar across the hind neck rich buff, mottled with brown: crown, nape, hind neck, and ear-tufts rich blackish brown, more or less spotted with ochraceous; forehead, region above the eyes, and inner webs of ear-tuft feathers greyish buff, pencilled with brown: primaries darker brown than the secoudaries, mottled with ochraceous grey at the tips, with a series of dark-edged fulvous or buffy white spots on the outer webs and corresponding palish bars on the inner; secondaries crossed with mottled bars of ochraceous grey, changing into buff at the inner edge; tail with 5 or 6 wavy mottled bands of the same; lores concolorous with the forehead, the terminal portions of the shafts black.

Facial plumes and cheeks buff-grey, crossed with dark pencillings; chin buff-white, unmarked: general hue of throat, chest, and under surface buff, richest on the chest, and paling to whitish on the abdomen and lower thigh-plumes; the feathers of the ruff boldly tipped with blackish; the throat and chest with fine transverse pencillings of brown: feathers of the breast, flanks, and sides of abdomen with clearly-defined blackish-brown shafts, branching off into cross vermiculations of ochraceous brown; thighs and upper part of tarsi more or less marked with brown, lower part almost unmarked, in some examples the entire tarsus devoid of marking; under wing-coverts very variable in some, buff, unmarked, in others, tipped and spotted with sepia-brown; edge of the wing whitish buff.

Obs. The hue of the head and hind neck varies in depth; some examples, fully mature, have these parts but little darker than the back; it may, however, be laid down as a general rule that the oldest birds have the darkest heads, and this is usually accompanied by a richer tone in the buff markings of the upper surface and the ground-colour of the chest and flanks. Some birds have the breast-feathers crossed with yellowish-buff markings.

The large example above referred to as in the British Museum is very dark on the upper surface, and has the light markings but slightly tinged with buff; the under surface, which is conspicuous for its whitish ground-colour, has the breast-feathers very openly pencilled with transverse rays. Several examples which I have examined from the hill-districts are decidedly greyer in their light markings than low-country birds, and show an absence of the buff tinge on the breast, which is unusual in the latter.

Rufous varieties of this Owl are occasionally met with in Ceylon. A living specimen, which I had for some time at Galle, had the iris chestnut-brown, the bill fleshy brown with a dark tip; feet vinous brown. The portions of the upper surface which are buff and greyish in ordinary birds were rufous in this, and the under surface was also a rich rufescent buff.

Young. Iris in some brownish yellow, in others reddish yellow; bill dusky horn-colour, under mandible bluish; feet brownish grey. The iris darkens with age, and the bill loses its plumbeous tint, the bill and feet in yearlings resembling those of the adult, but slightly less olivaceous.

The nestling has the plumage fluffy and the upper surface whitish, closely and indistinctly barred with brownish, the amount of each colour being about equal; the head, however, is lighter than the back, being the reverse of the adult character: wing-coverts brown, with irregular buff markings, the brown hue pencilled with greyish, the greater wing-coverts darker than the rest; quills pale brown, crossed with wavy mottled bars of fulvous: the bars on the centre of the outer webs and those at the base of the inner webs unmottled; lores greyish, mottled with brown; beneath greyish white, entirely crossed closely with wavy lines.

A pair of nestlings from the same nest, which differed even in the colour of the irides, were clothed in this whitish garb. In other specimens the character of the plumage is buff or ochraceous. A bird of the year before me has the head, neck, chest, and breast buff-yellow, the crown and hind neck closely crossed with blackish; back and wing-coverts dark grey, much mottled with buff, the dark shaft-lines but little developed; lesser coverts dark brown, forming a dusky patch above the ulna; quills with the bars wider than in the adult, and the dark portions more mottled; tail as in the adult, the ground-colour towards the base not so dark; the face, throat, and fore neck with close transverse lines of brown. On the sides of the breast the adult dark-striped feathers are appearing: the abdomen and legs are whitish, and the under wing-coverts and edges of the wing are buffy white.

Obs. With regard to the specific name of this Owl, there can, I think, be no doubt that Forster's figure, in his 'Indische Zoologie,' refers to a Ceylonese bird, and is meant to illustrate his description in the text, notwithstanding that the drawing is not very like the original, and is given in a work which, though it professes to deal only with the ornithology of India, contains birds from other parts of the world. It is stated by the author that the Owl comes from Ceylon; and there is no other species in the island which could be meant to do duty for the present. In part the German description is as follows:—"Die Ringe von Federn um das Auge sind sehr hellgrau, und der äussere Ring bräunlich gelb. Der Rücken ist braun; die Brust ganz blass gelblich mit schwarzen pfeilförmigen Flecken besäet."

But that which clearly defines the Plate to be a representation of a Ceylonese bird is the faithful drawing of the well-known plant, the *Gloriosa superba*, which is depicted (not in conformity, however, with its nature) as entwining the dead tree on which the Owl is perched. In a footnote at p. 13 is contained a lengthy dissertation on the poisonous properties of the *Gloriosa superba* root, proving that Forster was dealing, both as regards bird and botany, with that which pertained to Ceylon.

Mr. Hume considers the Ceylon bird identical with the Indian species entitled *S. griseus* by Jerdon, and which is united with *S. malabaricus* of the same author by Mr. Sharpe.

Distribution.—This Scops Owl, which, next to the Fish-Owl, is the commonest of its family in Ceylon, is widely diffused throughout the island. It is, however, located more numerously in the western and southern portions than elsewhere, and wherever it is found is commoner near the sea-coast than in the interior. At Negombo, Colombo, and Galle, and the districts adjacent to these localities, it is, for an Owl, decidedly numerous. About the capital it is so common that it may be heard nightly, by those acquainted with its note, about the cocoanut-plantations in Colpetty and Slave Island, and even frequents the Fort. A little colony, which divided their quarters between the large trees in Queen's House Gardens and the old Suriah's which formerly stood in front of the new buildings in Chatham Street, several times came under my notice while I was stationed at Colombo. In the Fort of Galle it likewise frequented the most public places, being often seen in the rows of trees near the Master Attendant's house. Throughout the Galle district, as far inland as the Hinedun Pattu, it is common. In the South-east and Eastern Province it is less frequent, and further north, at Trincomalie, it is not at all common. Layard records it from the Jaffna peninsula, whence I have myself seen skins. Mr. Holdsworth found it at Aripu, in the Mannar district. In the lower hills about Kandy, and localities from there to about 3000 feet elevation, it occurs; but I do not think it has been noticed much above that height.

On the continent this species of Scops Owl inhabits various parts of the peninsula of India, being common in the south. It does not appear to extend eastward into Burmah and Malayana, where it is represented by the allied species *S. lempiji* of Horsfield. It is included by Captain Butler in his list of birds from the Mount Aboo district; but it does not appear to be found further south in the Khandala and Western Deccan region. In the northern parts of India it is abundant, according to Mr. Hume, who remarks, in 'Nests and Eggs,' "that it is widely distributed throughout the Punjaub, the North-western Provinces, Rajpootana, the Central Provinces, and Oudh." I observe, however, that Mr. Ball does not record it from Chota Nagpur, which province bounds the Central Provinces on the north-east.

Habits.—Forster's Scops Owl frequents cultivated country and the neighbourhood of towns and villages. It is found in the plantations of the natives, in their cocoanut-gardens, in low jungle, bamboo-thickets, and even in old buildings. About Colombo it is well known, being frequently heard from the verandahs of the Colpetty and Slave-Island bungalows uttering its monosyllabic note in the surrounding cocoanut-trees. In the country it takes refuge in thickets and low jungle, and is partial to the deep shade afforded by "bamboo choena." In this latter it roosts on the horizontal branches of the "bataliya," beneath a thick canopy of tangled branches; while in the vicinity of human habitations it hides in holes of trees or in old buildings. It sleeps heavily, and has but limited powers of vision by day, for it may be approached within a few yards before perceiving that it is observed; when thus roused it flies off swiftly, quickly realighting, and turning round its head in the direction of its disturber, erects its ears and regards him with a fixed stare. It is by no means shy at nights, allowing itself to be shouted at when seated in a shady tree, uttering its monotonous *whok* note before taking flight. It usually frequents thickly-foliaged trees at night, about which it captures moths and beetles, taking them, according to Layard, on the wing. I have found its stomach to contain lizards as well as Coleoptera. It is strictly nocturnal in its habits, not issuing forth from its hiding-place before dusk, and it then resorts to the shade of thick trees and utters its monosyllabic note for some time. Layard says that this is changed "when flying to *wāh-hā wāh-hā*, quickly uttered and mingled with a tremulous cry." A pair of nestlings, referred to above, which I had for some little time displayed several interesting habits. They huddled together in one corner of their box, and when awakened during the day made a rapid stamping with their feet, consisting of some half a dozen blows delivered with such rapidity that there was no appreciable

T

duration of time between them. During the night, when hungry I presume, they made a snoring and hissing noise, and continued it for hours at a time. I have heard this note in the early evening in the *Hibiscus*-trees in the Galle Fort, and infer that it is the result of hunger. When looking at me, both this pair and the rufous bird already referred to oscillated their bodies to and fro, and moved their heads awry with the most comical aspect. They held their food in their talons, two toes in front and two behind, in the same manner as the Ceylon Wood-Owl, and after nibbling at it, paused, as if considering the expediency of the measure, and then quickly bolted it whole. Mr. S. Bligh had a tame bird at Kandy that would follow him round the room, alight on his shoulder, and nestle itself in his beard.

Forster's name for this Owl, as Mr. Holdsworth has shown in his catalogue, was ill-chosen; the term *Bakamūna* applies to the large Fish-Owl, signifying that it is a fish-eater, which the subject of this article certainly is not. I have, however, myself heard Ceylonese villagers, perhaps without thinking, apply the name *Bakamūna* instead of *Bassa* to the smaller Owls; and some such mistake probably led to Forster's adoption of the name.

Nidification.—In the southern parts of the island this Scops Owl breeds in February and March. It nests in hollow trees or in holes made by Woodpeckers in palms. A nest found at Oodogamma during my stay at Galle was placed in the hollow between the frond and the trunk of a Kitool-palm (*Caryota urens*). A few leaves or grass-stalks usually line the hole in which the eggs are deposited. These are from two to four in number, spherical in shape, and of a pure glossy white, and average, according to Mr. Hume, 1·25 inch in length by 1·05 inch in breadth. Mr. Blewit is mentioned by Mr. Hume as having found nests in holes of trees, which were lined with leaves and straw. The parent bird is said to fight vigorously when her retreat is invaded.

SCOPS SUNIA.

(THE RUFOUS SCOPS OWL.)

Scops sunia, Hodgs. As. Res. xix. p. 175; Blyth, J. A. S. B. xiv. p. 182; Jerd. Ill. Ind. Orn. pl. 41 (1847); Sharpe, Cat. B. ii. p. 67 (1875).

Ephialtes sunia, Gray, Cat. Mamm. &c. Nepal, Coll. Hodgs. p. 51; Kelaart, Prodromus, p. 96, et Cat. p. 116; Layard, Ann. & Mag. Nat. Hist. 1853, xii. p. 106; Horsf. & Moore. Cat. B. Mus. E. I. Co. i. p. 70 (1854); Holdsworth, P. Z. S. 1872, p. 418.

Scops aldrovandi, Blyth, Cat. B. Mus. A. S. B. p. 36 (1849).

Ephialtes pennatus, Jerd. B. of Ind. i. p. 136 (1862); Hume, Rough Notes, p. 386 (1870); Jerd. Ibis, 1871, p. 347.

Ephialtes bakkamœna, Blyth, Ibis, 1866, p. 255; Jerd. Ibis, 1871, p. 347.

Scops pennatus, Gould, B. of Asia, part 22 (pt.); Hume, Nests and Eggs, i. p. 65 (1873).

The Indian Scops Owl, *apud* Jerdon; *Choghad Rusial, Sunya Rusal*, Nepal; *Chitta guba, Yeria chitta guba*, Tel. (Jerdon).

Bassa, Punchi-Bassa, Sinhalese.

Adult female (Ceylon).—Length (estimated from skin) 6·8 inches; culmen from cere 0·5; wing 5·0; tail 2·4; tarsus 0·75; mid toe 0·7, its claw (straight) 0·34.

The above measurements are taken from an adult specimen shot in the Kandy district. The dimensions of a "*reddish Owl*," shot in Haputale last January, concerning which Mr. Bligh writes me, which may appertain to this species (though more probably referable to the rufous phase of the next), are as follows :—Length 6·2 inches; wing 4·8; tail 2·25; tarsus 0·9. Weight 2¼ oz.

Iris " bright yellow; bill ' horny,' darker at tip " (*Whyte*); foot fleshy green; cere olive-brown.

Head, entire upper surface, sides, and lower part of throat fine rufous-chestnut; forehead and part of crown and ear-tufts with broad black mesial stripes, diminishing to narrow lines on the occiput and hind neck; back and lower scapulars with irregular black shaft-lines and transverse pencillings of the same hue; least wing-coverts with the concealed bases of the feathers blackish, the median and greater series with narrow pointed central stripes; lateral scapulars with most of the outer webs white, and a black terminal patch confined to the outer web; a few of the outer feathers of the median and greater wing-coverts with white lateral patches; edge of carpal joint white: outer webs of primaries and secondaries crossed with wavy black marks, developing on the inner webs into broader but ill-defined bars, the basal portions of the webs much mottled with blackish; tail crossed with fine, skeleton, wavy bars of blackish, those on the inner webs of the 3 lateral feathers being darker and more complete than the rest.

Most of the loral plumes and some of the feathers at each side of the forehead white at the base; terminal portions of the loral shafts black; chin fulvous; ruff-feathers with deep black tips, forming a prominent border; the uppermost series white at the centre; chest-feathers with broad mesial stripes; ground-colour of the breast and flanks white, each feather with a black mesial stripe of varying width at the centre, breaking up into transverse pencillings on a rufous ground patch, these markings so arranged as to have an incomplete and clear white bar across the centre of the feather; underlying abdominal plumes, thighs, and under tail-coverts buffy white; knees and front of tarsus rufous, posterior part of tarsus whitish; a few longitudinal dark marks on the tail-coverts; under wing mingled buff and rufous, the outer feathers marked with black.

Young. An immature male shot in the cinnamon-gardens in May last, and which is just beginning to acquire its yearling plumage, measured as follows in the flesh :—Length 6·0 inches; wing 5·0; tail 2·25; tarsus 0·75.

Iris bright yellow.

Above a paler rufous than the adult, the feathers of the forehead and crown with narrow mesial lines of black as in the adult, and all the feathers of the body with indistinct white terminal margins, preceded by a blackish but inconspicuous and narrow bar. The shorter lateral scapulars have the outer webs white, but not so pure as in the adult; the longer series with central blackish markings; the winglet with white spots on the outer webs of

T 2

the feathers, but the primaries *without them*, the entire feather being rufous, with skeleton black bars on the outer webs, and broad bands of the same on the inner; tail crossed with blackish lines.

Throat and fore neck paler rufous than the back, barred obsoletely with brownish; breast and lower parts mostly white, the feathers with mesial black lines and cross rays branching off from them; under wing white, the exterior feathers dashed with rufous and marked with black.

The least wing-coverts in this specimen are those of the full yearling dress and are dark rufous, contrasting as a band against the paler hue of the rest of the wing. Jerdon remarks of the young of Indian examples that they are duller red than the adults, with the feathers more black-shafted, and that there is much white on the lower surface.

Obs. The adult example above described, and which has lately been sent home to the British Museum by Messrs. Whyte and Co., corresponds well with the series of *Scops sunia* in the national collection, its distinguishing features being the uniform rufous upper surface and the striated head, which are characteristic of the specimens from Malacca and India with which I have carefully compared it. This marked uniformity of coloration prevents the Ceylonese specimens, now to hand, of this species from being confounded with hepatic examples of the next (*Scops minutus*), notwithstanding that there is but a very slight disparity in size between the two.

One of Hodgson's specimens from Nepaul measures, wing 5·5, tarsus 0·85; another from Madras, wing 5·2, tarsus 0·85; a third from Pinang, wing 5·5, tarsus 0·9. There are slight differences in these from the Ceylon bird, which it may be as well to notice here, and which are as follows:—Nepaul: tail less barred, the central feathers almost wanting the bars, and those on the remainder fainter and more widely separated than in the Kandy specimen; the upper breast uniform with the chest, the white ground-colour commencing lower down. Madras: tail and breast similar to the Nepaul bird. Pinang: tail almost the same as in the Ceylon bird, the rufous colouring of the chest extending further down the breast. The upper surfaces and facial markings of all correspond with those of the Ceylonese example.

Distribution.—This little Owl was first recorded as a Ceylonese bird by Dr. Kelaart, who, however, gives no particulars of its habitat beyond remarking, at page 96 of his 'Prodromus,' " *Scops sunia*, a very small, reddish-yellow, Eared Owl, is occasionally seen in the very highest parts of the mountains." Since the publication of this note, the species does not seem to have been identified with certainty until now. Layard did not meet with the bird, simply remarking, in his ' Notes,' that it was procured at Nuwara Elliya by Kelaart. Mr. Holdsworth, writing 19 years after, has but little additional evidence to adduce; he quotes Kelaart, and says, " I have some recollection of seeing a specimen from the hills which I believe was the bird Kelaart referred to, and I think the species may be included in the Ceylon list." The example from which my description is taken was caught alive at Kattakelle, near Kandy, and is the only adult bird that I myself have seen from the island. An immature bird, described in this article, was killed close to Colombo in that well-known locality the cinnamon-gardens, proving that the species is widely distributed in Ceylon, inhabiting the low country on the seaboard as well as the mountainous districts. I am unable to state with certainty that the example spoken of by Mr. Bligh in a recent letter to me belongs to this species; but I have no doubt it will be found to inhabit the Haputale as well as the Nuwara-Elliya district, in which latter Kelaart seems to have made its acquaintance.

Jerdon remarks that it is found in India in forests and well-wooded districts, but is not very common. He procured it at Madras, and likewise obtained it in the Eastern and Western Ghauts, but not in Central India. In the sub-Himalayan districts it is by all accounts a fairly common bird. Mr. Thompson found it so in the Gurwhal forests, and Captain Hutton met with it frequently in Northern India. It extends into the Malayan peninsula as far south as Pinang, of which island it is an inhabitant.

Habits.—The Rufous Scops Owl is said to be an inhabitant of wooded districts and the edges of forests. In the Kandyan Province of Ceylon it has been met with on all occasions, I believe, on the outskirts of the jungle, either resorting to the vicinity of bungalows in search of food, or, like other Owls which have been so killed, hiding in detached trees, having wandered during the night far from its accustomed habitat. Mr. Whyte writes me that his specimen was caught by a coolie perched in a mango-tree, its plumage so saturated with rain that it was quite unable to fly. I do not find much concerning its habits in the writings of Indian naturalists. Jerdon remarks that the first specimen he ever procured was found dead outside his house at Madras, and had probably been killed by the crows; he says that it has a low mild hoot, which is often heard soon after dark, and further that all he examined had fed on insects. Such is the food of most of our small Owls in Ceylon, the coleopterous class coming in for the greatest share of patronage; and doubtless the present species is as much an insect-feeder in the latter island as in India.

Nidification.—In India this Owl breeds, according to Mr. R. Thompson, from March until August, in holes of trees, usually at no great height from the ground. Unfortunately this gentleman never took the eggs, though he says the bird was common in the Gurwhal forests.

Captain Hutton states that it breeds in hollow trees, "laying three or four white eggs on the rotten wood," in March.

SCOPS MINUTUS.

(THE LITTLE SCOPS OWL.)

(Peculiar to Ceylon.)

Glaucidium malabaricum!, Whyte, Str. Feath. 1877, p. 201.
Scops minutus, Legge, Ann. & Mag. Nat. Hist. 1878, i. p. 175.
The Little Owl of Planters in Ceylon.
Punchi-Bassa, Sinhalese.

S. minimus : similis *S. malayano*, sed minor et saturatior, subtùs obscurior et brunneo magis vermiculatus : colore rufescente dorsi gulæ et præpectoris absente.

Adult. Length to front of cere (from skin) 6·0 inches; culmen from cere 0·55; wing 4·75 to 4·85; tail 2·1 to 2·3; tarsus 0·7 to 0·8; mid toe 0·75, claw (straight) 0·3; height of bill at cere 0·25. Weight 2½ oz.
Iris yellow; cere greenish; bill olivaceous brown; feet fleshy brown; claws dusky.
Above dark brown, the feathers of the head, back, rump, scapulars, tertials, and wing-coverts crossed at the centre with transverse spots of ochraceous, spotted finely and closely vermiculated on the rest of their surfaces with greyish and ochraceous grey, surrounding transverse irregular markings of blackish; feathers of the hind neck crossed with bold wavy markings of whitish, and margined with rufescent buff; outer scapulars white externally, with terminal black spots and oblique central bars of the same, edged with rufous; the primary and outer secondary coverts have their dark markings mingled with rufous patches, and set off with white spots near the tips of the outer webs; primaries and secondaries brownish rufous, mottled with blackish brown, and the inner webs banded broadly with the same; the outer webs of the first five primaries crossed with five white, blackish-margined bars, the tips paler than the rest of the feather and mottled with dark brown; tail brownish, washed with rufous on some of the feathers near the base, mottled with blackish brown and crossed with five or six bars of buff-white with black edges.
Ear-tufts concolorous with the head, and rufous at the base of the feathers; loral plumes black, with white bases; facial disk grey, pencilled with blackish; ruff pale rufous, the feathers edged and centred with blackish brown; chin whitish; fore neck and under surface, together with the flanks, closely stippled with iron-grey on a white ground, the feathers with broadish central stripes of blackish, and crossed on their concealed portions with fine, wavy, transverse, black marks; on the lower parts the stippling is more open, the under tail-coverts being chiefly white, with the markings confined to the tips; legs rufescent, with wavy brown transverse marks; under wing-coverts whitish, shaded with rufescent, and crossed with irregular markings of brown.

The above is a description of the type specimen in the British Museum. A second, killed near Kandy, is slightly larger, having the wing 4·85 inches. It has the markings both above and beneath bolder and more open on the back, the transverse white spottings are larger, and the black markings take the form of shaft-lines; the ruff is rich buff and much more deeply tipped with black, and the under surface from the breast downwards is whiter and not so closely stippled, the markings taking the form of open vermiculations, with bold mesial stripes on most of the feathers.

Another example in the Colombo Museum, kindly loaned to me by the authorities of that institution, is in a rufous phase of plumage; whether this is the result of youth or not, I am unable to say, as it has no signs of nestling attire about it. Wing 4·82 inches. Iris yellow.

Upper surface, in the distribution of its markings, similar in most respects to the second example above treated of, but the mesial striæ not pronounced, the tips of the feathers mottled with blackish grey and fulvous, and the webs across the centre rufous; the lateral scapulars have the outer webs chiefly white, tipped with mingled black and rufous, the anterior quills with rufous-white marginal spots on the outer webs; the lower ear-tuft feathers and those of the ruff a decided rufous, the latter tipped with black, anterior to which is a fulvous-white patch on each feather; the breast and flanks rufescent white; the feathers of the sides of the breast and flanks with mesial black stripes and blackish mottlings at the tips; some of the striæ with a rufous edge, and some of the feathers rufous at their bases.

Young. A young bird in nestling plumage, which I had in confinement for a short time at Trincomalie, appears to

belong to the present species; and should time and a more extended acquaintance than I have been able to cultivate with this little Owl prove that I am right, it will be apparent that some examples assume a rufous phase, and perhaps retain it through life.

The dimensions of this specimen prove, I think, that it is too small to belong to the last species.

Wing 4·7 inches; tail 2·0; tarsus 0·75.

Iris yellow; bill greenish horn; cere olivaceous; feet brownish.

The general hue of the upper surface is rufescent fulvous (the back and median wing-coverts more rufous than the rest), mottled throughout with greyish, and faintly cross-rayed or pencilled with blackish; forehead and crown with not very perceptible shaft-lines of black; lateral scapulars white, with black terminal patches; inner webs of the greater wing-coverts mottled with blackish, the outer webs of the foremost series indented with white; outer webs of the first five primaries deeply indented with white, with a black edge to each indentation; inner webs mottled and crossed with dark shadings, vermiculated with the rufous ground-colour; ground-colour of the tail rufous-grey, mottled and cross-rayed with black; outer web of the lateral feathers indented with white.

Terminal portion of loral plumes white; face and edge of forehead greyish, cross-rayed with dusky, and beneath the eye with fulvous; ruff-feathers rufous, with fine dark tips; throat and chest mottled with yellowish buff and dark grey on a white ground; breast, flanks, and lower parts white, with cross-pencillings of dark sepia-brown and rufous, the dark markings on the flanks developing into indistinct shaft-lines. Under wing buff-white, clouded with dark brown and rufous near the edge; under tail-coverts whitish, pencilled with dusky; exterior side of tibia and tarsus marked with transverse lines of rufous.

Obs. This little Owl, in its ordinary brown plumage, approaches nearer to *Scops malayanus* than to any other Asiatic member of the genus; in size and in its rufous phase it comes close to *Scops sunia*. The closely stippled under plumage peculiar to the present species does not exist in the Malayan bird, which likewise has the ground-colour of the back to a considerable extent rufous, as also the sides of the breast-feathers. It is a much larger bird, the wings of a male and female in the Norwich Museum measuring 5·7 and 5·3 inches respectively. Rufous examples of *Scops minutus* will, I think, always be distinguished from ordinary specimens of *Scops sunia* by their smaller size, and by the less uniform character of the upper-surface plumage.

Distribution.—This small Owl, which is peculiar to the island, appears to be widely diffused throughout the hills of the Central Province, while it occurs rarely in various parts of the low country. Numerically speaking it is a rare bird, very few examples having as yet been procured. It is not possible to say whether Kelaart ever met with it or not; in continuation of the paragraph I have quoted in the last article, referring to *Scops sunia*, he speaks, at page 96 of his 'Prodromus,' of "the allied species, *Scops pennata*, being a low-country bird," which seems to imply that he was acquainted with a second small Scops Owl, none of which genus inhabit the island, with the exception of the present one.

To Mr. Bligh must be given the credit of obtaining the first authenticated example, which is the type of the species now in the national collection. It was caught in the chimney of his bungalow at Kotmalie, at an elevation of nearly 4000 feet. He writes me to say that he has met with four examples in all, the most of which I know are referable to the Haputale district. In May 1874 another specimen, referred to by Mr. Whyte under the name of *Glaucidium malabaricum* (the Malabar Wood-Owlet), was shot by Mr. J. R. Hughes on the Kitlamoola Estate; a further individual was killed by Mr. Macefield on the Deltota Estate, in April last year; and some time previous to this another in the rufous stage was shot near Colombo, and preserved in the new museum. The natives who brought me my young specimen at Trincomalie stated that it was a well-known bird to them; but I am, of course, unable to say that their remarks may not have referred to the last species. In the early part of 1876 I once or twice observed a very small Owl frequenting the trees in the Queen's House Gardens, which may probably have been this species. It will be seen, therefore, that though this species inhabits the low country, it is evidently more partial to the hill-districts, affecting the higher ranges as well as the upland valleys round Kandy.

Habits.—This species appears to be an inhabitant of the outskirts of woods, gardens, isolated jungles, thickets, &c., in the vicinity of forest. Mr. Bligh, who has had more experience of it than any one else, has generally observed it in the neighbourhood of his bungalow. One example found its way into the chimney, and fell down into the fireplace stupified by the smoke. Another, to the best of his belief, took up its abode for many months near his house, testifying to its existence there by bringing into the verandah of the bungalow its quarry, and devouring it in that peculiar locality—the remains of Bats, Finches (*Munia kelaarti*), "Bush-creepers" (*Zosterops ceylonensis*), and even those of a Robin Flycatcher (*Erythrosterna hyperythra*) affording ample testimony to the meals the little depredator had silently consumed in the dead of night! It was at last shot, and at the time had taken up its abode in a thicket of passion-flowers, out of which it sallied each evening, and resorted to a neighbouring grove of tall trees.

Since this article was written, I have heard again from my friend concerning one of these interesting little birds. He writes, " I have had the pleasure of seeing another of these little Owls several times of late by the bungalow; it is no doubt the mate of the one I lately shot: it generally alights on a thick branch, and unless you see it move, you would take it to be only a knot of wood, and it keeps, as a rule, perfectly still for some minutes at a time. It has a very feeble call, different in compass to any of the smaller Owls which I am acquainted with, though similar in character; it is like a short and feeble ' woot,' as it were jerked out. It is by no means a noisy or shy bird."

Besides small birds, the food of this Owl consists of moths and Coleoptera. In confinement it has much the same manner as Forster's Scops Owl. I kept my bird in a box, and when I approached it, it threw its head back, and staring up at me oscillated its body to and fro with a low growl of alarm.

It is scarcely necessary for me to remark that the cleverly-drawn lesser figure in the Plate accompanying this article represents this little Owl. It is from the type specimen in the British Museum.

Head small, with the disk almost obsolete. Bill short, cere tumid. Nostrils oval, and near the anterior margin. Wings long and pointed, the 1st quill falling short of the 3rd, which is the longest, by more than the length of the tarsus. Tail moderately long. Tarsus stout, slightly longer than the anterior toe, scantily feathered to the base of the foot. Toes covered with hairy plumes to the tips, which are covered with one or two transverse scales.

NINOX SCUTULATA.

(THE BROWN HAWK-OWL.)

Strix scutulata, Raffl. Tr. Linn. Soc. xiii. p. 280 (1822).

Strix hirsuta, Temm. Pl. Col. i. pl. 289 (1824).

Athene malaccensis, Eyton, Ann. Nat. Hist. xvi. p. 228.

Athene scutulata, Gray, Gen. B. i. p. 85 (1844); Kelaart, Prodromus, Cat. p. 116 (1852); Horsf. & Moore, Cat. B. Mus. E. I. Co. i. p. 68 (1854).

Athene scutellata, Layard, Ann. & Mag. Nat. Hist. 1853, xii. p. 106.

Ninox scutulatus, Blyth, Cat. B. Mus. A. S. B. p. 38 (1849); Jerdon, B. of Ind. i. p. 147 (1862, in pt.); Hume, Rough Notes, p. 420 (1870); Armstrong, Str. Feath. 1876, p. 303.

Athene hirsuta, Bp. Consp. i. p. 41.

Ninox hirsutus, Bp. Rev. et Mag. de Zool. 1854, p. 543; Hume, Str. Feath. 1874, p. 151; id. 1875, p. 40.

Ninox hirsuta, Holdsworth, P. Z. S. 1872, p. 418; Ball, Str. Feath. 1874, pp. 333 & 383; Bligh, J. A. S. (Ceylon Br.), 1874, p. 66; Legge, Ibis, 1875, p. 279; id. Str. Feath. 1875, p. 368.

Ninox scutulata, Sharpe, Cat. B. ii. p. 156 (1875); Hume, Str. Feath. 1875, p. 285.

Ninox scutellata, Inglis & Hume, Str. Feath. 1876, p. 373.

The Brown Wood-Owl, Tickell, J. A. S. B.; *The Hairy Owl*, G. R. Gray.

Choghad besra, Hind.; *Kulpechak*, Beng., lit. "Death-Owl;" *Paini ganti vestam*, Tel.; *Tangki perchiok*, Lepch. (Jerdon); *Kheng-Boop*, Arracan; *Raja wali*, Malacca (Horsf.). *Bassa*, *Punchi-Bassa*, Sinhalese; *Anda*, Tamils in Ceylon.

Adult male and female. Length to front of cere 9·8 to 10·3 inches; culmen from cere 0·55; wing 7·5 to 8·3; tail 4·5; tarsus 1·0 to 1·1; middle or outer anterior toe 1·1, its claw (straight) 0·53; height of bill at cere 0·27.

The above measurements are taken from eight Ceylonese examples. The average length of wing is about 7·8; one example in my collection measures 8·0, and another of a pair in the Poole collection 8·3, quite an abnormal dimension.

Iris golden yellow; cere dusky greenish; bill blackish blue on the sides of the upper mandible, the tip and the lower mandible paler; feet dusky yellow, the soles richer yellow than the upper surface of the toes; claws blackish, bluish at the base.

Above glossy chocolate-brown, the hue of the head, back, and sides of neck darker than the rest and pervaded with a cinereous tint; lower scapulars, greater wing-coverts, and rump slightly paler than the back; the outer scapulars marked with a largish concealed white patch, chiefly on the outer webs of the feathers; the tertials barred on

U

their concealed portions with white, the terminal bauds generally showing; edge of the wing white; least wing-coverts darker brown than the rest; primaries, their coverts, and the secondaries rich deep brown, the margins of the longer primaries fulvous, and the inner webs of all the quills crossed with narrow bars, which, near the tip, are faintly lighter than the ground-colour, and near the base fulvescent buff; tail drab-brown, crossed with five deep brown bars and tipped pale, the basal bar being concealed beneath the coverts.

Edge of the forehead and base of the loral plumes white, shafts and tips of the loral and chin-plumes black; upper throat whitish, the feathers with dark shafts; sides of the face and ear-coverts concolorous with the head; chest, breast, and flanks rich chocolate-brown; the chest and upper breast-feathers margined laterally with fulvous-yellow; the centre of the breast, belly, flanks, and the lengthened tibial coverts crossed with a broad bar of white on the centre of each feather, and a patch of the same at the base: on the lower flanks some of the bars are usually interrupted at the centre; vent and under tail-coverts white, the latter sometimes barred or streaked slightly with dark brown; legs rufous-brown; the thighs spotted with buff; bases of the tarsal feathers whitish; bristles of the feet brown; lesser under wing-coverts chocolate-brown, spotted and margined with fulvous; primary under-coverts dark brown, scantily barred with buff.

Examples (even those which are adult) from Ceylon vary to a certain extent in the depth of the upper-surface colour, some being much darker on the back than others. The hue of the tail varies considerably, the oldest birds probably having the ground-colour less smoky or more cinereous than others. The yellow edgings of the chest-feathers extend down the sides of the breast in some examples, and the edges of the white bars on the under surface are conspicuously tinged with ochraceous yellow. The specimen mentioned below, from Maskeliya, and another I have seen from the Central Province are very dark above, and have the primary-coverts almost blackish brown: they are likewise very large birds.

Young [*]. A nestling, taken from the nest by Mr. MacVicar, is described to me as very like the old bird in general aspect, clothed with fluffy brown feathers above, and having brown-centred white-margined feathers on the lower parts. At two mouths its plumage greatly resembled that of an adult. There are, however, slight differences which will be noticed in the following description of a yearling bird in my collection:—Upper surface lighter brown than the adult, with the lower head and hind neck contrasting more with the colour of the back; upper tail-coverts with pale tips; greater wing-coverts paling into rufous-brown at the edges, which are very finely margined pale; longer primaries with white indentations at the outer edges; secondaries edged near the tips with whitish; tail light drab, deeply tipped with greyish, and barred with five bands of a lighter brown than in the adult; beneath the coverts there is a sixth band.

Cheeks paler than the crown; chin white; chest and breast pale chocolate-brown, the former margined with whitish, and the latter barred very broadly with white; the feathers of the lower breast and belly tipped with white; lengthened tibial plumes, vent, and under tail-coverts unmarked white, the latter of which are in the *fluffy* stage.

Obs. The Ceylonese Hawk-Owl was considered by Temminck, who described it in 1824 from the island, as distinct from the *Strix scutulata* of Raffles, from Sumatra. The specific name of *hirsuta* has accordingly been applied by most writers to the Ceylon species, as the type of *N. scutulata* was not forthcoming for purposes of comparison and discrimination. Mr. Sharpe, in his catalogue of the Owls, has given an exhaustive series of comparative descriptions of the Hawk-Owl from the Indian Peninsula, Ceylon, Malacca, Labuan, various parts of China, Japan, and Formosa, and considers them to be identical. The race from Sumatra was not then represented in the collections he examined, and this important link was wanting to complete the chain of evidence as to the widely-spread species being the *S. scutulata* of Raffles. Lord Tweeddale has, however, since received an example from Sumatra which may fairly be considered to represent Raffles's bird. It is said to correspond well with Malaccan birds, as Mr. Sharpe suggested would one day be found to be the case; and the latter I find are not separable from our Ceylonese race. In size they compare well with our birds, the wings of four which I have examined varying from 7·7 to 7·9; another individual from Rangoon, and two from Labuan, the latter slightly smaller (7·4 in the wing), are likewise not to be separated as regards plumage from Ceylon birds. The hue of the upper surface, the dark cinereous tinge of the head, and the barring of the flanks and sides of the abdomen are the same in all. Whatever the birds from Northern India, Cachar, China, and Japan may be (and it is not my province, in a work such as this, to go into the vexed question of this species), those from Southern India, Ceylon, Sumatra, the Nicobars,

---

[*] I find that I was in error, at page 280, 'Ibis,' 1875, in my description of the immature plumage; further investigation and experience have tended to show that the iris is variable in the adult, of which the bright fulvous edgings to the throat-feathers is also a frequent character.

Labuan, and Malacca represent the one species to which the oldest applied name of *scutulata* must be applied. With regard, however, to *Ninox lugubris*, which Mr. Sharpe separates from the present bird, I would remark that an immature specimen from Ceylon is quite as pale as any that I have seen of *lugubris*.

Distribution.—The Brown Hawk-Owl is widely distributed throughout the low country, and is also found in the mountain-zone at a considerable elevation. It is not uncommon in the wooded portions of the Western Province, extending from the Pasdun Korale northward through the Raygam and Three Korales to Kurunegala. It has been obtained as near Colombo as Kasbawa and Kotté. In the forest and jungle-clad country south of the Bentota river it prefers the vicinity of rivers to the interior of the woods, and on the banks of the Gindurah it is quite common. In the Wellaway Korale and throughout the Eastern Province it is pretty generally dispersed, frequenting the borders of most of the tanks and the forests beneath the Hewa-Elliya Hills. Near the sea, between Batticaloa and Trincomalie, I found it at most halting-places along the coast-road, particularly at the Virgel and Topoor. It is to be found throughout the northern forest-tract, but not so plentifully as in the Trincomalie district, appearing in the Eastern Province to be always more common near the sea-coast than in the interior.

From the hills I have it from Maskeliya, whence Mr. E. Cobbold has kindly sent me a specimen, killed at about 4000 feet elevation; in Kotmalie Mr. Bligh procured it, that being the only hill-locality this gentleman has found it in. I have never heard its hoot in the Upper hills, and infer therefore that it does not inhabit so great an elevation; it would therefore, on the whole, be considered a rare species in the Central Province, and especially as regards Dumbara and the vicinity of Kandy, which is an excellent locality for some of its family. It appears to have successfully eluded the pursuit of our energetic ornithological pioneer, Layard, for he did not meet with a single specimen until he had been nearly eight years in the island.

Elsewhere the Brown Hawk-Owl is found in various wooded districts throughout India, extending into Burmah and Siam, and down the Malay Peninsula to the Straits, taking in the Nicobar Islands in its range to Sumatra. To the south-east it is found in Labuan and the west coast of Borneo, onwards to Celebes and the Moluccas. Turning to the north again we find it, as the *N. japonica*, inhabiting China, Formosa, and Japan. Touching its distribution in India, Mr. Bourdillon records it from the Travancore, where it confines itself to an elevation of above 2000 feet; Jerdon, who combines it with *N. lugubris*, says he has seen it in the Carnatic, Malabar coast, and Central India, and that it is rare in the Deccan and North-west Provinces. Concerning these latter, Mr. Hume says that it is almost unknown there, as also in the Punjaub and Rajpootana. In Chota Nagpur, Mr. Ball records it as not common. Capt. Feilden writes of the specimens he sent Mr. Hume from Thayetmyo, and which Mr. Hume identifies as *N. hirsutus*, that it is not common in that place. The note, he remarks, is like the mew of a small kitten; but our Ceylon bird has no such cry as this. In the Irrawaddy Delta, at Elephant Point, Dr. Armstrong found it abundant amongst clumps of trees and thin jungle near the coast. His specimens are, however, larger than the true *scutulata*, and perhaps are the same as the Cachar birds, which Mr. Hume separates as *N. innominata*.

Habits.—This Hawk-Owl has a marked preference for the vicinity of water; it is an insect-feeding species, and finds an abundance of such food near the borders of tanks and on the banks of rivers flowing through forest. It takes up its abode by day in thick jungle, particularly that description which is found growing to a height of about 30 feet at the upper borders of tanks, and which is densely matted at the top, forming a most suitable canopy from the rays of a tropical sun. Here the Brown Owl roosts, and, sleeping with "one eye open," does not admit of an easy approach; directly his haunts are invaded, out he shoots as sharply as any shy diurnal bird, and, taking sometimes a considerable flight, retreats into the most suitable cover he can find. In the hills it seems to frequent the interior of the forest, as Mr. Bligh informs us (*loc. cit.*) that he found three sitting together on a branch in "dense jungle," proving that it is more than usually sociable for a bird of its ilke. It hoots in the evening just after sundown, and is much more loquacious on moonlight nights than when it is dark. About 10 o'clock, after feeding, it recommences its not unmelodious hoot, resembling *whoŏ-wuk, whoŏ-wuk*, and which Layard not inaptly likens to the lowing note of the Bronze-winged Pigeon (*Calcophaps indica*). On a fine night it may be heard at a long distance in the almost unbroken stillness of the Ceylon forest, accompanied occasionally by the deep bay of the Sambur deer or the

u 2

complaining cry of the Loris (*Stenops gracilis*). In the morning it calls until a late hour, appearing to be regardless of the scorching rays of an eight o'clock sun, at which time I have seen it in an exposed situation on the banks of the Gindurah giving out its last matutinal cry. These Owls feed almost exclusively on beetles, moths, and grasshoppers, and seem to take their food until retiring in the morning, the stomach of a bird I killed in the Wellaway Korale at two o'clock in the afternoon being filled with undigested Coleoptera. They capture insects on the wing, and have the movements of Goatsuckers while hawking. Mr. Davison records of the allied Andaman species (*N. affinis*), that he observed it at Camorta, Nicobars, *hovering* in front of a cocoa-nut-palm, taking short circular flights from its perch, from which it would now and then dart suddenly up to a height of 15 or 20 feet. The singular cries attributed by Tickell and Dr. Hamilton to the North-Indian species are not applicable to our Ceylon bird. These writers liken them to the noise made by a strangling cat or a hare when caught by hounds. Besides the well-known hoot which I have referred to above, it is possible that this species is the author of a singular note which I have heard in the north-east and south-east of Ceylon, but which I never succeeded in identifying. It may be likened to the syllables *tohok—chok-korok*, uttered in moderately slow and even time and repeated for a long interval.

Nidification.—This species breeds in the early part of the year. Layard records shooting a female in November with the ovaries distended with eggs; and a nest found at Kæsbawn in the first week of April by the taxidermist of the Colombo Museum, Mr. Hart, contained one egg. This was pure white, of course, round in shape, and measured 1·45 by 1·27 inch. Another nest, containing one nestling, was found by Mr. MacVicar in April 1873, near Bopé. It was situated in a hole in a mango-tree, about 15 feet from the ground; at the bottom of the cavity there were no materials, the chick reposing simply on the dead wood of the tree.

Of small size. Bill short, rapidly curved from the cere. Cere tumid. Nostrils circular. Facial disk obsolete, the loral plumes very long. Wings short, rounded, falling short of the tail by more than the length of the tarsus; 1st quill short, as in *Ninox*, 4th and 5th quills subequal and longest. Tail moderate. Tarsus stout, longer than the anterior toes, well feathered. Toes covered with hairy plumes, claws rather long and acute.

GLAUCIDIUM CASTANONOTUM.

(THE CHESTNUT-BACKED OWLET.)

(Peculiar to Ceylon.)

Athene castanoptera, Blyth, J. A. S. xv. p. 280 (1846, nec Horsf.).

Athene castanotus, Blyth, Cat. B. Mus. A. S. B. p. 39 (1849); Kelaart, Prodromus, Cat. p. 116 (1852); Layard, Ann. & Mag. Nat. Hist. 1853, xii. p. 105; Blyth, Ibis, 1866, p. 259.

Tænioglaux castanonotus, Bp. Rev. et Mag. de Zool. 1854, p. 544.

Noctua castanonota, Schl. Mus. P.-B. *Striges*, p. 34 (1862).

Athene castaneonotus, Blyth, Ibis, 1867, p. 295; Hume, Rough Notes, p. 412 (1870).

Athene castaneonota, Gray, Hand-l. B. i. p. 39 (1869); Holdsworth, P. Z. S. 1872, p. 418.

Athene castanonota, Legge, Ibis, 1874, p. 11.

Glaucidium castanonotum, Sharpe, Ibis, 1875, p. 259; id. Cat. Birds, ii. p. 215 (1875).

The Ceylon Chestnut-winged Owl, Kelaart, Prodromus; *Chestnut-winged Owl*, Europeans in Ceylon.

Punchi-Bassa, Sinhalese; *Sin-Anda*, lit. "Small Owl," Tamils in Ceylon.

Suprà castaneo-rufus, scapularibus extùs nigro et fulvo fasciatim notatis, interdum sed rariùs albo maculatis : tectricibus alarum extùs obscurè nigro transfasciatis, majoribus vix fulvo apicatim notatis : primariis fuscescenti-nigris, extùs rufescenti-fulvo indentatis, secundariis castaneis obscurè nigro transfasciatis : supracaudalibus et rectricibus nigricantibus, anguste fulvo apicatis et 9-fasciatis : pileo cum nuchâ et collo postico, colli et capitis lateribus nigricanti-brunneis, angustè rufescenti-fulvo transfasciatis, interscapulio quoque paullò nigricanti fasciato : torque collari albo indistincto, plumis quibusdam longitudinaliter et irregulariter albo notatis : plumis anteocularibus albidis, scapis elongatis nigris : plumis oculo circumdatis albidis : mento albo, utrinque ad genas anticas triangulariter extenso : maculâ jugulari magnâ albâ : plagâ latâ gulari nigricanti-brunneâ fulvescenti-albo transfasciatâ : pectore summo et lateruli nigricante, castaneo lavato et rufescenti-fulvo transfasciato : pectore medio, abdomine et subcaudalibus purè albis, his medialiter nigro notatis : hypochondriis rufescenti-brunneo longitudinaliter maculatis : tibiis et plumis tarsalibus albis brunneo maculatis et fasciatis : margine alari albido : subalaribus et axillaribus albis flavo vix lavatis, illis conspicuè subterminaliter nigro notatis : rostro pallidè olivaceo, rictu et cerâ plumbeis : pedibus olivaceis : iride lætè flavâ.

Adult male and female. Length to front of cere 7·0 to 7·4 inches; culmen from cere 0·6 to 0·62; wing 4·9 to 5·55; tail 2·3 to 2·6; tarsus 0·9 to 1·1; middle or outer anterior toe 0·8, its claw (straight) 0·55; height of bill at cere 0·3. Expanse 18·7 inches; weight 4 oz.

The above measurements are taken from a series of seven examples, in which the males exceed as a whole the females, although one of the former has the wing 4·9 inches.

Iris primrose-yellow, in some slightly mottled with brown at the outer edge; eyelid dark olive-brown; cere and gape dusky greenish; bill greenish horn-colour; feet olivaceous, in some "woody" green, soles yellowish; claws brown, pale at base. Layard erroneously describes the iris of this species as reddish brown.

Head, sides, and back of neck, down to the interscapular region, cheeks, throat, chest, and sides of breast dark brown, everywhere narrowly barred with whitish, more or less tinged with buff, particularly on the head and back of neck, where the markings, in some, are rufous-white; back, scapulars, upper tertials, and all the wing-coverts reddish chestnut, with indications, more or less distinct, of dark bars across the feathers; primaries, with their coverts, and secondaries brown, pervaded with a chestnut hue, and barred with the hue of the back, which, towards the base of the inner webs, turns into fulvous-buff and spreads over the feather; the outer terminal bars of the primaries pale fulvous; edge of the wing pure white; upper tail-coverts and tail brownish black, the former barred with buff-white or pale rufescent, and the latter tipped and crossed with seven narrow non-corresponding bars of white.

Loral plumes black, the lower ones barred with whitish; a patch beneath the cheeks, a large space in the centre of the fore neck, centre of the breast, and all the lower parts white, the feathers on the sides of the breast, belly, and thigh-plumes with broad shaft-streaks of rufous-brown, or in some blackish brown; under tail-coverts in some unmarked white, in others marked with a few dark streaks; legs whitish posteriorly, the thighs barred with blackish brown; bases of the tarsal feathers blackish, showing on the surface: plumes of the feet greyish; secondary under wing-coverts white, the feathers beneath the point of the wing spotted with dark brown and ochraceous; base of the primaries yellowish white.

Obs. Scarcely any two specimens of this Owl are marked alike, the amount of indistinct dark barring on the chestnut mantle differing in almost every example. Some birds have the feathers at the lowermost portion of the hind neck boldly barred with white, others have them marked with spear-shaped centres of white; and in some, again, the rufous hue of the back obscures the light bars for some distance up the hind neck. The most singular variation, however, exists in the casual occurrence of white feathers in the scapulars. I have observed this on two examples, one of which forms the subject of the figure in the Plate. In this the outer webs of the lateral scapulars are white, surrounded by a blackish-brown edging. In this characteristic the species shows an inclination towards *G. castanopterum*, the Javan Owlet, which has the outer feathers of the greater wing-coverts, as well as the lateral scapulars with the outer webs, white: the lower breast and flanks are likewise boldly dashed with broad longitudinal streaks of rufous-brown in this latter species. It is also a larger bird, measuring from 5·7 to 6·1 in the wing. It is worthy of remark that the example delineated in the Plate has the stripes of the lower parts exceedingly rufous, approaching in this respect also to the Javan bird.

Distribution.—This pretty little Owlet, one of our peculiar Ceylonese forms, was considered by Kelaart to be confined to the hill-zone. It was discovered by Dr. Templeton, and described by Mr. Blyth of Calcutta, from specimens forwarded to him by the Doctor in 1846. It is found chiefly in the mountains of the island and the low country of the western and southern portions. It is tolerably common in Saffragam and in the Hewagam, Pasdun, and Raygam Koralcs, and is not unfrequent near Colombo. I have obtained it at Galkissæ, and Layard speaks of it as being very common near Colombo in 1852, but remarks that for nine years previously no specimens had been procured in the neighbourhood. This was, perhaps, from want of search, for it breeds not far distant from there. It occurs in the Kurunegala district and also in the south-western wooded hills. It is found in the jungles at the base of the Haputale hills and on the north side of the hill-zone at the foot of the Matale ranges, but how far north it extends I am unable to say. I have never met with it in the northern forest-tracts, nor on the coast from Batticaloa northwards; it has been procured on the west coast as high up as *Madampe*, beyond which I am not aware that it has been traced. In the Kandyan Province it is a common bird and widely distributed, being well known in all the coffee-districts, among which may be mentioned, more particularly, Dumbara, Kalebokka, Haputale, and Muskeliya. It is not uncommon in the main range, in which I have met with it as high as Kandapolla, 6300 feet, and Dr. Kelaart has it in his list of birds from Nuwara Elliya.

Habits.—The Chestnut-winged Owl inhabits by choice forest and thickly-wooded country, but it by no means confines itself to jungle, for in the Western and Southern Provinces it is fond of the areca-palm and jack-tree groves, among which the Sinhalese build their habitations, close to the doors of which I have sometimes heard it, and on one occasion killed it. It perches in the top branches of tall trees and is very shy. It is crepuscular as well as nocturnal in its habits, issuing from the umbrageous retreat in which it has passed the day as early as four o'clock, and flying from tree to tree in its vicinity, calling continuously until sundown. Its note, which is a repeated guttural cry resembling the syllable *kraw*, is again heard in the morning shortly after daybreak, and is sometimes continued on gloomy days until 8 or 9 o'clock. . I have never heard the

Cuckoo-like call spoken of by Layard as belonging to this Owl, and am inclined to think that he, like myself, mistook the note of the Hawk-Owl (*Ninox scutulata*), which answers to his description, for that of this species. Its usual food consists of Coleoptera and lizards, the former of which it takes on the wing. My friend Mr. Forbes Laurie has seen these Owls in the Kalebokka district hawking at sundown about wooded streams, and capturing beetles. Higher game than either of these, however, is sometimes aspired to; for Mr. Cobbold, of Maskeliya, informs me that he has witnessed one of these birds attacking a squirrel, and others have known them to kill small birds, such as Finches (*Munia*) and the Hill White-eye (*Zosterops ceylonensis*). This little Owl sees well in broad daylight, and has a very acute sense of hearing.

Nidification.—This species breeds, in the west of Ceylon, during March, April, and May. It lays in a hole in the trunk or limb of a tree, the cocoanut-palm being sometimes chosen; the eggs are deposited on the bare wood, and are two in number. A pair which I examined, and which were taken from the nest by the taxidermist of the Colombo Museum, were oval in shape, pure white in colour, and measured respectively 1·41 by 1·15 inch, and 1·34 by 1·08 inch, showing a considerable disparity in size.

The right-hand figure in the Plate accompanying my article on *Scops minutus* represents the example above referred to, with more white on the scapulars than I have seen in any other. Mr. Keulemans's talented pencil has portrayed this Owl in an attitude very characteristic of the genus *Glaucidium*.

GLAUCIDIUM RADIATUM.

(THE JUNGLE OWLET.)

Strix radiata, Tickell, J. A. S. B. ii. p. 572.
Noctua perlineata, Hodgson, J. A. S. B. xi. p. 269.
Athene erythroptera, Gould, P. Z. S. 1837, p. 136.
Athene undulata, Blyth, J. A. S. B. xi. p. 457.
Athene radiata, Blyth, J. A. S. B. xv. p. 281; id. Cat. B. Mus. A. S. B. p. 39 (1849); Horsf.
    & Moore, Cat. B. Mus. E. I. Co. i. p. 67 (1854); Jerd. B. of Ind. i. p. 143 (1862);
    Hume, Rough Notes, ii. p. 409; id. Nests and Eggs, i. p. 70 (1873); Ball, Str. Feath.
    1874, p. 383; Hume & Butler, ibid. 1875, p. 450.
Tænioglaux radiata, Bp. Rev. et Mag. de Zool. 1854, p. 544.
Noctua radiata, Schl. Mus. P.-B. *Striges*, p. 34.
Glaucidium radiatum, Sharpe, Ibis, 1875, p. 259; id. Cat. Birds, ii. p. 217 (1875).
Glaucidium malabaricum, Legge, Str. Feath. 1876, p. 242 (first record of species from
    Ceylon).

The Barred Owlet of some; *Jungli Choghad*, Hind.; *Chagad*, Nepaul; *Chota Ralpencha*,
    Beng.; *Adavi paine gunte*, Tel. (Jerdon).

Adult female. Length to front of core 7·9 inches; culmen from cere 0·7; wing 5·1; tail 2·6; tarsus 1·0: mid toe 0·8, its claw (straight) 0·5; height of bill at cere 0·3.

The above measurements are taken from the only specimen procured in Ceylon. A series of North-Indian *G. radiatum* gives as follows:—wing 4·8 to 5·4 inches; tail 2·5 to 2·7.

Iris pale greenish yellow; bill and core dusky greenish, tip of the mandible yellowish; feet greenish yellow.

Entire head above, sides and back of neck, back, scapulars, wing-coverts, upper tail-coverts, and tertials a dark and somewhat ashen brown, closely and narrowly barred with rufescent white on the upper parts, and with white on the longer scapulars, tertials, and tail-coverts; the lateral scapulars with broad bars of white on the outer webs; lesser wing-coverts obscurely barred with rufescent; winglet, outer median and greater coverts dark brown, barred narrowly with rufous, the outermost median feathers with a broad white patch on their outer webs: primaries and secondaries rich hair-brown, crossed with bands of rufous, paling into whitish at the edge of the longer primaries and near the tips of the secondaries: tail darker brown than the primaries, tipped and crossed with seven narrow bars of white.

Lores blackish, the basal portions of the webs whitish; face and ear-coverts concolorous with the head, but more openly barred; chin whitish, the plumes tipped with black; beneath the cheeks a broad band of white running beneath the ear-coverts, and an extensive patch of the same on the lower part of the fore neck; across the throat a band of brown, narrowly barred with pale rufescent, blending into the markings of the fore neck; chest, sides of breast, flanks, and thigh-coverts a blacker brown than the upper surface, barred on the chest with fulvous-white, and on the lower parts with broader bands of white; down the centre of the breast, the abdomen, under tail-coverts, and legs white: outer side of thighs and upper portion of tarsus barred with brown: under tail-coverts marked with bar-like spots of the same; under wing-coverts fulvous-white, paling into white at the edge of the wing, and marked down the centre with a longitudinal band of rufous, the feathers composing it spotted with brown; under surface of base of primaries rufescent.

Obs. The example from which the above description is taken corresponds with a number of North-Bengal, Darjiling, and Nepaul examples in the collection of the British Museum, and is another of the singular instances, exemplified in *Spizaetus kelaarti* (the Ceylonese race of *S. nipalensis*), *Bubo nipalensis*, and others, in which a North-Indian bird is found to extend its range to Ceylon over the heads, so to speak, of the South-Indian and neighbouring species. Not being well acquainted while in Ceylon with either *G. radiatum* or *G. malabaricum*, I naturally assigned my bird ('Stray Feathers,' *loc. cit.*) to the latter, as it came tolerably close to it in description. It has, however, I find, no pretensions to a relationship with the southern form, which, besides its more rufous colouring, has a smaller wing—three specimens having, respectively, wings measuring 4·8, 4·9, 4·8 inches, and, as a rule, more white about the fore neck. The Ceylonese bird, however, in the less rufescent tint of the upper-surface bars, and in the somewhat blacker hue of the dark flank-bands, has some slight difference to Bengal birds, but no more than is generally the case with insular examples of northern forms. The amount of white on the outer secondary wing-coverts and the type of barring on the tail are identical in both; in fact a specimen in my collection procured by Mr. A. Anderson in North India is, with the exception of the slightly rufescent bars of the upper surface, the counterpart of the bird described in this article.

Distribution.—This curious Owlet is in reality not an uncommon bird, but it appears not to have been procured in the island by any one but myself. Guided solely by the clue to its range afforded me in its remarkable note, I think I shall not be in error when I say that it is widely distributed, but not so much so as the Chestnut-winged Owlet, being, for the most part, confined to the southern half of the island, extending up the eastern side, perhaps, to the termination of the heavily-wooded country to the south of the Virgel, and occurring in the Uva district of the Central Province. There is, however, no reason to infer that it may not exist in the northern forests, but I have never heard it in them. I first met with it in 1873, while encamped in the recesses of the extensive timber-forests in the hills on the south bank of the Gindurah. In the same year my acquaintance with its extraordinary call was renewed in several parts of the low country between Haputale and Hambantota, but no example was procured. In 1875 I came upon it again in various localities between Batticaloa and the base of the Hewa-Elliya range, and also heard its hoot in the jungles on Namooui-kuli mountain, near Badulla. From the number of birds I heard in the east of Ceylon, I infer that its head-quarters are in that part of the island; and, as a hill-bird, it may (in common with other species, which range from the eastern side into Uva only, without going west of Nuwara Elliya) be confined to the eastern portion of the mountain-zone, or, on the contrary, be found throughout the whole of it; for I have no doubt that it will some day be met with in the Peak jungles, which are similar in character and climate to those of the south-western district.

The habitat of this Owl on the mainland, according to Mr. Hume, is "chiefly the sub-Himalayan country and the lower ranges of the hills themselves as far west as Mandi." It is, however, found in widely-scattered districts throughout India. Though it does not appear to be found in Lower Bengal nor in the plains of the North-west and Central Provinces, yet Mr. Ball records it as not uncommon in Chota Nagpur. It has been procured in parts of the Madras Presidency and at Anjango on the Travancore coast. Captain Butler records it as an inhabitant of the woods at the foot of Mount Aboo, though it does not occur anywhere else in the Guzerat district, nor in the Kandhala region worked by Mr. Fairbank. Dr. Cantor has procured it at a place called Keddah in Malacca.

Habits.—The Jungle-Owlet frequents lofty timber-forests (the "Mukalana" of the Sinhalese), the dense jungle generally growing in the Eastern Province, luxuriant woods in the Park country, and even low scrubby jungle near the sea-coast. In the latter situation I met with it at Tevalāmune, on the Batticaloa lake. Its habits are more diurnal than any other Owl I am acquainted with, and its curious call attracts notice wherever it is to be found. This is, for the most part, uttered by day during dull mornings and afternoons, or at any time when the bird is disturbed in the forest by a sudden sound, such as the report of a gun or the bark of a dog; at such times its loud spasmodic call impresses the hearer with the suspicion of anger in the little "bird of ill-omen" at being disturbed in its sylvan retreats! The effect of diurnal gloom on its disposition seems very marked, as it hoots at the fancied approach of night as soon as the sun is overcast with the quickly-passing showers so common in the Ceylon jungles. The only example procured by me, after many attempts to satisfy myself as to the authorship of such strange notes, was shot in the banks of the Maha-oya, on the new Batticaloa Road, about 10 o'clock on a damp August morning, when drenching showers were following each other at intervals of five or ten minutes, causing the little fellow to shout with unusual frequency, and enabling me to track him through the dripping underwood. He was continually on the move, and when overtaken was seated on a high *Euphorbia* tree beneath a dense cluster of its massive leaves. The note commences with the syllable *kāōw* slowly repeated and gradually accelerated until changed to *kāōw-whap, kaōw-whap,* which

x

increases in loudness until it is suddenly stopped. Tickell remarked of it in India that it kept up its clamorous cries during the greater part of the day. Thompson likens the note to the syllables *too-roo-roo-roo*, which does not accord with my experience of the singular Ceylon cry. Its flight is quick and straight, performed with vigorous flappings of the wings, and is very un-owl-like in character.

Mr. Thompson, as quoted by Mr. Hume in his 'Rough Notes,' remarks, "Its flight is both rapid and strong, with closed wings like that of the Besrah. It kills and devours all kinds of small birds, even taking them in the daytime. I had one caught which came down at a chicken three times all by itself, and killed it in the broad daylight." Notwithstanding these rapacious propensities, insects doubtless form its chief food, as will be seen from the extract subjoined below; the stomach of my specimen was crammed with beetles, a favourite food with small Raptors in Ceylon.

In Mr. Hume's notice of this species in 'Rough Notes' is contained the following interesting account of its habits. He says (p. 410), "These birds in confinement tame readily and eat raw or cooked meat. I have seen them in the daytime, in the shady verandah in which they were kept, kill and eat crickets, ants, and butterflies. A pair of sparrows made a nest on the interior cornice of the enclosed end of the verandah in which they lived. At first the sparrows teased and bothered the owls the whole day long at intervals, the owls merely retreating inside their box, chattering angrily; but one night two of the three got loose, killed both sparrows, eating their breasts and entrails, and all the young ones, of which not a trace was left*. They did not attempt to leave the place (this was at Dehra), and I let the third loose, after which they gradually grew wilder (returning, however, for some weeks for the day to their box), and at last left the house altogether, although, when I gave it up, they were still hanging about the trees in the very jungly compound. They were excessively noisy birds, both by night and even at intervals by day, in fact, at times, a perfect nuisance. Dogs were their abomination; and the way in which, menaced by a puppy of mine, who evidently thought it famous fun, they would lower their heads, set out their wings and ear-coverts, and 'curse and swear' (a mixture of hissing and chattering utterly indescribable in words) was really quite 'edifying'"!

Jerdon says that it flies actively about during the day when disturbed, and testifies to having found it rarely in small flocks—probably a young brood with their parents.

Nidification.—In India this Owlet breeds from April until May, which is doubtless the season for its nesting in the south of Ceylon. It nests, according to Mr. Thompson, in holes in small trees. The eggs have not been procured; but the young ones, which have several times been taken, are from three to four in number.

---

* These were evidently bolted whole.—W. V. L.

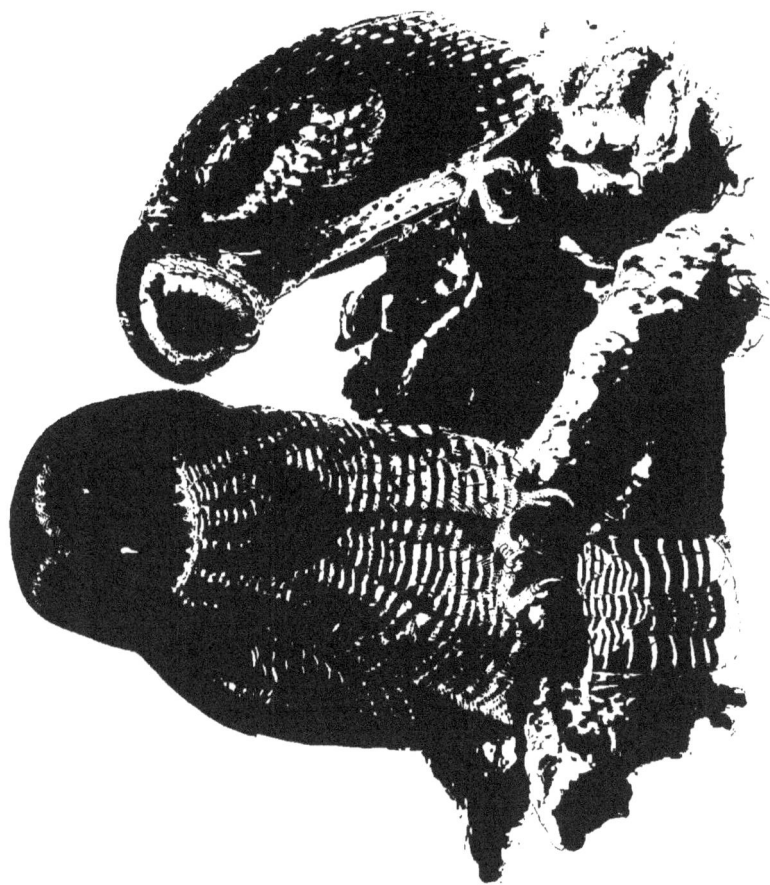

ACCIPITRES.

STRIGES.

BUBONIDÆ.

SYRNIINÆ.

Genus SYRNIUM.

Of moderate size. Bill stout, cere advanced; a well-developed facial disk incomplete above the eyes. Head without ear-tufts. Wings moderate, rounded; the 4th quill the longest, the 1st falling short of the 4th by the length of the tarsus and middle toe. Tail moderately long. Tarsus stout and thickly feathered. Toes in some thickly feathered, in others furnished with hair-like bristles, and sometimes bare; outer anterior toe longer than the inner; claws long and powerful.

SYRNIUM INDRANI.

(THE BROWN WOOD-OWL.)

Syrnium indrani, Blyth, Cat. B. Mus. A. S. B. p. 40 (1849, in part); Kelaart, Prodromus, Cat. p. 116 (1852); Layard, Ann. & Mag. Nat. Hist. 1853, xii. p. 107.

Syrnium indranee, Jerd. B. of Ind. i. p. 121 (1862, in part); Holdsworth, P. Z. S. 1872, p. 415; Hume, Str. Feath. 1873, p. 429; Legge, ibid. 1874, p. 342; Rainey, ibid, 1875, p. 332; Sharpe, Cat. B. ii. p. 282 (1875).

Bulacca indranee, Hume, Rough Notes, ii. p. 347 (1870).

Syrnium ochrogenys, Hume, Str. Feath. 1873, p. 431.

Brown Owl, Devil-bird, Europeans in Ceylon; Oulama Owl, Kelaart.

Ulama, Sinhalese.

Ad. brunneus, capite et dorso concoloribus: collo postico obscurè albido vel pallidiore brunneo fasciato: scapularibus eodem modo fasciatis, extùs latè albo transversim notatis: rectricibus alarum bruuneis ochrascenti-brunneo vel albido transfasciatis, majoribus extùs latiùs albo fasciatis: supracaudalibus bruuneis albido angustiùs transfasciatis: rectricibus bruuneis albo 12-fasciatis: areà faciali corvinà vel ochrascenti-fulrà inconspicuè brunneo fasciatin notatà: plagà superciliari albidà, scapis plumarum latè nigris: plumis oculis circumdatis et plagà antoeculari nigris: limbo faciali brunneo: corpore reliquo subtùs fulvescente vel albido et brunneo regulariter transfasciato: pectoris lateribus et plumis tibialibus et tarsalibus obscuriùs fasciatis: subalaribus pectori concoloribus: remigibus subtùs brunneis ochrascenti-fulvo intùs fasciatis, versùs basin latiùs notatis: rostro cœruleo-albicante: iride castaneà: pedibus corneis.

Adult male and female. Length to front of cere 17·0 to 18·0 inches; culmen from cere 1·0 to 1·1; wing 11·75 to 13·5 (average of seven examples 12·75), expanse (wing 13·25) 43·5: tail 6·5 to 7·5; tarsus 2·0 to 2·3; outer anterior toe 1·4 to 1·6, its claw (straight) 0·0; height of bill at cere 0·55. These dimensions are from a series of six Ceylonese examples.

Iris chocolate-brown, or a slightly reddish brown in some; pupil, in light, bluish; cere dusky bluish or olive: bill bluish near the cere, culmen darker than the sides, tip whitish horn-colour; toes dusky bluish; claws bluish horn, darker at the tips.

Head and upper surface glossy sepia-brown, palest on the secondary wing-coverts, longer scapulars, and rump, where the feathers are narrowly crossed with wavy bars of buffy white; lower tertials fulvous-brown, and barred similarly

x 2

to the adjacent coverts; least wing-coverts uniform, like the back; winglet and primary-coverts deep brown, barred with dusky fulvous; primaries paler brown than their coverts, secondaries somewhat lighter still, the whole deeply tipped with white and barred on both webs with dusky fulvous, paling into whitish at the inner edges, and into buff at the basal portions of the longer primary outer webs; first primary darker than the rest, and unbarred on the outer web; tail deep brown on the terminal portion, paling towards the base, tipped with white and barred with narrow non-corresponding bands of buffy white.

Disk rufous tawny, changing into whitish above the eyes, in some examples faintly barred with dark wavy lines; a circle of black feathers immediately round the eye, extending more or less to the loral plumes, which in some specimens are almost black, in others the basal part of the webs is whitish; upper part of ruff blackish brown, paling to dark brown beneath, and bounded externally by a zone of fulvous; beneath this the feathers of the throat are brownish, this colour usually taking the form of a zone across the fore neck; chest and under surface fulvous, closely barred with brown; under tail-coverts whitish, barred with darker brown than the breast; thighs, tarsi, and toes more ochraceous than the under surface, crossed with narrow, wavy, brownish bars; under wing-coverts concolorous with the breast, barred more closely and paling to buff-white at the edge of the wing; primary under wing-coverts blackish brown, paling to buff at the base; basal portions of the inner primaries and the secondaries beneath fulvescent white.

Examples from the upper hills (whether as a rule or not I cannot say) are darker on the disk, ruff, and lores than the low-country birds, and exhibit at the same time the facial barring which Mr. Hume found to be absent in his examination of the specimen on which he founded his Ceylonese race or subspecies *S. ochrogenys*. These birds have the barring of the under surface darker than the ochraceous-faced, paler-eyebrowed ones from the low country; but the ground-colour varies, being occasionally paler than in the latter.

Young. The nestling has the iris paler brown than the adult; cere and bill bluish leaden.

It is clothed with whitish down on the body, which gives place to the first or nestling feathers, which are edged with greyish buff, the scapulars, quills, and tail assuming from the first the brown hue noticed in the following description:—

Plumage on leaving the nest. Head, hind neck, scapulars, wing-coverts, lower back, and upper tail-coverts pale rufescent brown, the body-feathers broadly edged with whitish margins of a fluffy character; the scapulars and wing-coverts boldly barred with buff-white, and the greater coverts deeply tipped with the same; primaries and their coverts dark sepia-brown; the secondaries paler brown, the whole barred with pale ochraceous brown, and deeply tipped with whitish, the bars at the internal bases of the quills buff; tertials paler or more ochraceous brown than the secondaries, and narrowly barred with buffy white; back brownish; tail concolorous with the primaries, barred with narrow whitish marks, and tipped with white.

Lores and plumes between the eye and the forehead black; cere and bill rufous behind the eye; ruff and chin deep brown, the former edged with whitish; entire under surface buff-white, the feathers crossed with softened and indistinct rays of light ochraceous; under wing-coverts pale fulvous.

The above is a combined description from the example in my aviary (which was the subject of my article in 'Stray Feathers') and a second, shot in the Central Province; at this young stage even the face in the latter hill-bird is not so golden as was that of my tame, low-country one. The latter exhibited the following change of plumage during the first year:—After the lapse of a few weeks (about the 15th of June) the tips of the interscapular feathers next the scapulars and those of the lower part of the sides of the neck just above the point of the closed wing began to darken, and a V-shaped mark, having its apex about the middle of the back, was formed; this was the origin of the deep sepia-brown back of the adult. About a fortnight later the mature feathers (buff, barred with brown) began to appear on the tarsus, the fluffy plumage falling out to give place to them. In a short time the ground-colour of the tail-feathers deepened into blackish brown, and the adult feathers began to assert themselves elsewhere on the throat and parts of the breast. The tarsus and tibia took about three weeks to change, and by that time the whole of the interscapular region had become very deep sepia-brown; the downy feathers along the ulna commenced to fall out, and the deep brown edge to develop itself, while the wing-coverts kept pace with the rest, the whole wing rapidly becoming dark. In the mean time, while this moult was going on, the scapulars, quills, and tail-feathers darkened, assuming by a change in the feathers the hue of the adult. By the 31st of July the whole of the under surface was fully clothed with new feathers, the lesser wing-coverts were fully grown, the back had assumed the adult appearance, and the chin had become deep brown, the ruff extending beneath it by degrees. The facial disk had not altered at that time, but as the bird grew older it darkened into the normal yellow-rufous colour. The feathers of the head were the last to change, that part becoming dark brown about the middle of September; but it was not until the 30th November, when the bird was about 8 months old, that the last immature feather disappeared from above the right eye.

Obs. The Brown Wood-Owl, which has generally been associated with the species described by Col. Sykes from Southern India as *Syrnium indranee*, has of late been separated by Mr. Hume as *S. ochrogenys*, the grounds for so doing being that it was considered by him to have a more ochraceous disk than the Indian bird, and likewise to have that part not cross-rayed with dark lines. Sykes's type is not forthcoming now, nor are there any Southern-Indian birds in English collections, as far as I have been able to discern, from which it can be gathered what the species really is like. It is, of course, distinct from the Nepal bird (*S. newarense*), notwithstanding that some of the latter species are quite as small as Sykes's specimen was. His description, which applies well to Ceylonese examples, is in part as follows :—"Abdomine subrufo, brunneo graciliter fasciato ; regione circumoculari nigrâ ; disco rufo, brunneo marginato." With regard to the second point, concerning which it may be remarked that there is no evidence to show that it did not exist in the Indian bird, it will be seen that hill Ceylonese examples have the face more or less cross-marked with brown rays, though low-country birds have not as a rule. On the whole, therefore, in the absence of specimens from the districts where Sykes and Jerdon got them, it will be well to retain the Ceylon bird under its old title, until evidence is forthcoming to separate the Indian species, particularly as Mr. Hume lately writes me that he now considers the Nilgherry and Ceylonese species to be one and the same. In order to further the existing information concerning this interesting bird, and more especially for the benefit of my Ceylon readers, who are more or less interested in the so-called Devil-bird, it seems expedient to give a figure of the species, which I have accordingly done*.

Distribution.—The Brown Wood-Owl is distributed over the whole of Ceylon, inhabiting the low-country jungles of both the north and the south of the island, as well as the forests of the hill-zone up to the altitude of the Nuwara-Elliya plateau. In the Kandyan Province it is pretty generally found throughout all the coffee-districts, and is not at all uncommon in the neighbourhood of Kandy. In the upper ranges I have met with it at Kandapolla, and in the British Museum there are specimens from Nuwara Elliya. In the western parts of the low country it is a bird of local distribution, but in the wild jungles of the north and east I imagine it is everywhere to be found. I have myself met with it close to Trincomalie, and others have procured it in various parts of the Vanni. In the Colombo district it has been shot as near to Colombo as Kreshawa, and in the scattered jungles, commencing about 20 miles inland and extending more or less to the base of the hills, it is not unfrequent. More favourable to its nature are, however, the continued woods and forests clothing the country, further south, between the Kaluganga and Dondra Head, and there it is tolerably common. Between Kalatura and Agalawatta, in a comparatively maritime part of the country, I have heard several of these Owls on a single evening hooting within a short distance of each other.

Jerdon remarks that this species is found throughout Southern India, in Ceylon, and the Malayan Peninsula. He makes mention of it as follows :—" It frequents the forest only, and is most common at a considerable elevation. Col. Sykes found it in the dense woods of the Ghâts. I procured it first on the Nilghiris, and afterwards along the Western Ghâts in the Wynaad and Coorg. It has also been sent from Goonsoor." It does not appear to have been found north of the Deccan, and does not inhabit either Burmah or Tenasserim ; with regard to the Malayan Peninsula it has been procured in that region by Dr. Maingay, Lord Tweeddale being in possession of a skin sent home by that gentleman. On the authority of the late Mr. Swinhoe it has also been assigned to the island of Formosa ; the specimen was described in 'The Ibis,' 1863, p. 218, under the name of *Bubo caligatus*, and was supposed by Mr. Gurney to belong perhaps to this species ; but it was afterwards found to be *Syrnium newarense*, and is described as such in Swinhoe's " Catalogue of the Birds of China," P. Z. S. 1871, p. 344.

Habits.—This fine Owl, which has received the ill-omened name of Devil-bird, on account of the dire noises which the natives of the island have always ascribed to it, frequents shady forest-groves, woods of moderate extent, and portions of heavy jungle, near clearings and open places. I have met with it half a dozen times without being able to procure it, so sharp-sighted is it by day ; it was, on several occasions, being most thoroughly mobbed by the Jungle-Drongos (*Buchanga longicaudata*) in company with a host of Bulbuls, who were pursuing it from tree to tree with a chattering incessant enough to bewilder a wiser bird than even an Owl! On another occasion I witnessed its persecution, in a forest near Ambepussa, by two or three pairs of

* My Plate was drawn some months prior to working out my article, and the bird was styled by Mr. Hume's name *ochrogenys*, which I have now had altered.

Racket-tailed Drongos (*Dissemurus lophorhinus*) ; so that the "Devil-bird," notwithstanding its redoubtable sobriquet, does not appear to be much respected, by the King-Crows at any rate !

There is no bird in Ceylon to which so much interest attaches, both among the European and indigenous population, as the present. If the subject of ornithology be mooted in conversation, questions are invariably asked as to the "Devil-bird," What is it? have its direful notes been heard? and so forth. Very diverse opinions have always existed as to the identity of the bird, notwithstanding that the natives of the island, and consequently those who have worked at its ornithology and gathered much of their knowledge of the habits of its birds from them, have always attributed the discordant notes uttered by some nocturnal bird to the present species. Kelaart writes, "The shriek of the Devil-bird (*S. indrani*) is truly appalling. The superstitious natives listen to these dismal cries with great horror; some death or less misfortune is apprehended when an Owl sings (?) nightly over a hut or on a tree overshadowing it." Layard follows with the information that the Wood-Owl "utters the most doleful cries, which the natives consider the sure signs of approaching evil." Sir E. Tennent writes that the Sinhalese regard this Owl "literally with horror, and its scream by night in the vicinity of a village is bewailed as the harbinger of impending calamity;" and further that there is a "popular legend in connection with it, to the effect that a morose and savage husband, who suspected the fidelity of his wife, availed himself of her absence to kill her child, of whose paternity he was doubtful, and on her return placed before her a curry prepared from its flesh. Of this the unhappy woman partook, till discovering the crime by finding the finger of her infant, she fled in frenzy to the forest, and there destroyed herself. On her death she was metamorphosed, according to the Buddhist belief, into an *Ulama* or Devil-bird, which still at nightfall horrifies the villagers by repeating the frantic screams of the bereaved mother in her agony."

I have been assured by gentlemen in Ceylon that the Owl which makes these wonderful noises is a small, whitish bird, and some have told me that they have seen it in the act of uttering them. This description would seem to indicate the next species, a bird until lately quite unknown in the island. The author just quoted publishes, in a footnote at page 248 of his 'Natural History of Ceylon,' a letter from Mr. Mitford, late of the Ceylon Civil Service, and one who took great interest in the birds of the island, from which it will appear that this gentleman was doubtful as to the identity of the Devil-bird. He says, "The Devil-bird is not an Owl. I never heard it until I came to Kurunegala, where it haunts the rocky hill at the back of Government-house. Its ordinary note is a magnificent clear shout like that of a human being, and which can be heard at a great distance, and has a fine effect in the silence of the closing night. It has another cry like that of a hen just caught; but the sounds which have earned for it its bad name, and which I have heard but once to perfection, are indescribable, the most appalling that can be imagined, and scarcely to be heard without shuddering. I can only compare it to a boy in torture, whose screams are being stopped by being strangled. The only European that had seen and fired at once agreed with natives that it is of the size of a pigeon with a long tail. I believe it is a Podargus or Night-Hawk." I believe myself that there is no doubt about the bird being an Owl, as none of the Nightjars in Ceylon ever utter notes at all resembling these cries. The natives, however, who brought me my young specimens of the Wood-Owl at Galle did not seem to know that they were the birds accredited with these noises, but simply called them *Bakkamūna*, or "Large Owl." Mr. Holdsworth, who was of opinion, from the description given him by natives of the Devil-bird, that it was an Owl, was fortunate enough to hear its cries one night in the Aripu district, but was unable to discern the author of them. While watching at a waterhole for the purpose of shooting bears, he was suddenly alarmed by piercing cries and convulsive screams suddenly issuing from a small patch of bushy jungle about thirty yards from his hiding-place. He says, "My hunter at first thought a leopard was there, and told me to keep quiet; but the cries increased, and became so horribly agonizing, that it was difficult to believe murder was not being committed. . . . . Before I reached the place all was silent as before, and the idea of the Devil-bird flashed across my mind. This was afterwards confirmed by the hunter, who, however, did not care to talk much about it." My readers will gather from the above summary of evidence that there does exist in Ceylon some nocturnal bird which utters very singular notes, but that it is not quite clear what the species really is. The natives at different times and different places have given me the most contradictory answers concerning the delinquent; but in many parts of the island they believe that it is the Brown Wood-Owl, and from them Messrs. Kelaart and Layard received the idea it was so, and hence the general idea current among Europeans as to the supposed identity of this noisy bird of ill-omen. I endeavoured during my stay in Ceylon to discover whether these notes really were attributable to this bird or not; but, as regards my personal experience, I failed in finding out any thing satisfactory in the matter. My rearing up two of them (one of which I had in confinement for more than a year) did not assist me in my inquiries; for, as I stated in my article in 'Stray Feathers,' 1874, the only approach to any hoot which they made was a low growl, very seldom uttered, and a faint wheezy screech when they were very hungry; nor did they ever utter their far-sounding sonorous call, well known in the Ceylon hills, which resembles the syllables *to-whooo*, repeated at short intervals. Owls do not, as a rule, give vent to their natural calls while in confinement; and I therefore do not consider my evidence in this quarter very conclusive. Since writing on this species, however, I have been assured by a gentleman who kept a pair of these Owls, one of which is now in the Zoological Gardens, that in 1875, and during the month of March, which is about their breeding-time, the pair alarmed the inmates of his house by uttering the most dismal and wailing cries imaginable; and although these notes were not described to me as being so horrible as they have been depicted above, I think this testimony is much in favour of the idea that in the breeding-season this bird does utter loud and singular cries, which in the dead of night fall with more than their real harshness of sound upon the ear. It still remains, therefore, for some one interested in the ornithology of the island to persevere in shooting the bird in the act of making these noises, and so settle the matter once and for all. Whether it be the present or some other species it is doubtless the case that these peculiar notes are only uttered during the breeding-season. In a state of confinement this Owl is any thing but an unpleasant bird. It has the power of almost erecting its dorsal scapulars and pectoral feathers when under the influence of emotion or surprise, and looks much like a porcupine in appearance when so doing.

The habits of my tame birds were exceedingly interesting, their quaint manners, grotesque bearing, and familiar actions rendering them daily objects of admiration. I therefore take up room to subjoin the following extract from the article above referred to, in which, after referring to the singular habit of revolving their heads, with their eyes fixed on the object of their attention, and then hugging them forward in order to gain a better sight of it, is written as follows:—"When given any thing of no great size to eat, such as a *Calotes* or small bird, it invariably seized it in its foot, grasping it with the outer toe to the rear, and holding it up after the manner of a Parrot, nibbled at various parts with a view of tasting it, after which it would suddenly jerk it into its mouth, head foremost, and swallow it without any exertion whatever. On the 10th June, when only three months old, it swallowed entire a large *Calotes* lizard; but this feat, I consider, was outdone by its companion, which I reared the following year, and which bolted, at the age of six weeks, a *Dicæum minimum* and *Cisticola schœnicola* with as much ease as if they had been small pieces of meat. This peculiarity of holding its food in the foot was very interesting to witness, the bird at these times, under the influence of pleasurable emotions, presenting a highly grotesque appearance, opening and slowly shutting its large eyes, and tasting the dainty bit with every now and then an epicurean snap of its mandibles. This, by the way, is performed by pressing the under mandible against the tip of the upper, and then letting it go with a snap against the basal edges of the latter. He delighted in a good wash, and took his bath almost regularly every day, flying over to the 'chattie' generally in the forenoon, and squatting down in the water, which he would throw over him on all sides; his ablutions took sometimes more than five minutes to perform, after which it was his custom to mount on a high perch, and hang down his wings until he was dry, presenting the most ridiculous aspect imaginable. He remained sometimes more than an hour in this position, feathering and pluming himself until able to fly about. The process of feathering was performed in general with the eyes shut; and it was interesting to watch the manner in which he would seize one feather after another without ocular assistance, leading them out from base to tip, and working them with a quick movement of the under mandible. Their powers of vision were not good on a dark night, and when young this was particularly noticeable." One which I kept in a box insisted on perching on the side all day, where it slept in peace; when tired it would lower its body until its breast rested on the wood, and in this position, with its head stretched out, it would remain for half an hour at a time. At sunset it became lively, snapping its bill loudly when approached, and displayed then, as the light decreased and objects became more perceptible to its vision, the singular habit of revolving or rotating and then darting out its head in the manner already mentioned. Fish was a favourite article of diet with these birds; they bolted good-sized "sardines" whole, in the same manner that they treated birds and lizards.

Nidification.—This Owl breeds in February, March, and April, and nests in the hole of a large tree; one of my young birds was taken from a cavity in the hollow of a lofty Hora (*Dipterocarpus zeylanicus*), the rotten wood at the bottom of which formed the nest. Two eggs brought to me from Baddegama as belonging to this species were very round in shape and pure white like other Owls' eggs; but they measured only 1·45 inch in length by 1·25 in breadth, and were too small for those of this species.

The figure in the Plate accompanying this article is taken from a male specimen shot in the Kandy district, and exemplifies the less rufous and more striated disk observable in hill-birds, as distinguished from those inhabiting the low country.

Of smaller size than *Syrnium*; disk more perfect above, with a patch of stiff feathers on each side of the anterior portion. Wings rounded, reaching to the end of the tail. Legs feathered. Toes covered with hairs; inner anterior toe longer than the outer one; middle claw slightly serrated, as in *Strix*.

PHODILUS ASSIMILIS.

(THE CEYLON BAY OWL.)

(Peculiar to Ceylon.)

Phodilus badius, Hume, Stray Feathers, 1873, p. 429; Whyte, ibid. 1877, p. 353.
Phodilus assimilis, Hume, Stray Feathers (Notes), vol. v. p. 137.
The Bay Screech-Owl, apud Jerdon.
Bassa, Sinhalese.

P. similis P. *badio*, sed saturatior et tectricibus exterioribus nigricantibus: primariis intùs nigricantibus nec rufis: plagâ subalari tectricum majorum nigrâ nec rufâ: plumis pectoralibus nigro bipunctatis.

Adult, presumed male (*vide* Plate). Length (from skin) 10·5 inches; wing 7·1; tail 3·5; tarsus 1·55; middle toe 1·1; claw (straight) 0·65; outer posterior toe 0·15.

Adult, presumed female (British Museum). Wing 7·8 inches.

Female. "Length 11·5; wing 8·12, expanse 27·5; tail 3·5; tarsus 2·0 (?); mid toe and claw 1·5 " (*Whyte*).

"Iris dark brown; bill greenish white, with a dash of dark brown on edge of upper mandible, and dark spot on the nostrils; feet pale whitish green; claws pale ash, ridges of the scutæ of the toes of a darker green than the prevailing colour" (*Whyte*). Cere probably olivaceous.

Forehead and facial disk pallid reddish grey; loral plumes blackish at the base, the webs about the centres of the feathers rufous-brown; ruff-feathers white, very faintly tinged with rufous, and with a terminal black bar and the external tip rufous: crown and occiput with the back, scapulars, and lesser primary wing-coverts rufous, deepest on the head and slightly brownish on the other parts; feathers of the head with dark shafts and terminal black spots; a light buff patch on the centre of the occiput, on which the terminal dark spots are larger; back and sides of neck, inner webs of the scapulars and tertials, and the centre feathers of the median wing-coverts brownish buff; the buff feathers of the neck and wing-coverts with terminal brown spots, and the rufous portions of the back, together with the rump and upper tail-coverts, with a series of central, alternating white and black spots; least wing-coverts rufous-brown; outer webs of the primaries and secondaries, excepting the first primary, rufous, barred with black; inner webs blackish grey, barred with black; outer webs of the longer winglet-feathers and of the first primary white, barred with black; second and third primaries with the interspaces near the tip white; tail rufous, narrowly barred with eight wavy blackish bars, each feather with a white terminal spot enclosing a black one.

Throat and chest buff, changing on the breast, flanks, abdomen, and thighs into delicate rufous isabelline, each feather

---

* This genus has hitherto been associated with *Strix* in the family of Strigidæ. Professor Milne-Edwards has, however, lately pointed out, in an article in the ' Comptes Rendus,' Dec. 1877, that its affinities structurally are with *Syrnium*. The posterior margin of the sternum is deeply cleft, the structure of the tibia is similar to that of this latter genus, the clavicle is also similar, and the skull differs in its formation from that of *Strix*. Its external appearance, however, is that of this latter genus, and the claw of the middle toe is, like it, serrated.

with two central spots on a white ground-patch; tibia fulvous-buff; tarsus isabelline, like the breast; lesser under wing-coverts buff; a large patch of brown marked with rufous at the edge or beneath the metacarpal joint.

The above is a description of the bird figured in the Plate. The example in the British Museum has the point of the wing, lesser wing-coverts, and inner webs of primaries darker still; it likewise has the rufescent feathers of the forehead spotted with dusky grey. The specimen described by Mr. A. Whyte in 'Stray Feathers' (vol. v. p. 354) has 9 bands on the tail; and both have the peculiar buff occipital patch, looking like the remains of immature plumage.

Young.   The unfledged nestling is clothed with dusky grey down.

Obs.   This is a well-marked distinct race of the continental and Malayan *Phodilus*; the differences between the two are slight, but they are well pronounced and constant; Mr. Hume, besides noticing the dark wing-coverts and wing-lining patch and the blackish inner webs of the quills which characterise the Ceylonese bird, remarks that it is smaller. I do not know whether as a species it will prove to be so. The wings of three examples of *P. badius* in the British Museum are as follows:—India, 7·7 inches; Malacca, 8·0; Sarawak, 7·1. Jerdon gives the wing of the example he described as 8·5. In the Norwich Museum are the following:—Java, w. 7·5 inches; Borneo, w. 7·4; Borneo, w. 7·5; Java, w. 7·4; Java, w. 7·5. On the whole, therefore, the balance is in favour of the Ceylonese bird.   *Phodilus badius* differs from *P. assimilis* in having the head and back brighter rufous, in having the lesser wing-coverts, primary-coverts, and winglet concolorous with the back and not rufous-brown as in *assimilis*; these parts are not mottled with blackish; the inner webs of the quills are clear rufous, like the outer; the tail has only five or six bars; the feathers of the breast and belly have only one spot instead of two, and the patch under the wing is rufous instead of brown.

History and Distribution.—The present member of the interesting and little-known genus *Phodilus* is one of the most recently discovered of the peculiar Ceylonese birds. The first specimen on record was killed by a native about the year 1871, at a place called Lewelle Ferry, some three or four miles from Kandy. It was preserved by Messrs. Whyte and Co., of Kandy, and obtained from them by Mr. H. Neville, C.C.S., who sent it to Mr. Hume. To this gentleman is due the credit of discriminating our species, on the testimony of this example, from the Indo-Malayan bird *P. badius*. In November 1876 a second example (the skin of which, through the kindness of Mr. W. Ferguson, passed into my hands, and is now in the collection of the British Museum) was captured by a coolie on the Martinstown Estate, Kukkul Korale. It was taken from the nest together with three young ones; and Mr. H. M. Hector, to whom I am indebted for much information on the subject, and on whose estate the birds were caught, writes to me that the Sinhalese brought another bird of the same species to his superintendent, but there being no accommodation for it at the bungalow it was released. In February 1877 a third specimen was procured by Mr. Reeves of Ratota, and its capture recorded by Mr. A. Whyte at p. 201 of vol. v., 'Stray Feathers;' while at p. 353 of the same vol. Mr. Whyte notices a fourth caught in the following July on the estate of Mr. Weldon, Dickoya, who states, in his letter to Mr. Whyte, that it was the second of the kind which his cooly had caught. There appear, therefore, to have been, as far as I can ascertain, six examples* of this rare Owl shot or captured in Ceylon, showing that its range extends throughout the hill-regions of the island, and that the habitat, as far as is yet known, of the bird lies between 1500 and about 3000 feet elevation. Future research will, however, doubtless reveal its presence both in the low country and in the upper hills; and it is to be hoped that hereafter all examples met with will be both preserved and recorded with data of sex and measurements in 'Stray Feathers' or other ornithological publications.

Habits.—This recently discovered nocturnal denizen of our forests has come to such a limited extent under any one's notice, that it is not in my power to place on record much concerning its economy. It has shown itself to be an inhabitant of forest-jungle, out of which it evidently strays at nights in search of food, and, like many other Owls, when unable or too late to return to its usual haunts, hides where it best can on

---

* Mr. Hume writes me, since this was penned, that he has received two additional specimens from Ceylon.

estates, in isolated trees or in old buildings, and, owing to its completely nocturnal habits and imperfect day-sight, falls an easy victim to any one with eyes sharp enough to discover it. Mr. Reeves's specimen was taken in an old cooly hut. Mr. Weldon writes, as above mentioned, to Mr. Whyte :—"This bird was caught by a cooly in a tree in the daytime on my estate, and is the second of the kind he has caught here. It was put on a perch in a dark room, but refused to eat, and died after two or three days' confinement." The bird brought by natives to Mr. Hector's superintendent appears to have been taken the same way, being the third instance of capture by hand during the day. Mr. Hector, in a letter kindly written to me after my departure from Ceylon, throws some light on the nature of this Owl. He says, in speaking of the brood of young birds, "there were three apparently of different ages, as the largest very much exceeded the other two, which also differed considerably in size. The largest one was about the size of our ordinary Quail, with a flattish-shaped head. It seemed a vicious bird, as it used to peck the other two continually, and one day, I found, had pulled many of the feathers out of the smallest, and seemed to be trying to tear its flesh, so that I had to separate them." Unlike most Owls, it does not seem to thrive in confinement. Mr. Hector kept the parent bird of these young ones five weeks; but in Mr. Weldon's case, his lived but two days. Mr. Reeves writes me concerning his bird that it lived about a week and was fed on lizards and small fish caught in a neighbouring stream, and preferred the latter to any thing else. With regard to the note of this species, it may or may not be the author of the hideous sounds attributed to the Devil-bird; but I have no authentic information as to any of its cries. Of the allied species, *P. badius*, Messrs. Mottley and Dillwyn, as quoted by the late Mr. Blyth in the 'Ibis' for 1866, p. 252, state :—"It has only a single note, frequently repeated, and which is much like the first note of the common Wood-Owl's cry."

Nidification.—The Ceylon Bay Owl appears to breed at the latter end of the year, nesting in hollow trees. Mr. Hector writes me that "the nest was made in a hole in a tree and composed of dry twigs, moss, and feathers." The number of eggs in this was three, so that they may be inferred to vary from two to four, as in some other species of Owls.

For the loan of the specimen figured in the Plate accompanying the last article I am indebted to the kindness of Mr. Reeves and his brother-in-law Mr. J. C. Horsfall, in whose house at Altrincham it is mounted. I was unable to figure the example presented to the British Museum, as the tail is not perfect. Mr. Keulemans has drawn this Owl with the wings slightly drooped, in order to show the characteristic dark inner portions of the quills.

Fam. STRIGIDÆ.

"Hinder margin of sternum entire, with no distinct clefts; furcula joined to keel of sternum ; inner margin of middle claw serrated; inner and middle toes equal in length; between the anterior portion of the facial area a frontal patch of small stiff feathers always present and very broad." (*Sharpe*, Cat. Birds, ii. p. 289.)

Hinder margin of sternum entire, with no distinct clefts; furcula joined to keel of sternum. Head smooth. Bill straight at the base, compressed, feeble, with the tip much curved. Nostrils large, oval, and oblique. Facial disk complete and entirely surrounded by a ruff of stiff feathers. Wings long in comparison to the tail, pointed, with the 2nd quill the longest, and the 1st subequal to the 3rd. Tail even. Legs long; the lower part of tarsi clothed, as the toes, with bristles. Toes long and scutellate above; claws much curved, the inner edge of the middle serrated.

STRIX FLAMMEA.

(THE BARN-OWL.)

Strix flammea, Linn. S. N. i. p. 133 (1766); Gould, B. of Eur. i. pl. 36; Blyth, Cat. B. Mus. A. S. B. p. 41 (1849); Horsf. & Moore, Cat. B. Mus. E. I. Co. i. p. 81 (1854); Schlegel, Vog. Nederl. pl. 41 (1854); id. Mus. P.-B. Striges, p. 1; Gould, B. of Gt. Brit. i. pl. 18; Sharpe, Cat. Birds, ii. p. 291 (1875); id. in Rowley's Orn. Miscellany, pt. viii.

Strix javanica, Gm. S. N. i. p. 295; Jerd. Madr. Journ. x. p. 85; Blyth, J. A. S. B. xix. p. 513; Horsf. & Moore, Cat. B. Mus. E. I. Co. i. p. 82; Jerd. B. of Ind. i. p. 117; Hume, Nests and Eggs Ind. B. p. 59; Kelaart, Prodromus, Cat. p. 116; Layard, Ann. & Mag. Nat. Hist. 1853, xii. p. 107.

Strix indica, Blyth, Ibis, 1860, p. 251; Hume, Rough Notes, ii. p. 342; Gould, B. of Asia, pt. xxiv.; Holdsw. P. Z. S. 1872, p. 415; Hume, Str. Feath. 1873, p. 163, et 1875, p. 37.

The White Owl, Albin; Le petit Chat-huant (Brisson); L'Effraye, Buffon; The Screech-Owl; The Indian Screech-Owl (Jerdon). Lechuzo, Spanish.

Karaya, Karail, Hind., also Buri-churi, lit. "bad bird;" Chaao pitta, Tel.; Chaao kuravi, Tam. (Jerdon); Daris, Java (Horsf.); Serrak, Malays (Horsf.).

Adult male and female. Length to front of cere 13·8 to 14·0 inches; culmen from cere 1·0; wing 11·4 to 11·7; tail 4·3 to 4·7; tarsus 2·4 to 2·6; mid toe 1·3 to 1·4, its claw (straight) 0·75 to 0·85.

Obs. The above measurements correspond fairly with those of Indian and Burmese birds, and are taken from a series of specimens. The measurements of six examples from the above localities are:—Total length 11·0 to 14·8 inches; wing 11·0 to 11·8; tarsus 2·5 to 2·75. The expanse of Indian birds, according to dimensions contained in 'Stray Feathers,' varies from 38·0 to 30·7.

Iris black; bill fleshy white; cere flesh-colour; bare portion of tarsi and feet fleshy brown; claws brown.

General hue of upper surface, including the tail and wings, rich tawny buff, the visible portions of the feathers profusely and finely stippled with whitish and dusky grey, and each with a terminal white spot, "pointed" above with blackish brown; on the head and hind neck the white portion of the spot is inconspicuous, and the dark very much reduced in size; on the rump, tertials, and quills the tippings are more extensive than elsewhere; edge of the wing and outer web of the first winglet-feather white; inner webs of the primaries and secondaries whitish towards their edges, and crossed on the terminal half with broken-up brownish bars ; outer webs and tips mottled like the back with traces of bars corresponding to those of the inner webs ; a white, dark-edged spot at the tips of the quill-feathers, most conspicuous on the tertials ; tail crossed with four narrow, wavy bars of brown, mottled with whitish ; the tip white, mottled with brown, and terminated with a black-and-white spot.

Facial disk white, speckled in some with grey ; a rufous patch in front of the eye ; ruff (of stiff, erect feathers) glossy white interiorly, rich rufous at the tips of the exterior feathers, which are also pencilled round the edges with brown : sides of the neck concolorous with the back, the feathers with terminal greyish-bordered spots ; throat and under surface white, faintly tinged here and there, and more particularly on the flanks and tibia, with delicate buff ; feathers of the flanks and sides of the breast with dark triangular terminal spots ; under tail-coverts in some spotted, in others entirely white. Under wing-coverts and lower surface of quills white ; the lesser coverts with blackish spots and dashes of buff.

Obs. Some Ceylonese specimens are of a richer or more orange-buff than others, and in all that I have examined is the tail concolorous with the back. The spottings of the under surface are always present, in a greater or less degree, some examples having the belly and thighs as much marked as the breast. I have not seen any traces of zigzag markings in Ceylon birds, and thus they possess more affinity to the Indian than to the Malayan type of this variable species. The plumage of this Owl fades considerably with the age of the feathers and perhaps from exposure to the sun's rays. Such a specimen I possess in my collection, the appearance of which would suggest the idea that the bird had selected an exposed situation wherein to roost. The buff tint has entirely disappeared from the exposed portion of the upper-surface feathers, and the dark spots are very pale. The feathers in this example are much abraded throughout.

Young. In the European bird the nestling is covered with white down, the wing-feathers having the normal buff hue, with greyish and white mottlings.

Young bird on leaving the nest (Sharpe, Cat. Birds, ii. p. 293).—"General colour above orange, but profusely obscured with light grey, all the latter plumes vermiculated with ashy brown and having a distinct subterminal white spot margined both above and below with brown ; the head and hind neck coloured like the back, but more decidedly orange, especially the sides of the neck, which are bright orange with a few bright spots ; wing-coverts coloured like the back ; primary-coverts orange, mottled at the tips like the rest of the coverts, but much paler externally and inclining to whitish ; quills orange, mottled at the tips with greyish, and having distinct cross bars of grey mottled with white ; tail pale orange, barred with greyish and mottled with the same.

"Facial disk silvery white, the feathers rufous round the eye and especially in front of the latter ; ruff glistening white, the upper plumes washed with orange, and the lower ones also tipped with clear orange ; feathers of the under surface pure white, tinged with orange on the chest, but not spotted."

Distribution.—This cosmopolitan and well-known bird is an inhabitant of the north and north-west coasts of Ceylon. The natives have, my friend Mr. W. Murray informs me, a tradition among themselves that it was introduced by the Dutch. I will not venture to pronounce an opinion on this point, but will simply remark that its range is extremely limited. Layard only noticed it at Jaffna, where it used to be common about the fine old Dutch fort, living in the ramparts ; it was tolerably numerous there, Mr. Murray writes me, until about ten years ago ; since then its numbers, which were always limited, have been thinned by people collecting, and by the birds occasionally being caught abroad in the daytime by the Jaffna Crows, and quickly treated to lynch-law by these tyrannical citizens. It inhabits some old buildings in the neighbourhood of the town, and has been met with in the fine banyan-tree on the road to Chavakacheri. It no doubt frequents other ruined buildings along the coast to Manaar, in which district Mr. Holdsworth records it from Aripu. Further south it has been found at Puttalam, where Mr. R. Pole, of the Ceylon Civil Service, informs me he has seen it. I am not aware that its range extends lower down the coast than this latter place. On the opposite coast it is not known.

The Barn-Owl inhabits India, Siam, and Malayana, and extends eastward to Arabia. As regards the other continents of the globe it may be said (now that the many races hitherto recognized have been amalgamated into the one species by Mr. Sharpe) to be found in all of them. It is generally distributed throughout Europe and Africa, extending, in its race of *S. poensis*, to the island of Madagascar and to the Cape Verds. In America the western form *S. pratincola* is distributed throughout the North-American, West-Indian, and Neotropical Regions, as far south as Chili, and extends over to the Galapagos Islands. In the Australian Region it is represented by *S. rosenbergi* from Celebes, and by *S. delicatula* from Australia and a portion of Oceania, but not from Tasmania or New Zealand—the latter locality being without any member of the genus, while Tasmania is inhabited by the large race (*S. castanops*) of the Australian species *S. novæ-hollandiæ*.

Habits.—As in the old country, the Screech-Owl, or, as it is better known, the " Barn-Owl," frequents, in Ceylon and India, ruined edifices, forts, wells, and buildings of every description, preferring these architectural retreats to those afforded by old and hollow trees. It, however, sometimes takes up its abode in the latter. In Jaffna Mr. W. Murray, who resided there for many years, and had abundant opportunity of noticing its habits, writes me that it "lives in the small square drains leading from the silt-traps on the bastions to the moat.". . . . . " At dusk," he remarks, they sit at the openings overlooking the moat, and screech to one another for a good half-hour before starting on their foraging expeditions ; many feed about the fort, but some fly across the Jaffna lake to the islands in search of food." Mr. Holdsworth found them frequenting a store-house in his compound, " each regularly perching in a dark corner under the roof, at opposite ends of the long building, and apparently living in harmony with hundreds of Bats which hung from the roof and walls around." In India, Jerdon found them frequenting cells and powder-magazines in the vicinity of cantonments, and it therefore appears that in the East, as well as in Europe, it loves to haunt the habitations of man. In such localities it has opportunity of doing good in the capture of rats and other noxious vermin, enormous quantities of which it must destroy in a single year. Most people who take any interest at all in the natural history of birds now absolve the inoffensive and useful Barn-Owl from the sins which used to be laid at its door, and instead of accusing it of destroying birds, game, &c., are aware that it is a vermin-killer, and does far more good than harm. The old story is well known of the farmer, who, missing his pigeons one by one, laid in wait for the fancied robber, the Barn-Owl, and, having shot the unfortunate bird issuing from the dove-cot, was surprised to find a huge rat, the real depredator, in the bird's talons. Dr. Jerdon affirms, in his 'Birds of India,' that he has known it more than once fly into the room in which he was sitting with open doors and windows after a rat which had entered.

The note of the Screech-Owl, as its name implies, is a loud cry or scream, which it sometimes utters on the wing, in addition to which it is said by Indian observers to utter doleful wailings and sounds, such as are generally believed to be solely the voice of the Wood-Owl (*Syrnium*). Mr. J. H. Rainey, a writer in 'Stray Feathers ' (vol. iii. p. 333), relates that he has often been awakened " by cries which closely resembled two infants in distress," and on following the bird has shot what he identified as the Indian Screech-Owl. These occurrences took place at the end of the cool season, before the birds began to breed, and were, doubtless, says Mr. Rainey, their amorous calls or love-notes. It is the habit of this Owl to issue out from its roosting-place at dark ; but I have more than once seen English members of the species abroad before sunset even. In confinement they sleep throughout the whole day, which cannot be said of some Owls (see my remarks on the Ceylon Wood-Owl) ; and I have seen a caged bird outside a shop window, in one of the most crowded thoroughfares in London, fast asleep, totally unconscious of the din and roar going on around him. Mr. Holdsworth, who observed the habits of a pair that frequented a storehouse in his compound at Aripu, never observed them abroad until some time after sunset.

The habits of this Owl in confinement are very interesting. They are voracious in their appetites, and very fond of bathing. Mr. Blewitt, as quoted in 'Nests and Eggs,' p. 59, remarks of some he reared, that they would invariably *disgorge* the flesh of Hawks and Owls that had been given them to eat.

Nidification.—In Jaffna, I understand, this Owl breeds in June and July, nesting in the drains in the escarpments of the Fort ditch, without fear at that time of their nests being washed away. In India they breed from February to June ; and Mr. Hume observes that holes in wells are the favourite localities ; the nests are, however, often found in hollow trees, where there are no suitable buildings to be chosen. The eggs are usually laid on the bare surface of the cavity*, but sometimes a small stick-nest is made, which, says Mr. Hume, resembles that of a Pigeon. The number of eggs is variously stated as from three to seven, the latter being no doubt unusual. They are generally pure white, but sometimes have a

---

* In England I have found the eggs on the bare stone at the top of a barn wall.

creamy tinge. Mr. Hume says that the eggs of the Indian birds are more oval than those of the European. The average size is 1·69 by 1·28 inch. In Europe it is found that this Owl lays occasionally a second and third clutch of eggs before the first brood leaves the nest, these latter, as Professor Alfred Newton remarks, materially aiding the development of the unhatched chicks during the nightly absence of the parents in search of food*.

* Yarrell's 'British Birds,' 4th edition, p. 197.

Order PSITTACI.

Base of upper mandible covered with a cere, in which the nostrils are pierced, as in *Accipitres*; upper mandible vaulted, much curved, the tip overhanging the lower, which is short and rounded. Wings with ten primaries. Tail with twelve rectrices. Legs short. Feet zygodactyle.

Sternum large, much as in *Accipitres*, but narrower, and with an oval aperture in the posterior edge; cranium very large. Œsophagus dilated. Tongue fleshy. Of gay plumage.

Fam. PSITTACIDÆ.

Bill with the upper mandible wide at the base, suddenly compressed near the tip; margin with a well-pronounced lobe; under mandible short and obtuse. Wings moderately long. Tail variable. Tarsi covered with small tubercle-like scales.

Subfam. PALÆORNINÆ.

Bill moderate, the upper mandible moderately hooked. Tail very long and wedge-shaped: the central feathers far exceeding the rest, which are much graduated. Legs and feet proportionately small.

Of medium size.

PSITTACIDÆ.

PALÆORNINÆ.

Genus PALÆORNIS.

Bill rather short; upper mandible evenly curved from the base, the tip moderately produced; the margin furnished with a rounded indentation; the cere small, nostrils pierced close to the culmen; under mandible short. Wings moderately long; 2nd quill the longest, 1st and 3rd slightly shorter. Tail very long, the central feathers usually produced much beyond the next two. Tarsus very short and finely reticulated. Outer anterior toe longer than tarsus; outer posterior toe longer than the inner anterior one; claws rather straight and short.

PALÆORNIS EUPATRIUS.

(THE ALEXANDRINE PARRAKEET.)

Psittacus eupatrius, Linn. Syst. Nat. p. 140, ♀ (1766).

Psittaca ginginiana, Brisson, Orn. iv. p. 343, pl. 29. fig. 1, ♀ (1760).

Palæornis alexandri, Vigors, Zool. Journ. ii. p. 49 (1825); Jerdon, Cat. B. S. India, Madr. Journ. 1840, xi. p. 208; Blyth, Cat. B. Mus. A. S. B. p. 4 (1849); Kelaart, Prodromus, Cat. p. 127 (1852); Layard, Ann. & Mag. Nat. Hist. 1854, xiii. p. 262; Horsf. & Moore, Cat. B. Mus. E. I. Co. ii. p. 610 (1856); Jerdon, B. of Ind. i. p. 256 (1862); Finsch, Papageien, p. 11 (1868); Holdsw. P. Z. S. 1872, p. 425; Hutton, Str. Feath. 1873, p. 333; Legge, Ibis, 1874, p. 14.

Palæornis eupatrius, Walden, Ibis, 1873, p. 297; Hume, Str. Feath. 1873, p. 433; Ball, Str. Feath. 1874, p. 389; Legge, ibid. 1875, p. 199; id. Ibis, 1875, p. 282.

Palæornis magnirostris, Ball, Str. Feath. 1873, p. 66; Hume, Str. Feath. 1874, p. 176.

The Ring Parrakeet, Edwards, Glean. vol. vi. pl. 292 (1760); *The Alexandrine Parrakeet*, Latham; *Grande Perruche aux ailes rougeâtres*, Buffon, Hist. Ois. vi. p. 156; *Rose-band Parrakeet* of some.

Ra-i-tota, lit. "Royal Parrakeet," Hind. in south of India; *Pedda chilluka*, Tel.; *Peria killi*, Tamils in India and Ceylon; *Chundanon*, in Maunbhoom (Beavan).

Laboo girawa, Sinhalese.

Adult male. Length to front of cere 18·0 to 19·5 inches; culmen from cere 1·55; wing 7·8 to 8·2; tail 11·0 to 12·0; tarsus 0·75; outer anterior toe 1·0, its claw (straight) 0·45; depth of upper mandible at cere 0·73 to 0·78. The bill varies in depth; in old birds the lobe gets worn away, and in others I have observed that one side of the bill is higher than the other.

Iris yellowish white or very pale yellow, with a bluish-grey inner circle; eyelid dull reddish; bill deep cherry-red, paler on the lower mandible, and the tips of both somewhat yellowish.

Legs and feet greyish sap-green, or greenish plumbeous, or plumbeous grey; claws dusky. Above grass-green, brightening to emerald-green on the forehead and lores, and darker on the wings; a faint blackish stripe from the nostrils to the eye; occiput and cheeks pervaded with a greyish-blue tinge; a broad black mandibular stripe passing down and across the side of the neck, where it meets a rose collar which encircles the hind neck; on the secondary wing-coverts a dark red patch; 1st primary and inner webs of remaining quills dark brown, the former with a bluish and fine yellow edge; bases of secondaries washed with blue; central tail-feathers passing from the base into blue, and thence into yellowish at the tips; under surface of tail yellowish; beneath dingy or faded green, brightening on the lower flanks and sides of the abdomen; under wing and under tail-coverts pale green.

Female. Total length 17 to 18 inches; wing 7·5 to 7·8; tail 0·0 to 10·0; greatest depth of upper mandible 0·7. Iris dingy yellowish white, with darkish inner circle; bill, legs, and feet as in male.
The female wants the black mandibular stripe and rose collar.

Immature male. Similar to the female in plumage, but generally larger, attaining a total length of about 10 inches in the first year. In some the rose collar is present in an imperfect state; but these are probably birds of the second year. The iris is greenish white generally.

Obs. The Ceylonese race of this Parrakeet is, like many other representatives of Indian species inhabiting Ceylon, smaller than the continental; for although very large examples of males are sometimes met with, this sex is, as a rule, shorter in the wing and tail, and possesses a smaller bill than most members of it from India, while a still greater disparity exists between individuals of the other sex from the two localities. Three adult males from peninsular India and the N.W. Provinces in the national collection measure in the wing 8·3, 9·2, and 8·2 inches, and the tail in the second attains as much as 13·4, with the bill 0·85 in height at the cere; the mandibular stripe in some Indian individuals is very broad. Mr. Ball gives the wing-measurements of 2 males from Chota Nagpur as 8·05 and 8·5, and those of the tail 11·6 and 12·0; the corresponding dimensions of two females from the same district are 8·2, 8·35, and 12·2, 12·0. In the north and north-west of India a race exists with a glaucous blue tinge on the head, and likewise larger than the Ceylonese, which Mr. Hume considers deserving of subspecific rank under the title of *Pal. nivalensis* of Hutton. In Burmah and the Andaman Islands another is characterized by its larger bill as *P. magnirostris* of Ball. This latter is no larger in the wing than Indian examples, but as regards the bill the upper mandible attains, in some instances, the great height at cere of 0·90. Lord Tweeddale, however, received individuals from this locality smaller in the bill than Ceylonese; and though Mr. Hume remarks that these were probably females, yet they must have been compared with individuals of the same sex from Ceylon.

Distribution.—This fine Parrakeet is a common and widely diffused species in Ceylon. It appears to be as much entitled to the name of Alexandrine Parrakeet, in memory of the great Emperor whose voyagers brought it from the East, as the Indian bird; for it would be difficult to assign the true locality whence it was first procured in those days of yore. The old writer Willughby, in his 'Ornithology,' published in 1678, remarks of this species, which he calls the "Ring Parrakeet of the ancients":—"This was the first of all the Parrots brought out of India into Europe, and the only one known to the ancients for a long time; to wit, from the time of Alexander the Great to the age of Nero, by whose searchers (as Pliny witnesseth) Parrots were discovered elsewhere, viz. in Gagandi, an island of Ethiopia." Edwards says that his plate was taken from a specimen brought alive to London in one of the East-India Company's ships.

To return, however, to its distribution in Ceylon, it is found throughout the north of Ceylon, from Chilaw upwards, more particularly along the seaboard round to Batticaloa, where it is very abundant indeed. In some portions of this long line of coast its presence is notably wanting; for instance, at Trincomalie it is rarely seen, although 15 miles to the north of it and on the south of the Bay it is common. In the jungles of the interior it is locally distributed. In the south-east I found it tolerably plentiful in the Wellaway Korale, from which locality it ascends in the dry season to the Haputalo ranges. In the scrubby maritime district of Hambantota it is replaced by the next species. It occurs here and there in small numbers throughout the southern and western Provinces, and in the Kandyan district is not unfrequently met with in the dry season; and in Madulsima I have seen it as high as 3500 feet. Mr. Bligh has observed it on one occasion at Nuwara Elliya. It is tolerably numerous along the base of the Matale ranges from Dambulla to Kurunegala.

In the Peninsula of India it is found, according to Jerdon, in the forests of Malabar, in the hilly region of Central Indian, and in the northern Circars, and occasionally in parts of the Carnatic; in the extreme south it is not nearly so common as *P. torquatus*. In Chota Nagpur it is, according to Mr. Ball, by no means universally distributed, but in the Rajmahal hills is much more common. Mr. Fairbank does not record it from Khandala, and in northern Guzerat it is rare. In the sub-Himalayan region it is common as the *P. sivalensis* of Hutton, and in Burmah exists as the large-billed race *P. magnirostris*, which extends through Tenasserim to the islands of the Bay of Bengal.

Habits.—Large colonies of this species take up their abode in districts where cocoanut cultivation borders on forest and wild jungle, which afford an abundance of fruit-bearing trees, on the berries of which the Alexandrine Parrakeet subsists. It is also found in openly timbered country and in forest. It roosts in considerable numbers in cocoanut-groves, often close to a village, pouring in about half an hour before sunset in small swiftly flying parties from all directions, which, as their numbers increase towards the time for roosting, create a deafening noise in the excitement of choosing or finding their accustomed quarters. The fronds of the cocoanut afford them a favourite perch, on which they sleep huddled together in rows. At daybreak the vast crowd is again astir, and after much ado, flying from tree to tree with incessant screaming, small parties start off for their feeding-grounds, flying low, just above the trees, and every now and then uttering their full and loud note *ke-gar*; this sound is more long-drawn and not so shrill as that of the smaller bird, and can be heard at a great distance. Isolated birds have a habit of apparently leaving the rest of the flock and roaming off at a great height in the air, every now and then giving out a loud scream, which often attracts the attention of the traveller or sportsman for some little time before he is aware of the position of the Parrakeet, which is flying swiftly on far above his head. It is a shier bird than its smaller congener, and rather difficult of approach when not engaged in feeding or in the business of settling down for the night; at the latter time numbers may be shot without their companions doing more than flying out of, and directly returning to, their chosen trees. In the forests of the south-eastern part of the island I observed these Parrakeets resorting at evening to dead and sparsely foliaged trees, the bare branches of which afford them a similar perch to that of the palm-frond.

They feed on grain as well as on the fruits and berries of forest-trees; and I on one occasion captured a fine specimen which had become entangled in a species of vetch which covered the earthy portions of a rocky islet near Pigeon Island; it had been feeding on the seeds of the plant, and while extracting them from the pod had got beneath the tangled mass and was unable to extricate itself again. In confinement this species is possessed of the usual docility peculiar to the Parrot order, and is a very favourite pet in Ceylon with both Europeans and natives; I do not think it is as often taught to imitate the human voice as the next species, but I have heard it occasionally speak native words with a fair amount of distinctness. Indian writers say that it is taught with facility to speak; but I think that as a general rule in Ceylon it is kept more as an ornament than for its powers of talking, and when newly feathered, with its tail in perfect order, is a very handsome bird.

Nidification.—Layard writes that he was informed by natives that this bird laid two eggs, building, of course, as all Parrakeets, in a hollow tree. It excavates the hole in which it breeds, generally choosing a small limb, of which the hard shell to be cut through before reaching the interior cavity is not very thick. I have never succeeded in getting the eggs, and therefore can state nothing certain concerning their size. Mr. Hume gives the dimensions of one belonging to the Northern-Indian form, *P. sivalensis*, as 1·52 by 0·95 inch, a very unusual shape for the egg of a Parrakeet, which is generally round. Mr. Rainey writes in 'Stray Feathers,' concerning the breeding of the Rose-band Parrakeet in the Sunderbunds, that "they build their nests in the hollows"—of trees with light wood—"first scooping them down perpendicularly some two and a half feet, so that it requires a long arm to be able to remove the nestlings within . . . . The eggs are usually two or three and sometimes four in number, and are deposited in the end of the hollows, the scrapings of the wood being gathered below to form a soft bed for them and the young when hatched."

PALÆORNIS TORQUATUS.

(THE ROSE-RINGED PARRAKEET.)

Psittacus torquatus, Bodd. Tabl. Pl. Enl. p. 32 (1783).

Psittaca torquata, Brisson, Orn. iv. p. 323 (1760).

Psittacus alexandri, Linn. Syst. Nat. p. 141 (1760).

Palæornis torquatus, Vig. Zool. Journ. ii. p. 50 (1825); Sykes, P. Z. S. 1832, p. 96; Jerd. Cat. B. S. India, Madr. Journ. 1840, xi. p. 207; Blyth, Ann. & Mag. Nat. Hist. 1840, xii. p. 90; id. Cat. B. Mus. A. S. B. p. 4 (1849); Gray, Cat. Mamm. &c. Nepal Coll. Hodgs. p. 113 (1846); Kelaart, Prodromus, Cat. p. 127 (1852); Layard, Ann. & Mag. Nat. Hist. 1854, xiii. p. 262; Horsf. & Moore, Cat. B. Mus. E. I. Co. ii. p. 611 (1856); Jerdon, B. of Ind. i. p. 257 (1862); Holdsworth, P. Z. S. 1872, p. 425; Hume, Str. Feath. 1873, p. 170; id. Nests and Eggs, Rough Draft, p. 116 (1873); Ball, Str. Feath. 1874, p. 389; Legge, Ibis, 1874, p. 14, et 1875, p. 282; Butler, Str. Feath. 1875, p. 457.

La Perruche à collier, Buffon, Pl. Enl. p. 551; *Alexandrine Parrakeet*, Latham; *La Perruche à collier rose*, Buffon, Hist. Oiseaux; *The Rose-collared Parrakeet*, Kelaart; *Mango Parrot* in India; *Small Green Parrot* and *Ring-necked Parrakeet*, Europeans in Ceylon.

Tiya, Bengal; *Gallar*, Hind. in N.W. Provinces; *Tenthia suga*, Nepal; *Lybar*, Mussooree; *Ragoo* and *Kerah*, Mahrattas; *Lybar tota*, Hind. Shikarees in South; *Chilluka*, Telegu; *Killi*, Tamil; *Teea-tota*, natives of Maunbhoom.

Rana girawa, Sinhalese.

Adult male. Length to front of cere 13·5 to 16·1 inches; culmen 1·1; total length of the latter 1·72; wing 6·0 to 6·6; tail 8·2 to 9·8; tarsus 0·5; outer anterior toe 0·95, claw (straight) 0·38; greatest height of upper mandible 0·5. Adults of this Parrakeet vary extraordinarily in size, evidently attaining the maximun dimensions when several years old. The smaller of the above measurements, which are taken from a series of Ceylonese birds, are those of an individual in fully adult dress.

Iris yellowish white or white tinged faintly with yellow; bill, upper mandible deep red, margin at gape and lower mandible blackish; legs and feet dusky bluish slate or greenish olivaceous, often with a brownish wash on the tarsus.

Above and beneath grass-green, with a greyish wash on the hind neck and chest, and the lower parts more delicate than the back; a well-defined blackish streak from nostril to eye; a black mandibular stripe, meeting on the gorge, and curving round as in the last to meet a collar of pale red on the hind neck (the black and red overlap more than in the last species); occiput and hind neck above the collar with the upper edge of the black stripe at the cheek washed with delicate azure-blue; inner webs of quills brown, with a fine yellowish edging to both webs; the central rectrices and outer webs of next pair are blue, fading towards the base to green; the rest green, washed on the inner webs with yellowish; tips and under surface of all the feathers yellowish; under wing yellowish green; tertiary under-coverts yellow.

Female. Total length about 10·0 inches; wing 6·0. Soft parts as in male.

Similar in plumage, but wants the black stripe and rose collar, in place of which latter it has a narrow emerald-green band across the hind neck.

Young. The nestling is similar to the female, and males attain in the first year a size quite equal to adults of the other sex. There is a dusky indication of the black stripe, and on obtaining the dress of maturity the feathers above the green collar become edged with rose, and a narrow edging of black begins to appear where the broad black stripe eventually is put on. At the same time the hind neck becomes tinged with azure-blue.

Obs. Ceylonese examples of this Parrakeet average smaller than Indian. I have not met with a male which exceeded

6·6 in the wing, the usual size being 6·3, while several which I have examined from various Indian localities, such as Kamptee, Mysore, Hyderabad, &c., measure as much as 6·7 ; they are, however, no larger in the bill than insular examples, the measurements of this organ in several being 0·45, 0·43, 0·46 (height at front). The finest specimens I have seen in Ceylon were from the north ; I noticed, on the contrary, that Hambantota birds were smaller than those from other parts of the island.

Distribution.—This pretty Parrakeet is very abundant in the districts which it affects ; it is an inhabitant of all the dry low-country parts of Ceylon, and is more abundant on the seaboard and the adjacent maritime regions than in the interior. It is very partial to the cocoa-nut and palmyra districts on the east and north coasts ; commencing, therefore, at the Jaffna peninsula, where it is common, we find it more or less plentiful down the east coast and round the east corner of the island to the Girawa Pattu, or "Province of Parrots," beyond the western boundary of which it is rarely seen. From there up the west coast, as far as the district immediately to the north of Negombo, it is absent ; here it reappears again, and is very abundant about Chilaw, where it was noticed particularly by Layard, likewise at Puttalam and throughout the Seven Korales to the base of the Kurunegala and Matale hills, along which it is tolerably numerous. Along the west coast to Manaar, and thence northward to Jaffna, it is very common. It occurs in suitable localities throughout the northern forest tract, and in portions of the Park country, as well as at the base of the Medamahanuwara, Madulsima, and Haputale ranges, but I do not know of its ascending to any elevated patnas.

This species is common throughout all India, from the south of the Madras Presidency to the foot of the Himalayas. It is a denizen of the low-lying parts of the country as in Ceylon ; for I do not find it recorded from any elevation of consequence either in the north or south. Its range extends into the north-western parts of India. Captain Butler notices it as very common in Northern Guzerat, as well as on Mount Aboo, and Mr. Hume the same as regards Sindh. In Burmah it is likewise common, and extends down the peninsula to the latitude of Penang. It was introduced into the Andaman Islands by Col. Tytler, but Mr. Hume says it has now entirely disappeared. It is also found in North-eastern Africa and Senegambia.

Habits.—The Rose-ringed Parrakeet frequents openly-timbered plains, scrubby land in the vicinity of cocoanut cultivation, low jungle along the sea-coast, and, in fact, all localities where it can obtain an abundance of wild berries and fruit to subsist on. Like the last species it assembles in flocks, but of far greater number, to roost among the cocoanut-trees, often in the midst of a village, and even, as at Trincomalie, in the centre of a town. It commences to return from its feeding-grounds at an early hour ; and often about 4 o'clock in the afternoon I have watched little troops of a dozen or more glancing over the tops of the trees, and sweeping across open places in the jungle, or twisting through a palmyra-grove with surprising quickness, towards their evening haunt, their light green plumage glittering in the rays of the declining sun, while the foremost of the flock uttered his shrill but not unpleasant note, as if to cheer his companions on. In the early morning it is marvellous with what celerity they spread themselves over the whole surrounding country, branching off in little parties, probably the same which returned together on the preceding evening, as if they were resolved to reach a certain spot by a given time or they would find their breakfast vanished !

They feed for about three hours, and then towards 10 o'clock settle about in twos and threes in the thick foliage of shady trees, and remain silent, suddenly darting off with a scream when disturbed. They are very difficult to see when seated thus among leaves, and unless they were to fly off on the approach of man, would with difficulty be observed. In the evening they become, like the last species, regardless of a gun, and are often shot in large numbers by the natives, who wait beneath the trees as they return to roost. They feed chiefly on berries, but they can, as Jerdon remarks, be very destructive to grain. Burgess, as noted by the Doctor in his 'Birds of India,' remarks that they carry off the ears of corn to trees to devour at leisure. When looking for a tree in fruit, I have seen them, as Jerdon noticed, "skimming close to and examining every tree ; and when they have made a discovery of one in fruit, circling round, and sailing with outspread and down-pointed wings till they alight on the tree." I have often wondered at the skill with which flocks of this Parrakeet glance and twist between the trunks of a tolerably thick palmyra-grove, flying with arrow-like speed, and do not strike against them ; but it appears that sometimes they are not quick-sighted enough, for it is on record that they have flown against the walls of houses and been killed. The Shahin Falcon preys on this species, and some observers say that Owls kill them at night. Its note is shriller and shorter than that of the large bird, and is much uttered in the mornings and evenings. It is noteworthy that caged specimens in England always become noisy, even in the long summer evenings, about 5 or 6 o'clock, the exact time of going to roost in their native country.

Layard writes the following account of a large colony of these Parrakeets at Chilaw:—" Hearing of the swarms which resorted to the spot, I posted myself on a bridge some half a mile away, and attempted to count the flocks that came from one direction, eastward, over the jungle. About five o'clock in the afternoon straggling bodies began to wing their way homeward, but many of them came back again to pick up the scattered grains left on the fields near the village; about half-past five, however, the tide fairly set in, and I soon found I had no flocks to count—it was one living screaming stream : some high in the air winged their way till over their homes, when, with a scream, they suddenly dived downwards with many evolutions until on a level with the trees; others flew along the ground rapidly and noiselessly, now darting under the pendent boughs of some mango or solitary tree, now skimming over the bridge close to my face with the rapidity of thought, their brilliant green plumage shining in the setting sunlight with a lovely lustre.

" I waited at this spot till the evening closed in, and then took my gun and went to the coconut-tope which covered the bazaar. I could hear, though from the darkness I could not distinguish, the birds fighting for their perches; and on firing a shot they rose with a noise like the rushing of a mighty wind, but soon settled again, and such a din commenced as I shall never forget."

This is the most commonly domesticated of the Ceylonese Parrakeets, and is a great favourite with Europeans and natives; it learns to talk well, and is very often brought home to England as a pet.

Nidification.—This species breeds in holes in trees, often at a considerable height from the ground, and lays four or five white eggs on the dead wood at the bottom of the cavity. The mouth of the hole is, Mr. Hume remarks, very neatly cut, circular, and about 2 inches in diameter. The nesting-time is in March and April; and the hen bird is given to sitting very close, for Captain Butler writes that he had to push one off her nest with his hand, and even then she would not leave the hole, although there were no less than three entrances by which she might have escaped. The eggs, which are of course white, are devoid of gloss, and are broad ovals in shape; they measure as the average size, according to Mr. Hume, 1·2 by 0·95 inch.

PALÆORNIS CYANOCEPHALUS.
(THE BLOSSOM-HEADED PARRAKEET.)

Psittacus cyanocephalus, Linn. Syst. Nat. p. 141 (1766).

Psittaca bengalensis, Brisson, Orn. iv. p. 348 (1760); Gm. Syst. Nat. i. p. 325 (1786).

Psittacus indicus, Lath. Ind. Orn. i. p. 86 (1790).

Palæornis cyanocephalus, Blyth, Cat. B. Mus. A. S. B. p. 5 (1849); Kelaart, Prodromus, Cat. p. 127 (1852); Layard, Ann. & Mag. Nat. Hist. 1854, xiii. p. 264; Horsf. & Moore, Cat. B. Mus. E. I. Co. ii. p. 616 (1856).

Palæornis rosa, Jerdon, B. of Ind. i. p. 259, et Ibis, 1872, p. 6; Holdsworth, P. Z. S. 1872, p. 425; Legge, Ibis, 1874, p. 14; Gould, B. of Asia, pt. xxvi. (1874); Legge, Ibis, 1875, p. 282.

Palæornis bengalensis, Jerdon, Cat. B. S. India, Madr. Journ. 1840, xi. p. 208.

Palæornis purpureus, Hume, Nests and Eggs (Rough Draft), p. 116 (1873); Hume, Str. Feath. 1873, p. 433; Ball, Str. Feath. 1874, p. 390; Brooks, ibid. 1875, p. 232; Butler, ibid. p. 457.

La Perruche à téte bleue, Brisson, Orn. iv. p. 359, pl. 19. fig. 2; *Blossom-headed Parrakeet*, Latham and Gould, Birds of Asia; *Rose-headed Parrakeet*, *The Ashy-headed Parrakeet* (Kelaart).

Faraida, lit. "the plaintive or complainer," Beng.; *Tui-suga*, Nepalese; *Tiua-tota*, Hind. in the south; *Bengali-tota*, in the Punjaub; *Rama-chilluka*, Telegu.

Battoo girawa, *Malitchia*, Sinhalese; *Killi*, Ceylonese Tamils.

Adult male. Length to front of œre 12·0 to 13·0 inches ; culmen 0·8 ; total length varying from 13·0 to 14·0 ; wing 5·1 to 5·25; tail 7·0 ; tarsus 0·5; outer anterior toe and claw 1·0 ; depth of upper mandible at cere 0·37.

Iris white, pale yellowish white, or greenish white, with a dusky or greyish inner circle, which latter is divided sometimes from the pupil by a whitish ring ; cere olivaceous green ; eyelid olive-brown ; bill, upper mandible orange-yellow, variable in depth of hue, and in some with a dusky tip, lower mandible black or blackish brown*; legs and feet dusky sap-green, claws plumbeous with dusky tips.

Head, face, and nape covered by a cap of flame- or rose-red, which is bounded beneath by a narrow black collar and overlaid gradually from the crown and cheeks downwards with delicate blue; the black collar is concealed by the overlying cap on the hind neck and widens below the cheeks, passing up by the base of the bill to the gape ; below this collar the neck is encircled with verdigris-green, varying in extent, and passing into the yellowish green of the back and scapulars; wings, rump, and upper tail-coverts verditer-green, brightest on the latter parts ; quills brown internally and with a fine yellow outer margin; a dark red spot on the median wing-coverts : middle pair of tail-feathers blue, washed with green at the base, and with deep white tips, the rest green with the tips yellowish and bases of inner webs rich yellow; beneath yellowish green, more verdant on the lower parts and under tail-coverts; axillaries and under wing-coverts pale emerald-green.

Individuals vary in the hue of the rump and the depth of the white tail extremities; and many that have attained the adult cap, but have not arrived at the full age of maturity†, have a greenish-yellow semi-collar below the green ring, and a more yellowish hue on the back, tertials, and under surface.

---

* Like other Parrakeets this species usually has the bill so discoloured that it is difficult to tell what its colour really is.

† This is difficult to define, for, as Capt. Hutton remarks from observation of caged birds (Str. Feath. vol. i. p. 311), each subsequent year after the third "only adds to the richness of colouring."

Female. Less than the male; wing 4·9 to 5·1 inches. Upper mandible yellow; lower dusky or blackish. The cap is dull plum-blue, wanting the black collar and mandibular stripe, and bounded by a yellow ring clearly defined on the sides of the neck; back brownish green; wings wanting the red shoulder-spot; chest washed with yellowish.

Young. The nestling is clothed with green feathers; the bill is at first black, changing in the male, at about a fortnight old, into yellow.

The bird of the year has the bill greenish yellow, dusky along the culmen; iris white, tinged with green; legs and feet plumbeous green.

Plumage green throughout, brightest on the rump and lower back, paling slightly on the forehead, and with the hue of the hind neck rather light, contrasting somewhat with the dark green of the nape; the central tail-feathers are rather short and washed with blue, the tips being whitish.

In the next stage the forehead becomes paler and the head bluish, with a dusky edge bordering the lower mandible: the central rectrices are blue, as in the adult, but with less of the white colour at the extremities.

Lutinos of this Parrakeet are occasionally met with. A beautiful example, in perfect luteous plumage, was given to His Royal Highness the Duke of Edinburgh by the Mudliyar Jayetilke of Kurunegala.

Obs. Ceylonese examples of this Parrakeet are, as a rule, smaller than Indian, among which northern birds seem to be the largest. The wings of five males from India are as follows—Madras, 5·3 inches; Bengal, 5·4; Bengal. 5·5; Nepal, 5·7; Nepal, 5·3: those of two females—Bengal, 5·2; "India," 5·4. The coloration of the rose and blue cap corresponds with that in the insular bird.

The Burmese bird, the *Pal. rosa* of Boddaert, founded on plate 888, Pl. Enl., which was long confounded with this species, has the head less covered with the azure hue, the axillaries and wing-lining *blue*, and the female has the red wing-spot as well as the male. Mr. Blyth, who published a remark on the subject in the 'Ibis,' 1870, appears to have brought the fact of these differences to the notice of Mr. Gould. Both species are beautifully figured in Mr. Gould's great work on the 'Birds of Asia;' but unfortunately in the letterpress the specific names have, I conclude, by a *lapsus calami* become inverted; the Indian bird is called *rosa* and the Burmese *cyanocephalus*. He remarks, in the commencement of his article on the Indian bird headed *Pal. rosa*, that it should bear the name of *P. cyanocephalus*, founded on the "Perruche à tête bleue," Brisson, Orn. iv. p. 359, pl. 19. fig. 2; so that the mistake is apparent at a glance. The Burmese bird *Pal. rosa* is, I observe, styled *Pal. bengalensis* by writers in 'Stray Feathers,' this name being in reality a synonym of *Pal. cyanocephalus.*

Distribution.—This beautiful Parrakeet is abundant in many parts of the low country, and tolerably plentiful in the coffee-districts up to an elevation of about 4000 feet. It is not, as a rule, found very near the sea-coast. In the Galle district it is first met with about 15 miles inland, and is common from there up to the Morowak Korale, wherein the country and vegetation suit its habits. In the interior of the Western Province, from Aviswella to Ratnapura and through the Saffragam valley to the district lying to the south of Haputale, as also in the Pasdun and Raygam Korales, it is a common bird. I have seen it about 10 miles inland from Kalatura. Mr. Parker writes me that it is not found nearer Puttalam than Uswewa, and northward of this it keeps to about the same distance from the sea; further inland about Kurunegala and in most parts of the Seven Korales, as well as along the base of the hills to Dambulla, it is tolerably plentiful. Beyond Anaradjapura it becomes scarcer, being only found in certain suitable localities. In the Jaffna peninsula I have not seen nor heard of it. It appears not to be found near Trincomalie, but to the south of the Virgel I once met with it, and that, too, at no great distance from the sea. It is not uncommon in the Eastern Province and about Nilgalla. In the Magam Pattu it frequents the cheenas of the natives. As regards the Central Province, it is common in the Knuckles, Pusselawa, Deltota, Maturata, and other districts round Kandy. In the vicinity of Badulla and in Madulsima it is likewise tolerably plentiful.

On the continent Jerdon says that "it is found more or less throughout India, extending into the Himalayas." It is common on the Malabar coast and in the jungles of the Carnatic and in the Eastern Ghâts. Mr. Bourdillon does not seem to have met with it in the Travancore hill-region, but Mr. Fairbank says that it is common on the Palani hillsides up to 4000 feet; the same writer remarks that it is common along the hills in Khandala, and visits the Deccan at some seasons in flocks. Mr. Ball says that it is found in most parts of Chota Nagpur, but at the same time it is somewhat local; it is likewise common about the Sambhur

Lake; and throughout the entire north-western region of Mount Aboo and Guzerat it is found, not frequenting, however, the parched-up province of Sindh, which is only to be expected, seeing that it is a bird which loves a luxuriant country.   It inhabits the plains at the foot of the great Himalayan range, extending into those mountains up to 5000 feet, but giving place as it goes eastwards to the Burmese bird, *P. rosa*, which ranges as far westward as Nepal.

Habits.—In the Central Province this Parrakeet frequents chiefly the patnas on the hillsides and the vicinity of the paddy-fields of the natives in the valleys. In the low country it is partial to wooded lands near rice-fields, open glades, cheenas, and clearings generally in the jungle. It forms one of the most pleasing ornithological features of Ceylon, what with its gay plumage and its restless disposition, leading it to dash about in small parties, which glance with the swiftness of an arrow down the valleys and ravines of the verdant forests, and make these lovely spots re-echo with its musical whistle, while its bright green attire contrasts with the many-coloured foliage of the woods. It perches much on the very tops of trees, balancing itself on the smallest leafy twig, and remains perfectly motionless until started into flight by the approach of danger. Mr. Ball remarks that the way in which these birds conceal themselves in trees is a matter of surprise, and hints that it is apparently not only the colour but the position in which they perch that accounts for their similarity to the surrounding foliage. In hot weather in Chota Nagpur they choose the Sal tree (*Shorea robusta*), and one may approach within a few feet of the birds without being able to distinguish a single individual. I have myself observed the same thing with *Palæornis torquatus*, when perching or feeding in a tree with small light-green roundish leaves, the name of which I am not acquainted with.

The present species is very fond of dead trees, which usually stand in cheenas in the low-country jungles ; it climbs actively about the branches of these, using its bill, and shows its plumage off to advantage against the charred wood. Its flight is very swift indeed, and when shooting down a ravine it proceeds with an oscillating or side to side motion, its wings half-closed, at a speed surpassed by few birds in Ceylon. Its note is a clear and high-pitched musical whistle, which is usually uttered on the wing ; it is possessed of considerable vocal powers, and can be taught, in confinement, to whistle tunes, Captain Hutton recording an instance of one which whistled many familiar airs. It is not, however, kept as a caged bird in Ceylon to such an extent as either of the foregoing species.

It is most destructive to the grain fields of the natives, devouring enormous quantities of Kurrukan, which is in many wild parts of the forests the only edible seed they cultivate ; it also attacks the brinjals and small red cucumbers which are much grown on newly burnt cheenas. Large flocks take up their quarters in these localities and resist all attempts to drive them away, returning immediately after being shot at, and settling on the tops of their favourite dead trees until they can again with safety renew their pillaging on the vegetables of the unfortunate half-starved cultivators.

Nidification.—This Parrakeet breeds from February until May in the western parts of Ceylon, nesting in holes in the smaller limbs of dead trees. I once found its nest in a mere sappling but a few inches in diameter ; at the bottom of the cavity were a number of dry pellets of earth, which made it apparently rather uncomfortable for the 4 young ones which were huddled together in it. They bore their own nest, choosing a partially decayed piece of wood, which they follow up into the centre of the branch, making the egg-cavity larger than the entrance. The eggs are laid on the dead wood, and the female is a very close sitter. The eggs are usually four in number and are pure white, the shell being devoid of gloss ; they average 1·0 by 0·81 inch

PALÆORNIS CALTHROPÆ.

(LAYARD'S PARRAKEET.)

(Peculiar to Ceylon.)

Palæornis calthropæ (Layard), Blyth, J. A. S. B. 1849, xviii. p. 800; id. Cat. B. Mus.
A. S. B. p. 340 (1849); Kelaart, Prodromus, Cat. p. 127 (1852); Layard, Ann. & Mag.
Nat. Hist. 1854, xiii. p. 263; Bonap. Rev. et Mag. de Zool. 1854, p. 263; Gray, List
Psittacidæ Brit. Mus. p. 22 (1859); Schlegel, Mus. P.-B. *Psittaci*, p. 83 (1862); Finsch,
Papageien, p. 53 (1868); Holdsworth, P. Z. S. 1872, p. 426; Layard, P. Z. S. 1873,
p. 204*; Legge, Ibis, 1874, p. 14; Hume, Str. Feath. 1874, p. 18; Gould, B. of Asia,
pt. xxvi. (1874); Legge, Str. Feath. 1875, p. 200; Walden, Ibis, 1874, p. 288.

Palæornis cathrapæ, Blyth, Ibis, 1867, p. 294.

Palæornis girronieri, J. & E. Verr. Rev. et Mag. de Zool. 1853, p. 195.

Psittacus viridicollis, Cassin, Pr. Philad. Acad. 1859, p. 373.

Layard's Purple-headed Parrakeet (Kelaart); *The Ceylon Parrakeet, The Hill-Parrot*,
Europeans in Ceylon.

Alloo-girawa, Sinhalese, Central Province.

♂ *ad.* suprà grisescens flavido adumbratus, capite puriùs pallidè purpurascenti-griseo, fronte, loris, regione paroticâ et
faciei anticâ viridibus: regione paroticâ posticâ pileo concolori, genis posticis nigris ad fasciam brevem cervicalem
nigram productis: collo postico intè smaragdineo fasciam collarem conspicuam formante: alâ saturatè viridi, remigibus intùs nigris, tectricibus alarum interioribus flavicanti-viridibus: dorso postico, uropygio et supracaudalibus
pallidè purpureis, his vix flavido lavatis: caudâ sordidè purpureâ, rectricibus flavo terminatis, exterioribus quoque
intùs sordidè viridibus, extùs purpureis: corpore subtùs latè viridi, gutture imo et præpectore smaragdineo lavatis:
subcaudalibus et caudâ subtùs flavis, illis viridi lavatis: subalaribus smaragdineis, majoribus et remigibus infrà
cinerascentibus: maxillâ scarlatinâ, mandibulâ saturatè rubrâ, apicaliter flavâ: pedibus fuscescenti-viridibus:
iride flavicanti-albâ.

♀ *ad.* mari similis, sed mandibulâ nigrâ distinguenda.

Adult male and female. Length to front of cere 10·6 to 11·0 inches; culmen 1·05; total length of the larger about
12·1; wing 5·4 to 5·6; tail 5·8; tarsus 0·6; outer anterior toe and claw 0·98 to 1·0; greatest height of upper
mandible 0·47.

Male. Iris yellow-white, or white, or greenish white, with a bluish-grey outer circle; cere dusky greenish; bill, upper
mandible coral-red, yellowish at tip, lower dusky red: legs and feet dusky greenish, plumbeous green, or plumbeous.
Head, nape, and cheeks faded leaden blue, changing into green on the region round the eye and at the base of the upper
mandible, and blending into a broad black border below the cheeks, which extends across the throat (more so in
some specimens than in others); across the hind neck a broad emerald-green collar, blending into the green of
the chest and under surface; the hind neck below the collar, interscapulars, and scapulars greenish grey, with a
bluish cast, changing into pale grayish blue on the back, rump, and upper tail-coverts, the latter part tinged
with green on the longer feathers.
Wings green, the median secondary coverts pale at the extremities; lesser coverts just above the forearm bluish; 1st
primary and inner webs of all the quills brown; tail French blue, tipped deeply with yellow, the inner webs of all
the lateral feathers washed, and the outer webs finely edged with green; the tail beneath yellowish; under tail-
coverts yellowish, tipped and edged green; under wing-coverts green.

---

* I am informed that the correct orthography of this Parrakeet's specific name should be *calthorpæ*, as the family
name was Calthorp, and not Calthrop. However, Layard says here, "I have to thank Mr. Holdsworth for restoring
the true reading of this name." I therefore leave the matter as it is.

2 A

The colours of this Parrakeet appear to be much affected by the sun's rays: the head and back lose their brightness, and become pervaded with a greyish hue, altering much the delicate character of the plumage. Specimens likewise fade after preservation.

Female. Bill, upper mandible black, lower blackish, tinged with reddish. The green on the lores and orbital region less in extent, and the centre of the back (as far as I have observed) more brilliant than in the male. As the coloration of the female's bill has been the subject of some controversy, I may remark that Mr. Holdsworth first pointed out that it was black. Adults of both sexes have sometimes, when in rich plumage, a slight cobalt-hue wash on the forehead and cheeks.

Young. Iris whitish; bill (male) pale orange, (female) upper mandible dusky black, lower reddish: foot and legs plumbeous.

Birds of the year are dull green above and yellowish green beneath; the head darker than the back and sometimes with a bluish tinge; there is an indication of the green collar on the hind neck: the back and rump cobalt-blue (brighter than the adult); tail green, washed with blue, tipped and edged internally with yellow towards the extremities, lower feathers of the upper tail-coverts green; some individuals have the tail bluer than others.

Obs. This species comes nearer the South-Indian Parrakeet (*Pal. columboides*) than any other, but has not even much in common with that. There is, however, a slight general resemblance in the two birds, which is in accordance with the relationship displayed between the avifaunas of the regions in question. *Pal. columboides* has the wing 5·7 to 6·0, and is therefore a larger bird, with a correspondingly longer tail. The black ring in this species completely encircles the neck, the under surface is slaty instead of green, the rump is green instead of blue, and the primaries and their coverts obscure blue.

Distribution.—The Ceylon Parrakeet was discovered by Layard, who writes thus concerning it in the 'Annals and Magazine of Natural History' for 1854 :—"My first acquaintance with this lovely bird was at Kandy, where I killed a male and female at one shot from a flock flying over my head; I took them for the common *P. torquatus* until I picked them up, and then great was my delight to find such an elegant new species. It proves to be the common Parrakeet of the hilly zone, and I have traced it to all parts of it." As Layard remarks, this species is distributed throughout the hills of the Kandyan Province; but it is singularly local as regards some parts, small districts here and there appearing to be surrendered almost entirely to the last species. Of such I may mention portions of the Pusselawa, Hewahette, and Kalebokka valleys, as well as parts of Dumbara, where, in the mouth of November, I have met with numbers of *P. cyanocephalus*, to the almost entire exclusion of the present Parrakeet. It is numerous about Kandy and Peradeniya, and also Deltota, in Upper Hewahette, Poondoloya, most parts of Uva, Madulsima, and Haputale, while it is still more abundant in Maskeliya, Dickoya, and throughout the Peak forests at intermediate altitudes. In the south it is numerous in the Morowak Korale, and very abundant indeed in the higher parts of the Kukkul Korale, notably in the Singha-Rajah forest, concerning which region I wrote, in my paper "On the Distribution of Birds in the Southern Hill-region of Ceylon" ('Ibis,' 1874), that I considered it more abundant there than in any other part of the island, a conclusion to which I still adhere.

It was thought for many years to be an inhabitant only of the hills, an idea which obtained on account of the very imperfect exploration of the forests round the base of the central zone, and the repeated working of naturalists over certain beaten tracks. In 1870 I first met with it in the low country, down in the valleys adjacent to the Hinedun Kanda or Haycock Hill, and was somewhat surprised at finding it there, while I had not seen nor heard it in the Oodogamma or Opaté forests, a district lying higher than the one in question. My next meeting with it was in the park-like woods lying between "Westminster Abbey" and Kollupitiya, on the new Batticaloa road, and which are studded with those remarkable rocky hills so characteristic of the Eastern Province. I subsequently found it about Nalanda, and all round the base of the Matale Hills, from Dambulla to Kurunegala : beneath the Ambokka range it is abundant. The greatest extent of low country, however, over which it is spread lies in the Western Province, between Ruanwella and Pelmadulla. I found it close to Ukawatta, about 26 miles from Colombo, where it was frequenting the tall timber-forests; it was also very common in the Kuruwite forests, and thence up to Gillymally, as well as in other parts of the valley of

Saffragam. Beyond the Karawita hills, which lie to the south of the Kaluganga, I again met with it in the forests of the northern or lower part of the Kukkul Korale, and traced it into the Pasdun Korale as far as the remote and sequestered village of Moropitiya. Nearer the sea than this locality I did not find it. In the south its coast-wise limit appears to be the vicinity of the Haycock, and in the east that of "Westminster-Abbey" hill.

From the above remarks it will be seen that this Parrakeet spreads into the low country at all points connected with an adjacent forest-covered range, in which it is numerous.

As regards the altitude to which it ascends, I have seen it between 5000 and 6000 feet above Maturata, at a similar height in the Wilderness of the Peak and in Haputale, and Dr. Kelaart records it from Nuwara Elliya, though neither Mr. Holdsworth nor Mr. Bligh met with it there.

Habits.—Layard's Parrakeet frequents the outskirts and open places in the interior of forests, patna-woods, wooded gorges, and glades in the vicinity of hills; it associates in moderately-sized flocks, and is a very noisy and restless bird, uttering its harsh "crake" on the wing, as it dashes up and down the magnificent valleys and forest-clad glens of the Ceylon mountains, and enlivens these romantic solitudes with its swift and headlong flight. It is entirely arboreal in its habits, settling in flocks among the leaves of its favourite trees, and silently devouring the fruit-seeds and buds on which it subsists. It is very partial to the wild fig, the fruit of the Kanda-tree (*Macaranga tomentosa*), the wild cinnamon-tree, and the flowers of the Bomba-tree. After feeding in the mornings it becomes garrulous, assembling in small parties in shady trees, and keeping up a chattering note almost similar to that of *Layarda rufescens*; towards evening it commences to feed again, and before going to roost roams about in small flocks, constantly uttering its loud harsh note, and settling frequently on the tops of conspicuous and lofty trees. In the Singha-Rajah forest their presence at evening was more conspicuous than that of any other bird; they darted up and down the deep gorges and across the small Kurrakan clearings in the forest, keeping up an incessant din; now and then they rested on the top of some dead tree standing in the cheena, and then suddenly glanced off, shooting with arrow-like speed between the trees of the forest, again to appear as they swept up the valley and away over the top of the gloomy jungle.

Its flight is bold and swift, but not of that glancing character peculiar to the last species; and this, together with its harsh cry, which can be heard a long way off, seems to distinguish it easily from *Pal. cyanocephalus.*

Nidification.—The breeding-season commences in January. It nests in holes in large trees; but I have never been able to procure the eggs, although I have more than once discovered the nest. I have seen one situated in a Hora-tree (*Dipterocarpus zeylanicus*); the old birds, on flying to it, clung to the bark outside the opening, and then pulled themselves into the hole, using the beak to assist them in entering. Layard writes that he was informed by natives that they laid two eggs, which, like those of other members of the family, would be pure white. In the Peak Wilderness they breed in the decaying trunks of dead Kitool-trees.

The figures on the Plate are those of an adult female in the foreground, with a slightly abnormal amount of black below the cheeks, and a young male from Kaloday, Eastern Province, in the background, which should have been drawn with the back turned to the front, so as to show the peculiarly light blue on the rump of immature males. Unfortunately, however, the requirements of the author and the tastes of the artist are sometimes at variance. I had wished that these birds should be figured on the "Jambu," a sketch of which, by Sir Chas. Layard, I furnished my artist with; but it was not found suitable, and he has introduced the common fig-tree of Europe instead.

PSITTACI.

Fam. TRICHOGLOSSIDÆ.

Bill with the upper mandible long, compressed gradually from the base, the tip rather straightened and acute; under mandible *longer than it is high*, the tip less obtuse than in the *Psittacidæ*.

Genus LORICULUS.

Bill with the upper mandible long, gently curved, compressed; margin with a slight lobe near the base; cere rather advanced; under mandible shallow and considerably elongated. Wings long, with the 2nd quill the longest, and slightly exceeding the 1st and 3rd. Tail short, rounded, scarcely exceeding the closed wings. Tarsus very short. Toes long, the outer anterior much exceeding the inner, which is about equal to the tarsus; claws stout, long, and well curved.

LORICULUS INDICUS.

(THE CEYLONESE LORIKEET.)

(Peculiar to Ceylon.)

Psittacus indicus, Gm. Syst. Nat. i. p. 349 (1788).
Psittacus asiaticus, Lath. Ind. Orn. i. p. 130 (1790).
Psittacula indica, Briss. Orn. iv. p. 390 (1760).
Psittacula conlaci, Lesson, Tr. d'Orn. p. 202 (1831).
Loriculus asiaticus, Blyth, J. A. S. B. 1849, xviii. p. 801; Kelaart, Prodromus, Cat. p. 127 (1852); Layard, Ann. & Mag. Nat. Hist. 1854, xiii. p. 261; Horsf. & Moore, Cat. B. Mus. E. I. Co. ii. p. 628 (1856).
Loriculus indicus, Bonap. Rev. et Mag. de Zool. p. 155 (1842); G. R. Gray, List Psitt. Brit. Mus. p. 55 (1859); Schlegel, Mus. P.-B. *Psittaci*, p. 132 (1864); Walden, Ibis, 1867, p. 467; Holdsworth, P. Z. S. 1872, p. 426; Legge, Ibis, 1874, p. 15.
Loriculus edwardsi, Blyth, Ibis, 1867, p. 295; Nevill, J. A. S. (Ceylon Br.), 1870–1, p. 32.
Coryllis indica, Finsch, Papag. ii. p. 714 (1868).
The smallest Red-and-green Parrakeet, Edwards, Glean. pl. 6 (1743); *Red-and-green Indian Parrakeet*, Lath. Synopsis (1781), also *Red-rumped Parrakeet*, Lath. Gen. Hist. (1822); *Das ceylonische Papageichen*, Finsch, Papag.; *The Small Ceylon Parrakeet*, Kelaart; *The Ceylon Love-bird* of travellers.
Gira-malitchia, Pol-girawa, lit. " Flower Parrakeet," Sinhalese.

Similia *L. vernali*, sed capite et occipite rubris, nuchâ flavo adumbratâ diversus. *Juv.* capite viridi dorso concolori distinguendus.

Adult male and female. Length to front of cere 5·1 to 5·2 inches; culmen 0·65; total length averaging 6·7 to 5·8; wing 3·6 to 3·8; tail 1·7; tarsus 0·4; outer ant. toe and claw 0·75.

Iris white; bill light orange-red, paler at the tip, lower mandible paler than the upper; legs and feet dusky yellow; cere yellow.

Lower hind neck, back, and wing-coverts leaf-green; forehead and front of crown rich deep red, gradually becoming overcast with an orange hue on the nape, and fading into the green of the hind neck; the upper part of the back is more or less pervaded with a dull golden cast; rump and upper tail-coverts deep red, outer webs of quills and the tail dark green; inner webs of primaries above dark hair-brown, 1st quill with a fine greenish-blue edging; beneath the inner webs of the quills and the lower surface of tail verditer-blue. Cheeks, region round the eye, and entire under surface pale green, washed with bluish across the fore neck.

Young. Iris dull grey or olive; bill dusky yellow; legs and feet olivaceous yellow, claws blackish. Head above green, with the forehead pale, and an aureous cast on the crown; rump and upper tail-coverts as in the adult; fore neck without the bluish tinge. Birds of the year are full-sized.

Lutinos of this species are not uncommonly met with. A description of a beautiful example is given by Mr. Nevill, of the Ceylon Civil Service (*loc. cit.*), as follows:—"Crown of the head and rump brilliant scarlet, shading into metallic orange on the rump; back vivid golden yellow, dappled with emerald-green, and tinged in places with orange: wings green, mottled with bright yellow; quills of the normal colour, tipped with yellowish white; beneath bright but paler yellow than the back, mottled with bright pale grass-green; throat yellowish; cheeks rufescent; under wing-coverts mottled green, yellow, and straw-colour."

Obs. *Loriculus apicalis,* from the Philippines, is very close to this Lorikeet: a specimen in the British Museum, from Mindanao, is scarcely separable in any other point but the coloration of the head, which is pale or yellowish red; the hind neck wants the aureous wash, and the throat has only a very faint wash of blue on it. *L. indicus* also resembles the Indian and Andaman species, *L. vernalis,* in most points, differing from it chiefly in the head. The latter bird has the head grass-green, concolorous with the back, with the forehead brighter than the crown, and the hind neck wanting the aureous colour of *L. indicus*; the red on the rump does not extend so high up the back; the coloration of the tail and wings is almost identical with that of the insular bird. The wing varies from 3·5 to 3·75 inches, or much the same as in *L. indicus.*

Distribution.—This pretty little bird, so well known as a caged pet to travellers who touch at Point de Galle, by whom it is generally styled the "Love-bird," is widely distributed throughout the low country of the island, and is commonly located in the hills up to an elevation of 3500 feet. In the south-west of the island it is extremely abundant, frequenting the cocoanut-groves close to the port of Galle, as well as the entire semi-cultivated interior of that district. Further up the west coast it is not common near the sea, but in the openly wooded and partly cultivated portions of the Western Province it is abundant; and in the Ratnapura and Kurunegalla districts is quite as numerous as about Galle. To the north of the Seven Korales it is less plentiful; but I have met with it here and there throughout all the forest-tracts of this part of the island, and in the N.E. monsoon have seen it in the woods near Fort Ostenburgh, Trincomalie. I have noticed it again in many parts of the Eastern Province, but I do not think it is as generally distributed there as in the west. Layard found it abundant about Hambantota, but I did not observe it at all in that district during two visits I made to it; in the north of the Magam Pattu I found it, but not on the scrubby sea-board near Hambantota. In the Central Province it is common about the patnas in Dumbara and Pusselawn and in many parts of Uva, and during the dry weather prevalent in the N.E. monsoon ascends above an altitude of 4000 feet. Mr. Thwaites, of Hakgala, informs me that he has seen it in the gardens at that season of the year.

This little bird is not very aptly styled *indicus*; but Gmelin, who named it from the figure in Edwards's plate, did not know from what exact locality he received his specimen, as all the information which Edwards could give about it was contained in the words, "brought from some Dutch settlement in the East Indies." When the bird became better known it was apparent that this settlement was Ceylon.

Habits.—The Ceylon Lorikeet frequents woods, detached groves of trees, compounds, native gardens, patnas dotted with timber, and, in fact, any locality which is clothed with fruit-bearing trees or those whose flowers afford it its favourite saccharine food. It is a most gluttonous little bird, constantly on the wing in active search for its food, darting with a very swift flight through the woods, uttering its sibilant little scream, its bright plumage flashing in the rays of the tropical sun. When it reaches a tree which attracts its attention it instantly checks its headlong progress, and alighting on the top, actively climbs to the fruit which it has espied, or should the tree prove barren, after giving out its call-note for a short time, darts off, perhaps in the opposite direction from which it came. It is excessively fond of the "toddy" or juice which exists in the Kitool or sugar-palm (*Caryota urens*), and feeds on it to such an extent that it becomes stupified and falls an easy captive to the natives, who cage it in large numbers for sale at Point de Galle.

While in a state of captivity they are fed on sugar-cane, of which they are very fond, but they do not live for any length of time should the supply of cane come to an end. It feeds so gluttonously on the beautiful fruit of the Jambu-tree that I have seen bird after bird shot out of one tree without their companions taking the slightest notice of the gun or the death of so many of their little flock*. When held up by the legs, after being shot, the juice of this fruit pours from their mouths and nostrils. The flowers of the cocoanut-tree come in for a large share of its patronage, as do also those of other trees, on the "cups" or calyces of which it subsists, biting them off in a pendent attitude. Layard writes that "at Gillymally they were in such abundance that the flowering trees were literally alive with them; they clung to the bright scarlet flowers head downwards, or scrambled from branch to branch, while the forest echoed with their bickerings. They bit off the leaves (which fell like scarlet snow upon the ground) to get at the calyx; and when this dainty morsel was devoured they flew off to the banana-trees, down the broad leaves of which they slid and fastened upon the ripening clusters of fruit or the pendent heart-shaped flower."

When roosting at night they sleep hanging by their feet from the perch.

The figure in the Plate facing my article on *Palæornis calthropæ* is that of an adult bird, and that on the Plate of *Xantholæma rubricapilla* an immature or yearling individual.

---

* I have observed the same of *Trichoglossus pusillus* in Australia, which it is sometimes impossible to drive from a tree laden with ripe cherries otherwise than by vigorously shaking the stems!

Order PICARIÆ.

Distinguished from the *Passeres* by the presence of a double notch in the posterior margin of the sternum. Bill not toothed. Wing with ten primaries. Feet in the Scansorial section arranged in pairs, two in front and two behind; in others simple, and arranged three in front and one behind.

Fam. PICIDÆ.

Bill straight, pointed or wedge-shaped at the tip. Head generally crested. Tail of twelve feathers, with the shafts rigid and stout. Feet zygodactyle, one of the posterior toes sometimes wanting.

Skull very strong. Tongue very long and extensile, and furnished at the tip with barbs. Sternum with a double notch in the posterior edge.

Subfam. PICINÆ.

Bill moderate or long, compressed, with the tip wedge-shaped, the ridge acute, and the lateral ridge just above the nostrils and *very* pronounced; the gonys long. Feet large; the outer posterior toe longer than the outer anterior one; the inner posterior toe always well developed.

Genus PICUS.

Bill of medium size, culmen nearly straight and sharp, lateral ridge parallel to the culmen and continued forward till it meets the margin. Nostrils concealed by a tuft of feathers. Wings with the 1st quill very short, the 2nd considerably shorter than the third, and the 4th and 5th the largest. Tail with the shafts stiff and decurved, the tip forked. Tarsus longer than the anterior toe, which is shorter than the versatile or long posterior toe; claws strong and curved.

PICUS MAHRATTENSIS.

(THE YELLOW-FRONTED WOODPECKER.)

Picus mahrattensis, Latham, Ind. Orn. Suppl. ii. p. 31, female (1790); Sykes, P. Z. S. 1832,
p. 97; J. E. Gray, Ill. Ind. Zool. pl. 32 (1830–2); Gould, Cent. Him. Birds, pl. 51
(1832); Blyth, J. A. S. B. 1845, p. 196; id. Cat. B. Mus. A. S. B. p. 62 (1849); Layard
et Kelaart, Prodromus, Cat. Append. p. 59 (1853); Layard, Ann. & Mag. Nat. Hist.
1854, xiii. p. 448; Horsfield & Moore, Cat. B. Mus. E. I. Co. ii. p. 674 (1856); Malh.
Mon. Picidæ, i. p. 108, pl. 28. figs. 1–3 (1863); Jerdon, B. of Ind. i. p. 275 (1862);
Holdsworth, P. Z. S. 1872, p. 426; Hume, Nests and Eggs, p. 122 (1873); Legge, Ibis,
1875, p. 283; Ball, Str. Feath. 1874, p. 390; Hume, ibid. 1875, p. 58.

Picus hemasomus, Wagler, Syst. Avium, gen. *Picus* (female), no. 30 (1827).

Picus aurocristatus, Tickell, J. A. S. B. 1833, p. 579.

Dendrocopus mahrattensis, Jerdon, Cat. B. S. India, Madr. Journ. 1840, p. 212.

Liopipo mahrattensis, Cab. et Heine, Mus. Hein. v. p. 44 (1862).

The Mahratta Woodpecker, The Black-spotted Woodpecker of Europeans.

Käralla, Sinhalese; *Tatchan-kuruvi*, Tam., lit. " Carpenter-bird."

Adult male and female. Length 6·8 to 7·0 inches; wing 3·7 to 3·95; tail 2·5; tarsus 0·7; outer ant. toe 0·55, claw
(straight) 0·33; hind toe 0·6; bill to gape 1·05. Females are usually smaller than males.
Iris variable, red or dull red; bill dusky bluish, culmen and tips dusky; legs and feet plumbeous, claws darker than toes.

Male. Forehead, front of crown, lores, and region round eye pale shining yellowish or straw-colour, extending further
back in some specimens than in others, and changing on the crown and occiput into pale crimson. Chin, face,
and a continued stripe from the ear-coverts down the sides of the hind neck, throat, centre of the chest, and
breast white, the ear-coverts slightly dusky in some; sides of neck and chest, down the centre of hind neck, hair-
brown, darkening on the upper part of the back into the brownish black of the upper surface, wings, and tail;
feathers of the back with white basal and lateral stripes; the wing-coverts, quills, and tail with large marginal
white spots, taking the form of bars on the inner webs of quills; the 1st (small) quill with the outer web unspotted;
sides of the breast, flanks, and under tail-coverts white, with very wide dark centre-stripes; centre of the breast
and belly pale crimson; tail-spots yellowish beneath. The brown of the neck and chest is very pale in some
specimens, probably the effect of the sun's rays, as the brown hue in most Woodpeckers is affected by them.

Female. Has the yellow of the forehead continued over the top of the head to the occiput, which wants the crimson hue;
ear-coverts duskier than in male; sides of the chest and flanks somewhat more covered with brown than in the male.

Obs. Ceylonese examples seem to be smaller than Indian. An individual in the national collection, from the N.W.
Provinces, has the wing considerably above 4 inches. Mr. Ball gives an extensive table of measurements,
loc. cit., from specimens shot in Chota Nagpur, by which it appears that males there have the wing exceeding
4 inches and females from 3·9 to 4·0 inches. The character of the black and white markings is similar in
specimens from India and Ceylon. The Burmese race has been separated by Blyth as *P. blanfordi*, with an
expressed doubt, however, as to its being really separable. Mr. Hume does not consider it to be so, and writes,
with reference to the alleged greater development of the white markings, that Indian specimens " vary much
inter se. An example from Wynaad is very dark; one from Kutch, again, very similar to *blanfordi*, and one from
Naubhur is undistinguishable from Thayetmyo birds: in the wings there seems to be no appreciable difference."

Distribution.—This little Woodpecker has a wide distribution in Ceylon, but is, notwithstanding, by no
means plentiful, and is rarely met with except by those who explore the wilds of the woods and forests. Mr.
Holdsworth says that it is common in the Aripu district. I have procured it in the Magam Pattu and soon it
close to the sea at Kirinde; in the Wellaway Korale it is also to be found. In the southern and western portions
of the island it is not found, as far as I am aware; and I have not seen it in the Trincomalie district, nor

in the interior of the country between there and the central road. Further north, however, it is found, for there the jungle is more suited to its habits. Mr. H. E. Hayes, of the Ceylon Public Works Dept., writes me that he has met with it at a place about 22 miles from Mullaittivu, called by the euphonious Tamil name of "Manawalempattumuripu!" Layard found it in the Northern Province and considered it to be confined to that part. I have seen it in the scrubs to the south of Kottiar, and all that densely clothed low jungle country lying between there and the Tamankadua Pattu is a most likely district for it. In the drier parts of the Central Province it is not unfrequent, inhabiting the secluded patna-nullahs, which are dotted here and there with clumps of wood interspersed with its favourite tree the *Euphorbia*. In such places I have seen it in the Hewahette district and also in Uva, in which latter part I once shot it on the Logole-oya, at an elevation of about 2500 feet.

In India the Mahratta Woodpecker is dispersed pretty well all over the peninsula, being found, according to Jerdon, in "almost every district up to the foot of the Himalayas, except in lower Bengal, though common in the Midnapore jungles." Particularizing the localities which it inhabits, we find it recorded by him as rare on the Malabar coast, but plentiful in the gap of Coimbatore. In the Palani hills it is not uncommon up to 5000 feet, a very considerable elevation for a heat-loving bird as it evidently is. It does not appear to be found in the Travancore hills, but Mr. Hume has received it from the Wynaad. In the Deccan and Khandala district it is widely dispersed, but not abundant. Further north, about the Sambhur Lake and in the Guzerat region, it is well known, though it appears, according to Mr. Hume, not to be found in Sindh. In Chota Nagpur it is distributed through the Province, though not very common. In Upper Pegu it again appears as the *P. blanfordi* of Blyth, and is, according to Captain Feilden, "found everywhere from the low grounds of Thayetmyo to the tops of the highest hills." Mr. Oates says it is common near the banks of the Irrawaddy, but was not observed by him far inland, showing that in Burmah as well as in other parts it is local.

Habits.—This species frequents low jungle and scrub, particularly that in which the *Euphorbia* grows; it is very partial to this tree; in fact every example I have met with in Ceylon was either actually on or in the vicinity of one. On the patnas I have usually observed it among scattered trees searching the trunks and branches with great agility, keeping chiefly to the underside of the latter, and working them out nearly to their extremities. It is a shy bird and difficult to procure, taking itself off with a short flight to an adjacent tree as soon as it perceives any one approaching it. It is usually a solitary bird, shunning the company of its species except in the breeding-season. It has a weak trill, not unlike that of the Pigmy Woodpecker, but of course louder; and Jerdon remarks that it also has a squeaking note. Layard observed it chiefly about *Euphorbia* trees, and Mr. Holdsworth noticed it on old fences as well as dead wood.

In India it keeps to particular trees—*Babool* in the Mount Aboo and Sambhur districts, and the Pulas-tree (*Butea frondosa*) in Chota Nagpur. Captain Feilden has observed it descending a tree tail foremost with great ease. Its food, according to my observations, consists mainly of small insects and ants; but Mr. Oates found small beetles in the stomach only.

Nidification.—The nest of the Yellow-fronted Woodpecker has never, to my knowledge, been found in Ceylon. It is almost sure, however, to nest in the *Euphorbia* tree. In India it breeds from March until April, nesting in a hole in a partially decayed branch, choosing, when it can, a Babool tree. Mr. Hume records the finding of a nest at Etawah, the hole being cut on the underside of a Babool branch about 1·5 inch in diameter, and leading to the excavated egg-cavity about 15 inches below it; the eggs were laid on chips of the wood made in excavating the hole. The eggs are three in number, less spherical than, but in size resembling, those of "the Lesser Spotted Woodpecker of Europe." Before being blown they are a delicate pink, turning glossy white after being emptied of their contents. They measure 0·87 by 0·68 inch, this being, according to Mr. Hume, the average of a large series.[*]

---

[*] I have not included *Picus macei* in this work. It was mentioned by Kelaart as having been procured in the island; but it is more than probable the bird was not correctly indentified. It is a North-Indian species, and could not well have occurred in Ceylon, as Woodpeckers are not birds which stray from their usual habitat.

2 B

Of small size.

Bill much as in *Picus*, short, widened at the base and conic; gonys quickly ascending. Wings longer than in *Picus*, the secondaries long in proportion to the primaries; tail much as in that genus, the outer feathers not so rigid; tarsi and feet the same, with the versatile toe longer than the anterior.

YUNGIPICUS GYMNOPHTHALMOS.

(THE PIGMY WOODPECKER.)

Picus gymnopthalmos, Bl. J. A. S. B. 1849, xviii. p. 804; id. Cat. Mus. A. S. B. p. 64 (1849).

Yungipicus gymnopthalmus, Bonap. Consp. Vol. Zygod. p. 8 (1854).

Picus otarius, Malh. Mon. Picidæ, i. p. 152, pl. 35 (1863).

Yungipicus gymnopthalmos, Kelaart, Prodromus, Cat. p. 128 (1852); Layard, Ann. & Mag. Nat. Hist. 1854, xiii. p. 448.

Yungipicus gymnophthalmos, Jerdon, B. of Ind. i. p. 279 (1862); Holdsworth, P. Z. S. 1872, p. 427; Jerdon, Ibis, 1872, p. 8; Hume, Str. Feath. 1873, p. 433; Legge, Ibis, 1874, p. 15; id. Str. Feath. 1875, p. 365; Bourdillon, ibid. 1876, p. 389.

Bæopipo gymnophthalma, Cab. et Heine, Mus. Hein. v. p. 59 (1863).

Little Black-and-white Woodpecker, Europeans in Ceylon.

Mal-kǣralla, Sinhalese.

Adult male and female. Length 4·7 to 4·9 inches; wing 2·8 to 3·0; tail 1·3; outer anterior toe 0·4 to 0·45, claw (straight) 0·25; bill to gape 0·6 to 0·7.

Iris white, greyish white, yellowish white, or reddish white (varies much): bill brownish olivaceous, somewhat paler beneath; eyelid and orbitar skin dull mauve or purplish; legs and feet greenish plumbeous.

Head above, centre of nape, and hind neck, back, wings, and tail very dark sepia-brown; back broadly barred, and the wing-coverts, quills, and tail spotted with white; on the rump and upper tail-coverts the white predominates, reducing the brown to bars; 1st quill and outer web of second unspotted brown; a broad white stripe passes from behind the eye to the nape; below this the cheeks, ear-coverts, and sides of neck are brown as the back; a narrow line of vermilion-red above the white stripe and partially concealed by the brown of the head; throat and entire under surface murky white; under tail-coverts striped and centered with brown; under wing-coverts white, barred with brown.

Female. Wants the vermilion superciliary stripe; otherwise as the male.

In some specimens the flanks and sides of lower part of breast show obscure brownish striæ.

Obs. Mr. Hume remarks of this species that "Ceylon specimens are absolutely identical with those from the Malabar coast." A female shot by Mr. Bourdillon in the Travancore hills measured—length 4·87 inches; wing 2·87; tail 1·25. This Woodpecker is very closely allied to the commoner Southern Indian race *Y. hardwicki*, which is, according to Jerdon, "brownish or sooty brown above, banded with white on the back; head pale rufescent or yellowish brown, scarcely deepening posteriorly." The Ceylonese bird, it will be observed, differs from it in being darker on the head and back; it is likewise smaller, the wing of *Y. hardwicki* averaging, according to Mr. Hume, 3 inches; he writes, *loc. cit.*, that typical examples of *Y. gymnophthalmos* have the whole head and back darker: but many specimens from Anjango, in the south of India, differ from some of *Y. hardwicki* from the north *only* in the much darker occiput and nape.

Distribution.—This Pigmy Woodpecker is tolerably plentiful in some parts of Ceylon, and has a wide range, being diffused over nearly all the low country, except perhaps the extreme north of the Vanni and the Jaffna peninsula, where it may also possibly occur. It is in the south-west of the island and in the Eastern Province where it is most abundant; in the latter part it is particularly seen about the dead trees standing in the beds of all the newly finished tanks. In the Galle district it is a common bird in localities suited to its habits; and about Colombo it is not uncommon, having been procured by myself as near that town as the cinnamon-gardens of Morotuwa. Layard states that he discovered it near the capital in the year 1848. About Uswewa, near Puttalam, Mr. Parker writes me it is common; beyond this in the Northern Province it is sparingly distributed, as far as I have been able to trace it, but, being difficult of discovery on account of its small size, it may often escape observation in that jungle-clad region. It occurs in the Central Province up to about 3000 feet. I have met with it in Pusselawa, Nilambe, Deltota, and parts of Uva, and I have no doubt it is to be found on the Dimbulla and Dickoya side as well.

In India it has been found, as far as present experience proves, only in the south, and even there it has escaped observation until rather recently. Jerdon had evidently seen it, though he had not procured it before the publication of his work, for he remarks, at p. 279, vol. i., " I have reason to believe that another and darker-coloured species is found in the Malabar forests; but whether this may prove identical with one of the Himalayan species or with the Ceylon bird in particular, I cannot now ascertain." Subsequently he satisfied himself of the question; for in his supplementary notes, contained in ' Ibis,' 1872, he writes that the Ceylon species occurs in the extreme south of Malabar and Travancore, and is the bird alluded to in the above-mentioned paragraph. Mr. Bourdillon has procured it in the latter district, whence also Mr. Hume has received numerous specimens, and Mr. Fairbank obtained it in the lower Palanis.

Habits.—This little bird, which, but for the frequent utterance of its shrill little note, would often completely escape observation, lives generally in pairs, and frequents the uppermost branches of trees, often perching *across them* for a short space of time. I have observed it settle thus on a mere twig, and then after a moment's pause sidle down to an adjoining branch. It works much at the broken tops of small dead branches, picking out worms and grubs from the rotten wood. In Rugam tank I observed it breaking off comparatively large pieces of dead surface-wood and searching beneath them for food. It is very fond of the jack-tree; and in the south of Ceylon I have often seen it in the "Dell" or wild bread-fruit trees (*Artocarpus nobilis*), which stand in low cheena wood, having been spared the axe for the sake of the timber. In the Northern Province I have usually observed it in large trees near rivers and tanks, and in the Kandyan country at the edges of coffee-estates or patnas. Its powers of flight, afforded by its long wings, are considerable, and its note, which is a prolonged trill, is audible at some distance, even when uttered at the tops of the loftiest trees.

Mr. Bourdillon's remark on this species, as observed by him in Travancore, is that "it lives in the tops of high trees, and is as difficult to observe as to shoot."

Nidification.—In the Western Province this Woodpecker breeds in February and March, nesting in holes in small branches. A nest which Mr. MacVicar found in the Colombo district, near Poré, was in a dead branch with an opening leading to it of about 1 inch in diameter. There were three young birds in it just hatched, and the egg-fragments were shining white.

Bill very strong, lengthened, the tip wedge-shaped ; lateral ridge very prominent, and parallel to the margin near the tip. Nostrils apart. Neck small. Wings much as in *Picus*. Tail moderately long and cuneate, the four central feathers subequal. Feet very strong, the hind toe considerably longer than the anterior ; claws stout, long, and much curved.

CHRYSOCOLAPTES STRICKLANDI.

(LAYARD'S WOODPECKER.)

(Peculiar to Ceylon.)

Picus ceylonus, Jerd. (*nec* Forster) Ill. Ind. Orn. pl. 47 (1847).
Brachypternus stricklandi, Layard, Ann. & Mag. Nat. Hist. 1854, xiii. p. 449 ; Jerdon, B. of Ind. i. p. 298.
Indopicus carlotta, Mahl. Rev. et Mag. de Zool. 1854, p. 379 ; id. Mon. Picidæ, pl. 67 (1863).
Brachypternus ceylonus, Jerdon, B. of Ind. i. p. 278.
? *Brachypternus rubescens*, Kelaart, Prodromus, Cat. p. 128 (1852).
Chrysocolaptes stricklandi, Cab. et Heine, Mus. Hein. v. p. 100 (1862) ; Blyth, Ibis, 1867, p. 297 ; Holdsworth, P. Z. S. 1872, p. 427 ; Legge, Ibis, 1874, p. 15, et 1875, p. 283 ; id. Str. Feath. 1875, p. 200.
Brachypternus erythronotus, Reich. Handb. Spec. Orn. pl. 629. fig. 4186 (1851).
Red Woodpecker, Jerdon's Illustrations ; *Hill-Woodpecker*, Europeans in Ceylon.
Kœralla, Sinhalese.

♂ suprà coccineus, pileo cristato : rectricibus alarum dorso concoloribus : alâ spuriâ et tectricibus primariorum et primariis nigris, his ad apicem pallidioribus et intùs albo ovate trimaculatis : primariis interioribus secundariisque coccineis, intùs nigris albo trimaculatis : uropygio celatim minutè albido maculato : supremaudalibus caudâque nigris : loris et frontis basi brunneis : regione paroticâ, colli lateribus et collo postico nigris, hôc ovato albo maculato : facie laterali gutturreque albidis, fasciis tribus nigris notatis, unâ supragenali, alterâ mystacali, et tertiâ medianâ gutturali : corpore reliquo subtus sordidè rufescenti-albo, plumis nigro marginatis, pectore et jugulo quasi gutturalis, abdomine quasi striolato : margine alari brunneo : tectricibus alarum reliquis nigricantibus albo maculatis : iride flavicanti-albâ ; rostro olivaceo ; pedibus olivaceis.

♀ capite nigro, albo punctato.

Adult male and female. Length 11·5 to 11·8 inches ; wing 5·8 to 6·1 ; tail 3·4 ; tarsus 1·0 to 1·1 ; outer ant. toe 0·9, claw (straight) 0·55 ; outer posterior toe 1·1, claw (straight) 0·55 ; bill to gape 1·9 to 2·1.

Females, though quite as large as, and equal in wing to, males, appear to have shorter bills *as a rule*. The claws of this species are very strong and deep.

Iris yellowish white or very pale buff ; bill brownish or olivaceous at the base, changing at the centre into greenish white, the tip assuming a dusky hue ; legs and feet plumbeous green or greenish slate.

Head, nape, back, rump, wing-coverts, outer webs of secondaries, and tertials crimson, dusky on the interscapulars : bases of the head-feathers brown, and those of the back and rump with concealed white spots ; region round the eye, ear-coverts, and down the sides of the neck blackish brown, paling on the lores and at the base of the upper mandible ; a line of white spots up the hind neck, continued above the ear-coverts to the eyes ; primaries and inner webs of secondaries and tertials black-brown, with a series of white inner-margined spots ; upper tail-coverts and tail black ; throat, cheeks, and a line above the gape to the nostril, centres of throat, chest, and upper breast-feathers,

and lower parts white; a mesial black line from the chin to the fore neck, and two down each cheek; fore neck and chest-feathers very broadly edged with blackish brown, gradually narrowing towards the lower parts, where it almost disappears; under tail-coverts white, crossed by angular dark bars: under wing-coverts barred white. The spots on the first primary vary; two is the normal number.

Female. Has the top of the head and the nape black, with round white spots: lores, sides of the neck, and ear-coverts blacker than the male and concolorous with the head; the longer under tail-coverts blackish brown.

Young. The nestling bird has the distribution of the markings the same as in the adult, but they are, together with the ground-colour, less pronounced. A young female before me has the head dull blackish, the spots on crown and forehead *sullied* white, while those of the crest are pure white. The white markings and spottings on the throat are likewise sullied white, the dark edgings are brownish black.

Birds of the year have the bill browner at the base than adults and shorter, measuring, on the average, about 1·85 inch to gape; the iris has a faint tinge of reddish, with a brownish-red outer circle. In some examples the primaries are tipped and crossed with white. Mr. Holdsworth alludes to an example which had the lower part of the back black, faintly barred with white, with crimson feathers appearing among the others.

Obs. Many individuals of this Woodpecker are met with in the low country of Ceylon with the feathers remarkably faded, those which are thus affected being chiefly the primaries at the tips, the coverts at the point of the wing and above the metacarpal joint, as well as on the hind neck; these I have found to be a dun-brown in some, and others a whity brown or greyish colour. The specimens were fully adult; and this singular feature could only have been the result of the action of the sun's rays on the plumage, the birds having frequented exposed situations.

This species is the Ceylonese representative of the South-Indian *Ch. delesserti*; but the latter bird has the back, scapulars, and wing-coverts golden red, and the bill is not so pale. Though first described as a new species by Layard in the 'Annals of Natural History,' 1854, it was previously known to Jerdon from specimens sent from Ceylon, and it was figured by him in his 'Illustrations of Indian Ornithology,' to face his article on *Brach. erythms*. It is very closely allied to the Philippine-Islands species *C. haematribon*, which differs from it in having the bill brownish, with the base of the under mandible pale.

Distribution.—This Woodpecker, the finest of its tribe in Ceylon, is widely distributed. It has been assigned hitherto to the hills alone, its range not having evidently been worked out; and I am at a loss to understand in what manner its presence in so many parts of the low-country forests has been overlooked by ornithologists collecting in the island. It is found throughout the Central Province from the altitude of the Horton Plains and the Pedro range downwards, but it is, as far as I have been able to trace it out, more plentiful in the higher than in the intermediate forests on the Kandy side. In Uva, however, it is to be found in most forests, following its way down the wooded passes into the low country. It is spread throughout the Eastern Province and the forest-region lying between the Haputale ranges and the south coast, and seems to thrive as well there as in the damp cool regions of the Nuwara-Elliya plateau. I have procured it within a few miles of Kirinde, on the banks of the river there. It is found through all the forest-tract to the north of Dambulla, and inhabits the open woods close to the coast near Trincomalie. Within a few miles of that place I have shot it in an overgrown cocoanut-compound, together with *Brachypternus ceylonus* and *B. puncticollis* ! In the Vanni it is common, and extends through the Anaradjapura district and the Seven Korales to Kurunegala and Puttalam, its numbers decreasing as it approaches the damp climate of the Western Province. South of the Deduru-oya it is much rarer. I have met with it in forest near Ambepussa, between Avisawella and Ratnapura, in the Pasdun Korale, and once near Baddegama in the Galle district, the precise locality there being the Government forest reserve of Kottowe.

I believe its numbers to have much diminished in the coffee-districts by the felling of the forest; but, notwithstanding, it seems to be local in its tastes. During several days' wanderings in the Peak forests, a most likely locality for it, I seldom heard its well-known trill, and again in the Knuckles forests I remember to have found it rare.

Layard procured the specimen from which he took his original description at Gillymally near Ratnapura, and mentions Mr. Thwaites getting a large number near Kandy, in which district it was evidently more common then than it is now. Mr. Holdsworth found it "abundant at Nuwara Elliya and in all tree-jungle in that district."

Habits.—Layard's Woodpecker is chiefly an inhabitant of tall forest and timber-jungle, but it is likewise found in tangled woods and groves of jungle which happen to be interspersed with large trees which it principally affects. In the south-east I invariably found it in the tall forest which lined the rivers flowing through that wild region; it shunned the thick thorny jungle clothing the arid land, and resorted to the more luxuriant belts which grew within the influence of the water. I generally found it in similar localities, or near the borders of tanks, in the northern part of Ceylon. In the Central Province it invariably affects the heavy jungle, either above the coffee-estates or in the valleys which have not yet been denuded of their beautiful clothing. It is very shy, always evincing a fear of man, and its habits escape observation by all except those who are much in the jungle. It is very active, working the tallest trees right to the top, and when sounding a hollow branch uses its powerful head and beak in dealing a "rattle" of blows with such inconceivable rapidity that the movement of its head cannot be discerned with the human eye!

This startling sound is produced by the Common Red Woodpecker; but it has not such a loud effect as when executed by the present species. I once watched one of these birds sounding a branch at the top of a lofty Keena-tree in the Lunugalla Pass, and observed that it held its head on one side and listened attentively each time before striking its rattle on the hard wood in order to force the frightened insects from their lair, in doing which it produced a noise which resounded through the forest. These Woodpeckers are usually in pairs not far distant from one another; and when two are running up the same trunk they keep on opposite sides of it, appearing not to wish to interrupt one another, each one suddenly vanishing round the bole on the appearance of the other, which has the effect of a game of "hide and seek." A single bird will work a tree from side to side, crossing and recrossing the trunk rapidly. Its feet and legs are very powerful, and it never seems tired of hunting for its food, which chiefly consists of ants. Its flight is swift, but not sustained for long. I have occasionally seen small parties in company, consisting of young birds with their parents; and on one occasion met with a pair near the Maha-oya, Eastern Province, which were searching about a huge fallen trunk, running along its horizontal surface as they would have climbed a standing tree. Its note is a weak trill, uttered in a high key and prolonged considerably; the voice of one bird is invariably answered by its mate, if within hearing distance.

Nidification.—I know nothing of the eggs of this species; but can state that in the hills it breeds at the beginning of the year, as I once found a nest at Elk Plains in January. It was situated in a hole in rather a small limb high up in a large tree, and the birds by their gestures appeared to have young.

The front figure in the Plate accompanying this article is that of a male shot at the Maha-oya, while the female represents an up-country bird killed at the Horton Plains.

CHRYSOCOLAPTES FESTIVUS.

(THE BLACK-BACKED WOODPECKER.)

Picus festivus, Bodd. Tabl. Pl. Enl. 696 (1783).
Picus goensis, Gm. Syst. Nat. i. p. 434 (1788).
Dendrocopus elliotti, Jerdon, Cat. B. S. India, Madr. Journ. 1840, xi. no. 208.
Chrysocolaptes melanotis, Blyth, J. A. S. B. 1843, xii. p. 1005.
Chrysocolaptes festivus, Gray, Gen. B. iii. App. p. 21 (1845 1); Blyth, Ibis, 1866, p. 355;
    Holdsworth, P. Z. S. 1872, p. 427 (first record from Ceylon); Jerdon, Ibis, 1872, p. 8;
    Adam, Str. Feath. 1873, p. 378; Ball, ibid. 1874, p. 391; Butler et Hume, ibid. 1875,
    p. 458; Legge, Ibis, 1875, p. 283.
Chrysocolaptes goensis, Blyth, Cat. B. Mus. A. S. B. p. 55 (1849); Jerdon, B. of Ind. i. p. 282
    (1862); Reich. Handb. Spec. Orn. p. 400, pl. 655. fig. 4359.
Indopicus goensis, Malh. Mon. Picidæ, ii. p. 82, pl. 66. figs. 1, 2 (1863).
Marram tolashi, Tamils in India (Jerdon).
Kĕralla, Sinhalese.

Adult male. Length 11·5 to 12·0 inches; wing 5·8 to 6·0, expanse 19·5; tail 3·0 to 3·6; tarsus 1·05 to 1·2; outer anterior toe 0·9, claw (straight) 0·5; outer posterior toe 1·0, claw (straight) 0·32; expanse of foot with claws 3·0: bill to gape 1·93, height at base 0·42.

The above measurements are from two Ceylonese specimens.

Iris (variable) in one example brownish, in the other crimson-orange; bill dull blackish or leaden horn-colour, darker at the tip; legs and feet greenish slaty, claws bluish horn or brownish ochraceous.

Head and crest bright but pale crimson, bordered by a broad blackish superciliary stripe, commencing at the nostrils and encompassing the occiput; forehead joining the superciliam brownish, bases of the head-feathers black; a broad stripe from the eye to the nape and thence spreading over the hind neck and interscapular region: throat, fore neck, lower part of face, and lower half of lores white; back, rump, upper tail-coverts, scapulars, least wing-coverts, and on each side of the white, passing up the side of the neck to the eye, brownish black; tail black: primaries and their coverts, inner webs of secondaries, and tertiaries blackish brown, with large round marginal spots to the quills, and corresponding greyish markings on the outer webs; throat with a dark mesial stripe, and two more down each cheek as in the last; beneath white, feathers of the throat and chest broadly edged with blackish brown, which diminishes to a narrow margin on the lower parts; under tail-coverts white with dark centres, the lower feathers entirely brown.

Female. Indian examples (I have not met with a Ceylonese specimen) have the crown and occipital crest light yellow, of a more orange hue than the colour of the wing-coverts: the forehead is spotted with white as in the last species; the wing-coverts a duller yellow than in the male. Blyth remarks that some females have the yellow crest tipped with crimson.

Young. A young male, shot by Mr. Parker in the Puttalam district, has the crest-feathers yellow, tipped with orange-red: the superciliary feathers brown and black, and those of the forehead black, marked or spotted with white, the latter colour predominating near the base of the bill.

Obs. Ceylonese examples appear to be altogether smaller and less robust than, and with the bills not so stout as in, Indian specimens; the black and white markings about the neck and throat are more open or bolder in the Indian bird, and this is especially noticeable in the lateral stripes leading down from the chin, in the black patch on the ear-coverts, and in the white stripe over the ears; the forehead, in the continental males, is conspicuously white, and the white centres of the chest-feathers more pronounced. A male in my collection, from Raipoor, measures

0·1 in the wing, bill to gape 2·2 ; another from the North-west Provinces, in the national collection, 6·1, the bill to gape 2·2 ; a female from the same locality 6·3, bill to gape 2·15, height at base 0·5. Mr. Ball records a male from Chota Nagpur as 6·3, and a female 6·1. The Raipoor specimen has the forehead all white, the feathers having black bases, and the crimson of the crown and crest is deeper than in my southern Ceylon example.

Distribution.—This is one of our rarest Woodpeckers, for though it is not uncommon in one or two districts, yet the localities that it has been hitherto known to frequent are few and far between. The first examples procured in the island were sent to Lord Walden in 1865, and were a male and female, obtained in a locality called Cocurry, the whereabouts of which I have been unable to determine ; but Mr. Holdsworth is of opinion that the birds were shot somewhere in the north-west of the island, as they were part of a collection made not far from Aripu. Further south, in the forests between Puttalam and the Seven Korales, it is, I am informed by Mr. Parker, not unfrequent, he having seen more than a dozen specimens in the jungles round Uswewa. I have never met with it but once, and that was on the Kirinde Ganga, a few miles from Tissa Maha Rama, in the south-east of Ceylon. I there procured a male of a pair which I saw in March 1872. The species should be looked for by future collectors in the forest on the banks of the Kattregama-oya, Koombookam Aār, and other rivers of the Park country, as this district is one which abounds in Woodpeckers. Mr. Parker procured a male near Uswewa in February 1876, and another in July 1877.

In India the Black-backed Woodpecker has a tolerably wide distribution. Jerdon remarks of it that it is "found in various districts of the peninsula and Central India, being rare in most parts and common in a few localities." He found it in the Eastern Ghats, in parts of Mysore, between Bangalore and the Nilghiris, in the Vindhyan Mountains near Mhow, and in the hilly and jungly districts of Nagpore, between that and the Nerbudda. Referring to our excellent Indian journal, 'Stray Feathers,' we find Mr. Bourdillon omitting it in his Travancore list, and likewise Mr. Fairbank from the Palani-hills birds. Jerdon also states that it is not found in the Malabar forests ; and therefore its place would seem to be taken in these regions by the common species *C. delesserti,* the above-mentioned districts all lying to the north of latitude 10°. In Chota Nagpur, Mr. Ball met with it on one occasion in the Palamow subdivision, and again in the Satpura hills. From the Central Provinces I have an example mentioned above ; and to the north-west of this district, besides inhabiting the Vindhyan mountains, it is found in the Sambhur-Lake region, concerning its distribution in which Mr. Hume writes, " Dr. King shot this species in the jungles at the foot of Aboo. I got it in similar jungles further up the Aravallis ; and Adam obtained it again near Koochamun, which is near the north-west extremity of the Sambhur Lake. It is quite foreign to the plains region (Guzerat), and is unknown in Sindh Cutch, Kattiawar, and Jodhpoor."

It was originally, as its name implies, sent from the Goa district, near which it has also been procured in the southern Mahratta country.

Habits.—This species frequents forest and jungle-clad country like the last, and is similar to it in its general habits. It is found working on the trunks of both large and small trees, and is very active in its movements, appearing likewise, from my small experience of it, to be shy in its nature. The note which I heard it utter was a weaker trill than that of Layard's Woodpecker, and much resembled the voice of the little Mahratta Woodpecker. I am indebted to Mr. Parker for several notes on its habits, one of which relates to its cry, which he says is not so loud nor so prolonged as that of *Brachypternus ceylonus* ; it would therefore seem to have two distinct calls, like this last-mentioned and other species. One of his specimens was shot in the act of fighting with the Common Red Woodpecker for the possession of a hole for (I presume) breeding-purposes. He writes me that they frequently fight with this species, whose aggressive propensities necessitate it ; he thinks that the great numbers of *Brachypternus ceylonus* in the north-western forests may perhaps prevent the Black-backed Woodpecker from spreading over the country, for these latter have " to fight pretty nearly every day before they can call their house their own, and must find their life a burden to them. With their powerful bills " (he remarks) " and well-formed muscular bodies, they are more than a match for the Red Woodpeckers ; but the latter do not hesitate to attack them, when the two species chance to meet in the same tree." The stomach of an example he shot contained insects and seeds, two of which latter were as large as peas. Mr. Ball saw one of them feeding on the ground where jungle and grass had recently been burnt, and writes that he saw and shot three which were busily engaged in searching the branches of a cotton-tree (*Bombax malabaricum*).

Nidification.—I am unable to give any information concerning the breeding-habits of this species. Mr. Parker observed a pair in July, engaged in making a hole in a tree standing in a submerged tank-bed, but was unable to get the eggs.

Subfam. GECININÆ.

Bill wider at the base than in the last subfamily; culmen more curved, the lateral ridge slight, in some genera absent; gonys short. Feet with the outer posterior toe shorter than the outer anterior one : hind toe small, obsolete in some.

PICIDÆ.

GECININÆ.

Genus GECINUS.

Bill rather short; upper mandible widened at the gape; culmen curved, lateral ridge near and parallel to it; nostrils partially concealed. Wings with the 1st quill short, and the 4th and 5th subequal and longest. Tail moderately long and pointed. Feet strong, with the anterior longer than the posterior toe; claws very strong and deep.

GECINUS STRIOLATUS.

(THE STRIATED GREEN WOODPECKER.)

Brachylophus squamatus, Jerdon, Cat. Birds S. India, Madr. Journ. 1840, xi. p. 213. no. 210.

Picus striolatus, Blyth, J. A. S. B. 1843, xii. p. 1000; Sundevall, Consp. Av. Picidæ, p. 60 (1863).

Picus squamatus, Jerdon, 2nd Suppl. Cat. B. S. India, Madr. Journ. 1844, xiii. p. 138. no. 210.

Brachylophus xanthopygius, Hodgson, Cat. Nepal Birds, p. 85 (1845).

Gecinus striolatus, Blyth, Cat. B. Mus. A. S. B. p. 57 (1849); Horsf. & Moore, Cat. B. Mus. E. I. Co. no. 962 (1854); Jerdon, B. of Ind. i. p. 287 (1862); Legge, Str. Feath. 1873, p. 488 (first record from Ceylon); Ball, ibid. 1874, p. 391; Hume, ibid. 1875, p. 68; Butler, ibid. p. 458; Inglis, ibid. 1877, p. 26; Fairbank, ibid. p. 396; Ball, ibid. p. 413.

Gecinus xanthopygius, Bonap. Consp. Gen. Av. p. 127 (1850).

Chloropicus striolatus, Malherbe, Mon. Picidæ, pl. 77, p. 134 (1862).

The Lesser Indian Green Woodpecker, The Small Green Woodpecker, Indian authors.

Adult male and female. Length 10·5 to 10·9 inches; wing 5·1 to 5·3 (a female measures 5·2); tail 3·8; tarsus 1·0; outer anterior toe 0·8, claw (straight) 0·42; outer posterior toe 0·75; bill to gape 1·45 to 1·6. Weight of male 3½ oz. Iris reddish, with a frosted silver outer circle; bill blackish, the upper mandible with a pale edge, lower mandible yellow, with the tip dusky; legs and feet dusky greenish. Mr. Oates describes the eyelid in Burmese specimens as bluish grey.

Male. Forehead, crown, and occiput dull crimson, bounded by a black line passing from the upper part of the lores over the eye to the nape, where it spreads out into a crest in continuation of the red of the hind head; below this line a white streak passing from above the eye to the nape; lores and cheeks dusky whitish, the latter with a black stripe formed by the centres of the feathers; ear-coverts greenish grey; back and wings dull green, changing on the rump and upper tail-coverts into yellow, with a wash of orange on the centre of this part; primaries, inner webs of tertials, and all the secondaries, except the green external portion, dark brown, with a series of external white spots on the primaries, and inner white marginal bars towards the bases of all the quills; secondaries with pale indentations at the inner edge of the green portions; tail blackish brown, with interrupted or marginal bars of greenish grey on the central feathers, the remaining feathers with dusky bars and pale edges.

Beneath greenish grey of different depths, darkest on the chest and palest on the lower parts, each feather with a sub-edging of brown, forming a lanceolate mark; on the flanks and parts of the breast there is a central stripe as well, and tho markings ou the throat are confined to these mesial lines; under wing-coverts with arrow-shaped bars, and tho bases of the under tail-coverts with a central spear-shaped mark.

Female. Has the crown as well as the nape black, the ear-coverts darker, and the throat perhaps duskier, as a rule. than in the male; the frontal feathers are pale-edged, the central portions only being black.

Obs. The under surface of this Woodpecker is variable in appearance, owing to a discoloration of the feathers; it is only in new plumage that the green hue of the chest and breast is pure; it soon becomes sullied, and scarcely any two specimens (at least according to my experience of a tolerably large series of Indian and Ceylonese individuals) have the lower parts of the same hue, some being completely brownish, the green tint of the central portions of the feathers being only perceptible on close examination.

Ceylonese examples are identical with Indian in plumage, and are quite equal to the general run of these in size. From an examination of a series in the national collection, from Nepal and other districts, I find that the wings in the males vary from 4·9 to 5·2 inches, and in females from 4·8 to 5·0. Mr. Ball's tabulation of Chota-Nagpur specimens shows the wing in 3 males as 5·05 inches, and in a female as 4·95; the bills in these specimens are remarkably long, varying from 1·6 to 1·7. A female from the Palanis measures—wing 5·0, bill from gape 1·4. Burmese examples are large: wing 5·3 to 5·35.

This species is tolerably closely related to three others, viz. *G. viridanus* from Burmah, *G. squamatus* from the Himalayas, and *G. dimidiatus* from Java. It is most nearly allied to the first named, which Jerdon calls "a duplicate of it." *G. viridanus*, however, is a larger bird: the wings in a male measure 5·4 and in a female 5·6 inches. It has a greener under surface, the quills are much darker, and the rump is not so yellow as in *G. striolatus*; the black superciliary line is bolder in the male, and the black moustachial band broader, with the feathers conspicuously white-edged. In the female the forehead is uniform black, and the cheek-band much more pronounced: while the quills and rump present the same distinction as in the male.

G. squamatus is also considerably larger than *G. striolatus*, and has the scale-like markings of the under surface confined to the lower breast and abdomen. The forehead in the female is again uniform black, and not edged with whitish. as in *G. striolatus*.

G. dimidiatus is about the same size as *G. striolatus* (wing 5·0 inches), with the bill perhaps shorter as a rule. The fore neck and chest is uniform green, the breast and lower parts with conspicuous, blackish, scale-like markings; rump not so yellow as in *G. striolatus*; in the female, as in the last, the forehead presents the same peculiarity, being quite black.

Distribution.—This Woodpecker has a restricted range in Ceylon, being, as far as is yet known, quite a hill-bird. Until late years it escaped all observation, and had no place in the Ceylon lists, which was owing to the imperfect exploration of the patnas in the Central Province, to which it is almost entirely restricted.

The first specimen brought to the notice of the ornithological world was killed in 1872 in the Pusselawa district, and was recorded by me, *loc. cit.* Mr. Laurie procured a female example about the same time in the Knuckles district. Subsequently Mr. Bligh obtained a pair in the Haputale ranges, the shooting of which was recorded in the 'Observer.' These were killed at an elevation of 4500 feet, and others have since been shot by him at the same elevation near the Catton estate. It is more plentiful in the Uva patna basin (*i. e.* the great stretch of grassy scrub-covered hills extending across from Udu Pusselawa to the northern slopes of the Haputale hills) and in the district beyond Badulla than in any other part, save perhaps the similarly featured country below Hangrankette. I have shot it near Lunugalla, and on the Logole-oya in Madulsima, and likewise in the valley in Lower Hewahette; and I once met with it in the low-lying patnas at the foot of the Howa-Elliya range at an elevation of about 1000 feet.

It is not improbable that it will be found in the Rakwana district, and perhaps on the Karawita hills; and in the Central Province it may possibly extend considerably down the valley of the Mahawelliganga, where the country is open, grassy, and dotted with scattered timber.

On the mainland this Green Woodpecker enjoys an extensive range, being found in Southern and Central India and in the Himalayas. Jerdon remarks that he has seen it in Malabar in low jungle near the sea, in bushy ground on the Nilghiris and on the Eastern Ghâts, and also that it occurs rarely near Calcutta.

In the Palani hills Mr. Fairbank procured a single specimen at 4000 feet elevation; but he remarks that it is absent from the Khandala district. Mr. Ball's experience of it is that it is rare in Chota Nagpur and more abundant in the Satpura hills, and that it occurs sparingly throughout the coast-region lying between the Mahanadi and Godavery rivers. In the north-west it is local; Captain Butler procured it in the jungles at the foot of the Arawalli range, and this is the only locality in all that region in which it is, according to Mr. Hume, to be found.

It is found in the Doon and in Kumaon, as well as in the sub-Himalayan tracts; and Mr. Inglis writes that in Cachar "it is very common during the cold weather, and also often seen in the rains." It extends thence to Assam and Upper Pegu, where it is, according to Mr. Oates, very common.

Habits.—The Striated Green Woodpecker frequents stunted trees dispersed about the patnas and bare hillsides in the Central Province, and being of a retiring, shy disposition, resorts mostly to the numerous ravines with which these districts are cut up. Jerdon remarks that it not unfrequently descends to the ground and feeds there; this I have seen it do myself, and have more than once observed it searching about the stems of quite small bushes. When flying off to a tree it generally alights at the bottom of the trunk and works the whole tree to the top, devoting most of its time to the small branches, from one to the other of which it flies before going to another tree. It is very active and also shy, being a most difficult bird to procure. When aware that it is being pursued it flies quickly from tree to tree, and leads the collector such a chase as soon leaves him breathless on the steep patnas up which he has been toiling under a blazing sun, baffled in his pursuit and listening to the restless Woodpecker's singular "*queemp*" cry as it disappears over the brow of the nearest rise! It is constantly uttering this note, and by it I have always discovered its whereabouts. It is nearly always in pairs; but on one occasion I discovered four in a wood in Madulsima, which were probably a young brood with their parents. I have reason to believe that it roosts *perched across* a branch, as I once shot at one in the dusk of the evening that had flown into the top of a tree above my head and had taken up that position. Mr. Bligh has met with it in coffee-plantations frequenting dead stumps of trees; but it does not appear to reside in such localities, merely visiting them from the neighbouring patnas. I have found its diet to consist almost entirely of black ants, which abound in the trees in the Central Province. It is very local, dwelling, I believe, in one spot, for it may be heard day after day in the same place.

Nidification.—In India this species lays from March until May, building, according to Mr. Hume, in holes in trunks or branches of trees. The eggs are four or five in number, "pure china-white and very glossy;" they vary from 1·02 to 1·1 inch in length, and from 0·74 to 0·85 inch in breadth.

Bill shorter than in *Gecinus*, widened at the base, with the lateral ridge well defined and parallel to the culmen, which is curved. Nostrils concealed by hair-like plumes; gonys short. Tail long and cuneate, the central feathers considerably attenuated. Feet with the anterior toe longer than the posterior; claws strong and well curved.

CHRYSOPHLEGMA XANTHODERUS.
(THE SOUTHERN YELLOW-NAPED WOODPECKER.)

Picus mentalis, Jerd. (*nec* Temm.) Cat. B. S. India, Madr. Journ. 1840, xi. p. 214. no. 211.

Chloropicus xanthoderus, Malh. Brit. Mus. 1844, et Rev. Zool. 1845, p. 402, et Monog. Picidæ, pl. 75, vol. ii. p. 114.

Picus chlorigaster, Jerd. Cat. B. S. India, Madr. Journ. 1845, xiii. p. 138. no. 31.

Gecinus chlorophanes, Blyth (*nec* Vieill.), Cat. B. Mus. A. S. B. p. 59 (1849); Kelaart, Prodromus, Cat. p. 128 (1852); Layard, Ann. & Mag. Nat. Hist. 1854, xiii. p. 448.

Chrysophlegma chlorophanes, Jerd. B. of Ind. i. p. 290; Holdsworth, P. Z. S. 1872, p. 428; Bourdillon, Str. Feath. 1876, p. 390; Fairbank, ibid. 1877, p. 396.

Picus xanthoderus, Sundev. Consp. Av. Picid. p. 58 (1863).

Green Woodpecker, "*Ground-Woodpecker*," Europeans in Ceylon.

Pachcha kūralla, Sinhalese.

Adult male and female. Length 8·7 to 9·2 inches; wing 4·5 to 4·75; tail 2·0 to 5·5; tarsus 0·75; outer anterior toe 0·7, claw (straight) 0·38; posterior outer toe 0·65; bill to gape 1·0 to 1·05. Females appear to average smaller than males.

Iris sombre red or brownish red; bill blackish, with the sides of the lower mandible and margin of the upper next the gape yellow; legs and feet olive-greenish or dusky sap-green.

Head, crest, and a patch on the lower part of cheeks crimson, the feathers of the nape below the crest rich yellow; bases of forehead and crown-feathers greenish black; upper surface with the wing-coverts bright olive-green; face, throat, neck, and under surface dull green; lores blackish, round the eye and on the cheeks the feathers are somewhat dusky; outer webs of inner primaries, secondaries, and tertials next the shaft orange-red; greater wing-coverts with their bases orange; inner webs of quills brown, with distant white spots, and the terminal portion of the outer webs of the primaries with whitish spots: tail black; beneath, the sides of the breast, flanks, and lower parts are barred with white; bases of throat-feathers white, showing more or less on the surface; under wing-coverts marked with greenish white.

Female. Has the forehead and head deep green, the bases of the feathers dark brown; the occipital feathers and yellow nape-patch as in the male, but the red cheek-stripe absent, that part being green like the sides of the neck.

Young. Birds of the year have the forehead dark green, and the feathers tipped with the crimson hue of the occiput; the fore neck and throat are brownish, and the dorsal feathers to a certain extent tipped with a pale hue; flanks and lower parts more barred with white than in the adult.

Obs. Ceylonese specimens correspond well with those from Madras. A male from this locality measures 4·7 inches in the wing, bill to gape 1·1; a female 4·6 in the wing, bill to gape 1·1. This species is closely allied to *Ch. chlorophus*, its northern representative, which inhabits Bengal, Assam, and parts of the sub-Himalayan hills. This is a larger bird, two males from Nepal measuring 5·5 and 5·6 inches in the wing, and a female 5·4; it has the hinder part of the head green, a band of red passing across the front of forehead over the lores and eyes to the occiput, where it occupies the terminal half of the nuchal feathers, while the nape and upper part of the hind neck are light saffron-yellow, richer in hue than in *Ch. xanthoderus*, and the wing-coverts are not so much washed with red as in this latter species.

The female wants the red stripe across the forehead and over the eyes ; the under surface is greyish, and less washed with green than in *C. xanthoderus*.

This Woodpecker has been styled by Jerdon and Blyth, and consequently by Layard and other writers on Ceylon ornithology, *Chrysophlegma chlorophanes*, owing, apparently, to a mistake made by Blyth in quoting Vieillot as the author of the species at page 59 of his Catalogue of the Birds in the Asiatic Society's Museum, Calcutta : he there gives as a synonym of *Ch. chlorophanes*, "*Picus chlorophanes*, Vieillot." Vieillot, however, gave no such name in his 'Dictionary,' as Malherbe remarks in his article on the present species and on *Ch. chlorolophus*. I have myself examined the pages in his vol. xxvi., devoted to the Woodpeckers, and cannot find any reference to any other Green Woodpecker from India but that relating to the *Pic à huppe verte* from Bengal. The species, then, in reality wanted a name until Malherbe met with specimens of it in the British Museum, sent there by Jerdon from South India, and described it under the above title. Blyth, by an error, quoted this name as a synonym of *Ch. chlorolophus* at page 58 of his Catalogue.

Distribution.—The "Ground-Woodpecker" is found throughout most of the low country, except the northern parts, where, as far as I am able to ascertain from report and my own observation, it has not yet been detected. As it is, however, nowhere very abundant and is of a retiring nature, it may have been passed over in the north of the Vanni, and it will be for future explorers to extend its limit to that part of the island. It is not unfrequent near Colombo, and is diffused generally throughout the Western Province, being perhaps most common in parts of Saffragam and in the Raygam and Pasdun Korales. In the south-west it is not uncommon both in the hill-region and the wooded country lying between the "Haycock" and Galle. On the eastern side of the island I have found it in the Friars-Hood and other districts ; I met with it also in the Wellaway Korale, and it most likely inhabits most of the Park country between there and Batticaloa. In the Kandyan Province it is found in the valleys intersecting the coffee-districts, but more particularly on the Uva side, where I have seen it at an altitude of 4000 feet. It was not uncommon about Lunugalla, inhabiting the jungle on the pass down to Bibile. Kelaart says that it is not unfrequently seen at Nuwara Elliya ; but I know of no one else who has seen it there.

In the south of India this Woodpecker is not uncommon. Jerdon writes that it is "found in the forests of Malabar, more especially far south, as in Travancore." In this district Mr. Bourdillon says it is very common ; and Mr. Fairbank obtained it in the Palani hills at a considerable elevation. It seems to be restricted to the extreme south of the peninsula; for Jerdon did not find it in the Eastern Ghâts nor in Central India, and Mr. Fairbank does not record it from the Deccan.

Habits.—This species affects the edges of forest and also the interior of the jungle, being partial to wooded ravines through which streams run, near the banks of which I have more than once met with it. It is also found in scattered jungle and low thickets, and may often be surprised on the ground in dense underwood. But though it is found so much near the ground, tapping about the roots of trees and searching for food on fallen timber, it nevertheless often betakes itself to the very tallest trees of the forest, and has a habit of mounting up to the very topmost branch and there remaining motionless for some time, uttering its loud monosyllabic note, which somewhat resembles that of the Bay Woodpecker. It is when not feeding or on sallying out the first thing in the morning that it utters its note ; and sometimes when flying across an open glade or cheena, as I have noticed it in the Eastern Province, it gives out its plaintive pipe ; but otherwise it is not a very noisy Woodpecker. When disturbed in the thick jungle, if it be on the ground, it decamps from tree to tree with a loud fluttering of the wings, and clings to the trunks near the roots. When on the wing for any distance its flight is performed with quick beating of the wings and long intermediate jerks, by which it progresses with considerable speed.

Layard writes that he has seen it on the ground " in pairs, breaking into the dried masses of cow-dung in search of Coleoptera. On being alarmed it takes refuge in the nearest tree or bush, and displays all the arboreal activity of its tribe, climbing round the branches and evading the eye by carefully keeping on the opposite side of the limbs."

Besides feeding on coleopterous insects, it is very fond of ants, with which I have found its stomach crammed.

In the breeding-season, which in Travancore apparently is in March, Mr. Bourdillon says that the "plaintive monotonous call of these birds (which somewhat resembles the breeding-call of the common Pariah Kite) may be heard at all hours of the day as they cling motionless to the topmost bough of some tall forest tree."

I know nothing of its nidification.

Bill short, wide at the base ; culmen much arched or curved, lateral ridge almost obsolete and close to the culmen ; gonys straight, its angle sharp. Wings much as in other genera of the family, but with the secondaries long. Tail rather short, broad at the base. Tarsus about equal to the anterior toe, which is longer than the versatile one ; claws much curved.

Of chestnut plumage.

MICROPTERNUS GULARIS.

(THE MADRAS RUFOUS WOODPECKER.)

Picus badius, Jerdon, Cat. B. S. India, Madr. Journ. 1840, xi. no. 214.

Micropternus gularis, Jerdon, 2nd Suppl. Cat. B. S. India, Madr. Journ. 1845, xiii. p. 139 ;
Blyth, Cat. B. Mus. A. S. B. p. 61 (1849); Kelaart, Prodromus, Cat. p. 128 (1852);
Jerdon, B. of Ind. i. p. 294 ; Holdsw. P. Z. S. 1872, p. 428 ; Hume, Str. Feath. 1873,
p. 434 ; Legge, ibid. 1875, p. 201 ; Hume, ibid. 1877, p. 477.

Micropternus phaioceps, Layard, Ann. & Mag. Nat. Hist. 1854, xiii. p. 450.

Phaiopicus jerdoni, Malherbe, Rev. Zool. 1849, p. 535 ; id. Mon. Picid. pl. 47. figs. 1–4 (1862).

The Bay Woodpecker of some ; *Brown Woodpecker*, Europeans in Ceylon.

Kæralla, Sinhalese.

Adult male. Length about 9·5 inches ; wing 4·5 to 4·7 ; tail 2·75 ; tarsus 0·75 ; outer anterior toe 0·75 to 0·8, its claw (straight) 0·4 to 0·42 ; outer posterior toe 0·65 to 0·7 ; bill to gape 1·15 to 1·3. The bill, considering its small size, is somewhat variable in length.

Iris chestnut-brown in some, brownish red in others ; bill black, with a slate-coloured or sometimes a greenish line at the sides of the lower mandible ; legs and feet " slaty " or blackish plumbeous.

General plumage rufous-bay, with a dusky hue on the under surface : head, region round the eye, and cheeks infuscated with brownish ; the feathers extending from the gape beneath the eye to the ear-coverts tipped with crimson, and occasionally those in front of the eye faintly pointed with the same ; the feathers of the lower part of the hind neck and all the upper surface beneath that part crossed with bars of brownish black, narrowest on the back and broadest on the inner webs of the quills and tertials ; tail with the central feathers deeply tipped with blackish, and the remaining bars five in number ; the three lateral feathers with the subterminal bar the same width as the rest : chin- and throat-feathers crossed with blackish-brown subterminal bars and tipped with whitish ; flank, sides of belly, and under tail-coverts barred with a lighter brown than the back ; first three primaries with a brown patch on the inner webs ; under wing-coverts crossed with narrow bars of brownish black.

Though the extent of the crimson tipping on the cheeks varies, I have not yet seen a Ceylonese specimen with it above the anterior angle of the eye, as is the case with the closely allied *M. badiosus*.

Female. Slightly smaller ; wing 4·5 inches ; bill to gape 1·1 to 1·25.

The rufous plumage paler throughout than in the male, at least in most specimens that I have examined ; cheeks wanting the crimson colour.

Young. In what appears to be an immature male bird, the feathers of the head are edged with rufous-bay, and the crimson cheek-patch is very small in extent : the chest and breast have the feathers crossed with crescentic bands of brown.

Obs. This species is said to vary in size from different parts of South India : I find no appreciable difference between western, southern, and northern specimens in Ceylon ; they average, as is the case with most Indo-Ceylonese forms, smaller than the continental birds. Mr. Hume, in his exhaustive notice of the genus ('Stray Feathers,'

1877), gives the wings of a series taken at random from the "Nilghiris, Ceylon, and Travancore as 4·72, 4·85, 4·75, 4·68, 4·71, 4·0, 4·78, 4·85, 4·8, 4·7;" all but three of these dimensions exceed the maximum of Ceylon birds. He remarks that the tail-bands are usually six in number: I take it for granted that the black tip, 0·6 to 0·8 inch in depth, is not included in this number; and if so, most Indian birds must have an extra band on the caudal feathers. Lord Tweeddale records an instance of a Malabar specimen having the crimson "points" quite round the eye; this appears to be a characteristic distinction of *M. badiosus* from Borneo. Ceylonese females are quite as pale as South-Indian.

Perhaps no genus of Woodpeckers has its members so closely allied as this; the different species have a general resemblance to one another, but yet possess certain nice points of distinction peculiar to types from certain regions which serve to assign them to specific rank. *M. phaioceps* from Bengal is a larger bird than ours: wings 5·1, 5·3, 4·7. It is paler on the head, and has the white-margined feathers of the throat concolorous with the fore neck and chest. *M. brachyurus* (or *M. badius*), according to Mr. Hume, has the white-tipped throat-feathers banded with dark brown like *M. gularis*; but they extend on to the cheeks, whereas in the latter they do not surmount the rami of the lower mandible; the head is paler than in *M. phaioceps*, and the crimson dotting of the face the same in extent, or not extending above the angle of the eye: this species inhabits Java, and to it Mr. Hume unites the bird from Tenasserim. The fourth species (*M. badiosus*) differs solely in the red points extending round the eye; but this would seem to be the case, in isolated instances, with some individuals from Malabar.

Distribution.—This bird has hitherto been considered rare in Ceylon, and likewise of local distribution. It is, however, widely distributed, for I have met with it in all my wanderings through the low country. It is less common, I think, in the north than elsewhere; but yet I have seen it in many parts of the forest-clad country from Tamblegam to the neighbourhood of Anaradjapura. It is found within four miles of Colombo, and is pretty evenly diffused throughout the Western Province: in the south I have met with it chiefly near the Gindurah, and in the south-east found it at Tissa Maha Rama and other places; it is not uncommon in parts of the "Park" country, and I have met with it near Nilgalla; but in all these districts of the eastern portion of the island it is likely to be passed over unless the collector be well on the alert, for it is found usually in the wilder parts of the forest, where the jungle is thin and scattered or interspersed with open glades. In the Seven Korales it is pretty common, and Mr. Parker writes me that it is numerous about Uswewa. It occurs in the valley of Dumbara; but I do not know that it ascends much higher than that. Mr. Bligh has seen it in Haputale up to about 2000 feet. In the peninsula of India it is found "in the forests of Malabar, both above and below the Ghâts, from the extreme south to north latitude 16°." At the latter extreme it is rarer than further south. Mr. Fairbank records it from Khaudala and Mahabaleshwar, where it inhabits the western slopes of the hill-ranges. Further north than this I do not think it has been met with. It is, I imagine, more common on the Nilghiris and the adjacent Malabar coast than in the extreme south, for I do not find it recorded either from the Travancore or Palani hills.

Habits.—The Bay Woodpecker is an active and restless bird, astir the first thing in the morning, making its loud note, *queemp-queep*, heard before many other birds have begun to think about their morning rambles! It is found in thick forest, in compounds filled with cocoanut, bread-fruit, and jack trees, at the borders of jungle-begirt paddy-fields, and in detached woods. It usually mounts to the top of a tree, and selecting some dead branch, taps away at it, diligently listening in the intervals until its luckless prey is discovered. It may be approached easily when thus engaged, and when disturbed does not fly far. It goes more on the ground, I think, than the last species, for I have several times surprised a pair breaking up dried cattle ordure; and on one occasion, in the north of Ceylon, came on one busily attacking a stream of black ants as they filed in close order, a dozen abreast, across a jungle-path. This insect, the short black ant (*Formica esrundans*?), forms the Bay Woodpecker's favourite food in the forest districts; it attacks the large black pendent nests which it constructs and entirely consumes its numerous inhabitants. Mr. Parker writes me that he once descried an individual issuing from a round hole in a large nest, and found, on examining it, that the interior was completely hollowed out. When flushed from the ground it rises with a loud flutter to the nearest tree, and often flies suddenly from branch to branch, and so decamps to another place of safety. I have more than once found its breast smeared and discoloured with some viscous substance, which must be the gum from the bark of certain trees. Its flight is very jerky and not swift, being performed with alternate beating and closing of the wings. I am unable to furnish any information concerning this bird's nesting.

2 D

Bill moderately long, with the culmen curved, wide at the base, with the nostrils apart; gonys short and straight. Wings and tail moderate; the latter cuneate, the central feathers slightly exceeding the next. Tarsi rather short; anterior toe longer than posterior; *hind toe minute*, and its claw *rudimentary*.

BRACHYPTERNUS CEYLONUS.

(THE RED WOODPECKER.)

(Peculiar to Ceylon.)

Picus ceylonus, Forster, Naturf. xiii. pl. 4.

Picus erythronotus, Vieill. N. Dict. d'Hist. Nat. xxvi. p. 73 (1818).

Picus sonneratii, Less. Traité d'Orn. p. 221 (1831).

Brachypternus erythronotus, Strickl. P. Z. S. 1841, p. 31.

Brachypternus ceylonus, Blyth, J. A. S. B. 1846, p. 282; id. Cat. B. Mus. A. S. B. p. 56 (1849); Kelaart, Prodromus, Cat. p. 128 (1852); Layard, Ann. & Mag. Nat. Hist. 1854, xiii. p. 449; Cab. et Heine, Mus. Hein. v. p. 171 (1863); Blyth, Ibis, 1867, p. 297; Holdsworth, P. Z. S. 1872, p. 428; Legge, Ibis, 1874, p. 15, et 1875, p. 284; id. Str. Feath. 1875, p. 202.

Brahmapicus erythronotus, Malherbe, Mon. Picidœ, ii. p. 90, pl. 69. figs. 1–4 (1863).

Red Woodpecker, Cocoanut-Woodpecker, Europeans in Ceylon; "*Toddy-bird*," natives in south of Ceylon.

Pastru carpentaru, Portuguese in Ceylon, lit. "Carpenter-bird" (from its habit of tapping trees).

Kæralla, Keberella, Sinhalese; *Tatchan-kuruvi*, Tamils in Ceylon.

♂ ad. suprà coccineus, colli et dorsi postici plumis basaliter nigris: uropygio nigro sordidè coccineo lavato: supra-caudalibus caudâque omnino nigris: tectricibus alarum coccineis, basaliter nigris, et pallidè coccineo apicaliter maculatis: remigibus nigris, albo fasciatim maculatis, primariis extimis pogonio interno tantùm notatis, secundariis extùs dorso concoloribus: plumis superciliaribus nigris albo minutè punctatis: facie laterali fulvescenti-albidâ nigro striolatâ: genis et gulâ fulvescentibus nigro maculatis: gutture imo nigro: corpore reliquo subtùs fulvescente, plumis nigro marginatis: hypochondriis, subcaudalibus et subalaribus transversim nigro fasciatis: rostro saturatè corneo, mandibulâ cyanescenti-corneâ: podibus sordidè viridibus: iride rubrâ.

Adult male and female. Length 11·4 to 11·75 inches; wing 5·2 to 5·85; tail 3·5 to 4·0; tarsus 1·0; outer anterior toe 0·8, its claw (straight) 0·45; outer posterior toe 0·7; bill to gape 1·5 to 1·6.

Iris red, dull red, or reddish; bill blackish, base and sides of under mandible leaden; legs and feet murky greenish, olivaceous green, or dusky sap-green.

Male. Head and crest, back, scapulars, and wing-coverts, with the outer webs of secondaries, tertials, and inner primaries crimson, brightest on the crest and back, and merging into the black of the rump; bases of the head-feathers black, those of the forehead being pointed or tipped with black; throat, fore neck, space behind the eye, hind neck, upper tail-coverts, and tail black; a stripe from behind the eye to the nape, and a broader one from the gape down the cheeks and sides of the neck to the chest white; ear-coverts striated, and throat closely spotted with bar-shaped marks of white; primaries, inner webs of secondaries and tertials, primary-coverts, and point of wing blackish brown; a series of bar-shaped spots on the inner webs of all the quills and corresponding marginal whitish spots on the outer webs of the primaries; secondary-coverts with terminal reddish-white spots;

beneath, in continuation of the throat, the feathers of the chest have white centres and broad black margins, which latter coalesce lower down into bars, most conspicuously on the flanks and under tail-coverts ; under tail- and under wing-coverts black, barred with white.

Female. Has the forehead and crown black, with terminal, circular, white spots, the occiput and nuchal crest being crimson.

Young. Birds of the year have the iris brown ; bill dark horn-colour, light bluish at the base beneath, and varying in length from 1·2 to 1·5 inch (tip to gape).

The forehead and crown in both sexes are black ; the *male* has the feathers on the latter part faintly tipped with reddish, which colour seems to spread to the frontal feathers at the end of the first year, and probably by moult ; the *female* has the forehead and front of crown unmarked at first, and the white-spotted feathers appear by moult at the age of about four months ; the face and throat are less spotted than the adult, the white markings being roundish and small ; on the chest there is much more black, the white spaces being broken in the centre by the black of the outer portions of the feather, while on the breast the black margins are broader and extend to the tip of the webs ; the pale terminal spots on the wing-coverts are absent or very faintly indicated.

With age the markings of the chest open out into broad white mesial stripes.

Obs. This species is not very distantly related to the next, bearing, as Blyth remarks, the same relationship to it as *Chrysocolaptes stricklandi* does to the South-Indian *Ch. delesserti*. In the dark race of *B. puncticollis*, as found in the forests of Ceylon, there is a still greater approach to the present species, for the well-matured male of it is almost as red on the back.

Distribution.—This Woodpecker is the most abundant species of its family in the island, and being such a common bird was known to the old naturalist Forster.

It is diffused throughout the entire island, with perhaps the exception of the extreme north of the Vanni and the Jaffna peninsula. It is abundant in the Western and Southern Provinces, and equally so in the forest-clad country lying to the south of the Haputale hills, in the interior of the Eastern Province, and scarcely less so in the jungles between Matale and Trincomalie and in the N.W. Province. Mr. Holdsworth did not observe it in the Aripu district, nor did I meet with it there nor in the island of Manaar ; some distance inland from Mantotte it is, I am informed, not uncommon, as also further north in the Vavonia Valankulam district. In the Kandyan Province it is not rare in the Knuckles district, in Pusselawa, Nilambe, Hewahette, Dimbulla, and Uva, being perhaps most numerous in the latter part. In his ' Prodromus,' Dr. Kelaart records it as very abundant at Nuwara Elliya ; but this remark, doubtless, really refers to Layard's Woodpecker, which might easily be mistaken, by an unpractised eye, for the present species. I have never seen it above 3500 or 4000 feet ; but there is no reason why it should not range higher than that elevation. It is found likewise in the hills of the Southern Province, for it is not uncommon above Morowaka and in other localities in the Rakwana district.

I did not notice it in the scrubby districts along the south-eastern seaboard, not meeting with it nearer the coast than about 10 or 15 miles north of Hambantota ; not so, however, on the western coast, where it frequents the coconut-plantations close to the sea-beach, being the first Woodpecker which the newly arrived collector meets with in his trips to Mount Lavinia or through the cinnamon-gardens to the villages about Kotté.

Habits.—Partial as the Ceylon Woodpecker is to coconut-groves and compounds containing jack, bread-fruit, and other cultivated trees, it is nevertheless found, in the wilder districts, in forest and jungle of all sorts. It is a fearless bird and very active, running up and round the stems of trees, searching flowers and nut-stalks at the heads of palms, and in a general way perpetually cramming itself with its favourite food, red ants (*Formica smaragdina*). Its usual note is the loud harsh call well known to most people in Ceylon, besides which it delivers a loud "trill" while searching for food ; and on many occasions I have observed a pair working about the roots of large trees in the forest going through a little parlance or conversation quite unlike the common notes. Its manners while feeding are quaint, striking loud blows and twisting its head attentively on one side with a view of finding out the whereabouts of its intended victim. It is also highly

2 D 2

interesting to a lover of nature to witness a pair of these birds carrying on their courtship, as they jerk to and fro, round and up a bare cocoanut-trunk, hammering and alternately cocking their heads on one side to listen, then feeding each other, and playing hide and seek round the bare stem, uttering the whole time a low love-chattering. The rattle which this Woodpecker performs when sounding a hollow branch for insects is quite as rapid as that of Layard's Woodpecker, but not so powerful. I have observed it sound a branch many times, twisting its head into a listening attitude after each series of strokes before it gave up the task as unsuccessful. Its harsh call above mentioned is uttered while the bird is in flight, which is, as Layard mentions, sustained "by short rapid jerks repeated at considerable intervals."

This species is very fond of searching about the flowers of the cocoanut-palm, which abound in various insects on which it feeds; and this habit has caused the natives to think that it resorts to the tops of the cocoanuts for the purpose of feeding on the toddy !

Perhaps the most remarkable feature in this bird's economy is its extraordinary pugnacity. As mentioned in the preceding article, it is addicted to fighting with the Black-backed Woodpecker, disputing with it the right of entrance into the holes which the latter has perhaps excavated for its nest. It is, however, not less amiable towards its own kin ! Mr. Parker writes me an account of a combat which he witnessed once, and comments on the disposition of the bird as follows :—"I think the Red Woodpecker is one of the most fearless (amongst his fellows) of any bird I have seen. One day, when examining a tank, I heard a tremendous screaming in a large tree, and I found there two Red Woodpeckers fixed *vertically* on opposite sides of a small *horizontal* branch hammering away at each other as they would do at a dead tree. They were far too busily engaged to take any notice of me, and after watching them for 10 minutes or a quarter of an hour I left them still screaming and fighting." I have observed that they do not live on very good terms with the Racket-tailed Drongo, *Dissemurus malabaricus*; but in this case it is the latter that I have always noticed to be the aggressor, flying at and driving the Woodpeckers from the trees in which they, the Drongos, may be sitting.

The skin of the Red Woodpecker is tough and very thick, but not so much so as that of either of the foregoing *Chrysocolaptæ*; its neck is thicker in proportion to the head than in those birds.

Nidification.—In the south of Ceylon the Red Woodpecker breeds from February until June, and not unfrequently nests in the trunk of a dead cocoanut-tree, cutting a round entrance and excavating the decaying part of the tree for some distance below it. I have never been able to procure the eggs, although the bird is so common.

The figures in the Plate accompanying this article represent a male and female of this Woodpecker.

BRACHYPTERNUS PUNCTICOLLIS.

(THE SOUTHERN GOLDEN-BACKED WOODPECKER.)

Brachypternopicus puncticollis, Malh. Rev. Zool. 1845, p. 404 (♂ adult).
Picus chrysonotus, Malh. Rev. Zool. 1845, p. 404 (♀).
Brachypternus micropus, Blyth, J. A. S. B. 1845, xiv. p. 194.
Brachypternus aurantius, Kelaart, Prodromus, Cat. p. 128 (1852); Layard, Ann. & Mag.
  Nat. Hist. 1854, xiii. p. 448.
Brahmapicus puncticolli, Malh. Mon. Picidæ, vol. ii. p. 92, pl. 70 (1–4), 1861.
Brachypternus chrysonotus, Horsf. & Moore, Cat. B. Mus. E. I. Co. ii. p. 656 (1856); Jerdon
  (*nec* Lesson), B. of Ind. p. 296 (1862).
Brachypternus puncticollis, Holdsworth, P. Z. S. 1872, p. 428; Hume, Str. Feath. 1876,
  p. 457; Fairbank, ibid. 1877, p. 396.
Brachypternus intermedius, Legge, Str. Feath. 1876, p. 242; Whyte, ibid. 1877, p. 201.
Yellow-backed Woodpecker, Europeans in Jaffna district.
Pastru carpentaru, lit. "Carpenter-bird," Portuguese in Ceylon.
Tatchan-kuruvi, Ceylonese Tamils; *Kærralla*, Sinhalese.

Adult male. Length 10·3 to 10·75 inches; wing 5·3 to 5·5; tail 3·5; tarsus 0·8; outer anterior toe 0·8, claws
(straight) 0·45; posterior outer toe 0·7; bill to gape 1·4 to 1·55.

Adult female. Wing 5·1 to 5·35 inches.

Male. Iris red; bill blackish or very dark plumbeous, edges of upper mandible paler; legs and feet dull sap-greenish,
claws blackish leaden.

Red back.—Male. Occiput and crest pale crimson, the feathers black at the base and with a narrow pale stripe down
the centres; forehead and crown black, each feather "pointed" with crimson; a white streak passing from the
nostril under the eye and expanding on the sides of the neck, where it meets another passing from above the eye
and over the ear-coverts; the latter white, edged with black; lower back, tail, primaries and their coverts, and
the inner webs of the secondaries black; interscapular region, middle of the back, scapulars, and adjoining greater
wing-coverts orange-yellow on the centres of the feathers, crimson at the tips, and olivaceous yellowish at the
bases; the extent to which the dorsal and scapular feathers are terminated with crimson varies much; outer
webs of secondaries and tertials dusky orange; some of the outer median wing-coverts with a whitish central
spot near the tips; quills barred with white, the secondaries on their inner webs, and the primaries, all but the
first, on both; the latter feathers with one white spot on the inner web; chin, throat, lower part of cheeks, and
fore neck black; the feathers of the chin and throat with a terminal triangular white spot and a bar of the same
across the bases; the feathers of the fore neck with only the terminal spot; chest, breast, and flanks white, with
broad, lateral, black margins, decreasing in width towards the abdomen; lower flanks barred with black; under
tail-coverts barred and tipped with the same; under wing-coverts and edge of the wing white, the feathers
margined with black.

The extent to which the crimson coloration is developed in some birds from the forests is shown in the figure in the
Plate; in this the back and scapulars are almost entirely crimson, with a yellowish hue about the centres of the
feathers, the latter colour being almost entirely overcome by the red; the outer webs of the secondaries, the
tertials, and the wing-coverts are reddish orange, with a yellowish hue slightly developed near the shafts; the outer
wing-coverts are somewhat tinged with yellow.

Female. Iris duller than the male. Forehead and crown black, the feathers with terminal spots of white. The crimson
of the occiput not so bright; the back and scapulars orange, or in some yellowish orange tipped with crimson.
Some have the back uniformly orange, while others have the feathers yellow and the red coloration confined to
the tips; but in all cases there is less of the latter hue than in the males.

PALE OR YELLOW RACE.—*Male.* Markings the same; the white wing-covert spots larger and continued more on the inner feathers; the ear-coverts with less black; the back and scapulars golden yellow; outer webs of secondaries, tertials, and the secondary wing-coverts dusky golden.

In one specimen from the Jaffna district, which presents an abnormal development of white in the spots of the primaries and those on the wing-coverts, the terminal spot on the feathers of the throat is connected with the basal bar by a mesial stripe, and this imparts the appearance of a specimen of *Brachypternus aurantius.*

Female. Does not differ from the male in the yellow coloration, the head and forehead being, as in the other race, black with white spots.

The older the bird in both sexes the greater the amount of white on the breast and chest, the white portions having evenly defined lateral edges and not indented with the marginal black as in birds not thoroughly mature.

Young. A female in nest-plumage has the dark portions brownish black instead of jet-black; the feathers of the forehead and crown with faint fulvous-white tips; the spots on the throat and fore neck very small, the basal marking in the form of a spot and not of a bar; the white central portions of the chest-feathers small and round in form, the black portions deeper than in the adult and extending to the tips as well as the margins.

Obs. The light form of this species resident on the coast is evidently the bird referred to by Layard as *B. aurantius*, which is a North-Indian Woodpecker, and not found in Ceylon, although it must be remarked that specimens are rarely found among our Jaffna birds with an extraordinary development of the white on the throat-feathers, which nearly approach individuals of typical *B. aurantius*; their isolation, however, precludes their being considered any thing but abnormally-marked examples of this truly puzzling species. The present species is, in fact, one of the most difficult birds to deal with in the whole of this work. Its extreme variability of coloration, apparently dependent on the effect of climate and situation, and its somewhat doubtful connexion with its relations of South India (*Brach. puncticollis* of Malherbe), together with a want of access on my part to a good series of Indian specimens carefully recorded from the *forests* and open low-lying districts analogous to the north coast of Ceylon, make it almost impossible for me to come to a satisfactory conclusion in the matter. First, as to variability of coloration: Ceylonese specimens from the Jaffna peninsula and adjacent coast, as far south as Manaar, exhibit no variation in the pale golden hue of the back, which resembles that of *B. aurantius*; in the island of Ramisserum, however, which belongs to the mainland, we at once get a richer yellow-backed bird than the Jaffna and Mansar one, and some examples even have a faint tipping of crimson to the feathers of the back, whereas in some specimens from South India the whole coloration of the upper surface has a dull orange hue, a similar example to which I once shot near Trincomalie, forming a good link between the Jaffna and the forest bird. One such specimen from "Malabar," in the British Museum, has the wing 5·9 inches; it is a female, and the largest specimen I have ever seen.

Directly we enter the forests in Ceylon, we find the back and scapulars of an orange-yellow instead of a golden yellow, and the tips more or less "touched" with crimson, or the whole back of a uniform reddish-orange hue. Examples from Madras, and presumably from the forests, have much less of the crimson tippings than is exemplified in some from Ceylon, make from the former locality corresponding with females from the latter. Malherbe's description of these birds from the Nilghiris is very exact. He says, "Le dos et les tectrices alaires sont d'un jaune-orange lavé de *rouge vif*; les plumes de ces parties sont olivâtres à leur base, puis lavées de jaune-orange et terminées d'un rouge à reflets qui borde aussi la moitié de la plume." Blyth, in describing his *B. micropus*, which seems to me to have been an example of a *pale-backed* bird of this species, speaks of the black of the nape being continued lower upon the shoulders, and considerably contracting "the golden orange of the back." The expression *golden orange* seems to imply a uniformity of coloration, as would have been the case with a pale-backed individual. It is evident, from what I have adduced here, that this Woodpecker varies immensely in India in its coloration, but not so much, I am inclined to think, as in Ceylon. The extremely red bird was first of all considered by me to be distinct, and was named (*loc. cit.*) *B. intermedius*; but as I now find, from an examination of a more complete series, that there is every grade from the pale yellow to it, and having never seen another so dark, it becomes necessary to unite the two extremes. Rather, perhaps, does the pale golden bird need separation; for I question whether any specimens from India can compare, in this respect, to the Ceylonese coast race. The area of country which it inhabits is, I think, too small to allow of *its* being elevated into a race or subspecies; but if, when further investigation is brought to bear on the matter, it be found to be paler than any Indian examples, the entire Ceylonese group of Golden-backed Woodpeckers might well be separated as *B. intermedius.* It is just possible that the very dark bird which I figure here may be a hybrid between *B. ceylonus* and the present species.

As touching the synonymy of this Woodpecker, Malherbe's name seems always to have been considered to have priority over Blyth's, owing probably to the November number of the 'Revue Zoologique' having been published earlier than the July number of the 'Journal of the Asiatic Society of Bengal'; otherwise *B. micropus* is the older of the two names. The *Picus chrysonotus* of Lesson never referred to this species; it was simply the female of *B. aurantius*; for, remarks Malherbe, the southern bird did not exist in the Paris collection at the time Lesson gave his name.

Distribution.—The pale race of this Woodpecker inhabits the Jaffna peninsula and the adjacent coast down to Manaar; further south it occurs, but less plentifully, to Puttalam, although specimens appear to be generally tinged with orange in the latter district; on the east coast it is found as far south as Trincomalie, but is not at all common on that side of the island. I have noticed a golden-backed Woodpecker south of Kottiar Bay, but I am not able to say to which race it belonged. In the forests the orange bird is found throughout the northern half of the island. I have procured it a few miles inland from Trincomalie; it is common at Anaradjapura, and throughout the Seven Korales down to the Puttalam district, where Mr. Parker has seen it in the jungles near Uswewa. I fully expected to find it in the jungles of the Eastern Province, but did not succeed, although I was shown a specimen by the late Dr. Gould which he had procured in the "Park" country while on a trip to that part of Ceylon.

I think I may safely say that directly this species enters the shady forests of Ceylon it alters its coloration, assuming the orange hue; no pale-backed bird has ever, to my knowledge, been shot in the interior, and no orange-backed one at Jaffna.

In Ramisserum Island the Southern Golden-backed Woodpecker appears to be very common. My native collector brought me a series of specimens from it, and said it was abundant there. Jerdon says that it is found in "various parts of Southern India, in the Carnatic, and in Malabar." From the latter district I have seen skins; and Mr. Fairbank writes that it is common in heavy forest on the lower Palanis; he has also met with it so far north as in the Khandala district near the Goa frontier.

Habits.—This handsome species frequents, on the sea-coast and in the maritime districts, cocoanut- and palmyra-groves, native gardens, compounds, and scattered jungle in the vicinity of the forest, while in the interior it is found throughout the forests, affecting the heaviest timber and the densest jungle. It has the same jerky flight and a similar loud note of alarm to the last species, and usually consorts in pairs, which do not keep close company, but generally follow each other about, sometimes working on the same tree, but more often searching for their food at a little distance from one another. It runs actively up the trunks of the cocoanut-trees, and when it has reached the top disappears into the head and searches about among the roots of the fronds and the dead flower-stocks, where there are generally numbers of ants to be found. It is very early astir, and when the day has scarcely dawned its loud note is to be heard among the cocoanut-groves in the Jaffna district. It is then very restless, flying from tree to tree before finding a suitable quantity of ants to attack; and a considerable time elapses before it settles down steadily to work, vigorously tapping and listening attentively for the result of its morning salutation to the varied insect inhabitants of the fine old tamarind- or jack-tree into which it has perhaps betaken itself. In the forests I have seen it devoting much attention to the huge bosses and knarled excrescences of the fine Koombook- or Mee-trees which one so often finds near the remote village tanks. It has a trill note, somewhat louder than that of Layard's Woodpecker.

Nidification.—I know nothing concerning the nesting of this Woodpecker; but Layard says that it excavates large holes in the male palmyra-trees, the wood of which is softer than that of the female.

The red-backed figure in the Plate represents the type specimen of my *B. intermedius*, described *loc. cit.*, and which was presented to the Colombo Museum by the late Governor of Ceylon, Sir Wm. Gregory. The pale bird is from the Jaffna peninsula, and the female in the background is an orange-backed bird from the forests near Trincomalie.

PICARIÆ.

Fam. CAPITONIDÆ.

Bill large, wide at the base, conic, inflated at the sides, the margins toothed in some; the culmen curved, the base of the upper mandible continued backwards to the gape; the tips of both mandibles acute; base of the bill furnished with bristles. Tail short and soft, of ten feathers. Feet zygodactyle. Sternum with the keel low and the posterior edge with two emarginations on each side.

Subfam. MEGALÆMINÆ.

Bill with the margin of the upper mandible smooth, variable in length; shorter than the head in some, longer in others.

Genus MEGALÆMA.

Bill conical, stout, wide at base, compressed towards tip; upper mandible overlapping the under at the gape, which is wide; culmen more or less arched. Nostrils exposed and in a basal groove parallel to the culmen, protected by long bristles pointing forwards. Lores, gape, and chin furnished with similar tufts. Wings short, the tertials comparatively long; 1st quill short, 4th and 5th subequal and longest. Tarsus longer than the long anterior toe, scutellated before and behind. Feet zygodactyle, with stout scales; the anterior toes syndactyle.

MEGALÆMA ZEYLANICA.

(THE BROWN-HEADED BARBET.)

(Peculiar to Ceylon.)

Bucco zeylanicus, Gm. Syst. Nat. i. p. 408 (1788).
Capito zeilanicus, Vieill. N. Dict. d'Hist. Nat. iv. p. 499 (1816).
Bucco zeilanicus, Cuv. Règne An. p. 457 (1829); Blyth, J. A. S. B. xv. 1846, pp. 13, 282; Hartlaub, Rev. Zool. 1841, p. 387.
Megalaima caniceps, G. R. Gray, Gen. Birds, ii. p. 429 (1846); Blyth, Cat. B. Mus. A. S. B. p. 66, 1849 (in part), et J. A. S. B. 1851, xx. p. 181 (in part); Layard, Ann. & Mag. Nat. Hist. 1854, xiii. p. 446; Cassin, Orn. Rep. U.S. Exp. Japan, p. 242.
Bucco kottorea, Hartlaub, Rev. Zool. 1841, p. 387.
Bucco viridis, Bonap. Consp. Av. i. p. 144 (1850).
Megalaima zeylanica, Blyth, J. A. S. B. 1851, xx. p. 181; Kelaart, Prodromus, Cat. p. 127 (1852); Layard, Ann. & Mag. Nat. Hist. 1854, xiii. p. 46; Horsf. & Moore, Cat. B. Mus. E. I. Co. ii. p. 638 (1856); Jerdon, B. of Ind. i. p. 311 (1862); G. R. Gray, Cat. B. Brit. Mus. *Capitonidæ*, p. 13 (1868); Holdsworth, P. Z. S. 1872, p. 420; Legge, Ibis, 1874, p. 15.

Megalæma zeylanica, Blyth, Ibis, 1867, pp. 297, 311; Marshall, Mon. Capitonidæ, pl. 40 (1871).

Le Kottorea, Levaill. Barbus, pl. 38; *Le Cabezon kottorea*, Vieill. N. Dict. d'Hist. Nat.

The Large Barbet, Kelaart; *Woodpecker*, Europeans in Ceylon.

Kotoruwa (so called from its note), Sinhalese; *Kootoor*, Ceylonese Tamils; *Kootooroo*, Portuguese in Ceylon.

Similis *M. canicipiti*, sed minor et capite et collo postico brunnescentioribus, et striis medianis minùs conspicuis: tectricum alarum maculis pallidis minùs conspicuis.

Adult male and female. Length 9·5 to 10·0 inches; wing 4·2 to 4·5; tail 2·5 to 2·7; tarsus 1·2; outer anterior toe and claw 1·15; posterior outer toe 1·1; bill to gape 1·6 to 1·8.

Iris reddish brown, with a pale outer circle, sometimes brownish buff; bill dull orange or fleshy red; legs and feet sickly yellow or pale olivaceous yellow; orbital skin dull yellow.

Bristles round the bill black: head, hind neck, throat, and chest umber-brown, passing on the lower part of hind neck into the grass-green of the back, wings, and tail; the brown parts with pale striæ, yellowish and most conspicuous on the lower part of hind neck, throat, and chest; wing-coverts with yellowish terminal spots: some of the tertials and rump-feathers with an occasional wash of bluish; outer primaries brown, with yellowish-grey edgings towards the tips; inner webs of remaining quills brown, with pale yellowish inner margins: chin obscure slaty grey (this hue not always discernible); ear-coverts brownish yellow; beneath, from the chest, light green, paling gradually into the brown of that part; under wing-coverts yellowish, tinged with greenish; under surface of tail bluish.

Young. The young quickly assume the plumage of the adult, being at first paler about the head and hind neck.

A young female, with the wing measuring 4·1 inches, in my collection, has the head, face, and hind neck pale brown, with the striæ whitish, in which colour they are continued to the green feathers of the interscapular region; the throat and fore neck are paler brown than the head, with the striæ whitish and blending gradually into the ground-colour: upper breast very slightly suffused with green. This example might well pass for a small specimen of *M. caniceps*.

Obs. This species is very closely allied to its representative in Central and Southern India, some specimens being scarcely separable were it not for their constantly smaller size. The wing in this Barbet, *Megalæma caniceps*, varies from 4·6 to 4·9 inches, the average length being, I imagine, about 4·75. It has the head, hind neck, and throat paler than in *M. zeylanica*, the stripes are broader and are continued down on to the interscapular region; the wing-coverts have the pale central spots more pronounced.

Megalæma viridis, from Malabar, Travancore, and other Southern-Indian hill-districts, is very nearly related to the last mentioned, but is smaller than it and less even than the Ceylonese bird. The wing varies from 3·7 to 4·4 inches. Jerdon's description of it is :—"Very similar to *Mrg. caniceps*, but smaller, the brown of the head and nape scarcely lineated; that of the under parts pale, becoming whitish on the throat; there are no pale specks on the wing-coverts, nor any traces of pale streaks on the green of the back." Another species from Southern India is the *Megalæma inornata*, Walden, which was, until lately, confounded with *M. caniceps*. It is readily distinguished from that species by the "absence of the broad pale median streaks on the pectoral plumage." It has the "chin, throat, breast, and upper portion of the abdominal region uniform pale brown; each feather has the shaft very faintly paler. The plumage above closely resembles that of *M. caniceps*; but the terminal spots on the wing-coverts and tertiaries are almost altogether wanting." In the uniformity of the throat it differs from all other Barbets.

Distribution.—This noisy well-known bird, commonly called a "Woodpecker" or "Woodcutter" by the Eurasian population and many Europeans, is very abundant in most parts of the low country, except close to the seashore or in large tracts of damp forest such as clothe much of the face of the southern half of the island. It is likewise an inhabitant of the Kandyan Province up to an altitude of about 2500 or 3000 feet in the western and northern parts, and to about 4000 feet in the drier district of Uva. Those parts in which it is numerous are the cultivated portions of the west and south-west, parts of the Eastern Province (in which it is locally distributed), portions of the flat forest-clad country lying between Lemastota and the S.E. coast, and the north-east of the island. It is found in the Vanni and throughout most of the country lying immediately to the north of Dambulla, wherever the jungle is of an open character. In the Seven Korales the same may be said of it ; and Mr. Parker writes me that it is common about Uswewa. Mr. Holdsworth does not record it from Aripu ; but it avoids such dry scrubby districts on the seaboard, being similarly absent from the brushy country about Hambantota.

As regards the Central Province it is not uncommon in Dumbara and in the valleys of Hewahette, Maturata, and other basins of the hill-tributaries of the Mahawelliganga. In the glens or steep ravines intersecting the great expanse of hilly patnas between Fort MacDonald and Haputale it is likewise found, and is now and then seen at a considerable altitude on the pass leading up to Hakgala. Near Banderawella I have met with it at about 4000 feet elevation.

Habits.—The Brown-headed Barbet inhabits compounds, open wooded country, dry jungle, and scanty forest where fruit-bearing trees are plentiful, on the seeds of which it principally feeds.

There is perhaps no bird better known than this one is to sportsmen or any others who are induced to visit or reside in the cultivated interior of the Western and Southern districts ; taking up their abode in some shady compound encircling the native cultivator's house on the nearest rise to his ancestral paddy-fields, these noisy birds commence early in the morning to call to one another, and make the woods resound with their guttural cries. Its loud scale-notes, commencing in measured time and increasing in rapidity and loudness, must be known to every European in the low country, and give rise to its native name of *Kotorawa*, which has a slight resemblance to some of the syllables in the scale ; they much remind one of the commencement of the laugh of the Great Brown Kingfisher, or " Laughing Jackass," of Australia. The food of this Barbet consists of every sort of tree-fruit, seed, and berry ; nothing seems to come amiss to it, for there is no tree that bears fruit that it may not sometimes be found in. It is not as gregarious as the next, or as the two smaller Barbets, but, on the contrary, is unsociably inclined towards its fellows, and more than two or three are seldom found in the same tree. It is active in its movements, seizing fruit that may be firmly attached to the stalk, and swinging its body from its perch, wrenches off the coveted morsel ; fruit and berries are swallowed whole, and in the north the favourite food is the berry of the banyan or the luscious seed of the Palu or iron-wood tree, of which the Ceylon bear (*Prochilus labiatus*) is so fond. It perches with the body inclining to the horizontal and the head thrust forward in an attitude of watchfulness, unlike the smaller Barbets, who sit bolt upright and twist the head stupidly from side to side. Coleopterous insects are likewise devoured by it ; and in captivity this Barbet has been known to exhibit, as some Toucans do, a carnivorous tendency. An interesting account of a caged bird is contained in Layard's "Notes on the Ornithology of Ceylon." At page 447, Ann. & Mag. Nat. Hist. 1854, he writes :—" One kept in a large aviary in Colombo destroyed all the little Amadinæ placed with it. Not content with snapping them up when within his reach, he would lie in wait for them behind a thick bush or the feeding-trough, pounce upon them unawares, and after beating them a little on the ground or perch, swallow them whole. When this cannibal came into my possession he was confined in a smaller cage than that in which he had at first been secured ; this seemed to displease him, and he went to work to find some means of escape ; he narrowly examined every side and corner to discover a weak spot, and having detected one, applied himself vigorously to bore a hole through it, as a Woodpecker would have done ; grasping the bars with his feet, he swung himself round, bringing his whole weight to bear upon his bill, which he used as a pickaxe, till the house resounded with his rapid and well-aimed blows. On being checked from exercising his ingenuity in this manner, he became sulky and refused to eat or offer his call of recognition when I approached him ; in a day or two, however, he apparently thought better of the matter, resumed his labours upon another spot, and fed as voraciously as ever, devouring huge slices of bananas, jungle fruits, the bodies of any small birds I skinned, &c. I hoped he would have lived long with me, but found him dead one morning ; and as he was fat and well-favoured, I presume he died a victim to the solitary system."

The flight of the *Kotorawa* is performed with quick beating of the wings, and is somewhat laboured, though by no means slow, owing to the amount of momentum which such a solid frame must naturally acquire.

Nidification.—This bird breeds from March until July. The latter month is rather late, I imagine ;

but at that date I found a nest with four young ones near Minery. It hollows out with its powerful bill a hole in a rotten tree just large enough to allow of its entering the egg-cavity, which is some distance down the trunk or branch. It does not use the same nest twice, but having found a tree with wood suited to its work, perforates it each year for the new nest, as many as 8 or 10 holes being sometimes visible in a tree by a jungle roadside. It is only when sounding wood before making its nest that these birds tap with their bills, the blows being very slowly repeated with perhaps an interval of 10 seconds between each. There are generally a few bents and grass-stalks collected for the eggs to lie on, but scarcely worthy of the name of nest. The eggs are three or four in number, pure white, glossy, and rather round in shape; they measure about 1·1 by 0·9 inch.

The upper figure in the Plate accompanying this article represents a male of this species from the Western Province.

MEGALÆMA FLAVIFRONS.

(THE YELLOW-FRONTED BARBET.)

(Peculiar to Ceylon.)

Bucco flavifrons, Cuv. Règ. An. i. p. 428 (1817).

Bucco aurifrons, Temm. Pl. Col. texte (1831).

Megalæma flavifrons, Bonap. Consp. Gen. Av. i. p. 143 (1850); Blyth, J. A. S. B. 1852,
   p. 179, et Ibis, 1866, p. 227; Gray, Cat. B. Brit. Mus. *Capit.* p. 8 (1868); Marshall,
   Monog. Capit. pl. 30 (1871).

Megalaima flavifrons, Kelaart, Prodromus, Cat. p. 127 (1852); Layard, Ann. & Mag. Nat.
   Hist. 1854, xiii. p. 447; Holdsworth, P. Z. S. 1872, p. 429; Layard, P. Z. S. 1873,
   p. 204; Legge, Str. Feath. 1875, p. 365.

Cyanops flavifrons, Jerdon, B. of Ind. i. p. 314 (1862); Blyth, Ibis, 1867, p. 297; Legge,
   Ibis, 1874, p. 15.

Le Barbut à front d'or, Levaill. Barbus, pl. 35; *The Yellow-headed Barbet*, Kelaart; *The
   "Shouter*," Europeans in planting districts.

Kotoruwa, Sinhalese.

♂ *ad.* suprà prasinus, interscapulio obscuriùs viridi, occipitis nuchæ et colli postici plumis clarè flavo medialiter
   striatis : remigibus nigris, extùs prasinis, intùs flavo marginatis : caudâ prasinâ : fronte et verticæ aurato-flavis :
   loris, fasciâ superciliari, facie laterali, gulâque totâ cyaneis : genis anticis aurato-flavis, fasciam mystacalem parvam
   formantibus : corpore reliquo subtùs pallidè viridi, juguli et pectoris plumis prasino marginatis : subalaribus
   pallidè ochrascenti-fulvis, obscurè viridi lavatis : rostro viridescenti-corneo, mandibulâ pallidiore : pedibus pallidè
   viridoscentibus : iride dilutè rubrâ.

♀ mari similis.

Adult male and female. Length 8·3 to 8·9 inches ; wing 3·45 to 3·7 ; tail 2·2 to 2·3 ; tarsus 0·9 to 1·0 ; outer anterior
   toe and claw 0·95 to 1·05 ; outer posterior toe 0·85 ; bill to gape 1·15 to 1·3 ; height at front of nostril 0·32 to
   0·4. *Females* average slightly smaller than males, and the bills of both sexes vary in size.

In this species the bristles at the gape and chin are slight, and the lores more feathered than in the preceding.

Iris light red, or pale brownish red, a pale outer circle often present ; bill greenish horn, slightly dusky at base of
   culmen, lower mandible paler ; tarsi and feet sickly green, the tarsi in some bluish ; soles yellow, claws dusky.

Lores, a superciliary stripe, cheeks, ear-coverts, and throat pale verditer-blue, lightest on the latter part : a spot beneath
   the gape, forehead, and front of crown amber-yellow, passing on the head into brownish green, and from that into
   the grass-green of the back, wings, and tail ; nape, sides, and back of neck marked with light striæ, yellowish on
   the former, and greenish white on the neck ; outer primaries and inner webs of quills brown, margined internally
   with yellowish ; longer primaries outwardly edged light towards the tips ; beneath pale green ; bases of abdominal
   feathers whitish ; chest and sides of breast with crescentic margins of brownish green.

Obs. The coloration of this Barbet is peculiar, inasmuch as it forms a link between the *Cyanops* group and that
   comprised of the members of the genus *Megalæma*. Although now classed with the latter, it has, as I have
   pointed out above, a slight dissimilarity in the less amount of facial bristling and more feathered lores, besides
   which its bill is shorter in proportion to its width at the base. Some variation in its plumage is observable ; the
   extent of the frontal yellow varies, in some specimens it ceases abruptly, while in others it passes back almost to
   the occiput. The striæ on the hind neck are sometimes broad and almost white in colour, individuals so marked
   having the lunulations on the chest very pale : this is, perhaps, a sign of immaturity. There is no difference in
   size between low-country and hill birds, some Mahara specimens in my collection being as large as, if not larger
   than, any others.

Distribution.—This Barbet has long been known as a peculiar Ceylon bird. Levaillaut described it in his great work among the Barbets, from a specimen in the Paris Museum, and Cuvier afterwards gave it its Latin title of *flavifrons*. Its head-quarters in Ceylon are the hills of the Kandyan Province and those of the southern group lying in the Kolonna, Morowak, and Kukkul Korales, downwards from all of which it spreads into the low country and has there a somewhat peculiar distribution. It is very abundant throughout all the Kandyan Province, ranging up to the forest of the main range, but not nearly in such numbers as it inhabits the coffee-districts. I have met with it as high as the Kandapolla woods, 6400 feet, but not at Nuwara Elliya or on the Horton Plains, although it is found just beneath the latter, at the foot of the "World's End" precipice. In the coffee-districts of Itakwana and the Morowak Korale it is numerous, but it is far more abundant in the Singha-Raja forests of the Kukkul Korale. As regards its dispersion through the low country, commencing in the south, we find it in the Opaté, Oodoguuma, and other fine timber-forests on the banks of the Gindurah, and in the dry season in the forest of Kottowa, near Galle. In the forest-region of the south-east I never met with it. In the Western Province it is common in some localities in Saffmgam, and is numerous in parts of the Pasdun Korale, whither it finds its way down from Kukkul Korale. It inhabits the hills stretching from Ambepussa to Avisawella, and thence spreads down the river to Kaduwella, and northwards to Mahara and Heneratgoda; in the south-west of the Raygam Korale it is not uncommon, and is numerous about Kæsbawa and other places in the Hewagam Korale. It extends from the Ambokka range into the Seven Korales, in which I have found it on the western slopes of the Doolookanda hill; but further out than this I was unable to trace it. I do not think it ranges much to the north of Dambulla, or I should most likely have met with it on the slopes of the isolated mountain of Rittagalla. In the Eastern Province its distribution is equally local; for it is met with in some forests near Kumberuwella, about 25 miles from Batticaloa, and also in the Friars-Hood forests, but thence through a wide expanse of forest-country to the foot of the Madulsima range it does not appear to be found.

I observe that Layard (P. Z. S. 1873) is of opinion that it did not frequent the low country of the Western Province in his day, but that it has spread outwards of late years. I think, however, that the above "distribution" will demonstrate to any one knowing the interior of Ceylon that its range is very peculiar, some districts coming in for a share of its patronage, while others adjacent to them are altogether passed over.

Habits.—The voice of this bird is one of the chief ornithological characteristics of the Ceylon hills; the notes which constitute it have somewhat the character of those of the larger bird, but differ chiefly in the "roll" with which they begin; they are commenced early in the morning, and continued for many hours, until the persistent Barbet, judging by the tone of his cries, becomes hoarse, and then there is a cessation, much to the relief of the wearied planter over whose bungalow the "shouter" has perhaps been calling to his mates away up at the forest's margin for the past hour! Mr. Bligh tells me that he observes a very perceptible decrease in this bird's loquacity as soon as it has begun to breed, although it has, of course, been more than usually noisy during the season of courtship. It delights in perching on the top of a tree growing at the brink of some dizzy precipice, from which its note swells far and wide over the beautiful coffee-planted gorge beneath; but still more curious is the manner in which the monosyllabic sound *quiŏk, quiŏk*, ascends audibly from the edge of the patanas far beneath the bungalow, and falls on the ear as distinctly as if it were issuing from a tree close at hand. In the low country it is found chiefly in forest, but sometimes about paddy-field woods, as at Mahara, Kaduwella, Ambepussa, and other places; in the timber-jungles of the south-west it is next to impossible to procure, as it keeps to the tops of the highest Hora- and Keena-trees, and would never be discovered were it not for its perpetual shouting. It is a gluttonous feeder, collecting in dozens among the branches of any tree in fruit, climbing intently about and wrenching off the berries with its powerful bill, at the same time letting much fall to the ground. In the Singha-Raja forest I found it feeding greedily on the berry of the Dang-tree (*Syzygium caryophyllæum*). Towards evening, after digesting its morning food, the Yellow-fronted Barbet begins its clamour again, and after feeding becomes silent before dusk. It is noticeable to what a great extent these birds answer one another; as soon as one commences its note, the refrain is taken up by another not far distant, and then by a third, and so on until the whole wood resounds with the not unmelodious but rather wearying sounds. I have not unfrequently heard from my friends in the coffee-districts that the continuous cry of this bird near the bungalow of a sick person has a most wearisome effect.

Nidification.—This Barbet has apparently two broods in the year, for the season of its breeding lasts from February until September. It selects usually a soft-wood tree, such as the cotton (*Bombax malabaricum*), and cuts a round hole into the heart of the branch or trunk, in which it excavates a cavity for its eggs some distance down from the entrance. The eggs are two or three in number, and are laid on the bare wood; they are pure white, rounded in form, with a smooth texture; they average 1·11 by 0·81 inch. When the trunk of a tree is chosen, several holes are sometimes commenced before a soft-enough place is found to excavate the nest.

The lower figure in the Plate accompanying the preceding article represents a fine specimen of this bird from the Southern Province.

Bill shorter and wider at the base than in *Megalæma*; culmen more arched; loral bristles very long. Wings with the 2nd quill the longest, the 3rd only slightly less than it. Legs and feet as in the last genus.

XANTHOLÆMA RUBRICAPILLA.

(THE LITTLE CEYLON BARBET*.)

(Peculiar to Ceylon.)

Bucco rubricapillus, Gm. Syst. Nat. i. p. 408 (1788); Blyth, J. A. S. B. 1847, pp. 386, 464.

Bucco lathami, Gm. Syst. Nat. i. p. 408 (1788); Lath. Ind. Orn. i. p. 205 (1790); Cuv. Règ. An. i. p. 45 (1829).

Capito rubricapillus, Vieill. N. Dict. d'Hist. Nat. p. 449 (1816).

Capito lathami, Vieill. N. Dict. d'Hist. Nat. p. 449 (1816).

Megalaima rubricapilla, G. R. Gray, Gen. Birds, p. 429 (1846); Kelaart, Prodromus, Cat. p. 127 (1852); Layard, Ann. & Mag. Nat. Hist. 1854, xiii. p. 448; Goff. Mus. Pays-Bas, *Buccones*, p. 26 (1863).

Megalaima lathami, G. R. Gray, Gen. Birds, p. 429.

Megalæma rubricapilla, Blyth, Cat. B. Mus. A. S. B. p. 68 (1849); Bonap. Consp. Av. p. 144 (1850).

Xantholæma rubricapilla, Horsf. & Moore, Cat. B. Mus. E. I. Co. ii. p. 646; Blyth, Ibis, 1867, p. 297; Marshall, Monogr. Capit. pl. 44 (1871); Holdsworth, P. Z. S. 1872. p. 430; Legge, Ibis, 1874, p. 15, et 1875, p. 284.

Le Barbet à couronne rouge, or the *Red-crowned Barbet*, Brown, Ill. xiv. (1776); *Le Cabezon à couronne rouge*, Vieill. N. Dict. d'Hist. Nat. p. 497; *The Rose-crowned Barbet*, Marshall; *The Red-fronted Barbet*, The "*Copper-smith*," also *Bell-bird*, Europeans in Ceylon.

Mal-kotoruwa, lit. "Flower-Barbet" (from its gay colours).

♂ ad. prasinus, plumis quibusdam cyaneo lavatis: remigibus saturatè brunneis, extùs dorsi colore lavatis: caudà viridi: narium plumis flavis: lineâ angustâ frontali nigrâ: verticie et fronte scarlatinis posteà nigro marginatis: strigâ supra- et infraoculari flavâ: facie laterali genisque viridi-cyaneis: gutture toto letè flavo, maculâ jugulari scarlatinâ: corpore reliquo subtùs viridi flavo lavatâ: subalaribus flavidis: rostro nigro, ad basin schistaceo: pedibus saturatè corallinis: iride rufescenti-brunneâ.

Adult male and female. Length 6·0 to 6·2 inches; wing 3·0 to 3·15; tail 1·4; tarsus 0·75; outer anterior toe and claw 0·75; posterior toe 0·65; bill to gape 0·85 to 0·9.

---

* This species has usually been styled the "Red-fronted Barbet;" but the next, though not so called, has also a red forehead; and therefore, taken in reference to Ceylon ornithology, the present name will, I think, be better.

Iris brown or reddish brown; bill black, pale beneath at the base; legs and feet opaque coral-red, claws blackish; orbital skin dull red.

Forehead and a spot on lower part of throat crimson, a black border at the base of culmen, and another above the crimson patch passing behind the eye to the cheeks; a superciliary stripe, cheeks, chin, throat, and round the crimson neck-spot shining gamboge-yellow, bases of throat-feathers black; from the black coronal band to tail, including the wings, dark green, tinged with bluish on the crown; outer primaries and inner webs of all the quills blackish brown, margined internally with yellowish; wing-coverts and back in many specimens edged bluish green; a patch of pale blue over the ear-coverts and side of neck, passing up into the bluish edgings of the crown; beneath, from the chest (which is washed with the yellow of the throat) pale green, with bluish edgings on the sides of the breast in some.

The amount of black on the crown varies, the band being narrowest in newly-plumaged examples, the black bases of the head-feathers amalgamating with it in abraded dress.

Young. Bill blackish; iris brown; legs and feet bluish brown. Forehead green, somewhat paler than crown, no trace of red band; throat yellowish, and the yellow cheek-spot present; crimson throat-spot wanting; green of upper and under surface as in adult. This is the plumage on first merging from the nest. Shortly afterwards the red throat-spot and frontal band are acquired.

Obs. This little Barbet is allied, but not very closely, to its South-Indian representative, *X. malabarica*, which also has the forehead and the space round the eyes, as well as the chin and throat, crimson; the occiput is black passing into blue; cheeks and sides of neck dull blue. In size it is similar to the Ceylonese bird; wing about 3·2 inches.

Distribution.—The Little Ceylon Barbet inhabits almost all the low country except the hot scrubby districts on the sea-board in the south-east and north-west of the island; but it is much more common in the southern than the northern half. In the Galle district it is very abundant, extending into the southern ranges to an altitude of 2500 feet; it is almost equally so all through the Western Province, and extends through the N.W. Province (beginning to be less abundant at Chilaw) into the northern forest tract, in some parts of which it is more plentiful than the next species, which is essentially a northern bird. About Trincomalie and along the north-east coast to Mullaittivu it dwells chiefly in the jungle some miles inland, while *Xantholæma indica* is found near the coast as well as in the interior. Mr. Holdsworth did not observe it at Aripu, which is a region unsuited to its habits; but it frequents the interior towards the Central Road, and is also found in the Jaffna peninsula.

In the Kandyan Province it is common in Dumbara and about Pusselawa, Hewahette, and other localities, but is less so in Uva than the next species. From this region it is found at intervals in the Eastern Province out to the east coast; and in the forest country from the base of the Haputale range to the edge of the scrub or "brush" country near Hambantota it is fairly common.

Habits.—This Barbet chiefly frequents cultivated country, scattered woods, the edges of paddy-fields, native gardens, compounds, and cocoanut-plantations; but in the wild districts of the north and east it is partial to luxuriant forest, in which it usually takes up its quarters near some spreading banyan-tree or other source of frugivorous supplies. It is one of the most noticeable birds about native villages, taking up its abode among the bread-fruit and jack-trees, and uttering its curious note, which has gained for it, as well as for the next species, of which the voice is somewhat similar, the name of "Copper-smith." It sits perfectly upright on the top of a tree, being very partial to the *Bombax malabaricum*, and jerks out its monosyllabical cry *wok, wok, wok,* slowly repeated, with a bob of the head at each note, and then breaks forth into *wok wok wok wok,* as if it had suddenly become impatient at the result of its parlance with its inattentive mate. It is usually solitary, or if accompanied by a mate appears not to dwell in very close fellowship with it, except, of course, during the breeding-season, when it may be seen in pairs in the same tree. It lives entirely on fruits and seeds like the rest of its congeners, but does not congregate in such flocks as the next species. The flight of this Barbet is tolerably swift, but of necessity somewhat laboured; it is performed with quick beatings of the wings, with now and then a long dipping motion.

Nidification.—The breeding-season of this little bird lasts from March until June, and it usually nests in the decayed branches of living trees, the bread-fruit (which is generally much encumbered with small, dead, top branches) being a favourite resort with it. It plies itself to the task of excavating the hole with great assiduity, first of all slowly tapping the wood all over until it has found what it imagines is a soft place; very often, after working in for an inch or so, it will find that the wood is too hard for its capabilities, and will then try another spot in the same branch. A nest I once found was in the topmost branch of a bread-fruit; the habitation was an old one, but close to it were one or two essays at making a fresh hole; the wood had evidently proved too hard and it had returned, perhaps reluctantly, to the old nest. The branch was about 4 or 5 inches in diameter, and the hole entering the cavity 2 inches and perfectly round; the nest was about 6 inches below the aperture, and the young, which were three in number, reposed upon the bare wood without any nest-lining whatever. The eggs are glossy white, rather spherical in shape, and measure about 0·9 by 0·65 inch.

In the Plate accompanying this article the figure of the young bird represents the nestling after quitting the nest.

XANTHOLÆMA HÆMACEPHALA.

(THE CRIMSON-BREASTED BARBET.)

Bucco hæmacephalus, Müll. Syst. Nat. Suppl. p. 88 (1776).

Bucco flavigula, Bodd. Tabl. Pl. Enl. p. 30 (1788).

Bucco philippensis, Gm. Syst. Nat. i. p. 407 (1788).

Bucco indicus, Lath. Ind. Orn. i. p. 205 (1790).

Capito philippensis, Vieill. N. Dict. d'Hist. Nat. iv. p. 498 (1816).

Megalaima philippensis, G. R. Gray, Gen. B. ii. p. 429 (1846); Blyth, Cat. B. Mus. A. S. B. p. 68 (1849); Kelaart, Prodromus, Cat. p. 127 (1852); Layard, Ann. & Mag. Nat. Hist. 1854, xiii. p. 447.

Xantholæma indica, Horsf. & Moore, Cat. B. Mus. E. I. Co. ii. p. 644 (1856); Jerdon, B. of Ind. i. p. 315 (1862); Holdsworth, P. Z. S. 1872, p. 430; Legge, Ibis, 1875, p. 284.

Megalaima hæmacephala, G. R. Gray, Cat. B. Brit. Mus. *Capit.* p. 10 (1868).

Xantholæma hæmacephala, Marshall, Mon. Capit. pl. 42 (1871); Hume, Nests and Eggs (Rough Draft), p. 131 (1873); id. Str. Feath. 1873, p. 458; Ball, ibid. 1874, p. 466; Hume, ibid. 1875, p. 77; Armstrong, ibid. 1876, p. 311.

Le Barbut des Philippines, Brisson; *Le Cabezon à gorge jaune*, Vieill.; *The Crimson-gorgeted Barbet*, Marshall; *Copper-smith*, Europeans in Ceylon; *Kat-khora*, Hind., or *Tambayat*, lit. "Copper-smith"; *Chota bassant bairi*, or *Chota Henebo*, Beng.; *Tokoji*, Telegu.

Kotoruwa, *Mal-kotoruwa*, Sinhalese; *Kokoorupan*, Tamil (Layard).

Adult male and female. Length 6·0 to 6·1 inches; wing 3·0 to 3·15; tail 1·5; tarsus 0·8: outer anterior toe and claw 0·75; bill to gape 0·9 to 0·97.

Iris reddish brown, with a pale or pearly-grey outer circle; eyelid red; bill black; legs and feet coral-red.

A broad frontal band and a patch across the lower part of throat glistening crimson; lores, the top and sides of head behind the eye, ear-coverts, and cheeks black; chin, throat, and a stripe above the eye, and a patch on the cheek sulphur-yellow; hind neck, back, and wings sap-green, slightly pervaded with bluish on the occiput; tail and outer webs of quills bluish green; outer primaries and inner webs of all the quills blackish brown, margined inwardly with whitish yellow; below the chest-patch green, washed next the crimson with yellow; breast and lower parts whitish, with broad dark green centres to the feathers, darkest on the flanks, and fading on the centre of the belly; bases of wing-coverts blackish, under wing yellowish.

The extent of black on the occiput varies, and specimens are likewise often seen with the back and wing-coverts edged yellowish green.

Young. Iris dark brown; legs and feet yellowish red. In the fully-plumaged nestling the crimson forehead and chest are wanting, the former being concolorous with the crown, which is dusky green; the yellow throat- and facial-spots are not so bright as in the adult; the lower part of the face and space just behind the eye only are blackish: the upper surface is pale-edged, and that part of the chest which is crimson in the adult is dull green: under surface much as in the adult, but the centres of the feathers are paler. Traces of the black crown are perceptible in the blackish bases of the feathers there.

Obs. Ceylonese specimens are identical with Indian in character of marking &c., but the latter may perhaps average somewhat larger. An individual from Kamptee has the wing 3·2 inches, another from "North India" 3·25. In some specimens the pale portions of the breast-feathers are strongly tinged with yellowish. Birds from the Burmese countries, says Mr. Hume, are not different from Indian; and an example from Acheen (Sumatra) is indistinguishable, as regards colour, from Indian examples, although somewhat smaller and shorter in the bill.

Distribution.—The little "Copper-smith" is diffused throughout all the dry region of Ceylon, commencing in the south a few miles west of Tangalla, and extending round the east side of the island (including the interior from the coast to the eastern slopes of Madulsima) to the extreme north. From the Jaffna peninsula it inhabits the west coast as far south as Madampe, the limit of its range extending thence across the country to Kurunegala, where it is very common. From Kurunegala it is found all along the base of the west Matale hills to Nalanda, and round to Bintenne; while all through the forests of the interior, stretching north of Dambulla, it is common. From the lowlands below Madulsima it ascends into Uva, in which region it is the prevalent small Barbet; but it does not cross the Hakgala ridge, and, in fact, its numbers decrease gradually to the west of the Badulla valley, and it is not very plentiful in the Uva patna basin. From Badulla it extends round the base of the hills, being found up their slopes to a height of about 2000 feet to Maturata and Hewahette, in which valley I have seen all four species together in the same ravine. In the north-east monsoon it strays in small numbers into Kandy, and as far even as Peradeniya, in the gardens at which place I have heard it in February. It is resident a little to the east of Kandy, namely at Hangeranketto. It will be seen that the range of this bird in Ceylon is entirely determined by climate, and is one of the most interesting of such cases to be found in the whole list of Ceylonese birds: the lower portions of the Kandy country towards the east are dry, and there this little Barbet establishes itself, and in the dry season penetrates to the west almost until it meets its fellows permanently residing in the low country of the North-west Province. The distance between Peradeniya, the most westerly point at which I have observed these birds coming from the east, to Kurunegala, where western birds are common, is not 20 miles in a direct line. I should not wonder if it be found in this intervening space, should naturalists take the trouble to look for it. I have never heard it myself at the back of Allegalla peak, and I do not know the low-lying chenna-hills between it and Gallagedera.

Beyond the confines of Ceylon the Crimson-breasted Barbet has a wide range. Jerdon speaks thus of its habitat:—" It is found throughout all India, extending into the Burmese countries, Malayana, and the isles." In some of the latter regions it is perhaps as common as it is in India. Capt. Feilden and Mr. Oates speak of it as common throughout Pegu, and Dr. Armstrong found it in abundance in thin forest-jungle in the Irrawaddy delta. In Tenasserim it is recorded as common; and Col. Tickell states that there appear to be two races of it in that Province, one of which inhabits the dense lofty forests, and the other the open country and villages, the two differing somewhat in voice. Mr. Davison procured it at Aeheen.

In Southern India I find that Mr. Bourdillon does not state it to be an inhabitant of the Travancore Hills; but Mr. Fairbank found it common up to 4000 feet in the Palanis, which form an eastern spur of the former range, and are, I have no doubt, much drier. In Central India and Bengal it is widely distributed, extending westwards into the Guzerat district, but not as far as Sindh, nor is it found in the Punjab or the Himalayas.

Habits.—This quaint little bird, being an inveterate fruit-eater, is found in all localities where trees affording it its favourite food are to be found. In the hills it affects scanty jungle and wooded ravines and hollows; but in the low country it is found, in addition to jungle, woods, groves, &c., in the gardens of the natives and the grounds surrounding the bungalows of the Europeans. It was a constant resident in the Fort at Trincomalie, and there I had much opportunity of observing its curious habits and manners. It appears not to indulge much in its powers of wing, but is a quiet retiring little bird, taking up its abode in the shady banyan or other such fruit-bearing monarch of the forest, and flying from branch to branch as it gorges itself with the ripe berry. If disturbed it flies off a short distance, and sits on the top of a neighbouring tree, twisting its head about and looking intensely stupid, until it suddenly remembers that its mate must be somewhere near, and it then commences its singular metallic-sounding call, resembling the syllable *wonk-wonk-wonk.* This is slowly repeated, and sounds like the striking of a hollow copper vessel; it is very distinct from the quicker sharper *wok-wok-wok* of the Ceylonese Barbet. In the breeding-season it delivers this note from morning to night, continuing it to a most monotonous extent without cessation: the pair sit close together, and utter it in concert, each note being accompanied by an odd-looking combined forward and sideward jerk of the head; and as of course both birds do not move together, the sound appears to come from different directions. I find that Sundevall takes another view of this curious effect; he writes, as quoted by Jerdon, " the same individual always utters the same note, but two are seldom heard to make it exactly alike. When, therefore, two or more birds are sitting near each other, a not unpleasant music arises from the alternation of the

2 F 2

notes, each sounding like the tone of a series of bells." The difference in sound, as I have already remarked, is produced by the alternate twisting of the birds' heads, that of one being directed towards, while that of the other is turned away, from the listener while the note is being delivered. Jerdon, I remark, advocates the same reason.

This Barbet has been stated to run up the trunks of trees; this it assuredly does not; it may be seen clinging to the bark of a tree at the commencement of the breeding-season, tapping the wood in order to find some soft or hollow place to make its nest in, but it has no power of proceeding up the surface of the trees. It congregates in large flocks, in company with Pigeons, to feed on the fruit of the Banyan-, Bo-, and Palm-trees, and quickly returns to the feast after being frightened away.

Nidification.—This species breeds from January until June, April being, I imagine, the month in which most young are reared. It generally nests in small decayed branches, boring them on the lower side when they happen to be slanting. As is the case with the former species, it selects, if possible, a branch that is hollow, and cutting its neatly-made round entrance, lays its eggs at the bottom of the cavity. Should the branch not be hollow, however, it will excavate to a depth of 6 inches or more, and will even continue to deepen it year after year. An instance of this is given by Jerdon, who had a pair breeding year after year in the cross beam of a vinery in his garden; the cavity was lengthened annually until "the distance from the original end was 4 or 5 feet." Another entrance was made from the underside, as was the first, and about 2½ feet from the nest. A pair that bred in a tree opposite my bungalow in the Fort at Trincomalie took from a fortnight to three weeks to construct the entrance and a short internal cavity. The opening was on the underside of a branch inclined at about 30°. The birds took it by turns to work, and the assiduity with which they laboured at the solid branch was extremely interesting. The little "carpenter" clung to the bark beneath the orifice, and swinging its body sideways and backwards would bring the whole of its strength to bear on the blow which it delivered with its stout little beak. I observed that the tail was seldom used as a support unless when a very vigorous blow was about to be dealt. When tired he would fly to an adjacent branch and look at the work with a contented aspect, and after a rest commence anew.

Mr. Parker writes me that he once watched one working at a hole in a most sedate manner. "After swaying his body sideways a little, whilst he was selecting a suitable chip to attack, he very gravely gave two or three sharp taps with his bill and detached a piece of wood. He then, *after looking round him*, proceeded in the same way to select another chip and detach it, and so on, as if he intended to spend his whole existence at the work."

Mr. Adam describes a nest which was made in the fork of a dead branch lying by the side of a thoroughfare, and so small that it could easily be lifted by the hand.

The eggs are usually three in number and are of an elongated shape; they are pure white, and have a pinkish tinge before being blown; they average 0·99 by 0·69 inch (*Hume*).

PICARIÆ.

Fam. CUCULIDÆ.

Bill more or less slender, curved, and compressed; the nostrils exposed and variable in position; gape wide. Feet zygodactyle; the outer anterior toe longer than the outer posterior one. Tail more or less long and broad.

Subfam. CUCULINÆ.

Nostrils swollen; the head sometimes crested. Tail variable. The tarsi feathered anteriorly, and the thigh-feathers long, hiding the tarsus. Stomach villous. (*Sharpe*, P. Z. S. 1873, p. 579, in part.)

Genus CUCULUS.

Bill moderately slender; gape wide, the culmen gently curved. Nostrils round, apert and basal. Wing tolerably long and pointed; the 3rd quill the longest, and the 1st shorter than or subequal to the 7th. Tail graduated, generally long. Tarsus not longer than the middle toe, feathered above at the front, the lower part covered with broad transverse scales. Inner anterior toe much shorter than the outer, and not so long as the outer posterior one.

CUCULUS CANORUS.

(THE COMMON CUCKOO.)

Cuculus canorus, Linn. Syst. Nat. i. p. 168 (1766); Sykes, P. Z. S. 1832, p. 98; Gould, B. of Europe, pl. 240 (1837); Jerdon, Cat. B. S. India, Madr. Journ. 1840, xi. p. 219; Blyth, Cat. B. Mus. A. S. B. p. 71 (1849); Layard et Kelaart, Prodromus, Suppl. Cat. p. 60 (1853); Layard, Ann. & Mag. Nat. Hist. 1854, xiii. p. 452; Horsf. & Moore, Cat. B. Mus. E. I. Co. ii. p. 702 (1856); Blakiston, Ibis, 1862, p. 325; Jerdon, B. of Ind. i. p. 322 (1862); Holdsworth, P. Z. S. 1872, p. 430; Gould, B. of Gt. Britain, vol. iii. pl. 67 (1873); Hume, Nests and Eggs (Rough Draft), p. 133 (1873); Ball, Str. Feath. 1874, p. 393; Bligh, J. A. S. (Ceylon Branch), 1874, p. 67; Hume, Str. Feath. 1875, p. 78; Butler, ibid. 1875, p. 460; Scully, ibid. 1876, p. 134; Hume, ibid. p. 288; Blakiston and Pryer, Ibis, 1878, p. 227; Dresser, B. of Europe, pt. 69 (1878).

Cuculus hepaticus, Sparrm. Mus. Carls. iii. pl. 55 (1787).

Cuculus borealis, Pall. Zoogr. Rosso-Asiat. i. p. 442 (1811).

Cuculus indicus, Cab. et Heine, Mus. Hein. iv. p. 34 (1862-3).

Cucu, Spanish (Saunders); *Phu-Phu* in Dehra Doon; *Kukupho*, Lepchas; *Akku*, Bhotan (Jerdon); *Kako*, Japanese (Blakiston); *Kakkok*, Turkestan (Scully).

Adult male (Kotmalie). Length 13·5 inches; wing 8·6; tail 7·2; tarsus 0·85; outer anterior toe and claw 1·1; bill to gape 1·17.

These are the dimensions of a very fine example shot in Ceylon. I subjoin others from specimens procured in different parts of the world, beginning with those contained in Dr. Scully's very complete notice of Eastern-Turkestan Cuckoos.

	Length. in.	Expanse. in.	Wing. in.	Tail. in.	Tarsus. in.	Bill from gape. in.	Weight. oz.
(1) Five males. Yarkand	12·8 to 14·0	23·0 to 24·0	8·5 to 8·9	7·0 to 7·7	0·8 to 1·0	1·15 to 1·3	3·2 to 4·5
Two females, juv., rufous.							
Yarkand	12·3 to 13·2	22·8 to 23·5	8·0 to 8·5	6·6 to 6·9	0·9	1·15 to 1·2	2·8 to 3·8
(2) Examples in Brit. Mus.:—							
Athens	13·0 (from skin)	..	8·0	6·8	0·8		
Germany	..	..	8·5	7·0	0·85	1·2	
Persia	..	..	8·6	7·0			
N.W. Province	..	..	8·9	7·0	0·9	1·1	
♀. "India"	..	..	8·3				
♂. Sweden	..	..	8·9	7·5	0·9	1·2	

Iris yellow, pale yellow; bill, upper mandible and tip of lower blackish, base of under mandible greenish yellow; gape and eyelid yellow; inside of mouth red; legs and feet yellow; claws dusky.

Above dusky bluish ashen, with a slight greenish gloss generally on the interscapular region; the rump and upper tail-coverts more bluish than the back; quills plain brown, crossed on the inner webs with pointed marginal bars of white, reaching to within about 2 inches from the tips of the longer primaries; winglet and primary-coverts darker than the quills; edge of the wing beneath the winglet white; tail blackish, with a slaty hue; the tips of the feathers white; the five lateral feathers on each side with central white spots, sometimes limited to one in number, and marginal indentations of the same on the inner webs and sometimes on both; the outermost feathers more spotted than the rest.

Throat and fore neck delicate ashen, blending on the sides into the darker hue of the hind neck; from the chest downwards white, crossed with narrow wavy bars of blackish, broadest on the flanks; vent and base of under tail-coverts unmarked; under wing-coverts white, marked as the chest; a wash of slate-colour along the under edge of the wing.

Young. Variable in plumage; more or less marked with rufous.

Above greenish brown, the green lustre very strong in some; the head, hind neck, back, and lesser wing-coverts with white tips to the feathers; quills and primary-coverts mostly barred with rufous, the inner parts of the bars whitish; tail more boldly spotted than in the adult, the white spots running into rufous adjacent patches; under surface buffy white, with bolder and darker bars than in the adult, which extend also to the throat, where they are closer together than on the breast. There is often a white nuchal patch.

After moulting the nest-plumage, specimens have the lower part of the throat and also the chin washed with buff, and the lower parts often retain the buff tinge; the bars are darker and sharper-edged than in old birds.

Rufous phase of young. Barred more or less on the whole upper surface with rufous bands, occasionally very broad, and predominating on the rump and upper tail-coverts over the slaty-brown ground-colour; the under surface is white as in others, barred with bold black bands.

Obs. Although the moderately close character of the barring which distinguishes adults of this Cuckoo from others is the same in all specimens, yet I notice a considerable variation in the bars themselves, consisting in their width and in the appearance of the edges, some being more softened off than others.

Cabanis and Heine separated the Indian Cuckoo, alleging that it was a smaller bird than the European; but the specimens which they had to deal with were doubtless those of *C. himalayanus*, now recognized to be quite distinct from *C. canorus*.

Distribution.—There have been two instances of the occurrence of the Common Cuckoo in Ceylon; and it is only to be wondered that a bird in which the migratory instinct is so powerful as in this interesting species has not been oftener met with in the island, particularly as on the other side of the Bay of Bengal it has been found as far south as Timor, lat. 10° S. Layard obtained the first example in the old Botanical Gardens at Kew, Colombo; and Mr. Bligh the second, which he shot on the Harangolla Patnas, Kotmalie, on the 7th October, 1873. This was at an elevation of about 4000 feet, lower than which it is not likely that the Cuckoo would reside in Ceylon during its stay. Layard's specimen was killed, of course, *en passant* to the hills.

Our English harbinger of spring can therefore only be looked upon as a mere straggler to Ceylon; but notwithstanding, as it is a bird which recalls home recollections to many of my readers who perhaps feel themselves exiled to the beautiful island of Lanka, I feel constrained to say more concerning its distribution, habits, and strange career as a nestling than the limits of this work on a local avifauna would otherwise warrant. During the breeding-season the Cuckoo inhabits more or less the whole of the Asiatic continent north of the Himalayas, extending its range as far north as the limit of forest-growth, considerably within the Arctic circle, and extending westwards from Japan right across to the neighbouring continent of Europe, over which it is entirely diffused, being of course, as regards the various districts in which it has been noticed in both regions, locally common and locally scarce: to the south of the Himalayas many birds remain and perhaps breed as low down as the latitude of Calcutta. Within its ordinary breeding-limit, however, it is to some parts only a visitant; Mr. H. Whitely records it as such to Hakodadi, in Japan. In China, says Swinhoe, it "occurs in the mountains of the south in spring, extending northwards to Pekin. During its migration we met with it on the plains." Mr. Blakiston notes it as common on Fujisan, one of the Japanese islands.

In Eastern Turkestan, writes Dr. Scully, it arrives on the plains about the middle of April (this is from the south of course), and leaves about the beginning of August. In Persia, Mr. Blanford says that it abounds; he heard its note frequently in the Baluchistan hills in February and March, and he is of opinion that it breeds in the Persian highlands, for he met with it in May in the wooded hillsides and valleys of Fárs. To Palestine it is also a summer visitant from the south; Canon Tristram ('Ibis,' 1866, p. 285) did not observe it before the 30th March; it was generally spread over the country, and was particularly abundant in the Jordan valley. As above remarked, it is spread over the whole of Europe to the extreme north. Messrs. Alston and Harvie Brown record it as very abundant at Archangel; in Sweden and Norway it is likewise common; and as regards the British Isles it travels to the extreme limit of the Shetlands, arriving in the south at the end of April, and laying, according to Mr. G. Dawson Rowley's observations, as early as the 1st of May. It does not appear to remain in Spain during the summer, merely passing through on its northward migration from Africa: Mr. Saunders did not find it laying anywhere in the country.

In North Africa it is a spring and autumn visitor, passing through on its way to the north from more southerly latitudes. Captain Shelley has shot it as early as the 30th April; and Von Heuglin states that it arrives from the south in March, and lingers on its way north until May, returning so soon again as August. These must be, in all probability, birds that have bred in the south of Europe. In Lower Nubia, Professor Hartmann heard it in May, and again in September and October. In Tangier it is common, arriving in spring from the south. On the west coast it has been procured in Fantee by Governor Ussher, in Damara Land, South-west Africa, by Mr. Andersson, and in Natal by Mr. Ayres. In South Africa, however, where it winters, it is evidently by no means common, as there are comparatively few instances of its capture there; where, therefore, the number of birds that pass through North Africa spend the winter has yet to be determined, and will most likely prove to be the upland regions of the continent, or the country of the great lakes so prominently brought before the world of late years by our great African travellers.

In Asia, where we have been discussing its summer habitat, its winter quarters are well known. In Bengal it is common, and thence is spread all over India to the extreme south, where it is rare. In the northwest of the latter country it is a spring visitant, passing through, according to Captain Butler, in May, and moving towards the hills; after the breeding-season it returns again, and is very plentiful in September. Neither Mr. Bourdillon nor Mr. Fairbank record it from the Travancore hills; and the latter does not speak of it in the Deccan, although Jerdon says that it remains two or three months in the spring in Central India, and that he heard its call at Goomsoor, Saugor, and Nagpoor in May and June. On the other side of the bay it is evidently a mere straggler, occurring in Pegu and perhaps in the Malaccan Peninsula and islands between there and Timor, which is its utmost limit to the south.

Habits.—The Cuckoo chiefly affects openly wooded or park-land, avenues of trees, the borders of woods, scrubby commons or wastes where a few trees are here and there interspersed among the low growth, in which its foster-parents usually nest. In the breeding-season, however, it wanders about so much that it may be found in the heart of large woods; and I observe that Mr. Blanford mentions hearing it in the jungles of the Persian hills. Shortly after arriving in the various localities where it intends to rear its young its welcome note may be heard from daybreak until late in the morning resounding merrily through the woods, which teem with numerous joyful songsters, not a few of which are perhaps destined to be the foster-parents of the Cuckoo's offspring, and to have their own ignominiously expelled by the unprincipled and unscrupulous little stranger! There is no bird in Europe about which so many strange theories have existed in the popular mind as the "harbinger of spring." Strange crimes and misdemeanours have been accredited to it from the earliest times; and among these, as is stated by the ancient writer and naturalist Pliny, none is so dire as its devouring its foster-parent; he remarks, "The young Cuckoo being once fledged and ready to fly abroad is so bold as to seize on the old titling and to eat her up that hatched her." Although we must absolve the Cuckoo from such a want of gratitude as is here depicted, yet the conduct of the young nestling, as will be noticed directly, is in the highest degree unnatural. It is believed by many that the old birds possess the power of fascinating the species in whose nest their egg is to be deposited, such a belief having obtained from the erroneous idea that the Cuckoo actually lays its egg in the nest it has chosen, which it certainly does not. A great difference can be detected in the sound of the Cuckoo's note as uttered by different birds, some giving it out as *wuk-koo*, the first syllable being very plainly pronounced. The Yarkandis syllabize it by the word *kak-kok*, which Dr. Scully says he thinks is a better representation of the note than ours. It is the love-call of the bird, and after the breeding-season, as is well known, ceases to be heard, causing many to think that the Cuckoo has left her accustomed haunt, whereas in reality it has only become silent. I have observed in England that it is usually heard before 9 or 10 o'clock, after which the bird is more or less silent until evening, when it again becomes as garrulous as it was in the morning. The Cuckoo's flight is powerful and very Hawk-like, being performed with regular beatings of the wing; it generally flies a moderate distance, mounting into an upper branch, where it alights and commences its note at once, which it continues for a little time and then becomes silent before moving on again. It is most noisy just at the time of laying.

Its diet is insectivorous and varied, consisting of caterpillars, grubs, worms, moths, and small insects. The stomach is clothed inside with a thick hairy or villous coating, which is, I believe, peculiar to all the subfamily Cuculinæ; at least all I have shot in Ceylon possess this character in a greater or less degree.

Nidification.—The chief amount of interest which attaches to the singular economy of the Cuckoo is naturally centered in its nidification, and the strange habit, as exemplified in the whole group of true Cuckoos, of fostering its young upon other birds. Connected with this are many points of great interest to the naturalist, such as its supposed polygamy, its instinct of laying eggs of a peculiar type to suit those with which it is deposited, its partiality for Warblers' nests, the fact of eggs of peculiar coloration prevailing in different localities, and the habit of conveying the eggs in the bill after laying them and then depositing them in the nest chosen to receive them, all of which justly tend to render the natural history of the Cuckoo one of the most interesting of any bird known.

In India the Cuckoo has been ascertained to lay during the latter half of May and the first half of June (Hume, 'Nests and Eggs'), usually choosing the nests of Pipits and Chats; among the former the Upland Pipit (*Heterura sylvana*) and Jerdon's Rock-Pipit (*Agrodroma jerdoni*), and among the latter the Indian Bush-Chat (*Pratincola indica*), the dark grey Bush-Chat (*P. ferrea*), the white-winged Black Robin (*P. cuprata*), and the Magpie Robin (*Copsychus saularis*) appear to be the favourite species. In the Almorah district Mr. Brooks says they lay in the nests of *P. indica* and *C. saularis*, and Mr. Thompson in the nests of Pipits. At Murree, Captain Marshall found the eggs in the nests of *A. jerdoni* and *P. ferrea*, and Mr. Hume obtained two eggs in the nests of *Heterura sylvana* near Kotegurh.

In Europe it has a great partiality for nests of Warblers; and in a long list of about a score of these birds given by Dr. Baldamus, in his exhaustive article in the 'Naumannia,' 1853, are mentioned the Blackcap, the Robin, the Garden-Warbler, the Whitethroat, the Lesser Whitethroat, the Reed-Warbler, the Wood-Warbler, the Marsh-Warbler, the Redstart, the Grasshopper-Warbler, the Nightingale, and the Willow-

Wren; other species given by the same author are the Common Wren, the Hedge-Sparrow, the Pied Wagtail, the Yellow Wagtail, the Marsh-Pipit, the Meadow-Pipit, the Skylark, the Yellow Bunting, the Butcher-bird (*Lanius collurio*), the Tree-Pipit, the Crested Lark, the Wood-Lark, the Reed-Bunting, the Brambling, the Crossbill, and the Linnet. Mr. Cecil Smith, in his 'Birds of Somersetshire,' mentions also an instance of a Blackbird's nest; in addition to which I may cite those of the Thrush, Great Tit, Turtle Dove, and Wood-Pigeon. It will be observed that the majority of these birds have much too small nests for the Cuckoo to be able to lay in, and that into some she of course could not enter, which fact alone would prove what used to be doubted by many naturalists, but is now universally accepted by all who have given their attention to the matter, viz. that the Cuckoo deposits her eggs in the nest by carrying them in her bill. Birds of late years have been killed with the eggs in their mouths; and I myself have seen one shot rising from an Essex meadow with an egg in its bill.

Females hang about certain localities for days, and having in the mean time discovered a nest which suits them, lay their eggs and, watching the opportunity when the rightful owners are away, convey them to their destination. A struggle not unfrequently ensues between the Cuckoo and the foster-parent, evidences of which are seen in broken egg-shells and other signs of a scuffle having taken place. The same bird, it has been ascertained, only deposits a single egg in one nest, and that generally after the rightful owner has begun to lay. Of this even the natives of Central Asia have cognizance; for Dr. Scully tells us, in his paper in 'Stray Feathers,' that the Yarkandis told him so, giving the nests in which the eggs were deposited as those of the Brown Shrike (*Lanius arenarius*), the Red-headed Bunting (*Euspiza luteola*), and the Indian Blue-throat (*Cyanecula suecica*). They say, he remarks, that all Cuckoos are of the female sex, and are not very particular in their choice of husbands, frogs being selected indifferently with birds! The latter strange idea emanates, no doubt, from the Cuckoo in Yarkand giving, according to Dr. Scully, "a prolonged sort of cry, somewhat resembling that of the toad (*Bufo viridis*), but somewhat louder." Dr. Baldamus contends that each Cuckoo lays "eggs of a certain colouring only, which corresponds with that of the eggs of some one species of Warbler, in the nest of which she deposits them;" but Mr. G. Dawson Rowley has found that this is not always the case.

The most remarkable feature, however, in the economy of the Cuckoo has yet to be noticed; and this is the extraordinary faculty in the young chick which prompts it, when newly born, and *before its eyes are open*, to eject its foster-brethren from the nest; and coupled with this is the scarcely less singular devotion evinced by the bereaved foster-parent for the little monster who has thus deprived her of the rearing of the rest of her offspring. With regard to the conduct of the young Cuckoo, it may not be known to all of my readers that a long account of it was published in the last century by Dr. Jenner, who gave the results of his observations in the 'Philosophical Transactions' for 1788. For a long time the Doctor's account of what he saw did not secure that amount of credence which it should have. The fact seems to have been known to the ancients that the young Cuckoo got rid of its fellow nestlings; but this, according to Pliny, was by the simpler method, perhaps, of devouring them, which somewhat rough treatment was, he considered, rather encouraged than otherwise by the unconscious foster-parent; for, writes he, "she joyeth to see so goodly a bird, and wonders at herself that she hath hatched and reared so trim a chick. The rest, which are her own, indeed, she sets no store by; yea, and suffereth them to be eaten and devoured of the other, even before her face."

Of late years the experience of Dr. Jenner has been verified by the observations of a lady in Scotland devoted to the subject of natural history, and who, in a little book on the Pipits, gave a sketch of what she saw. She was afterwards requested to publish an account of the proceeding in detail in 'Nature,' which she did. I quote here in part from Mrs. Hugh Blackburn's story as follows:—"The nest (which we watched last June, after finding the Cuckoo's egg in it) was that of the common Meadow-Pipit, and had two Pipit's eggs, besides that of the Cuckoo. It was below a heather bush, on the declivity of a low abrupt bank on a Highland hill-side in Moidart. At one visit the Pipits were found to be hatched, but not the Cuckoo. At the next visit, which was after an interval of forty-eight hours, we found the young Cuckoo alone in the nest, and both the young Pipits lying down the bank, about ten inches from the margin of the nest, but quite lively after being warmed in the hand. They were replaced in the nest beside of the Cuckoo, which struggled about till it got its back under one of them, when it climbed backwards directly up the open side of the nest, and

2 G

hitched the Pipit from its back on to the edge. It then stood quite upright on its legs, which were straddled wide apart, with the claws firmly fixed halfway down the inside of the nest, among the interlacing fibres of which the nest was woven; and, stretching its wings apart and backwards, it elbowed the Pipit fairly over the margin so far that its struggles took it down the bank instead of back into the nest. After this the Cuckoo stood a minute or two, feeling back with its wings, as if to make sure that the Pipit was fairly overboard, and then subsided into the bottom of the nest.

"As it was getting late, and the Cuckoo did not immediately set to work on the other nestling, I replaced the ejected one, and went home. On returning next day, both nestlings were found dead and cold, out of the nest. I replaced one of them; but the Cuckoo made no effort to get under and eject it, but settled itself contentedly on the top of it. All this I find accords accurately with Jenner's description of what he saw. But what struck me most was this: the Cuckoo was perfectly naked, without a vestige of a feather, or even a hint of future feathers; its eyes were not yet opened, and its neck seemed too weak to support the weight of its head. The Pipits had well-developed quills on the wings and back, and had bright eyes, partially open; yet they seemed quite helpless under the manipulations of the Cuckoo, which looked a much less developed creature. The Cuckoo's legs, however, seemed very muscular, and it appeared to feel about with its wings, which were absolutely featherless, as with hands, the 'spurious wing' (unusually large in proportion) looking like a spread-out thumb. The most singular thing of all was the direct purpose with which the blind little monster made for the open side of the nest, the only part where it could throw its burthen down the bank. I think all the spectators felt the sort of horror and awe at the apparent inadequacy of the creature's intelligence to its acts that one might have felt at seeing a toothless hag raise a ghost by an incantation. It was horribly 'uncanny' and 'grewsome'"*.

Comment upon this extraordinary feat is unnecessary, suffice it to say that the testimony of other observers is forthcoming to prove that the young Cuckoo ejects its companions when still in a perfectly unfledged state, thus displaying a more wonderful instinct than perhaps exists throughout the whole range of the bird creation!

Concerning the attachment of the foster-parents to their tyrannical offspring, I quote as follows from Mr. Gould's admirable article in the 'Birds of Great Britain':—"How wonderfully solicitous are the little birds for its welfare, and with what spirit do the foster-parents defend their nurtured Cuckoo. If its removal be attempted they display the greatest uneasiness. Wagtails will even fly in the face of the person who thus teases them; and if it be returned to them they will evince their joy by fondling and dancing around it, leaping over its back, and exhibiting many other demonstrations of delight. Yet in a few days their charge will wing its way to the leafy branch of some tree in the forest, and there sit uttering most strange, piercing, bat-like notes, varied occasionally by others resembling the syllables *chat-chat*." The affection displayed by the Wagtail in particular for the young Cuckoo, inciting it to feed it when grown to three times its own size, is well delineated in Mr. Gould's magnificent plate, in which a Pied Wagtail is drawn standing (as it was actually seen) on the back of a Cuckoo seated on a fence, and depositing a caterpillar in its upturned and gaping mouth. Touching the habits of the young, I subjoin from the same article the following interesting paragraph:—" A young Cuckoo, which was taken from the nest of a Wagtail at Formosa (Berkshire), exhibited many strange actions, which very strongly reminded me of a rattlesnake. If the hand was put towards it, it raised itself on its legs, protruded its neck, puffed out its feathers, and threw its head forward with a quick and determined stroke, precisely like a snake or viper, struck the hand with the open mouth, just as a snake would do, and immediately drew the hand back in readiness for another stroke. On the second day after it was taken, the bird was sufficiently reconciled to me and my daughter to take small pieces of raw beef or mutton and caterpillars from the hand, but continued to utter its piercing shriek whenever we approached it. Does not this peculiar electrifying shriek attract the attention of the smaller birds when it requires food? A delicate ear will hear this sound for thirty or forty yards, and it is probably heard at a still greater distance by the smaller birds."

The two types common in Cuckoo's eggs are the red and the grey. The ground-colour is whitish in some,

---

* 'Nature,' No. 124.

streaked and spotted with brownish red and purple ; in others it is stone-coloured or pale reddish, blotched with brownish grey, yellowish brown, or brownish red.

They average about 0·9 by 0·7 inch.

For permission to give the accompanying woodcut I am much indebted to the kindness of Mr. Gould. It is a copy from his facsimile of Mrs. Blackburn's sketch in the ' Pipits.'

CUCULUS MICROPTERUS.

(THE INDIAN CUCKOO.)

Cuculus micropterus, Gould, P. Z. S. 1837, p. 137; Blyth, J. A. S. B. 1842, xi. p. 902;
    Kelaart, Prodromus, Cat. p. 129 (1852); Layard, Ann. & Mag. Nat. Hist. 1854, xiii.
    p. 452; Jerdon, B. of Ind. i. p. 326 (1862); Swinhoe, P. Z. S. 1871, p. 395;
    Holdsworth, P. Z. S. 1872, p. 430; Legge, Ibis, 1874, p. 16; Hume, Str. Feath. 1875,
    p. 79.
Cuculus striatus, Blyth, Cat. B. Mus. A. S. B. p. 70 (1849); Horsf. & Moore, Cat. B.
    Mus. E. I. Co. ii. p. 703 (1856); Cab. et Heine, Mus. Hein. iv. p. 37 (1862).
Cuculus affinis (A. Hay), Blyth, J. A. S. B. 1846, xv. p. 18.
Great-billed Cuckoo, Blyth; *Ashy Mountain Cuckoo*, Kelaart.
Bou-kotako, Bengalese; *Takpo-pho*, Lepchas.

Adult male (Ceylon). Length 12·2 inches; wing 7·75; tail 5·8; tarsus 0·7; outer anterior toe 0·8; bill to
gape 1·24.

An example from Sumatra measures—wing 8·2 inches; tail 6·8; bill to gape 1·11. A male from Pegu—length
13·3; expanse 23·5; wing 8·25; bill from gape 1·35: a female—length 12·4; wing 7·6; bill from gape 1·3.
(*Oates.*)

Iris brown; bill dark horn at base of upper mandible, the tip blackish; under mandible fleshy, with a dark tip; gape
and orbital skin yellow; legs and feet ochre-yellow, claws dusky.
Back, wings, and tail brownish ashy, with a metallic or bronze lustre; the head and hind neck dusky slate-colour,
blending imperceptibly into the hue of the back; wings light ash-brown, the inner webs of the primaries and
secondaries crossed with marginal bars of white, except at the tips, the secondaries white at the base of the inner
webs; tail light cinereous brown, with tips and a series of shaft-spots of white, a blackish subterminal bar and
blackish shaft-streaks between the white spots, outermost feathers barred with white.
Lores, face, throat, and fore neck pale ashy, the cheeks darker than the rest; from the chest downwards white, with
distant blackish, clear-margined bars: under tail-coverts with the longer feathers only barred; under wing-coverts
buff-white, irregularly marked with black bars.

An example not quite adult has the concealed portions of the hind-neck feathers barred with rufous, and the chest-
feathers tipped and transversely marked with a paler hue of the same; the lateral upper tail-coverts, which lie
concealed, have the outer webs barred with rufous and white, with the interspaces dark brown.

This Cuckoo is at once recognized at a glance from *C. canorus* by the presence of the dark caudal bar.

Young. A specimen in nest-plumage from Darjiling has the upper surface a lustrous ruddy brown, all the feathers
more or less deeply tipped—on the head and hind neck with buff-white, on the back and wing-coverts with
rufous, the extreme tips being whitish, and on the rump and upper tail-coverts with dusky rufous, the quills tipped
and their inner webs barred with rufous; both margins of the central tail-feathers indented with rufous, and the
tips of all the feathers fulvous-white, the outermost pair barred with rufous, and the next two pairs with the
inner webs only barred with the same; there is no bar, but the terminal inch of all the feathers is unmarked,
imparting the appearance of a band: under surface buff, barred heavily with blackish brown; the markings of the
chest not so regular as on the breast and flanks.
An immature bird shot at Nalanda, Ceylon, is glossy grey-brown above, with a subdued ashen hue on the hind neck
and head; there is a lightish stripe above the eye and a narrow dark edge in front of it; greater secondary wing-
coverts tipped with rufous, the primaries and secondaries tipped with white, and the latter indented outwardly
with rufous; tail much as in the adult, but with the margins of the feathers indented with rufescent; chin and
lower part of chest washed with rufescent: a dark brownish patch on the sides of the chest, and the central
portion barred with blackish brown; under parts with the bars broader than in the adult.

Obs. The synonymy of this species is in a somewhat confused state, owing to Drapiez (Dict. Class. Hist. Nat. vol. iv,
p. 570) having described a Cuckoo from Java in 1823, of a cindery-brown colour ("brun-cendré") above, and

12 inches in length, which some unite with Gould's bird, discovered many years after in the Himalayas, but which others join with *C. himalayanus* of Vigors, a perfectly different bird, and not belonging to the terminal-bar-tailed group at all. It is scarcely possible to affirm what the *Cuculus striatus* of Drapiez really was; it was evidently an immature bird, as the outer primaries were indented with rufous; the dimensions of the wing were unfortunately not stated: taking all things into consideration it appears to me to have belonged to the brown bar-tailed section and not to the ashy one, of which *C. canorus* is the type. *C. himalayanus* is a miniature of this latter. Mr. Seebohm procured it on the Yenesay river; it migrates to China and Japan, and goes down to the Malay archipelago in winter; but so does the present species. In the British Museum is a specimen from Sumatra labelled *C. affinis*, with the wing 8·2, bill to gape 1·1 (this is identical with a Ceylonese example), and another from the Himalayas labelled *C. micropterus* (this has, perhaps, the bars on the lower parts broader, and is slightly darker on the throat and chest than the Ceylon bird; the bill across the gape is 0·75 inch, while the latter measures 0·71). Mr. Oates measured a male shot in Pegu as, wing 8·25 inches, bill from gape 1·35; a female, wing 7·0, bill from gape 1·3. The bills are very large in these, and Mr. Hume considers *C. micropterus* to refer to these large-billed birds. Perhaps there are two races of this Cuckoo in the Himalayas; but we do not know whether Gould's type had an exceedingly large bill or not. The Ceylonese birds which I have seen certainly are not so large in the bill as these latter specimens; but they evidently migrate from the Himalayas, and they most decidedly are not *C. himalayanus*. What the *C. affinis* of Lord A. Hay was is not quite clear. I cannot therefore apply his name to our bird, nor can I Drapiez's, if his species is to be considered the same as Vigors's (*C. himalayanus*, an altogether different type of bird), and therefore I must allow it to stand under Gould's name as heretofore.

Distribution.—This Cuckoo arrives in Ceylon during the month of October; but apparently its numbers are extremely limited, as but comparatively few examples have ever been recorded from the island. Kelaart speaks of it as a mountain species of rare occurrence and found in Dimbulla; Layard did not meet with it. Holdsworth writes that "the only two examples he met with were obtained in half-cultivated land in low country near Colombo." These were probably in migration to the hills at the time they were killed. I have shot it in the Kottowe forest near Galle, and have seen it in the same district on another occasion. It probably affects the subsidiary hills in the south-west of the island as much as any other part of the low country. I met with a Cuckoo, which I did not procure, but which I identify as belonging to this species, in the forests between Anaradjapura and Trincomalic; and Captain Wade, of the 57th Regiment, killed an immature individual at Nalanda at the north base of the Kandyan ranges; in addition to which I have seen it in the collection of Messrs. Whyte and Co., the specimen having been procured in Dumbara. It is doubtless a commoner species in reality than it appears to be, but, being a denizen of the forests, escapes nearly all observation during the period of its visit.

There is, I think, no doubt that this species migrates to Ceylon *viâ* the south of India from the Himalayan region; it is evidently very rare in the Peninsula. I notice that Messrs. Bourdillon and Fairbank do not record it in either of their lists from the southern hills; the latter notes it from Ahmednagar, but makes no comment as to its scarcity or otherwise. Jerdon found it rare on the Malabar coast and in the Carnatic, but "tolerably common in the jungles of Central India, as at Nagpore, Chanda, Mhow, and Saugor."

Taking the large-billed race to be only a local variety of the species which visits Ceylon, we find Mr. Hume recording this Cuckoo as "common throughout Lower and Eastern Bengal, and even up into the lower valleys of the Himalayas, in Sikkim, Bhootan, and Assam." In Pegu, according to Mr. Oates, it is numerous everywhere, but less so in the plains than in the hills. From Burmah it finds its way eastwards to China, where Swinhoe found it on the Upper Yangtsze; southwards it migrates in the cool season through the Malaccan peninsula to the archipelago, whence it has been procured in Java and Sumatra, and probably will some day be obtained in Borneo, if it has not been already met with there. Lord Tweeddale refers with doubt four examples procured in the Andamans by Lieut. Ramsay to this species; but the measurements of the wings, viz. 7·0 and 7·37 inches, are almost too small for *C. micropterus*.

Habits.—The Indian Cuckoo frequents high jungle and forest, particularly that on the sides of hills. It is a shy bird and keeps, as far as I have observed, to the tops of tall trees. It is very Hawk-like in flight, having much the appearance of a small Accipitre as it wings its way from the summit of one lofty tree to another. I noticed it in the Kottowe forest fly out of the upper branches of an enormous Hora-tree, and after proceeding a short distance alight on the very top of an equally high dead trunk. Its habit of keeping to the uppermost branches of these giants of the forest leads to its being seldom procured. Jerdon writes that it "repeats its call more frequently than other Cuckoos; this," he remarks, "is a double note of two syllables each—a fine, melodious, pleasing whistle, which the natives of Bengal attempt to immitate by their name *Bokutako*." Mr. Oates says that its note is double and very melodious, and that it selects the topmost bough of a tree (generally a dead one) and remains calling there for a quarter of an hour or more. Its loquacious habit, like that of the Plaintive Cuckoo, is evidently confined to the breeding-season; I never heard it, on the several occasions I have seen it in Ceylon, utter a note.

Its stomach is highly villous, and its principal food consists of caterpillars.

Its eggs have not yet been identified; but some suppose that it lays in the nests of Babblers (*Malacocerci*).

CUCULUS POLIOCEPHALUS.

(THE SMALL CUCKOO.)

Cuculus poliocephalus, Lath. Ind. Orn. i. p. 214 (1790); Blyth, J. A. S. B. 1842, p. 904, et
    Cat. B. Mus. A. S. B. p. 71 (1849); Horsf. & Moore, Cat. B. Mus. E. I. Co. ii. p. 704
    (1856); Jerdon, B. of Ind. i. p. 324; Hume, Nests and Eggs, p. 135 (1873).
Cuculus himalayanus, Gould, Cent. Him. Birds, pl. 54 (1832).
Hierococcyx poliocephalus, Bp. Consp. Gen. Av. i. p. 204 (1850).
Cuculus bartlettii, Layard, Ann. & Mag. Nat. Hist. 1854, xiii. p. 452 (juv.).
Cuculus lineatus, Less. Traité d'Orn. p. 152.
Cuculus tamsuicus, Swinhoe, Ibis, 1865, p. 108.
Cuculus ——?, Blakiston and Pryer, Ibis, 1878, p. 227.
The Hoary-headed Cuckoo of some Indian writers.
Daugham, Lepchas; *Pichu-giœpu*, Bhootias (Jerdon).
Hototogisu, Japanese.

Adult male and female. Length 10·0 to 10·75 inches; wing 6·0 to 6·2; tail 5·2 to 6·0; tarsus 0·75 to 0·85; outer
    anterior toe and claw 0·9 to 1·0; bill to gape 1·0 to 1·1. Expanse 17·3.

The above dimensions are from three examples procured in Ceylon. A Japanese specimen measures—wing 6·3 inches;
    tail 5·7.

Iris brown or brownish grey; bill, upper mandible and tip of lower blackish, gape, base of under mandible, and eyelid
    yellow; inside of mouth the same, but the base of the palate orange-red; legs and feet yellow, tarsus washed with
    brownish; claws brownish yellow.
Above almost uniform bluish ashen, illumined strongly with greenish, mostly on the scapulars; upper tail-coverts more
    bluish than the back; primaries slaty brown, with a greenish tinge, barred with white; tail blackish, tinged
    slightly with green, tipped with white, with a series of white shaft-spots and marginal indentations of the same;
    throat and fore neck pale fulvous, shaded with an ashen hue, and which colour blends softly into the grey of the
    sides of the neck; beneath, from the neck downwards, white, with the vent and under tail-coverts pale buff;
    breast, flanks, and thigh-coverts crossed with narrow softened-edged bars of blackish.
The above description is taken from a well-preserved Japanese example in Mr. Seebohm's collection, which is identical
    with Indian specimens.

Young. Above ashy brown; the feathers of the head more or less tipped with white, these markings being often confined
    to the superciliary region and occiput. Upper back, scapulars, and wing-coverts tipped and barred with whitish
    or pale fulvous; the lower back and upper tail-coverts marked with a series of white central transverse spots, the
    latter more or less barred with rufous as well; primaries and secondaries barred on the outer webs with rufous,
    and on the inner with white, changing somewhat into rufous near the tips; tail spotted as in the adult, and the
    central feathers barred with rufescent; chin and throat fulvous, barred with pale brownish; under surface as in
    the adult.
Individuals vary much *inter se* in the markings of the upper surface, some specimens being banded with rufous instead
    of white. An example shot in March at Colombo is acquiring the adult plumage, having the attire of the head,
    back, and rump mixed with bluish-ashen feathers.

Rufous phase. This species commonly assumes a rufous phase. Two individuals from Nepal which I have examined
    in the British Museum are entirely rufous above, with the head, hind neck, scapulars, back, and wing-coverts
    banded with blackish slaty, having a perceptible greenish lustre; in one the rump and upper tail-coverts are
    almost unmarked, the feathers only having terminal bar-like spots; the wings are greenish brown, barred with
    yellowish rufous; the tail glossy dark brown, barred with incomplete angular rufous bars, the feathers all tipped
    with whitish; chin and throat yellowish rufous, narrowly barred with blackish; breast and lower parts white,
    crossed with widely-separated blackish bars; edge of under wing rufous, the rest of it white, barred with black.

Obs. I have examined a specimen of the Small Cuckoo from Madagascar, the *Cuculus rochii* of Hartlaub (P. Z. S. 1862, p. 224), and which is kept distinct from *C. poliocephalus* by Mr. Sharpe in his admirable paper on the Cuckoos of the Ethiopian region, on account of its darker upper surface and the somewhat different banding of the under parts, the dark bars, according to him, being broader and the white interspaces wider. It is entirely the same as the Japanese specimen above described; the upper surface has the same hue, and the breast and lower parts barred the same, the under tail-coverts having precisely the same buff hue. I think that the two species will have to be amalgamated; and if so, the great range which the Small Cuckoo will then acquire will be only second to that of *C. canorus*.

Distribution.—The present species was described in the 'Annals and Magazine of Natural History' by Layard as new, under the title *C. bartletti*. His specimen was in immature plumage; and he writes of the bird that he obtained many examples of it both at Pt. Pedro and Colombo. Mr. Holdsworth does not seem to have identified it while he was in Ceylon, but speaks of a Cuckoo, closely resembling *C. canorus*, which he saw in an English garden in Colombo; and this I imagine, though it is very much smaller than the latter, must have, in reality, been this Cuckoo. It is, of course, migratory to Ceylon, and appears as isolated individuals on the west coast in October. Some years it is not seen at all, and during others several examples may perhaps come under the notice of collectors. Not a few were seen in the neighbourhood of Colombo in October 1876, one of which I procured at Borella, and another was shot near Kotté and preserved in the Colombo Museum. In December 1869 I obtained an example (immature, as are all which I have seen from Ceylon) on some trees at the lake side of the Galle face. It does not seem to have been noticed anywhere but in the Jaffna peninsula and about Colombo. It probably leaves the island in April.

On the continent it appears to enjoy a wide range; but is found more often in Northern than in Southern India, which makes its occasional occurrence in Ceylon somewhat noteworthy. It is known from the Nilghiris, but less so from the low country in the south of the peninsula. Mr. Fairbank records it from Ahmednagar, and Jerdon procured it as far south on the east coast as Nellore. He says that it is found throughout the Himalayas, migrating sparingly to the plains in the cold weather. "At Darjiling," he remarks further, "it is tolerably common, beginning its call still later in the season even than *Cuculus himalayanus*, this being rarely heard before the end of May, and continuing till the middle of July." Dr. Stoliczka procured it in Ladak, and to the eastward of the Himalayas it extends all the way to China and Japan, in the latter of which countries it is not uncommon. Swinhoe received specimens from Amoy and Szechuen and from North-west Formosa. In Java Mr. Wallace procured it, his specimens being, according to Blyth, similar to those from "the Himalayas and the Nilghiris," and, he adds, "from the mountains of Ceylon." It is not clear how he identifies it from the latter locality, for, according to my knowledge, it does not affect the hill-region at all. A specimen from Morty Islands, in the British Museum, is identical in plumage with other examples of this Cuckoo which I have examined, but is much longer in the wing, measuring 6·8 inches.

Habits.—The Small Cuckoo frequents low trees and stunted jungle near open places, and appears to be a tame bird, being stupidly heedless of observation, and allowing a near approach before taking wing. Jerdon remarks of it, " It is a very noisy bird, and has a loud, peculiar, unmusical call, which it frequently utters both when seated on a branch and when flying from tree to tree." . . . . . "The Bhootias," he adds, "attempt to imitate this in their name (*Pichu-giapu*) for the species."

It appears to feed much on caterpillars, one which I shot in my compound at the Colombo Lake being in the act of taking them from a plaintain-tree at the time.

Nidification.—Mr. R. Thompson says this species lays in May and June. An egg, which Mr. Hume believes to belong to this species, was taken by Mr. Brooks from the nest of a Warbler (*Reguloides superciliosus*), and is described as being an elongated, cylindrical ovate egg, and pure white and glossy; it measured 0·81 by 0·57 inch.

CUCULUS SONNERATI.

(SONNERAT'S CUCKOO.)

Cuculus sonnerati, Lath. Ind. Orn. i. no. 24, p. 215 (1790); Blyth, J. A. S. B. 1842, p. 906; id. Cat. B. Mus. A. S. B. p. 72 (1849); Kelaart, Prodromus, Cat. p. 129 (1852); Layard, Ann. & Mag. Nat. Hist. 1854, xiii. p. 452; Jerdon, B. of Ind. i. p. 325 (1862); Holdsworth, P. Z. S. 1872, p. 430; Legge, Ibis, 1874, p. 15, et 1875, p. 284.

Cuculus himalayanus, Jerdon, Cat. B. S. India, Madr. Journ. 1840, xi. p. 220.

Polyphasia sonnerati, Horsf. & Moore, Cat. B. Mus. E. I. Co. ii. p. 699 (1856).

Penthoceryx sonnerati, Cab. et Heine, Mus. Hein. iv. p. 16 (1862); Walden, Ibis, 1872, p. 367.

Le petit Coucou des Indes, Sonn. Voyage aux Indes, ii. p. 211 (1782); *Sonnerat's Cuckoo*, Lath. Syn. Suppl. p. 102; *The Banded Bay Cuckoo*, Jerdon; *Rufous Cuckoo* of some; "*Fine-weather Bird*," Planters in Ceylon.

Punchi koha, lit. "little Cuckoo," Sinhalese.

Adult male. Length 9·5 to 10·0 inches; wing 4·0 to 5·1; tail 4·8 to 4·0; tarsus 0·7; outer anterior toe 0·65, claw (straight) 0·25; bill to gape 1·1.

Iris brownish red, paling at the outer edge to slaty and in some to yellowish; bill blackish, gape fleshy yellow or reddish: base of lower mandible bluish, in some yellowish; inside of mouth orange-reddish; legs and feet brownish slaty, or bluish leaden in some, the soles yellowish, claws dusky blackish.

Forehead, top of the head, hind neck, upper surface, and wings hair- or nut-brown, with a green lustre, barred on the head, body, and wing-coverts with rufous-bay; feathers of the forehead with white bases, showing as spots on the surface; on the hind neck the bars almost monopolize the feather and are lighter; the upper tail-coverts have marginal spots or indentations of, and are tipped with, rufous; the quills and primary-coverts are unbarred, but are rufescent whitish inwardly, and are externally finely edged with rufous; tail deep brown, tipped white and edged or indented with rufous-bay, the inner webs of all but the centre feathers rufous with dark bars next the shaft, these latter have rufescent tips sometimes and at others want the light extremities altogether; entire under surface and feathers above the eye and down the side of the head between the nape and ear-coverts white, with narrow wavy blackish bars; ear-coverts darkish; the under tail-coverts and flanks, and in some specimens (probably young) the lower parts, tinted with fulvous; edge of wing white.

Female. Is, according to my experience, generally a smaller bird than the male. Length 9·5 inches; wing 4·5 to 4·0; bill to gape 1·0 to 1·05.

Iris hazel or reddish, with a yellowish outer circle; bill lighter than that of the male.

Has the upper-surface bars paler than in the male, and the under tail-coverts pure white or less coloured than the other sex.

Scarcely any two specimens of this Cuckoo are barred above precisely alike; with age the transverse marks seem to reduce themselves.

Young. Birds of the year are said to be more coarsely barred with paler bands than the adult, and to have the lower parts more tinged with fulvous.

Immature birds cannot be confounded with the rufous phase of *Polyphasia*, being, first of all, stouter or more massive, the bill much wider; and, secondly, they are more narrowly barred, and the under surface is all white, whereas in the latter the throat, chest, and generally the breast and abdomen are rufous.

Obs. This handsome little Cuckoo is closely allied to the Malayan species *C. pravatus*, Horsf., which inhabits Malacca and many of the islands of the Archipelago, including Sumatra, Java, and Borneo. This is a much smaller bird and more neatly barred, and wants the green gloss on the upper surface. Two individuals which I have examined in the national collection from Malacca measure in the wing 4·45 and 4·1 inches, and another from Sumatra 4·2. Lord Tweeddale gives the wings of three examples as follows :—Candeish, 4·88 inches ; Malabar, 4·75 ; Maunbhoom, 4·88. An individual from Tenasserim in the British Museum, which is scarcely separable from an immature bird from Ceylon, has the wing 4·6, appearing to be intermediate between the true *C. sonnerati* and *C. pravatus*.

Distribution.—The Bay Cuckoo is a resident in Ceylon, and scattered pretty freely over the island, but is nowhere very common, except in the Eastern Province. In this part it is frequent in many localities. I found it, particularly in the tank-district, affecting the open country between the Friars-Hood hills and the sea, and also cheenas in the vicinity of the tanks. In the north-eastern districts I have observed it chiefly in the north-east monsoon. In the South-west and in the Western Province it occurs in isolated places. I have either met with or procured it at Wackwelle near Galle, at Kaduwella near Colombo, in the Kuruwite Korale, at Ambepussa, and one or two other spots. Mr. Parker records it from Uswewa, and I have heard it in the North-central Province.

During the north-east monsoon it appears to ascend the hills, and is not uncommon in many parts of the Kandy country and also in Uva ; it is styled by the planters in some coffee-districts the " Fine-weather Bird," from its habit of calling before fine weather sets in.

Elsewhere this species is found almost only in the Southern and Central parts of India. Mr. Hume (' Stray Feathers,' 1875, p. 79) speaks of Captain Feilden procuring specimens of a Bay Cuckoo in Pegu which corresponded with Jerdon's description of *C. sonnerati*; and I have seen an individual from Tenasserim, as mentioned above, which could scarcely be separated from a Ceylonese specimen. Whether these will eventually prove to be the true *C. sonnerati* or not, I am unable now to say ; but if they should, it will much extend the range of the species.

Jerdon writes, " This elegantly marked little Cuckoo is found in the forests of Malabar and Travancore, where it appears tolerably common, also on the sides of the Nilghiris and in the Wynaad, and more rarely on the Eastern Ghâts, about the latitude of Madras." Of late neither Mr. Fairbank nor Mr. Bourdillon have procured it in the above-mentioned localities ; but the former records it from Khandalla, and Lord Tweeddale likewise from Maunbhoom, which is the most northerly locality from which I have heard of it.

Habits.—This bird frequents open places in the jungle, the edges of tanks where there are dead trees, sparsely-timbered country, and cheenas. It is very shy, and chiefly affects the tops of trees, where it remains motionless for a long time, piping its curious far-sounding whistle, which may be syllabized as *whī-whip, whiwhip—whī-whip, whiwhip.* It is particularly noisy in the morning before 9 or 10 o'clock, and in the evening just before and at sunset, calling for a considerable time without intermission, and consequently making its presence known wherever it has taken up its abode. When in forest it is difficult to find, being a small bird and generally seated across some horizontal branch near the top of the tree; but should there be an isolated tree standing in the open, near the edge of the forest-clearing or cheena, there the Banded Bay Cuckoo is sure to post itself, and then can easily be seen. In the Eastern Province I have come upon three or four in as many separate trees standing close together ; they do not seem to care about cultivating any close intimacy, though they are not unfrequently found in scattered company. Their call-notes are different from the whistle just mentioned ; commencing in a low key they suddenly change to a higher, and then die away into scarcely audible sounds. When approached they fly off to an adjacent tree, and commence calling anew. The diet of this species consists chiefly of Coleoptera, Mantidæ, and caterpillars.

Nothing seems to be known of the nidification of this species. Mr. Hume, it is true, mentions that an egg taken from the oviduct of one of the birds shot by Captain Feilden was bluish grey ; but it does not seem quite certain to what species these specimens belonged.

CUCULUS PASSERINUS.

(THE INDIAN PLAINTIVE CUCKOO.)

Cuculus passerinus, Vahl, Skriv. af Nat. Selsk. iv. p. 57 (1797).

Cuculus niger, Blyth, J. A. S. B. 1842, xi. p. 908.

*Polyphasia tenuirostris, Hodgs. Cat. B. Mus. E. I. Co. ii. p. 698 (1856, in part).

Cuculus tenuirostris, Kelaart, Prodromus, Cat. p. 129 (1852); Layard, Ann. & Mag. Nat. Hist. 1854, xiii. p. 453.

Polyphasia nigra, Jerdon, B. of Ind. i. p. 333 (1862).

Ololygon passerinus, Gray, Hand-list Birds, ii. p. 217 (1871); Hume, Nests and Eggs, p. 136 (1873); Ball, Str. Feath. 1874, p. 394; Butler, ibid. 1875, p. 461; Ball, ibid. 1876, p. 235.

Polyphasia passerina, Jerdon, Ibis, 1872, pl. i.; Holdsw. P. Z. S. 1872, p. 431; Legge, Ibis, 1875, p. 284.

Narrow-billed Cuckoo, Kelaart; *Pousiya*, Mahrattas; *Chinna katti pitta*, Telugu. *Koha*, Sinhalese.

Adult male and female. Length 8·75 to 9·2 inches; wing 4·4 to 4·5; tail 4·5; tarsus 0·65; outer anterior toe 0·65, claw (straight) 0·23; bill to gape 0·9.

Iris light red or yellowish red, in some red with a well-defined yellowish outer circle; bill blackish, often with a reddish tinge, the base of lower mandible slightly paler, inside of mouth orange-red; legs and feet (very variable) light reddish brown or greyish brown with a yellowish tinge, in some dusky reddish and in others yellowish; soles yellow, claws blackish.

Above dark ashy, blending on the sides of the neck into the uniform pale cinereous of the throat, chest, and breast; upper back, scapulars, and wing-coverts glossed with greenish; rump and upper tail-coverts more bluish than the head, the former edged with white at the base; quills plain brown; tail dark ashy blue, deeply tipped, barred on the inner webs, and edged outwardly towards the base with white; under tail-coverts, vent, and lower part of belly white, blending into the hue of the breast.

In some specimens there is more white on the abdomen than in others; and at times the under tail-coverts even are sullied with grey. Some individuals, otherwise in the normal adult plumage, have the tail scarcely tipped, and the inner edge of the feathers only slightly indented with white; and occasionally the tail is devoid of white markings.

Young. Birds of the year vary considerably in their coloration; but their prevailing character is to be marked with rufous, and nearly always on the tail (this has the mesial spots and marginal bars rufous), and the chin and throat with more or less of the same colour, while the under surface is marked with whitish or fulvescent bars. In some the caudal bars are white near the tips of the feathers and rufous at the base.

Hepatic phase. This species assumes frequently a rufous plumage analogous to that in which the common Cuckoo is often found. An example shot in March at Colombo has the upper surface, wings, tail, sides of the neck, and throat bright rufous; the feathers of the head and hind neck with a few terminal bars of blackish; the back, scapulars, and wing-coverts barred with greenish black; the terminal portion of the quills and the entire outer webs of the first primaries dull brown; shafts of the tail-feathers and a subterminal spot black; tips of all but

---

* The trifling differences in the bill and plumage of the members of this genus are not, I consider, sufficient to separate it from *Cuculus* as restricted. The structure of wing and tail is nearly similar in both. The same may be said of the next species, which differs from *Cuculus* mainly in the metallic lustre of the plumage. The tail in the bronze Cuckoos is, as a rule, less graduated than in *Cuculus*; but it is variable, scarcely any two species being exactly alike.

the central pair whitish ; under surface, under tail-, and under wing-coverts white, blending on the chest into the rufous of the throat and barred with wavy bands of blackish brown, which on the under tail-coverts are far apart ; some of the tibial feathers rufous.
In other examples the breast and flanks, as well as the chest, are rufous, these birds being probably in a younger stage than those which have a considerable amount of white on the lower parts.
These rufous individuals, I imagine, remain so throughout life, perhaps losing the bars on the upper surface entirely, while the quills would remain more or less brown.

Obs. The allied species, *P. tenuirostris* of Gray, which replaces the present bird in Burmah and the countries to the east of India generally, is very similar to it on the upper surface and throat, but has the breast, belly, and under tail-coverts rufous, darkest on the latter, which is consequently the very opposite character to that displayed by the Plaintive Cuckoo. It has a rufous phase ; but this differs slightly from that of the present species, the lower parts being banded more boldly, and the tail wanting the white tips.

Distribution.—The observations taken by various naturalists in Ceylon on the movements of this little Cuckoo tend to show that it does not make its appearance in all parts of the north of the island at the same time, the truth, doubtless, being that it arrives in one district and then wanders thence over the country, its distribution being materially influenced by climate. In the north-east, about Trincomalie, I have known it appear in the beginning of October, at which time it has scarcely done breeding in the south of India ; it was common enough in suitable places in the interior long before Christmas. Layard, however, remarks that it appeared about Jaffna in February, a time when it should have been assembling for its return northward, and under which conditions he most probably saw it there. Mr. Holdsworth's experience is again scarcely less noteworthy ; he did not notice it in the Aripu district before the beginning of January, from which it would appear that it visits the west coast considerably after its arrival on the other side of the island. It would not, however, be safe to assign to it any general period of arrival on the evidence of one or two seasons, as no doubt its appearance, as is the case with most migratory birds, varies considerably according to the kind of season and prevailing weather at the time at which it should be expected.
In the Galle district I have met with it in December, and in the Western Province have seen it about the same date. In these latter districts it does not occur in any great numbers, being a lover of dry climate. In the Hambantota country and all round the south-east coast it is very numerous. From the north down to Chilaw it is common, and in the Seven Korales and along the base of the Matale hills towards Kurunegala I have found it abundant in March. It has, I believe, been found in Dumbara, but I am not aware of its visiting any higher parts of the Kandyan country than that : in Uva it probably occurs at a greater elevation.
Concerning its distribution in India, Jerdon writes, " The Plaintive Cuckoo is found all over India in wooded countries. It is most abundant on the Malabar coast, in the Wynaad, and on the warmer slopes on the top of the Nilgherries, save in the Carnatic, but found here and there in jungly places and on the Eastern Ghauts ; rare in Lower Bengal, and up to the foot of the North-west Himalayas." Its distribution seems to be rather peculiar in some districts ; Mr. Ball says that it occurs rather sparingly in Chota Nagpur, that Captain Beavan procured a specimen in April in Maunbhoom, and that he himself got another in Sirguja in the same month. In the coast-region to the eastward he found it not uncommon in Orissa, but did not see or hear of it in travelling southward till he reached the western part of Raipur on the road to Nagpur.
Captain Butler writes, " The Indian Plaintive Cuckoo is not uncommon at Mount Aboo ; it arrives about the beginning of June, and its mournful ventriloquistic note soon makes one aware of its presence." Mr. Hume follows with the observation that it is found nowhere else throughout the whole region round about Aboo. From these remarks it appears that this species moves about in India to a considerable extent, migrating in the northern parts to the westward during the breeding-season.

Habits.—The Plaintive Cuckoo certainly does not lay much claim to such a title in Ceylon, for there it is one of the most silent of birds, which fact leads one to the inference that its notes are chiefly uttered in the breeding-season. It frequents open scrubby lands, plains dotted with jungle, bushy wastes, and such like ; when disturbed it flies from one low shrub to another, and perches generally upon the topmost branches. It is seen moving about a good deal in the early morning, and in the evening, in districts where it is numerous, assembles in small parties and roosts in thick bushy trees. I found it in considerable numbers once on the Kimbulama-oya, an affluent of the Dodura-oya, flying in and out of the trees growing on the banks; the birds were very wary, and it was with difficulty that I could get within shot of them. It is usually not very prone to allow of a near approach, being of a restless disposition; but when met with alone is not nearly so shy as when associating together. It feeds on caterpillars, Coleoptera, and other large insects, and may often be seen taking them on the ground; its stomach is villous in a high degree.

Concerning its note, which is so well known in India, Jerdon writes that it is "a plaintive call of two syllables, the last one lengthened out, which Mr. Elliott made *whi*, *whew—whi whew whew*, and which may be written as *ku-reer*, *ka-vee-eer*, and to which the bird, by pointing his head in different directions as he sits calling, gives a most ventriloquistic effect." I would remark, it is by a similar means that the Hawk-Cuckoo imparts such a singular sound to its call.

Nidification.—This little Cuckoo lays its eggs in the nests of Wren-Warblers, the Yellow-eyed Babbler (*Pyctorhis sinensis*), and also in that of the Grey-backed Shrike (*Lanius erythronotus*). Miss Cockburn, a lady who has done much towards furthering our information on the oology of the South of India, is, according to Mr. Hume, the only person who has identified its eggs, having found them in the nest of the Common Wren-Warbler (*Drymoipus inornatus*) on the slopes of the Nilghiris. I subjoin the following note from her, which Mr. Hume gives in 'Nests and Eggs of Indian Birds':—"On the 17th of September, 1870, the nest of a Common Wren-Warbler, which had two small eggs, and a third, which was much larger, but of something the same colour. A few hours after another Common Wren-Warbler's nest was found, which also contained two small eggs, one of which was broken, and a large egg. These two nests were not far from each other; I took them both. On the 22nd September another nest of the same Warbler was found, which also contained a large egg and two small ones.

"The same day one of my servants, seeing a Plaintive Cuckoo sit very quietly on a hedge, shot it. On examination it was found to contain an egg ready to be laid, of the same colour and spots as those found in the little Warblers' nests. On the 26th September, a Common Wren-Warbler's nest was found, which had only a Cuckoo's egg in it. The Cuckoo was seen near the nest, and the little Warblers in a great fright; for the appearance and flight of the Cuckoo very much resembles that of a small hawk. On looking in the nest there was the egg. It was left for two or three days; but on going to the spot the nest was found to be deserted, so the Cuckoo's egg was brought away.

"On the 5th October, 1870, another Common Wren-Warbler's nest was found; but this time it was occupied by a young Plaintive Cuckoo, which entirely filled the wee nest, and had the boldness to peck at my finger every time I tried to touch it. The nest had no young Wren-Warblers. Whether the young Cuckoo had pushed the little Warblers out, or whether no other egg, except the Cuckoo's, was hatched, it is impossible to say. I regret not having seen the nest till at this stage of the young Cuckoo's existence. A week after it had left the nest, but was caught among the bushes close by. Considering the smallness of a Common Wren-Warbler's nest and one of the Warbler's eggs having been found broken in one of the nests, as mentioned above, there can, I think, be little doubt but that this bird, like its European namesake, must carry her egg in her mouth and drop it into the nest."

The eggs thus found were of "a delicate pale greenish blue, blotched and spotted boldly but sparsely, and almost exclusively towards the large end of the egg, with reddish or purplish brown and pale reddish purple. The markings seem generally to form a very imperfect and irregular, but still more or less conspicuous, zone round the large end."

In size they varied from 0·78 to 0·81 inch in length, and from 0·53 to 0·57 inch in breadth.

CUCULUS MACULATUS.

(THE INDIAN EMERALD CUCKOO.)

Trogon maculatus, Gm. Syst. Nat. i. p. 404 (juv.) (1788).

Chrysococcyx lucidus, Blyth, J. A. S. B. 1842, xi. p. 917; Jerdon, 2nd Suppl. Cat. B. S. India, Madr. Journ. 1845, xiii. no. 225.

Chrysococcyx smaragdinus, Blyth, J. A. S. B. 1846, xv. p. 53.

Cuculus (Chrysococcyx) xanthorhynchos, Layard et Kelaart, Prodromus, Cat. Suppl. p. 60 (1853).

Chrysococcyx hodgsoni, Horsf. & Moore, Cat. B. Mus. E. I. Co. ii. p. 705 (1856); Jerdon, B. of Ind. i. p. 338 (1862); Swinhoe, P. Z. S. 1871, p. 394.

Lamprococcyx smaragdinus, Cab. et Heine, Mus. Hein. iv. p. 13, note, no. 6 (1862).

Lamprococcyx maculatus, Holdsworth, P. Z. S. 1872, p. 432.

Chalcites hodgsoni, Gould, B. of Asia, pl. xxi. (1877).

Le Curucui tacheté, Brown, Ill. Ind. Zool. pl. 13.

Angpha, Lepchas (Jerdon).

Adult male and female. Length 6·0 to 6·4 inches; wing 4·0 to 4·3; tail 2·7 to 2·9; tarsus 0·5; outer anterior toe (without claw) 0·62; bill to gape 0·75.

Iris brown or reddish brown; bill yellow at the base, with the terminal portion brown; legs and feet reddish brown.

Above brilliant emerald-green, with more or less of a coppery tinge, most prevalent at the margins of the feathers; the back and scapulars with a golden lustre when viewed in some lights: quills metallic brown-green, the inner webs of the primaries rufous at the centre and white at the base; tail much tinged with coppery, the outermost feathers barred with white and the interspaces blackish green; throat metallic green; under surface white, crossed with bold bands of bronzed green.

Young. The immature bird has the back, wings, and rump metallic green, more or less overshot with a coppery gloss, and the feathers barred terminally with rufous and dusky green: the head and hind neck rufous, with a strong coppery lustre, the feathers barred with blackish brown, and sometimes with whitish as well; tail green, the feathers rufous externally, the outermost feathers mostly white, barred with black-green or blackish, the next two pairs barred with blackish and tipped with white; barring of the under surface duller than in adults.

Immature examples vary much in the extent and character of their rufous coloration. An individual in my collection has the back, scapulars, and wings brilliant emerald-green as in the adult, with the head and hind neck rufous, strongly illumined with coppery; the crown and nape barred with whitish and brown, the former across the centres of the feathers; the outermost feathers are white externally and rufous internally, barred with greenish black, the penultimate almost entirely rufous, with green cross bands on the inner webs and a broad subterminal bar of the same, the extreme tip being white; the next pair are rufous on the outer webs, barred with green.

Obs. This species is not very aptly named *maculatus*. It was figured by Brown from a young bird with spotted wing-coverts sent from Ceylon by Governor Loten; he named it the "Spotted Curucui," from which Gmelin gave it its title of *maculatus*, looking upon the bird, however, as a Trogon.

C. lucidus, with which this Cuckoo has been occasionally confounded, is found over most of the Australian continent, and differs from the Indian bird in being of a paler, more coppery, and less lustrous green on the upper surface, and the whole of the under parts are barred with metallic greenish copper-colour. The young bird is brownish above, the green colour being confined to the back and tail; the throat and chest tinged and barred with pale brownish. The wing in this species varies from 4·1 to 4·3 inches.

Distribution.—The fact alone of Brown recording his specimen of the "Spotted Curucui," figured in his 'Illustrations of Indian Zoology,' as having been sent from Ceylon by Governor Loten, entitles this species to a place in our lists. It has not, to the best of my knowledge, since been met with or heard of even in the island; Layard knew nothing of it, and I conclude entered it in the catalogue of Ceylon birds by himself and Kelaart, published in the Appendix to the 'Prodromus,' solely on the authority of Brown.

It has not as yet been detected in Southern India, and Jerdon says it has been rarely procured even in the central part of the Peninsula; its habitat is essentially the sub-Himalayan region, and (according to Blyth) Arakan and Tenasserim. Jerdon obtained it at Darjiling, and Hodgson procured it in Nepal. Its occurrence in Ceylon can only be accounted for on the supposition of its having migrated southwards in the usual manner, following thus the example of all the true Cuckoos which visit Ceylon.

Habits.—But little is known of the habits of this lovely little bird; but they may, I have no doubt, be considered to resemble those of other members of this beautiful group. Gould writes of *C. lucidus* that "while searching for food its motions, although very active, are characterized by a remarkable degree of quietude, the bird hopping about from branch to branch in the gentlest possible manner, picking an insect here and there, and prying for others among the leaves and the corners of the bark with the most scrutinizing care." The same interesting manners are doubtless possessed by the present species. Jerdon states that the food of the one he shot at Darjiling consisted of insects.

Genus HIEROCOCCYX.

Bill wide at the gape. Wings shorter than in *Cuculus*; the 1st quill short and the 3rd longer than the 2nd. Tail subeven.

Plumage Hawk-like in character, the young being striped beneath.

HIEROCOCCYX VARIUS.

(THE COMMON HAWK-CUCKOO.)

Cuculus varius, Vahl, Skriv. af Natur. Selsk. iv. p. 60 (1797); Strickland, Ann. & Mag. Nat. Hist. 1846, p. 398; Blyth, Cat. B. Mus. A. S. B. no. 339, p. 70; Layard et Kelaart, Cat. Ceylon B. App. Prodromus, p. 60 (1853); Layard, Ann. & Mag. Nat. Hist. 1854, xiii. p. 462.

Cuculus fugax, Horsf. Tr. Linn. Soc. xiii. p. 178 (1821); Jerdon, Cat. B. S. India, Madr. Journ. 1840, xi. p. 219.

Cuculus lathami, J. E. Gray, Ill. Ind. Zool. p. 34, fig. 2 (1832).

Hierococcyx varius, Horsfield & Moore, Cat. B. Mus. E. I. Co. ii. p. 700 (1856); Jerdon, B. of Ind. i. p. 329; Holdsw. P. Z. S. 1872, p. 431; Ball, Str. Feath. 1874, p. 393; Bligh, J. A. S. (Ceylon Branch) 1874, p. 67; Bourdillon, ibid. 1876, p. 392; Ball, ibid. 1877, p. 413.

Bychan Cuckoo; *Sokagu Cuckoo*, Latham, Hist. of Birds.

Kupak or *Upak*, Hind.; *Kokgallo*, Bengalese; *Kuttipitta*, Tel.; *Takkhat*, lit. "Custom-house Bird," in Deccan; *Irolan*, Malabar (*apud* Jerdon).

Adult male and female. Length 13·0 to 14·7 inches; wing 7·4 to 8·2 (*Hume*); tail 6·5 to 6·8; tarsus 0·9 to 1·0; outer anterior toe and claw 1·2 to 1·3; bill to gape 1·15 to 1·3.

Females are smaller than males. The above limit of the wing is that of a male, and must be exceptional. Several specimens I have examined from Ceylon and N. W. India vary from 7·5 to 7·8 inches, which I imagine is about the average limit.

Iris yellow; bill, upper mandible and tip of lower brown, base of under mandible and gape yellow; orbits bright yellow; feet gamboge-yellow, claws dusky at the tips.

Above dark ashen grey, darkest on the interscapular region and palest on the rump and upper tail-coverts; basal margins of the feathers on the hind neck more or less rufous, showing on the surface of the plumage; quills and winglet grey-brown; inner webs of primaries partly crossed from the edge with wide bars of white, more or less mottled with grey; extreme tips of the secondaries pale; tail brownish ashen, tipped with rufous and crossed with a broad subterminal band of blackish brown, above which are four narrow bars of the same, with an adjacent pale cross ray at the lower edge, which expands and is more conspicuous on the outer feathers; under surface of the light portions whitish.

Lores, cheeks, and ear-coverts bluish ashen; chin ashen, the extreme point darkest; throat and chest rufous, the centres of the feathers bluish grey in some, with the basal edges whitish, in others the whole basal portion of the feather is bluish grey; lower part of chest, breast, and flanks barred with the same on the rufous ground, which pales gradually into unmarked buff-white on the belly, vent, and under tail-coverts; under wing-coverts pale rufous or fulvescent, the greater series barred with bluish ashen; under surface of quill-bars buff-white.

When not fully adult the markings of the under surface are darker and the rufous is confined to the chest. A specimen shot by Mr. Bligh in Kotmalie has the lores whitish; the chin and cheeks dark slate, with the centre of the throat white; the chest is washed with rufous, this colour is barred with slate, which gradually darkens on the breast into brown on a white ground-colour, the bands being at the same time edged with rufous ; thigh-coverts, vent, and under tail-coverts pure white; under wing-coverts buff, cross-rayed with brown ; under surface of the quill-bars pure white.

Young. Above dark cinereous brown, barred on the lower part of the hind neck, back, scapulars, and wing-coverts with rufous ; on the hind neck the bases and margins of the feathers are of this colour : primaries and secondaries barred exteriorly with rufous, internally with buff, shading into rufous near the shaft ; tail-feathers tipped with rufous and white, the subterminal bar very broad, and the remaining four more developed than in the adult, the pale succeeding cross rays being rufous and the interspaces ashen ; on the three lateral pairs of feathers the cross rays are whitish on the inner webs ; forehead and crown ashen brown, scarcely marked with rufous : beneath buffy white, the throat and fore neck marked with broad mesial strix of cinereous brown : the feathers on the sides of the neck edged with rufous ; breast with angular transverse spots of the same, which become more bar-like on the flanks ; belly and under tail-coverts unmarked ; under wing-coverts rufescent, barred with brown.

Obs. The closely allied species, *H. nisicolor* of Hodgson, from the Himalayas, may, Mr. Hume writes, be distinguished from the present by the young not having any barring on the flanks or abdomen, and also by its darker upper surface at all stages. It is, however, not likely ever to occur in Ceylon, as the larger form, *H. sparverioides*, common in South India, and which even migrates to China, has not yet been detected in the island.

Distribution.—This noisy Cuckoo arrives on the shores of Ceylon about the beginning of November, and makes its way at once to the hills, taking up its abode in considerable numbers in the forests of the main range. It is common about Nuwara Elliya, Kandapolla, and the " plains " lying between the Sanatorium and Totapella. On the Horton Plains themselves it is no less numerous, frequenting the picturesque woods which dot this beautiful and lonely spot. Layard was the first to record it from Ceylon, and writes that he shot three specimens in the old Botanical Gardens at Kew, Colombo ; these were evidently new arrivals. Mr. Holdsworth met with it, as I did, at Newara Elliya, at the beginning of the year, and Mr. Bligh procured it in Kotmalie in the month of November. He writes me that it is not uncommon in the Haputale range, and that it was yearly to be found on the Harangolla patnas in considerable numbers, making itself heard by night as well as by day. Messrs. Whyte and Co. have lately sent home a specimen killed in the Kandy district, and I have no doubt it takes up its seasonal quarters on the slopes of the Knuckles range.

On the mainland the Hawk-Cuckoo is, says Jerdon, the common species " of the plains of India, being found throughout the whole country, though most abundant in wooded districts." Mr. Bourdillon writes of it that it is abundant in the semicultivated land of the plains of Travancore, penetrating the jungles at the foot of the hills to 1000 feet elevation, but that it does not ascend the hill-slopes to any height, though common in the low country. Mr. Fairbank's experience of it in the Palani hills is similar ; he found it at the base and on the sides of the range only ; it is singular, therefore, that it should resort to the very highest points in Ceylon. In Khandala it is common ; and concerning its distribution in Chota Nagpur, Mr. Ball says that it is found in the jungly parts of the Province and that it inhabits the Rajmahal hills. In Jaipur and the south of Raipur, he remarks that it occurs in such abundance that its cry is a " positive nuisance and source of irritation both by day and night !" More towards the north-west of India it appears to be only a seasonal visitant ; for Capt. Butler, in his very complete list of animal migrations to the Mount-Aboo district (' Stray Feathers,' 1877), records it as only remaining during the rainy monsoon—to wit, from June until October.

Habits.—Unlike many of the Cuckoos, which are silent in the non-breeding season, the present species is extremely noisy at all times ; it frequents the high jungle in the upper ranges of the Ceylon hills, and is partial to the vicinity of the open grassy spaces called " plains " on the Nuwara-Elliya plateau. Its singular scale-like call, which is uttered while the bird twists its head round, is very characteristic of this region. In January it may be heard the whole morning in the picturesque woods on the Horton Plains, literally throwing its peculiar high-pitched notes in all directions : at one moment they seem to be in the distance ; at the next, when it turns its head towards the listener, they swell with strange force on the ear, mounting higher and higher until the bird appears to be obliged to stop.

Jerdon writes of it as follows :—" It frequents gardens, avenues, groves, and jungles, and its loud crescendo

2 1

notes are to be heard in the breeding-season, from April till July in the south of India (but beginning earlier in Bengal, according to Blyth), in every garden or avenue. It sounds something like *pibuba*, *pibuba*, repeated several times, each time in a higher note than the last, till they become exceedingly loud and shrill. Mr. Elliott makes it *whi-wheeba*; Sundevall calls it *piripiu*. This author further remarks that each word is pronounced about twice, nearly in this manner in the musical scale, C B B A—A C C B—B D D C; and it thus mounts the scale of notes at every second cry, three or four times, till the note is as high as the bird can raise it, when it makes a short pause and begins anew. . . . . . . It lives both on caterpillars and other soft insects and on fruits, and it is very fond of the fig of the banyan and other *Fici.*" It is said by the natives in India to be good eating; but Mr. Fairbank says that he tried it, and found the flesh intolerably strong-flavoured, which is not to be wondered at, as, according to his investigations, it feeds on lizards and insects. Its flight is strong and swift, and it has been noticed to have the habit of darting suddenly into bushes, to the manifest alarm of small birds, who sometimes mistake it for the Shikra and pursue it accordingly. Mr. Bligh informs me that it calls at night; he found it frequenting the skirts of the jungle bordering the grassy wastes on the Harangolla patnas.

Nidification.—The eggs of this species have not yet been identified, as far as I have been able to ascertain. It is believed to deposit them in the nests of the *Malacocerci*, or Babblers. Jerdon saw these birds feeding a young one, which was following them about screaming; he writes that, "on one occasion, at least, there were two or three young *Malacocerci* in company; so that the young of this species of Cuckoo does not always eject the young of its foster-parent from the nest."

Bill much as in *Cuculus*, the nostrils very protuberant and situated near the margin. Wings moderate, with the 3rd and 4th quills subequal and longest. *Tail forked*, with outer feathers short, and the penultimate the longest and forming the fork. Tibial plumes very long. Tarsus partly feathered down the exterior side.

SURNICULUS LUGUBRIS.

(THE DRONGO-CUCKOO.)

Cuculus lugubris, Horsf. Trans. Linn. Soc. 1820, xiii. p. 179 (Java).

Cuculus albopunctatus, Drap. Dict. Class. d'Hist. Nat. iv. p. 570 (1823), juv.

Psendornis lugubris, Hodgs. J. A. S. B. 1839, p 137.

Pseudornis dicruroides, Hodgs. J. A. S. B. 1859, p. 136 (Mountains of Nipaul).

Cuculus dicruroides, Jerd. Cat. B. S. India, Madr. Journ. 1840, xi. p. 221; Layard et Kelaart, Cat. Prodromus, App. p. 60 (1853); Layard, Ann. & Mag. Nat. Hist. 1854, xiii. p. 453.

Surniculus dicruroides, Blyth, Cat. B. Mus. A. S. B. p. 72 (1849); Horsfield & Moore, Cat. B. Mus. E. I. Co. ii. p. 695 (1856); Jerdon, B. of Ind. i. p. 336 (1862); Swinhoe, Cat. B. of China, P. Z. S. 1871, p. 394; Holdsworth, P. Z. S. 1872, p. 431; David & Oust. Ois. de la Chine, p. 61 (1877).

Cacangelus lugubris, Cab. et Heine, Mus. Hein. iv. p. 17 (1862).

Surniculus lugubris, Walden, Ibis, 1872, p. 368.

The Fork-tailed Cuckoo, Europeans in Ceylon; *The Black Fork-tailed Cuckoo*, Jerdon *The Fork-tailed Drongo-Cuckoo*, Blyth.

Kurrioviyum, Lepchas (Jerdon); *Awon-Awon*, Java.

Adult male and female. Length 10·0 to 10·3 inches; wing 4·8 to 5·3; tail 5·4 to 5·7 (to tip of penultimate). middle feathers about 1·5 shorter; tarsus 0·55 to 0·65; anterior toe 0·6, claw (straight) 0·25; bill to gape 0·9 to 0·95.

Iris brown; bill black; gape and inside of mouth orange-red; legs and feet blackish or deep reddish black, the edges of the tarsal scales whitish; claws black.

Plumage above and beneath black, with a blue and a green gloss or sheen, brilliant above and subdued on the lower surface; the head and tail have the blue lustre the strongest, and the back and wings green (in some specimens there are one or two white feathers on the occiput): the lateral tail-feathers are tipped and crossed with slanting bars of white, the penultimate has a series of white spots adjacent to the shaft, and all the rectrices a fine whitish edge at the base; the under tail-coverts, which are glossed more highly than the breast, are tipped and banded with white, and there is a conspicuous white tuft on the outer thigh-coverts.

Young. Iris red-brown; legs paler than the adult. In the first plumage the upper and lower surface have white tips to the feathers; the wing-coverts and rectrices are similarly tipped, and some of the underlying upper tail-coverts are barred as well; the head, back, and wings are less glossed than the adult, and the under surface is brownish black; the tail is more barred, the penultimate being thus marked instead of spotted, and the next feather has a series of median white marks. In this stage the tail is rounded, the penultimate being shorter than the adjacent inlying feather.

With age the spots disappear from portions of the upper surface, remaining longest on the upper tail-coverts, and some birds, not quite mature, have fine white tips to the wing- and upper tail-coverts, and a greyish-white edging to the under-surface feathers.

Obs. The Indian species, *S. dicruroides* of Hodgson (which was described from a specimen from " the mountains " in Nepal), has generally been kept distinct from the much earlier described and reputedly smaller Javan species, *S. lugubris.* I notice, however, that so high an authority as Lord Tweeddale remarks (*loc. cit.*) that " Himalayan, Ceylon, Malaccan, and Javan individuals do not differ," and are all the same as an example from Borneo, which is the subject of his notes. It appears to be a very variable species as regards size. The wings of two adults from Java, as given in the note in question, measure 5·75 and 4·82 inches, one from Nepal 5·37, one from Darjiling 5·75. I have examined a good series of Ceylonese examples and have found none to exceed the limit given above (5·3), and the usual dimension is from 5·0 to 5·1 in fully adult black birds.

Distribution.—This singular Cuckoo is rather locally dispersed in Ceylon, being common in one district and absent in another adjacent tract of country. As regards the Western Province, it is occasionally found not far from Colombo, and is very common in the Three Koralès and country intermediate between that and Ratnapura, and it extends into the hills, above the latter place, to a moderate elevation, occurring at Gillymally. In the south-west it is less frequent; in the Kurunegala and Puttalam district it is fairly represented, and it occurs here and there throughout the northern forest-tract at all times of the year, from the latter place across to Trincomalie, where it is not uncommon in the forests. In the Eastern Province I saw many examples, but did not meet with it in the Kattregam and Hambantota districts. In Madulsima and Uva I have seen it up to 4000 feet elevation, and procured it once near the Debedde gap ; in the Kandy country it is found towards the Hangerankette side and in Dumbara valley. Layard mentions, in his notes, that Mr. Thwaites sent him numerous specimens from the neighbourhood of Kandy ; it is probably more plentiful there some seasons than others.

In India it is sparingly distributed throughout the country. Jerdon writes, " I have procured it on the Malabar coast, the Wynaad, in Central India, and at Darjiling. I have found it in other parts of the Himalayas, ... and in Tenasserim and Burmah." Mr. Hume records it as rare in Tenasserim. It has been procured in different parts of the Malaccan peninsula and in Sumatra at Lampong, and, as above noticed, was first described from Java, where, according to Horsfield, it is found " in districts of secondary elevation, which are diversified with extended ranges of hills and covered with luxuriant forests." To the east of that island it has been found in Labuan and Borneo ; and Mr. Swinhoe remarks that it was procured by him in Szechuen, China, in the month of May. In India, judging by the experience of collectors recently, it is less common than in Ceylon.

Habits.—The Fork-tailed Cuckoo frequents a variety of situations, inhabiting the interior of dry forests throughout the north, scrub and low jungle in other places, grassy patnas dotted with isolated trees, and last, but not least, burnt clearings and vegetable plantations in the woods of the interior. In the latter it is chiefly observed in Saffragam and at the base of the western ranges, delighting on perching on the charred stumps and saplings which remain after the first firing of a cheena. It is exceedingly docile in its disposition, sometimes alighting on a fence by the side of a jungle-path and flying tamely on in front of the traveller, and at others sitting on a stump until approached within a few yards. At a distance, its tame habit will always serve, in conjunction with its small-looking head and bill, to distinguish it from a Drongo, to which it bears an otherwise absolute resemblance. Its remarkably human-like whistle, which consists of six ascending notes (sounding as if some one were practising a musical scale in the wilds of the jungle), is, I think, uttered chiefly in the breeding-season. I have heard it always in the north during the north-east monsoon ; at other times, in July and August, in the Western Province, it is quite mute. Its diet is mixed, consisting chiefly of caterpillars and beetles, but often combined with various seeds.

When on the wing it is very different from a Drongo, flying along with a steady movement, and not dipping in its progress through the air.

Nidification.—Judging from my examination of various specimens shot in the north, the breeding-season of this species appears to be in the early part of the year ; it is most noisy then. I have no information as to its eggs, or the bird in whose nest they are deposited. Jerdon suggests that it may possibly lay in those of King-Crows, to which it bears such a wonderful resemblance. He writes, "One day, in Upper Burmah, I saw a King-Crow pursuing what at first I believed to be another of his own species; but a peculiar call that the pursued bird was uttering, and some white on his plumage, led me to suppose that it was a Drongo-Cuckoo, which had perhaps been detected about the nest of the *Dicrurus*. Mr. Blyth relates that he obtained a pure white egg in the same nest with four eggs of *D. macrocercus*, and which, he remarks, may have been that of the Drongo-Cuckoo." It is extremely probable, I think, that it was.

Genus COCCYSTES.

Head crested. Bill more curved and compressed than in the preceding genera. Nostrils ovate, basal, exposed, and placed near the margin. Wings rather short, rounded, the 4th quill the longest. Tail long, much graduated. Tarsus longer than in *Cuculus*, exceeding the inner anterior toe; the upper portion feathered, the rest covered with broad transverse scales; outer posterior toe considerably longer than the inner one.

COCCYSTES JACOBINUS.

(THE PIED CRESTED CUCKOO.)

Cuculus jacobinus, Bodd. Tabl. Pl. Enl. 872 (1788).

Cuculus melanoleucos, Gm. Syst. Nat. i. no. 35, p. 416 (1788); Lath. Ind. Orn. i. p. 211 (1790).

Oxylophus edolius, Jerd. (*nec* Cuv.) Cat. B. S. India, Madr. Journ. 1840, xi. p. 222.

Oxylophus melanoleucos, Blyth, Cat. B. Mus. A. S. B. p. 74 (1849); Layard, Ann. & Mag. Nat. Hist. 1854, xiii. p. 451.

Oxylophus serratus, Kelaart, Prodromus, Cat. p. 128 (1852).

Coccystes melanoleucos, Horsf. & Moore, Cat. B. Mus. E. I. Co. p. 694 (1856); Jerdou, B. of Ind. i. p. 339.

Coccystes jacobinus, Cab. et Heine, Mus. Hein. iv. p. 45 (1862); Holdsworth, P. Z. S. 1872, p. 432; Hume, Nests and Eggs (Rough Draft), p. 137 (1873); Sharpe, P. Z. S. 1873, p. 597; Legge, Str. Feath. 1874, p. 366; Ball, ibid. p. 394; Butler, ibid. 1875, p. 461; Legge, Ibis, 1875, p. 284; Morgan, ibid. 1875, p. 315; Hume, Str. Feath. 1876, p. 457.

Jacobin huppé de Coromandel, Daubent. Pl. Enl. pl. 872; *The Pied Cuckoo*, in India; *Papiya*, Hind., also *Chatak*; *Kola Bulbul*, Bengal.; *Gola Lokila*, lit. "Milkman Cuckoo," also *Tangada gorankah*, Telugu (Jerdon).

Konde koha, lit. "Crested Cuckoo," Sinhalese.

Adult male and female. Length 12·0 to 13·0 inches; wing 6·4 to 6·7; tail 5·4 to 5·5; tarsus 1·0: outer anterior toe 0·8, its claw (straight) 0·3: bill to gape 1·1.

Iris dark brown; bill black: legs and feet bluish slate, edges of scales whitish, claws blackish.

Head, cheeks, upper surface, tail, and wings glossy green-black, the crown-feathers lanceolate and rather stiff, forming a thin crest one inch in length; quills dull black; basal half of primaries, with the exception of that part of the outer web of the 1st and inner web of the last, white: central rectrices tipped white, and the terminal ⅓ inch of the rest the same hue; entire under surface and under wing-coverts sullied white, which passes up behind the ear-coverts on to the sides of the neck: greater lower primary-coverts blackish.

Young. Birds of the year have the bill pale at the base: the legs and feet paler than the adults. The upper surface is sepia-brown, with the nape, ear-coverts, and sides of neck blackish: the forehead paler than the head, and the lesser wing-coverts are edged with greyish; beneath fulvous-grey or buff-white, with the sides of the throat brownish from the chin to below the ear-coverts.

Obs. An individual from the hills in the north-west of India measures—wing 5·7 inches, tail 6·8, bill to gape 1·3 another from Pegu, wing 5·9; both are identical with Ceylonese examples. Mr. Sharpe unites the African species with the Indian. A specimen from Damara Land, described by him *loc. cit.*, had the wing 6·4 and the tail 8·0 inches.

Distribution.—The Pied Cuckoo, which is a showy species, is widely distributed over the low country of Ceylon, but is subject to a partial migration away from the wet regions on the western and south-western sea-board during the prevalence of the S.W. monsoon. It appears about Colombo in November and December, and, when first arrived, lurks in any thick cover that may be to hand. I have seen it in the trees on the borders of the Slave-Island lake, but it soon disappeared for the jungles of the interior. In the Galle district it arrives about the same time and frequents the low jungle in the cultivated portions of the country. In the scrubby jungles of the Girawa and Magam Pattus and throughout the Eastern Province, in the jungles between the Mahawelliganga and the coast, in the maritime portions of the north and west, as far south as Chilaw it is a resident species, and in some of these districts is abundant. It is partial to those dry districts which are covered with low scrub, such as the neighbourhood of Hambantota and many similar spots on the east coast, the Jaffna peninsula, the N.W. coast, and the island of Manaar, as also the Puttalam and Chilaw district. I have seen it occasionally in the interior of the northern division of the island, but it is scarcer there than in the maritime portion. It ranges into the Central Province to a considerable elevation, occurring in Uva up to 3000 feet; but in the western portions (to wit, the valley of Dumbara and adjacent districts) it is not found at such an altitude.

This Cuckoo enjoys a wide range on the main land. Jerdon sketches out its distribution as follows:—
"It is found all over India, being rare on the Malabar coast, common in the Carnatic, and not uncommon throughout Central India to Bengal, where it is only at all common in the rains. It is more abundant in Upper Pegu than anywhere else that I have observed it . . . . I have seen it on the Nilghiris up to 5000 feet."

It does not appear to be found on the hills of the peninsula, but is common in the low country, on the Madura coast, and in Ramisserum Island. In Chota Nagpur it occurs rarely, as also in the Sambhur district. As regards Mount Aboo and Northern Guzerat, Captain Butler says it is very common, arriving just before the monsoon. In Cachar, Mr. Inglis met with it but once, and that was in May; but in Upper Pegu I find that Captain Butler and Mr. Oates corroborate Jerdon in saying that it is common; further south I observe that it has not been actually procured in Tenasserim, though it is doubtfully included in Mr. Hume's first list of birds from that Province. In North-east Africa it is, according to Mr. Sharpe, probably a migrant, and has been found in various parts of that region from August to November. Mr. Blanford has procured it in the Anseba valley, Antinori on the Blue Nile, and Ehrenberg in Nubia. It has been met with on the east coast and in various parts of South Africa, in Natal, the Transvaal, and other localities, and in the south-west of the continent it has been obtained in Damara Land.

Habits.—Low scrub, thorny jungle round the edge of forest, and open plains dotted here and there with brush-wood are the localities chiefly frequented by this Cuckoo; but it now and then occurs in avenues of trees or isolated shady groves, particularly when newly arrived in a district and the first cover to hand is being eagerly sought after. It is tame and usually solitary, although now and then I have seen a pair together; and in Pegu Mr. Oates has observed five or six in company. It is commonly seen sitting on the top of a low bush, and when flushed takes a short flight, but does not seek concealment in the bushes to any great extent. It has a rather plaintive, not unmelodious call, uttered when perched on some low tree; but at the commencement of the breeding-season, Mr. Holdsworth writes, "they are very noisy and incessantly flying from one place to another, one or more males apparently chasing the female, and uttering their clamorous cries." Jerdon remarks the same fact, and says that the call which the males utter at this time "is a high-pitched metallic note."

Its diet is insectivorous, consisting of caterpillars and various larvæ, grasshoppers, Mantidæ, &c.

Nidification.—In Ceylon the Pied Crested Cuckoo lays its eggs during the N.E. monsoon, choosing the nest of the Mud-birds or Babblers (*Malacocercus*) to deposit them in. Mr. Holdsworth observed them fighting

with those birds at Aripu, and Layard records an instance at Port Pedro of a pair of these Babblers tending
a young Crested Cuckoo in a bush; and when he drew near they flew away before him, feigning lameness, and
endeavoured to draw off his attention from their fosterling.   An egg taken from the oviduct of a female
killed in the Puttalam district was of a pale greenish or faded greenish-blue colour, and measured 0·95 by
0·74 inch.

In Aboo, Captain Butler states that they chiefly lay in the nests of the Striated Bush-Babbler (*Chatarrhœa
caudata*) and also in those of the Bengal Babbler (*Malacocercus terricolor*). The eggs are highly glossy and
closely resemble, says Mr. Hume, those of the first-named species, so that they are well fitted for deposit in
Babblers' nests; in shape they are " round ovals . . . very glossy, and of a delicate full sky-blue," and
average 0·94 by 0·73 inch.

COCCYSTES COROMANDUS.

(THE RED-WINGED CRESTED CUCKOO.)

Cuculus coromandus, Linn. Syst. Nat. i. p. 171. no. 20 (1766); Lath. Ind. Orn. p. 216 (1790).

Cuculus collaris, Vicillot, N. Dict. d'Hist. Nat. viii. p. 229 (1816).

Oxylophus coromandus, Jerd. Cat. B. S. Ind., Madr. Journ. 1840, xi. p. 272; Blyth, J. A. S. B. 1842, p. 920; id. Cat. B. Mus. A. S. B. p. 74. no. 363 (1849); Kelaart, Prodromus, Cat. p. 128 (1852); Layard, Ann. & Mag. Nat. Hist. 1854, xiii. p. 451.

Coccystes coromandus, Horsf. & Moore, Cat. B. Mus. E. I. Co. ii. p. 693 (1856); Jerdon, B. of Ind. i. p. 341 (1862); Cab. et Heine, Mus. Hein. iv. p. 45 (1862); Holdsworth, P. Z. S. 1872, p. 432; Hume, Nests and Eggs (Rough Draft), p. 138 (1873); id. Stray Feath. 1875, p. 82; David & Oustalet, Ois. de la Chine, p. 61 (1877).

Coucou huppé de Coromandel, Buffon, Pl. Enl. p. 274; *Coromandel Cuckoo* of some; *The Collared Crested Cuckoo,* Kelaart.

Yerra gola Kokila, Telugu; *Tseben,* Lepchas (Jerdon).

Konde-koha, lit. "Crested Cuckoo," Sinhalese.

Adult male. Length 15·0 inches; wing 6·3; tail 9·3; tarsus 0·9; outer anterior toe 0·95, claw (straight) 0·35; bill to gape 1·3.

Adult female. Length 16 inches; wing 6·7; tail 9·7; bill to gape 1·4.

Iris hazel-brown; bill black, inside of mouth and nostrils coral-red; legs and feet bluish slate, claws black. Above, the head, including the lores, upper part of cheeks and ear-coverts, the hind neck, back, scapulars, tertials, and least wing-coverts black, with a bronze-green lustre on the upper parts, and a deep blue-green gloss on the upper tail-coverts and tail; the head is less glossed than the back and crest, which latter is 1¾ inch in length, and stands out boldly from the nape; the green of the tertials is paler or more overcast with a brownish lustre than other parts: a conspicuous collar of white across the hind neck; quills, greater and median wing-coverts rich chestnut, the primaries dusky towards the tips; tips of rectrices fulvous-white, the centre pair only edged with it: beneath, the throat, fore neck, and sides are yellowish ferruginous, paling into white on the breast and upper part of belly, the abdomen, vent, and thigh-coverts becoming dusky grey; under tail-coverts green-black, edged with fulvous-white, some of the feathers having pale centres.

Female. Differs slightly; somewhat less deep in hue above; tips of rectrices whiter; throat not so rich, the colour not extending to the chest, and the lower part of breast not so pure, the grey of the abdomen pervading it somewhat.

Young. In Hodgson's drawing of the "young, hardly fledged," the bill and eyelid are pale red; the iris is pale brown, and the legs and feet reddish fleshy.

Head, upper surface, and tail brown, the feathers of the head, back, rump, and scapulars with broad fulvous margins: the quills and wing-coverts more deeply margined with the same; tail-feathers edged outwardly and tipped with pale rufous; ear-coverts and entire under surface white.

The bird of the year has the green portions of the upper surface, including the upper tail-coverts, tipped and edged towards the extremities of the feathers with rufous, and the ground-colour brownish metallic green; throat whitish, washed with yellowish rufous; under tail-coverts and abdomen tipped with ferruginous.

Distribution.—The Red-winged Crested Cuckoo, one of the handsomest of its tribe in Ceylon, is a migratory bird to the island, arriving about October and departing again in April. Whether or not it leaves the extreme north of the island altogether, I have been unable to ascertain with certainty; but there is no question about its being a visitor to the southern parts of the west coast, for in October 1876, while I was at

2 K

Colombo, an individual was captured on a canoe, some miles from the coast, and on which it had alighted in an exhausted state. When it first arrives it is not unfrequently seen in the Western Province, and then disappears from the sea-board, taking up its quarters in the interior of the low country and ascending the hills to some altitude. It occurs sometimes in Dumbara, and in March 1877 Mr. Bligh saw an example near his bungalow on the Catton Estate at an elevation of more than 4000 feet; he informs me that they are very rare in the Haputale district, and, indeed, its numbers throughout the island are very limited. The island of Manaar and the adjoining coast may perhaps be considered an exception; in the former I saw a good many in March, and Mr. Simpson says it is found about Illepekadua, and in the interior between that place and Mahintale. Mr. Holdsworth does not record it from Aripu. Layard procured it at Ratnapura.

On the mainland the Coromandel Cuckoo enjoys a wide range, but seems to be nowhere numerous. Jerdon writes of its distribution:—" It appears to be a rare species everywhere, though generally spread through India and Ceylon, extending into Burmah and Malayana. It is said to be common in Tenasserim and the Malayan peninsula. I have seen it in Malabar and the Carnatic, and it is also found in Central India and not very uncommonly in Bengal; in the latter country only during the rains. I obtained it in Sikhim in the warmer valleys."

It has been procured by very few collectors of late years either in South or Central India. I find no record of it in 'Stray Feathers' from the Peninsula; but I am aware that it is not uncommon in Ramisseram Island, having received specimens from there, and it must consequently be found on the adjacent coast about Tuticorin. Concerning its range to the east of the Bay of Bengal, Mr. Oates writes that in Pegu it is widely distributed, but not common. Captain Feilden seems to have fallen in with it to a much greater extent; he says :—" This bird is the commonest Cuckoo at Thayetmyo; in the thicker parts of the jungle every bamboo valley contains one or more pairs. They arrive in the beginning of the rains, and the young birds do not leave until October." This is the period at which the species visits us in Ceylon, so that there would appear to be a regular migration north and south at the beginning and end of the rains. In Tenasserim Mr. Davison only found it at a place called Meeta myo, which is about the centre of the province. There is a specimen in the British Museum from Sarawak; it goes, as we know, to Celebes, and it probably occurs in intermediate localities, perhaps in Java, but from there I have not heard as yet of any specimens. It is very desirable that we should know more of the movements and seasonal distribution of this bird, as it is one of the most attractive of its tribe in India. Swinhoe procured it at Amoy.

Habits.—I have observed this species in thick scrub and thorny jungle. A specimen was shot by Mr. MacVicar in the cinnamon-gardens near Colombo, a locality decidedly favourable to its habits. It is very shy, flying quickly up from the ground on being surprised, alighting then on the nearest bush or low tree, and speedily threading its way through the branches to the other side, when it again takes wing. The stomachs of those I have procured contained beetles, grasshoppers, Mantidæ, and other large insects. Captain Feilden notices that they have a Magpie-like chatter usually, but that they utter a "harsh, grating, whistling scream when watching over their young;" and this, I imagine, would be their ordinary note of alarm.

Nidification.—The breeding-season appears to be during the rains, i. e. from June until October. Mr. Hume describes an egg, which was taken from the oviduct of a female shot in Tipperah, as being a broad oval and of a "fine and glossy texture; in colour it was a moderately pale, somewhat greenish blue, without any specks or spots."

Captain Feilden has reason to believe that it lays in the nests of Quaker-Thrushes (*Alcippe phayrei*?). He writes, "I have frequently shot the young bird from the middle of a brood of young Quaker-Thrushes; and, as far as I could see from the thickness of the jungle, the old Thrushes were feeding the young Cuckoo. An egg taken from the nest of a Quaker-Thrush, that I believe to have belonged to this bird, was very round and pale blue."

The dimensions of the egg alluded to above are 1·05 by 0·92 inch.

Bill stout, wide at the base, not so much compressed as in *Coccystes*. Nostrils oval, exposed, not so near the margin as in the last. Wings long; the 3rd and 4th quills subequal and longest, the 1st nearly equal to the innermost. Tail not so much graduated as in the last genus. Legs and feet stout. Tarsus about equal to the anterior toe, and shielded with stout, broad, transverse scutæ.

EUDYNAMYS HONORATA.

(THE INDIAN KOEL.)

Cuculus honoratus, Linn. Syst. Nat. i. p. 169 (female) (1766); Lath. Ind. Orn. i. p. 214 (1790).
Cuculus niger, Linn. Syst. Nat. i. p. 170. no. 12 (male) (1766).
Cuculus indicus, Lath. Ind. Orn. i. p. 221 (1790).
Eudynamys orientalis, Sykes (*nec* Linn.), P. Z. S. 1832, p. 97; Jerd. Cat. B. S. India, Madr.
    Journ. 1840, xi. p. 222; Blyth, J. A. S. B. 1847, p. 468; id. Ann. & Mag. Nat. Hist.
    1847, p. 385; id. Cat. B. Mus. A. S. B. p. 73 (1849); Kelaart, Prodromus, Cat. p. 129
    (1852); Layard, Ann. & Mag. Nat. Hist. 1854, xiii. p. 451; Horsf. & Moore, Cat. B.
    Mus. E. I. Co. p. 708 (1856); Jerdon, B. of Ind. i. p. 342; Irby, Ibis, 1861, p. 230;
    Legge, ibid. 1874, p. 16.
Cuculus (Eudynamys) honoratus, Blyth, J. A. S. B. 1842, p. 912 (female).
Eudynamys honoratus, Gray, Gen. Birds, ii. p. 464 (1845).
Eudynamys honoratus, Walden, Ibis, 1869, p. 327; Holdsworth, P. Z. S. 1872, p. 432; Hume,
    Nests and Eggs (Rough Draft), p. 139 (1873); Ball, Str. Feath. 1874, p. 394;
    Anderson, Ibis, 1875, p. 142; Hume, Str. Feath. 1876, p. 403.
Eudynamys koronata, Hume, Str. Feath. 1873, p. 173.
The *Black Indian Cuckow*, Edwards, Nat. Hist. Birds, p. 58 (male); *Brown and Spotted
    Indian Cuckow*, Edwards, tom. cit. p. 59 (female); *Coucou tacheté de Bengal*, Daubent.
    Pl. Enl. pl. 294 (female); *Black Cuckoo* of some; *Koel*, Hind. (female), sometimes
    Koreyala, lit. " spotted "; *Kokil*, Bengal.; *Kokila*, Tel. (male *Nalak*, female *Podak*).
Kaputa koha (male), *Gomera koha* (female).
Coosil and *Koel*, Ceylonese Tamils (*apud* Layard).

Adult male. Length 15·0 to 15·3 inches; wing 7·0 to 7·4; tail 7·0 to 7·5; tarsus 1·15 to 1·2; outer anterior toe
    1·25, its claw (straight) 0·45; bill to gape 1·55.
Iris crimson; bill pale bluish green, blackish or dusky round the nostrils; legs and feet london blue.
Entire plumage black, with a strong metallic green lustre; the scapulars, wing-coverts, and tail with a bluish sheen
    as well.

Adult female. Length 15·5 to 16·5 inches; wing 7·3 to 7·75; tail 7·8; tarsus 1·25; outer anterior toe 1·3, claw
    (straight) 0·45.
Iris red; bill faded bluish, dusky at base and round the nostrils; legs and feet dusky sinty green or faded bluish.
Above metallic brownish green; hind neck, back, and lesser wing-coverts spotted and barred with white, the markings
    on the first-named part are limited to spots, and the barring on the wing-coverts consists of interrupted bar-
    like spots; the wings vary considerably in the character of their markings: quills, upper tail-coverts, and tail

barred with white, tinged generally to a greater or less extent with fulvous; forehead marked with fulvous terminal spots or broad mesial stripes; beneath white, with basal or longitudinal blackish marks on the throat, angular or arrow-headed on the chest, gradually changing into wavy bars on the breast; the flanks and the under tail-coverts boldly barred with dark greenish brown; the lower parts from the breast downwards more or less washed with fulvous.

Young. Males in nestling plumage have the iris mottled red; bill greenish, dusky at the base; legs and feet plumbeous. Upper surface and wing-coverts dingy or brownish metallic green, the feathers with white terminal spots; under surface the same, with terminal bars of white; under tail-coverts and under wing-coverts barred with white: tail-feathers tipped and adjacently marked with whitish.

The female in immature plumage has the head, hind neck and its sides, face, and throat striated with rufous; the spottings on the back and wing-coverts, and the bars on the scapulars and tertials, tinged strongly with the same: the bars on the tail-feathers rich tawny, and the under surface washed with the same.

Young females vary in their coloration, as, in fact, the adults also do; scarcely any two female Koels are marked *exactly* alike, differing in the extent of the spotting and barring of the upper surface, and in the amount of rufous on the forehead, of which our birds seem to have more than Indian.

Obs. A comparison of a series of Indian examples with my Ceylonese specimens does not disclose any points of difference between the two races except in the above-mentioned respect. A male from Madras has the wing 7·2 inches, and is identical with examples in my collection; another from Central India is larger, wing 7·8 inches.

The several species of Koel which inhabit the region to the east of the Bay of Bengal and the Malayan archipelago are closely allied, the males being black, and the chief specific difference lying in the coloration of the females.

E. malayana, Cab. et Heine, from Assam, Burmah, Tenasserim, Malacca, and Sumatra, has a larger bill than our bird, is longer in the wing, and the females are boldly marked with rufous. I have measured examples in the national collection varying from 7·4 to 7·6 inches in the wing, but it is said to reach 8·0 inches.

E. ransomi, from Ceram and Bouru, is a very fine species, with the wing 8·0 to 9·0 inches, tail 8·0 to 9·0, bill to gape 1·4; the female is very handsomely marked, its coloration being likewise rufous.

E. orientalis. An example, male, from Lombok has the wing 8·1 inches: a female is greenish black, barred with fulvous and white; the forehead and a superciliary band yellowish rufous, centre of the crown dusky green; wings and tail barred with rufous: another female is rufous beneath, barred with black; the upper surface dark green, barred with black.

Distribution.—The Koel is found all over the low country. It is equally common in the northern and southern portions of the island, including the Jaffna peninsula. Mr. Holdsworth only observed it at Aripu from November till April, and inferred that it was migratory to Ceylon. It moves about a good deal according to the weather, leaving the sea-board of the Western Province for the interior during the wet windy months from May until October; but this is all: away from the sea I have seen it at all seasons. On the east side of the island it appears to be stationary, being at all times to be observed in that part; and this is likewise true of the north-east. It is numerous in the delta of the Mahawelliganga and on the coast in places to the north of Trincomalie. In the interior it is much rarer, and, in fact, is liable to be passed over in a cursory inspection of many parts of the northern half of the island, as it is local in its distribution there. I have not seen it from the hills, but have been given to understand that it has occurred in Dumbara.

On the continent this noisy bird is very common in most parts of the Indian peninsula. It is abundant in Ramisserum Island and on the south coast of India; in the Palani hills it likewise occurs; in the Deccan it is common and widely distributed. Mr. Ball says it is tolerably common in the eastern parts of Chota Nagpur, but is seldom met with in the western, more jungly districts. Further to the west it appears to be a visitant only in the breeding-season, from April until October. Mr. Adam remarks that during his stay at Sambhur it only visited the place once or twice during the rains; in Sindh it is likewise non-resident, and in the Mount-Aboo district it occurs during the above-mentioned period of the year. Its inhabiting the Laccadive islands is especially worthy of remark. Mr. Hume found it on every inhabited island that he visited; he writes that, "unless perhaps at Amini and one or two of the Cannanore Islands, where there are Crows, they can only be, as the people affirm, seasonal visitants, there being no bird in whose nests they could lay their eggs."

Habits.—This is one of the noisiest birds in Ceylon, making the woods and paddy-fields ring with its peculiar scale-like call. It frequents groves of trees, compounds, wooded knolls in paddy-fields, and jungle near water or bordering open ground. It is a skulking bird, loving concealment in thick trees and tangled bushes, and delighting in the shady foliage of trees which are matted at the top with creepers. It moves actively about when flushed and driven into a tree, hopping along the slanting limbs, and springing from branch to branch till it gains the other side, and then escapes to a further place of concealment. It is the male which utters the peculiar note *ku-il, ku-il*, or *koyo koyo*, which mounts each time higher and higher and increases in vigour until it fairly rings through the woods; he is usually perched in some thick tree, and when he has finished his vociferation one or two females may be seen issuing from their places of concealment and flying towards him. This cry may often be heard at night. Adult males seem usually to be in the minority, or else they do not move about as much as the other sex, many more of which may always be seen in the course of a day's ramble.

The Koel is almost exclusively a fruit-eating species, and feeds greedily on all sorts of luscious seeds and berries; from the stomach of a male I once took two entire nuts of the Kitool-palm: it is fond of the banyan-fruit, but in Ceylon does not much affect localities in which this tree grows. Blyth states that it ejects the large seeds of any fruit that it has eaten by the mouth: he syllabizes another note uttered by the male as *ho-whu-ho*; but I have not heard this. Both sexes are much more noisy in May and June than at other times, as this is the breeding-season in Ceylon.

Layard remarks that the natives so much admire the note of this bird, that their poets compare it to the voices of their mistresses, which, however, as he aptly continues, cannot be very soft, for the Koel can be heard a mile away!

Nidification.—In the Western Province this parasitic Cuckoo breeds in May and June, laying nearly always in the nest of *Corvus levaillanti* (the Black Crow), and not in that of the smaller citizen species, as in India, for the simple reason that the latter does not inhabit the jungle to which the Koel resorts to rear its young. I am indebted to Mr. MacVicar for many valuable notes on the nesting of this bird, a number of whose eggs he has taken in the Western Province, and more especially in the vicinity of Kœshawa. The following are the particulars of four nests found in the months of May and June, 1875:—

(1) Eggs: 4 Crow's; 4 Koel's.
Differed considerably, as if they had been laid by different birds. Two were of a pale green ground, spotted rather thickly with longitudinally-directed markings of olive-brown, confluent slightly round the obtuse end, and laid over numerous blotches of lilac or pale bluish grey: dimensions, 1·24 by 0·93 inch and 1·2 by 0·9 inch. The other two were of a light brown colour, covered with small reddish-brown and purple spots: dimensions, 1·35 by 1·1 inch and 1·34 by 1·0 inch.

(2) Eggs: 3 Crow's; 3 Koel's.
Ground-colour olivaceous green, blotched and marked (sparingly at the small end) with two shades of olive-brown over numerous smaller spottings of indistinct bluish grey; the markings almost confluent at the obtuse end.

(3) Eggs: 2 Crow's; 4 Koel's.
Two distinct types. Ground-colour of two olive-brownish grey, marked all over, but mostly at the large end, with reddish brown, over numerous smaller spots of bluish grey; at the obtuse end the spots are large, but all over the rest of the surface in the form of small specks: dimensions, 1·32 by 1·0 and 1·38 by 1·0. The other two were of a greenish ground-colour, speckled with purplish and brown spots, mostly towards the obtuse end, where the markings become confluent: dimensions, 1·36 by 0·95 inch and 1·25 by 0·96 inch.

(4) Eggs: 2 Crow's; 2 Koel's.
In shape very stumpy. Colour dark olive-green, spotted with dark reddish brown, confluent round the obtuse end. Dimensions, 1·18 by 0·92 inch and 1·15 by 0·95 inch.
The average size of these eggs was 1·31 by 0·95 inch.
In India the Black Crow lays too early for the Koel; and my lamented friend Mr. Anderson* remarks

---

* In Mr. Hume's 'Nests and Eggs of Indian Birds' will be found a lengthy extract from a paper by this observant naturalist on the nesting of the Koel.

that this is why the nest of the Common Grey Crow is chosen. These clever birds seem to know that they are imposed upon by the Koels, and consequently hold them in strong dislike, constantly attacking and pursuing them during the breeding-season. When the female Koel is about to intrude her egg into the Crow's nest she is accompanied sometimes by the male. It is supposed that the young Koel ejects the Crows from the nest, as in the case of the Common Cuckoo; for Mr. Hume found a young one in a nest with three Crows newly fledged, and a week later " the Crows were missing, and the young Cuckoo thriving." The young Cuckoos persistently follow the Crows for some time after they have " flown," and are even then fed by them with as much care as if they were their own offspring.

Subfam. PHŒNICOPHAINÆ.

Bill robust, in most species higher than wide; culmen much curved. Nostrils not swollen and more or less linear. Tail long and graduated. Tarsi robust and naked, or slightly feathered on the upper part.

Bill stout, wide at the base, and suddenly compressed, the tip well bent down; the upper mandible very high: the nostrils linear and close to the margin, which is lobed just beneath them. Face clothed with a short papillose substance. Wings rounded; the 5th quill the longest and the 1st the shortest. Tail very long, broad, and much graduated. Tarsus longer than the middle toe and its claw, covered with broad transverse scales; outer anterior toe considerably longer than the outer posterior one; claws short and much curved. Feathers of the throat with stiff shafts projecting beyond the webs.

PHŒNICOPHAËS PYRRHOCEPHALUS.

(THE RED-FACED MALKOHA.)

(Peculiar to Ceylon.)

Cuculus pyrrhocephalus, Forster, Ind. Zoologie, p. 16, pl. vi. (1781); Gmelin, Syst. Nat. i. p. 417. no. 40 (1788); Lath. Ind. Orn. i. p. 222 (1790).

Phœnicophœus pyrrhocephalus, Stephens, Gen. Zool. ix. i. p. 59 (1825); Blyth, J. A. S. B. 1842, p. 927; Bonap. Consp. Gen. Av. i. p. 98 (1850).

Phœnicophaus leucogaster, Vieill. N. Dict. d'Hist. Nat. xviii. p. 461 (1816).

Melias pyrrhocephalus, Less. Traité d'Orn. p. 131 (1831).

Phœnicophaus pyrrhocephalus, Blyth, J. A. S. B. 1845, p. 199; Gray, Gen. Birds, ii. p. 459; Blyth, Cat. B. Mus. A. S. B. p. 75. no. 369 (1849); Kelaart, Prodromus, Cat. p. 129 (1852); Layard, Ann. & Mag. Nat. Hist. 1854, xiii. p. 453; Legge, J. A. S. (Ceyl. Br.) 1870–71, p. 37; Holdsworth, P. Z. S. 1872, p. 433; Legge, Str. Feath. 1873, p. 346; id. Ibis, 1874, p. 16, et 1875, p. 285.

Phœnicophaus ceylonensis, Licht. in Mus. Berl.

Phœnicophaës pyrrhocephalus, Cab. et Heine, Mus. Hein. pt. iv. p. 68 (1862).

The Red-faced Cuckoo (*rothköpfige Kukkuk*), Forster; *Malkoha*, Pennant and Kelaart.

Mal-kændetta, Sinhalese, Western Province; *Warrelliya*, in Friars-Hood district.

♂ supra metallicè viridis, alis dorso concoloribus, primariis extùs vix cyanescentibus : rectricibus olivascenti-viridibus, latè albo terminatis : pileo colloque postico nigricantibus, plumis albido marginatis, quasi striolatis : facie laterali totâ nudâ, papillosâ, rubrâ : genis et regione parotica, mento guléque summâ striolatim albis : gutture reliquo nigro : præpectore et colli lateribus nigris albido striatis : corpore reliquo subtùs albo : tibiis fuscescenti-viridibus : subalaribus metallicè chalybeo-viridibus, remigibus quoque subtùs chalybeo nitentibus : rostro flavicanti-viridi, mandibulâ pallidiore : pedibus cyanescenti-schistaceis, unguibus brunnescenti-corneis : iride brunneâ.

♀ mari similis, sed iride albâ distinguenda.

Adult male. Length 17·8 to 18·2 inches; wing 6·0 to 6·2; tail 10·5 to 11·1 (lateral feathers only 5); tarsus 1·3 to 1·4; anterior toe 1·1, claw (straight) 0·35; bill to gape 1·5 to 1·6. Expanse 17·5.

Iris brown; bill apple-green, paling at the tip, and the lower mandible lighter than the upper; legs and feet bluish slate, claws brownish horn.

Female. Length 18·0 to 18·7 inches; wing 6·2 to 6·4; tail 11·0 to 11·3; bill to gape 1·5 to 1·65.

Iris white.

Whole face as far back as the ears, passing over the eye and across the base of the upper mandible, clothed with a short blade-like crimson substance, resembling a rudimentary feather; crown, back, and sides of neck greenish black, with the terminal margins of the feathers white; back and wings deep brilliant metallic green, blending into the hue of the neck; quills slightly darker, with a bluish lustre; tail metallic bronze-green, the terminal portion white, increasing from about an inch on the central feathers to two inches on the laterals, and separated from the green by a smoky-brown margin; throat and upper part of chest deep black, the feathers of the chin and of the space beneath the crimson cheeks white, with black shafts; breast and lower parts pure white, changing abruptly from the black of the chest, at the lower edge of which the feathers are tipped with white; flanks and thighs dark greenish black; under wing-coverts metallic green. The tips of the head- and neck-feathers are furcate, the shaft protruding from the fork.

The extent of the striations on the hind neck, and the amount of white tipping at the edge of the black chest, vary in individuals. Some examples, probably immature birds, have the thigh-coverts and lower flanks tinged with fulvous.

Obs. This remarkable genus has no representative in India. Jerdon speaks of *Ph. curvirostris*, an inhabitant of Burmah: but this bird has a very different shaped and situated nostril, on account of which Mr. Sharpe, and justly so it would appear, has made it into a new genus, *Rhinococcyx*. It has the same singular facial clothing, but not to so great an extent as in our bird.

The most singular feature in the economy of the present species is the difference in the colour of the eyes in the two sexes, as noticed in my description above. Layard probably procured a female and noted the colour as white: specimens sent to Lord Tweeddale, and a living bird which Mr. Holdsworth had, appear to have been males and had brown eyes. I was fortunate enough, on two occasions, to shoot a pair together, and was able to demonstrate the fact of the sexual difference.

Distribution.—The Malkoha is found in most of the forests and heavily-clad jungle-districts of the low country; but, notwithstanding, has always been considered one of our rarest species, an idea which naturally arose from the extreme difficulty of penetrating its haunts. It occurs sparingly throughout the south-western hill-region, or the tract of country extending from the Kaluganga, through the Pasdun and Hinedun Korales, to the eastern confines of the Morowak Korale. It is likewise to be found in most of the damp forests of the Western Province, particularly in the hills stretching from the neighbourhood of Avisawella to Kurunegala, and occurs even at Maham and Kotté, in the vicinity of Colombo. It occurs throughout the jungles of the great northern forest-tract, extending from the Western Province through the Seven Korales to the Vanni, the most northerly point in which I have seen it being the forests on the road from Trincomalie to Anaradjapura. In the Eastern Province, however, it is far more numerous than in the aforesaid districts, for I have met with it in flocks of ten or a dozen in the jungles at the base of the Friars Hood, and also near Bibile beneath the Madulsima range. Mr. Bligh has procured it at a considerable altitude in the Lemastota hills, into which it doubtless ascends from the Wellaway-Korale forests in the dry season. On the western side of the hill-zone Mr. Holdsworth has observed it in the Kandy district; but I have no evidence of its being found at a greater elevation than that.

This species is one of the earliest known Ceylon birds. Its gay plumage no doubt made it an object of attraction to the early travellers; and Forster described it in his ' Indische Zoologie ' so far back as 1781, giving a plate of it done in the crude style of that period. He, however, does not make any mention of the discoverer of this interesting Ceylonese form, which leads to the inference that natives first made it known to Europeans in the island.

Habits.—This handsome bird is a denizen of forest and heavy jungle, and is of such a shy and retiring disposition that it is but little known to Europeans, even those who are stationed in the wilds of the interior.

The natives of the Western and Southern Provinces, a part of the island in which the population is chiefly located in the cultivated districts, are less acquainted with it than with most birds ; but the inhabitants of the northern and eastern jungles, whose scanty villages are situated, for the most part, in the depths of those primæval wilds, recognize the *Mal-kändetta*, without hesitation, as a not uncommon bird. Layard, who considered its range to be limited to the mountain-zone, speaks of it as being eaten by the natives, and probably alludes to the Kandyans of the Dumbara district before it was denuded of forest, and when it contained this bird in much greater numbers than it does now. The natives of the " Friars-Hood " jungles, where it is commoner than in other parts of the island, call it " Warreliya," or " long tail."

The Malkoha is fond of tall or shady forest in which there is a considerable amount of undergrowth or small jungle, into which it often descends, after making a meal off the fruits of the lofty trees overhead. When flushed it invariably flies up into high branches and is difficult to come up with, as it quickly makes off, taking short flights from tree to tree. I have seen a flock of six or seven feeding among the topmost boughs of one tree, and noticed that they moved very quicky about among the leaves, sharply wrenching off the berries which they were seeking and devouring them whole. As a rule it is a silent bird, the only note with which I am acquainted being a rather low monosyllabical call like *kaa*, which it utters when flying about. Although I have occasionally found the remains of small insects in its stomach, it is almost exclusively a fruit-eating species, and its flesh is consequently by no means to be despised. It is tender and not unpleasantly flavoured ; and Layard remarks, with justice, that the natives consider it a great delicacy. I have known an individual persistently return to a tree, on the berries of which it had been feeding, a few minutes after being shot at.

Nothing is known of the nidification of this species.

In the Plate accompanying this article, the figure in the background with the white eye represents a female shot in the Vanni.

Bill more slender than and not so deep as in the last genus, not so inflated near the base; the gape more festooned. Nostrils ovoid, basal, and placed higher up than in *Phœnicophaës*. Eye surrounded by nude skin. Wings with the 4th and 5th quills subequal and longest. Tail, legs, and feet much as in the last. Shafts of the throat-feathers rigid.

ZANCLOSTOMUS VIRIDIROSTRIS.

(THE GREEN-BILLED MALKOHA.)

Zanclostomus viridirostris, Jerd. Cat. B. S. India, Madr. Journ. 1840, xi. p. 223, et Ill. Ind.
Orn. i. pl. 3; Blyth, J. A. S. B. 1845, p. 200; id. Cat. B. Mus. A. S. B. p. 76. no. 375
(1849); Bonap. Consp. Gen. Av. p. 99 (1850); Kelaart, Prodromus, Cat. p. 129 (1852);
Layard, Ann. & Mag. Nat. Hist. 1854, xiii. p. 458; Horsf. & Moore, Cat. B. Mus.
E. I. Co. ii. p. 690 (1856); Jerdon, B. of Ind. i. p. 346 (1862); Holdsworth, P. Z. S.
1872, p. 432; Legge, Ibis, 1874, p. 16, et 1875, p. 284; Hume, Str. Feath. 1876, p. 458;
Fairbank, ibid. 1877, p. 397.

Phœnicophaus jerdoni, Blyth, J. A. S. B. 1842, p. 1095.

Rhopodytes viridirostris, Cab. et Heine, Mus. Hein. pt. iv. p. 68 (1862).

The Small Green-billed Malkoha, Jerdon, B. of India.

Kappra-popya, Hind.; *Wamana kaki,* lit. " Dwarf Crow," Telugu.

Mal-kændetta, Sinhalese; also *Handi-koota (apud* Daniell); *Koosil,* Ceylon. Tamils (Layard).

Adult male and female. Length 15·0 to 15·75 inches; wing 5·1 to 5·4; tail 8·4 to 9·3; tarsus 1·3 to 1·35; outer anterior toe 0·9, its claw (straight) 0·3; bill to gape 1·2 to 1·4.

Iris deep brown; bill pale leaf-green; orbital skin in front of eye cobalt-blue, paling behind into pale bluish; legs and feet dusky green or greenish blue.

Above greenish grey, overcome with a strong green gloss from the hind neck down to the rump; lores, at the base of the bill, and round the orbital region shading into blackish; wings and tail deep metallic green, the tips of the quills dusky; terminal portion of tail-feathers white, deepest on the outer webs of all but the central pair, which are evenly tipped and with less white than the rest; throat blackish, with greyish or pale fulvous striæ, formed by the double tips of the feathers being of that colour, and exceeding the black shaft; on the chest the feathers gradually become fulvous-grey, and from that pure fulvous on the breast and abdomen; flanks, thighs, and under tail-coverts cinereous, the two latter washed with fulvous.

Some examples, probably immature birds, have the under surface paler than the above, and the upper surface less glossed with green; the striæ of the throat are less fulvous in some than in others.

The furcate formation of the throat-feathers is most singular, and was, it appears, first pointed out by Blyth, with his usual habit of minute and accurate observation.

Obs. On comparing Ceylonese with South-Indian examples, I find no appreciable difference; an individual from Madras measures as follows—wing 5·1 inches; tail 9·2; tarsus 1·35; bill to gape 1·23.

This species does not differ widely in plumage from the North-Indian *Z. tristis,* which has not got the under parts rufous, and has the throat whiter, with the nude skin round the eye crimson, instead of blue. The latter species, however, is much larger, the wing measuring 6¼ inches according to Jerdon, and it is consequently styled the " Large Green-billed Malkoha."

Distribution.—This Cuckoo is widely diffused throughout the low country of Ceylon, being most numerous in the northern half and south-eastern division of the island, including, as regards the former, the Puttalam and Chilaw districts and the Seven Korales.

It does not, as far as I am aware, ascend into the hill-zone to any considerable altitude, although it is found in the hilly country at the base of the Hewa-Elliya ranges, at an elevation of about 1000 feet. In the above-mentioned low-country districts it is dispersed throughout the forests and low jungle, being everywhere to be found by those who know what sort of locality it frequents; in the south and west, however, it affects only those spots which are suitable to its habits. It is found in tangled thickets here and there throughout the Colombo district, and in the south-west corner of the island is more local still; for instance, it frequents the thorny tangled brake covering the peninsula on the east side of Galle harbour, and is scarcely to be found anywhere else in the neighbourhood. Mr. Holdsworth records it as abundant at Aripu; and further north, as well as in the island of Manaar, it is equally so. It is found in the Jaffna peninsula.

Elsewhere this Malkoha is found only in the south of India. In Ramisseram Island it is common, and likewise on the mainland of the peninsula. In the Palani hills Mr. Fairbank procured it at the eastern base. Jerdon says that it is found as far north as Cuttack, where it meets the larger species. " In the bare Carnatic and the Deccan," he writes, " it is chiefly to be met with in those districts where the land is much enclosed, as in part of the zillah at Coimbatore, where large tracts of country are enclosed by thick and, in many cases, lofty hedges of various species of *Euphorbia*. . . . . Throughout the west coast, where jungle and forests abound, it is much more common, especially in those parts where bamboos occur, and where numberless creepers entwine themselves and hang in luxuriant festoons from every tree."

Habits.—The Green-billed Malkoha frequents dense low jungle, the tangled edges of forest, scrub near the sea-coast or surrounding large woods, thickets, and so forth. It is not particularly shy, but does not care to subject itself to long observation, making off with a stealthy flight, and threading its way quickly through the most tangled underwood. I have often noticed it in pairs, but just as frequently flushed it singly, its mate being probably not far distant. In the Northern Province and the jungles to the south of Haputale, where it is abundant, it may frequently be seen flying across the roads. Its diet consists of various fruits and berries and also insects; in the stomach of one I found a large locust almost whole. In India it is said to be almost entirely insectivorous. Jerdon writes that it "diligently searches the leaves for various species of *Mantis*, Grasshopper, and Locust, whose green colours and odd forms, though assimilating so strongly to the plants on which they rest, are but of little avail against its keen and searching eye." In his 'Birds of India,' he remarks that he never found it feeding on fruit; in Ceylon it is the exception to find that it has partaken of any thing else. It is difficult to flush a second time; for when thoroughly alarmed it skulks in the thickest underwood it can find, or escapes, by the use of its legs, among the branches forming its retreat. Its note is a low crake, sounding like *krāā*, generally uttered after it has been flushed; but it is usually of a silent habit.

Nidification.—We are indebted to Miss Cockburn for the only information yet to hand of this bird's nesting. She obtained one nest in March on the Nilghiris. It was large, and consisted of sticks, put together much in the style of a Crow-Pheasant's nest. It contained two white eggs.

Bill very strong, high at base, well curved throughout. Nostrils lateral, and protected by a membrane. Wings rounded, the 6th quill the longest. Tail long, wide, considerably graduated. Tarsi stout and shielded with broad transverse scales, longer than the anterior toe with its claw. Toes stoutly scaled; hind claw very long and straight. Feathers of the head, neck, and throat spinous.

CENTROPUS RUFIPENNIS.

(THE COMMON COUCAL.)

Centropus rufipennis, Illiger, Abhandl. Akad. Wiss. Berl. (1812) p. 224; Horsf. & Moore, Cat. B. Mus. E. I. Co. ii. p. 681 (1856); Jerdon, B. of Ind. i. p. 348 (1862); Holdsworth, P. Z. S. 1872, p. 433; Jerdon, Ibis, 1872, p. 15; Hume, Nests and Eggs (Rough Draft), p. 142 (1873); Legge, Ibis, 1874, p. 16; Morgan, Ibis, 1875, p. 315.

Centropus philippensis, Sykes, P. Z. S. 1832, p. 98; Blyth, J. A. S. B. 1842, p. 1099; id. Ann. & Mag. Nat. Hist. 1847, p. 385; id. Cat. B. Mus. A. S. B. p. 78 (1849); Kelaart, Prodromus, Cat. p. 128 (1852); Layard, Ann. & Mag. Nat. Hist. 1854, xiii. p. 450.

Centropus castanopterus, Steph. Gen. Zool. xiv. i. p. 215 (1826).

Centropus pyrrhopterus, Jerdon, Cat. B. S. Ind., Madr. Journ. 1840, xi. p. 224.

Centrococcyx rufipennis, Ball, Str. Feath. 1874, p. 394; Fairbank, Str. Feath. 1877, p. 397.

The Philippine Ground-Cuckoo, Kelaart; *The "Crow-Pheasant*," Europeans in India and Ceylon, also "*Jungle-Crow*," in Ceylon; "*Lark-heeled Cuckoo*," Jerdon's Catalogue.

Mahoka, Hind.; *Kuka*, Beng.; *Marmowa*, at Monghyr; *Jemudu-kaki*, lit. "Euphorbia Hedge-Crow," Tel.; *Kalli-kaka*, lit. "Hedge-Crow," Tam. (Jerdon).

Ætti-kukkula, Sinhalese; *Chembigum*, Ceylonese Tamils.

Adult male and female. Length 17·5 to 18·5 inches; wing 7·3 to 8·1; tail 9·5 to 10·0; tarsus 1·9 to 2·0; outer anterior toe (with claw) 1·75 to 1·9; hind toe 0·6, its claw 0·8 to 0·83; bill to gape 1·65 to 1·85.

Females appear to average larger than males.

Iris vermilion; eyelid blackish leaden; bill black; legs and feet black.

Entire plumage, with the exception of the scapulars and wing-coverts (which are glossy cinnamon-rufous) black, dull on the crown and throat, and with the hind neck and its sides, as well as the chest and upper breast, illumined with steel-blue edgings, blending into a greenish hue at the centres of the feathers; these hues are brightest on the hind neck; back, rump, and flanks moderately glossed with greenish; tail-feathers glossed with green, mostly on the four lateral feathers; forehead and chin brownish, gradually darkening into the hue of the crown and throat respectively; tips of the quills smoky brown; scapulars somewhat darker rufous than the wings; under wing-coverts shaded with blackish.

The gloss on the tail-feathers varies in individuals: in some the central pair have scarcely any, the ground-colour partaking slightly, if examined carefully, of a ruddy-brown hue.

Young. The yearling bird has the head and nape marked with rufous strim; hind neck barred with fulvous; scapulars and wings crossed with rather broad bars of blackish; tail barred with spear-shaped bands of dusky whitish; throat-feathers centered and barred with fulvous; breast and thighs the same.

The above description of the young is from an Indian individual; I have not had the opportunity of examining Ceylonese examples in the immature stage, but they have been described to me as similar to the one here noticed.

Layard procured an albino of this Cuckoo at Pt. Pedro, in which "the black and purple portions were changed to a dirty creamy white, the dark red portions to a light brown."

Obs. Ceylonese *C. rufipennis* differs from the Indian bird of this species in its paler forehead and throat, these parts, as a rule, being in the latter concolorous with the adjacent dark plumage. I say, as a rule, because I find that, as in Ceylon, so in India, examples vary *inter se* in this respect; an example from Kamptee and another from the North-west Province are so close to the insular bird that the latter cannot well be discriminated as a separate race. Mr. Swinhoe, when at Galle, shot a pair of Coucals, which he considered ('Ibis,' 1873, p. 230) distinct from the true *C. rufipennis*, on account of their smaller size and larger bills (the size of bill is not given), as well as their broader tail-feathers barred obscurely across. The wings measured 7½ inches, which corresponds with those of Indian specimens : the tails evidently point to the individuals being immature.

An example in the British Museum from Kamptee measures 7·9 inches in the wing, and four measured and recorded by Mr. Ball are as follows:—Sex ?, Gangpur, 7·7; ♂, Rajmehal hills, 7·2; ♂, Natpuras, 7·8; ♂ juv., Calcutta, 7·55. These would compare very well with five Ceylonese examples taken at random from a series. The tails of the first three here enumerated measure 10·8, 10·5, 10·5 inches respectively; this is longer than they ever attain to in the insular bird, and I have observed the same inferiority in this respect when comparing my specimens with those in the national collection.

The larger species (*C. eurycercus*) from Borneo, Labuan, Sumatra, Java, as well as from Tenasserim, Burmah, Nepal, Sindh, Sikkim, and other parts of India (if the continental species be the same), differs from *C. rufipennis* in having the back coloured red like the wings, which are a paler rufous than in the latter species ; likewise in the blue-glossed tail and the much more metallic blue lustre of the hind neck, and finally in the darker under wing-coverts : it is, in all its races, a larger bird than *C. rufipennis*. A Labuan specimen measures 8·3, a Sumatran 8·7, and a Borneau 8·6 inches in the wing : the Sindh and Sikkim birds vary from 9·0 to 9·5 according to Mr. Hume, and some I have measured from other localities 7·9 to 8·3.

Distribution.—The "Jungle-Crow," or "Crow-Pheasant" as it is popularly called, is found throughout all the low country, including the island of Mannar and the Jaffna peninsula, in which latter districts, as well as in most of the north of the island, it is extremely abundant. It ascends the hills, ranging up to 3000 feet throughout the year in some districts, and reaching the altitude of the Nuwara-Elliya plateau in the cool dry season. In June I have met with it in Upper Hewnheitte, and in January I have heard it behind Hakgala mountain and in the railway gorge.

It is very abundant in the south-west and west of the island, and is tolerably numerous in the Eastern Province and along the north-east coast. At Trincomalie it frequented the native gardens in the heart of the Bazaar. In forest-districts it is local, being chiefly found where the jungle has been cut down and low scrub grown up. It is common in Dumbara, and particularly about Kandy, Paradeniya, and generally along the banks of the Mahawelliganga. On the Uva patnas it is not uncommon ; and in Haputale Mr. Bligh has seen it above 4500 feet.

On the continent the Common Coucal inhabits chiefly the southern and central portions of India. It is common in Ramisserum Island and on the adjacent coast, and Mr. Fairbank observed it up to 3500 feet in the Palani hills ; it likewise inhabits and breeds in the Nilghiris. It is common in the Deccan and in the Khandala district especially. Mr. Ball writes, "The Crow-Pheasant is tolerably common throughout the Chota-Nagpur division," but "circumstances, which it is not easy to detect, seem to influence the distribution of this bird. In some portions of the district I have been for weeks without seeing a single specimen, suddenly then I come upon a tract in which I do not fail to hear or see several every day."

In the North-west Provinces it is also found, as well as in the plains of Upper India. Mr. Hume remarks that it is abundant along the banks of the larger rivers in Sindh, but that in lower Sindh it is less common than in upper. In the Sambhur-Lake district it is "very rare" (*Adam*).

Habits.—The Common Coucal inhabits almost every variety of situation except gloomy forest, the interior of which it shuns. In the south and west of the island it is found in low woods, cultivated lands, the outskirts of heavy jungle, compounds, native gardens, and the borders of paddy-fields, and is usually a shy bird, betaking itself, when flushed in the open, to the cover of the adjacent wood, and quickly climbing and making its way through the branches out of sight. In the north, particularly in the Jaffna peninsula, it is the very reverse of shy, walking about the native compounds sometimes close to houses, and exhibiting no concern with regard to the inmates. It may be that it finds its food scarcer here in the dry season than in the less parched-up districts in the south. It walks with an even and stately gait, or proceeds with long hops, and, when winged and pursued, runs with great speed through the jungle, and is exceedingly difficult to capture unless stopped with a second shot. Some of its habits are very curious; and Layard remarks with truth, "On being alarmed it scrambles rapidly to the summit of the tree in perfect silence, and glides away in a contrary direction to that whence the cause of its terror sprung." It resorts often to a favourite tree to roost, probably a shady "Jack," or, better still, an Areca-palm, of which it is very fond, and which generally stand in the vicinity of native houses. Into these it flies late in the evening, when it can take refuge in them unobserved, and then hides itself in the thickest part of the foliage. At daybreak the following morning its deep notes are heard issuing from the thick foliage and answered by the bird's mate, who is in another tree close at hand; but there is not a sign of either to be seen: this conversation goes on at intervals, and I have known it sometimes to last for twenty minutes before either of the Coucals stirs from the spot in which it has passed the night: when the time has come for a move, they hop out from their night's quarters, and fly away sometimes in opposite directions, and are seldom seen in close company during the day.

There is perhaps no bird-note in Ceylon so well known, nor one which strikes the new arrival from Europe with such astonishment, as the wonderful sound which this Cuckoo issues from its capacious throat. It is heard far and wide for miles on a still evening, and is so deep and weird-like that it is difficult to imagine it is produced by a bird, still less by so small a one as this. It consists of a single call quickly repeated, which may be syllabized as *hŏŏŏp, hŏŏŏp, hŏŏŏp*; and this is uttered with the mouth wide open and the bird's head thrust down sideways at each note, an exertion which appears necessary to bring out such a voluminous sound. The most lengthy description on paper would fail to give any idea of the nature of the voice of this and, still more, of the next species; but I am perhaps not wrong in maintaining that the luxuriant woods, the sequestered vales waving with verdant fields of rice, the forest-clad hills and shady palm-groves, all of which go to form the smiling face of nature in Lanka's isle, would lose no little of their charm for the ornithologist were they devoid of the Crow-Pheasant's resounding call.

It feeds on a great variety of insects and even reptiles, consuming beetles, slugs, scorpions, centipedes, lizards, and, I believe, small snakes sometimes. It pilfers birds' nests, and eats either eggs or very young birds. Mr. Parker informs me that he has seen one trying to get up the tube of a Weaver-bird's nest to attack the young in it, but in this it failed. Jerdon records the fact of a gentleman in the Indian Custom's department having seen one of these birds dragging along a young hedgehog by the ear, a task which it could not well have undertaken had it not contemplated making a meal off the unfortunate animal.

Nidification.—The species breeds from May until September. Its nest, which is not often discovered, is built in a low tree, generally in the midst of thick woods, and is a large globular structure, composed of twigs and small sticks, with an opening in the side near the top, and is fixed in a fork of a branch or among a mass of small thick boughs. One which I found close to the bungalow on the Gangaroowa estate was placed in a *Lantana*-thicket; it was near the top of a tangled mass of the branches of this well-known pest (*Lantana mixta*); the body of the structure rested in a large saucer-like foundation constructed by the bird of the branches of the *Lantana*, mixed with others brought to the spot; it was about a foot in external diameter, and the exterior was lined with roots. The eggs were two in number, stumpy ovals in shape, and of a chalky texture, although the surface was smooth; the colour was pure white in one and buff in the other, and they measured 1·54 by 1·14 inch and 1·45 by 1·16 inch.

In India it has been observed by Mr. Blewitt that the nests are not always domed, some that he has found being simply structures about the size of a large round plate, with a depression in the centre for the eggs; in some instances the nests are placed high up in large trees and in an exposed situation. Three appears to be the normal number of the eggs, although four or five are sometimes met with.

CENTROPUS CHLORORHYNCHUS.

(THE CEYLONESE COUCAL.)

(Peculiar to Ceylon.)

Centropus chlororhynchus, Blyth, J. A. S. B. 1849, xviii. p. 805; Gray, Gen. Birds, iii.
App. p. 22 (1845); Blyth, Cat. Birds A. S. B. p. 78 (1849); Kelaart, Prodromus,
Cat. p. 128 (1852); Layard, Ann. & Mag. Nat. Hist. 1854, xiii. p. 450; Cab. et
Heine, Mus. Hein. iv. p. 116 (1862); Blyth, Ibis, 1867, p. 298; Holdsworth, P. Z. S.
1872, p. 433; Legge, Ibis, 1874, p. 16.

Green-billed Jungle-Crow, Europeans in Ceylon.

Ætti-kukkula, Sinhalese, Western Province.

Similis *C. rufipenni,* sed rostro viridi et magis curvato: pileo et collo postico amethystino-purpureo nitentibus : inter-
scapulii plumarum apicibus scapularibusque concoloribus : remigibus terminaliter magis quam in *C. rufipenni*
infuscatis.

Adult male and female. Length 16·2 to 17·75 inches; wing 6·3 to 6·5; tail 9·0 to 9·5; tarsus 1·7 to 1·8: outer
anterior toe 1·35 to 1·5, its claw (straight) 0·5; outer posterior toe and claw 1·4, long posterior claw 0·7 : bill
to gape 1·6 to 1·75.

Iris deep red or dull crimson : bill pale apple-green, slightly pale along the margins; inside of mouth, except towards
the tips, orbital skin, and nostril-membrane black; legs and feet black; claws dusky, greenish at the base.

Entire plumage, except the wings, scapulars, and tips of interscapular feathers, black, glossed on the back of the head.
hind neck, upper part of interscapulary region, and the throat with purple, changing towards the tips of the
feathers into beautiful amethystine; the lower parts and upper surface of tail with blue, and the back with
obscure metallic green : the quills are dark chestnut, much more infuscated at the tips than the last species : the
wing-coverts and scapulars are darker still, or of a dull maroon, with the bases of the feathers blackish; under
wing-coverts blackish.

Young. The fledged nestling has the iris slate-grey ; bill dusky at base and along the culmen, with the apical portion
greenish ; legs and feet dusky flesh-colour.

Wings and scapulars red as in adult, black plumage the same : but the feathers of the head are encased in soft sheaths
or " pens," each of which terminates in a long white hair-like process, which in time drops off, the feather emerging
from the tip.

The *yearling* bird has the bill as in the adult, but with the tip of the lower mandible dusky. The upper plumage is
not so highly bronzed as in the old bird : wing-coverts obscurely barred with blackish, tips of quills more infuscated : inner webs of tertials concolorous with the tips.

Obs. This species is closely allied to the preceding, its most conspicuous distinguishing characteristic being its green
bill, which is also more curved than that of *C. rufipennis;* but the richer metallic hues and dark-tipped wings
would well suffice to separate it even were the bill of the same colour.

Distribution.—This handsome species was discovered by Layard in 1848 on the Avisawella road ; but one
specimen was then procured by him, which was forwarded to Blyth and described by this naturalist under its
present title. In 1852 Layard again met with it, securing another example at Hanwella and three more " in
the dense jungle near Pallabaddoola, at the foot of the Peak." These researches, therefore, gave but a very
small range, the extreme limits falling within forty miles. Mr. Holdsworth records the fact of seeing an
individual of the species once, but did not procure it. Mr. Neville, I understand, obtained several specimens
in the Western Province, probably between Ratnapura and Colombo, and was, prior to the date of my
acquaintance with it, the only collector who, besides Layard, as far as I am aware, ever procured it.

Instead of being so rare as was hitherto supposed, this " Jungle-Crow " exists in considerable numbers
throughout the tract of country which it inhabits. This consists of the south-west hill-region, ranging from

the many jungles near Galle up to the altitude of the coffee-districts of the Morowak Korale, the whole of the Western Province, and the strip of country lying between Kurunegala and Dambulla. In this latter region I do not think it extends into the Seven Korales beyond the influence of the hill rains. It is not uncommon on the Dedaru-oya and in the jungles between the Ambokka range and the outlying rocky hills, of which the Dolookanda forms the most conspicuous point; and I have met with it as far north as the Kimbulana-oya, where it is crossed by the direct road from Kurunegala to Anaradjapura *viâ* Ilambawe.

This portion of the Seven Korales is very dry, and this bird only inhabits there the heavy jungle on the borders of the seasonal rivers and streams. Whether it extends out to the north-west beyond the locality indicated I am unable to say; but near the hills I have traced it from Kurunegala up to the vicinity of Dambulla. To return to the Western Province, which is its head-quarters, this bird is there common in all the heavy forest and jungle, as well as in bamboo-cheena from Ambepussa to Ratnapura, inhabiting all outlying dense woods between this line and Colombo. About Hanwella, in the Ikkade-Barawe forest, in the jungles near Poré, and thence south to Horenue, its deep booming note may always be recognized by those who know it, and in the forest named it is abundant. I found it numerous in the Ratnapura district, and traced it up to Pallalmaddoola, which is high up (2500 feet) in the Peak forest. To this elevation, and perhaps somewhat higher, it doubtless ascends all along the western slopes of the Kandyan hills and round through the Peak jungles for some distance east of Ratnapura. Westward of this place I met with it through the Pasdun Korale to Agalewatta; and southward of this it will be found to occur sparingly in the jungles on either side of the Bentota river, and other heavily timbered localities between there and the Hinedun-Pattu hills. I have heard it near Denniya and in the Singha-Rajah forest. Near Galle it is met with in the Kottowe jungles. I have thus far taken pains to trace out the distribution of this little-known bird perhaps more minutely than may at first sight be thought necessary; but it seems expedient so to do, as it is so seldom seen that many who are not acquainted with its note would pass it over entirely did they not know in what districts to look for it. I cannot say how far eastward of Ratnapura it extends, nor whether it occurs on the eastern slopes of the Kolonna Korale; but in all probability future research will much extend its limits both in the south and probably also round the northern base of the Kandyan hills.

Habits.—Of all our forest birds perhaps the present species is the most wary and seldom seen, scarcely ever emerging from the almost impenetrable fastnesses in which it lives. The Ceylon Coucal almost defies all discovery except by those who have made themselves acquainted with its note and care to follow it into its retreat. It is a denizen of tangled thickets, underwood in forests and on the banks of rivers, dense bamboo-jungle (to which it is especially partial), ratan-cane brakes, and such like, and rarely shows itself in the open except by the side of a road passing through forest, to which it will drop for an instant from an adjoining tree on espying a grasshopper or other insect, quickly retreating again under cover before any but the quickest shot can secure it. In the early morning, when the bamboo-cheenas in the wild parts of the Western Province are resounding with its deep far-reaching call, it mounts up from the underwood into some creeper-covered tree, which is a favourite situation with it, and gives forth its sonorous, long-drawn *hoō—whoop*, *whoōp*, which can be heard with distinctness for many miles round, echoing far over the luxuriant glades and waving rice-fields into the distant beetling wooded crags, from which it is answered back by more than one of its lurking fellow mates; for, as is the case with its congener, one note thus given out is the signal for many more, called forth from all sides, until there is a sudden cessation, as if by common consent. As will be gathered from my remarks on its habits, it is an exceedingly difficult bird to procure; for years I had been seeking it in the jungles of Ceylon, knowing well that the loud peculiar Coucal-notes which I often heard in the damp forests of the west could not be those of any other bird, but was never able to procure a specimen, until one morning, in the Hewagam Korale, I penetrated into a dense bamboo-thicket towards a huge overgrown tree, in which one of these birds was sending forth an unusual number of its sepulchral calls, and succeeded in bagging it, thus identifying the species with its note and enabling me, by adopting this device, to procure many specimens, and to jot down in my notebook, on auricular testimony, its distribution wherever I went. Its habit is to call for several hours in the morning and evening, or after a shower of rain, when it mounts up into a tree to escape the dripping underwood and dry its plumage. When disturbed, or after re-alighting on being flushed, it has a very singular monosyllabic note, somewhat resembling the dropping of a stone into deep water, and which may be likened to the syllable *dhjöonk*; this is uttered by both sexes; but whenever I procured a specimen uttering the loud call in question it proved to be a male. Its diet consists of Coleoptera, spiders, snails, and grasshoppers, and in the stomach of one example I found a number of minute Ammonites. When winged it runs, like the preceding, very rapidly through the dense jungle and quickly escapes pursuit.

Nidification.—The breeding-season probably begins in April or May and lasts until July. In August I procured the nestling which forms the subject of the accompanying Plate, and which had not long left the nest. It was seated on a low branch in some dense underwood and uttered a sound resembling the note of the adult, but not so deep. On the first occasion that I heard it I was unable to find the bird, supposing it to be an old one which had flown away on my approaching it; but on passing the exact spot the following day I again heard the note, and succeeded in finding its author, which must have remained in precisely the same position during the 24 hours that had intervened. The nest and eggs are, in all probability, almost identical with those of the Common Coucal, the latter being perhaps somewhat smaller.

The figure in the Plate represents an adult bird, shot in the Seven Korales, feeding the nestling alluded to, which was procured in Mr. Chas. de Zoysa's fine forest at Kuruwite, where the species is abundant.

Genus TACCOCUA.

Bill higher than wide at the nostrils, the culmen much curved and hooked at the tip, the margin boldly lobed at the base. Nostrils exposed, basal, almost linear and pierced in a depression near the margin. Wings short and rounded, the 4th quill the longest. Tarsus as long as the inner anterior and outer posterior toes without their claws, covered with very broad scales.

Feathers of the fore neck and chest with very stiff shafts. Eyelid furnished with stout eyelashes.

TACCOCUA LESCHENAULTI.

(THE DARK-BACKED SIRKEER.)

Taccocua leschenaulti, Lesson, Traité d'Orn. p. 144 (1831); Blyth, J. A. S. B. 1845, xiv. p. 201; id. Cat. B. Mus. A. S. B. p. 77 (1849); Jerdon, B. of Ind. i. p. 352 (1862); Holdsworth, P. Z. S. 1872, p. 438 (first record from Ceylon); Hume, Nests and Eggs (Rough Draft), p. 145 (1873); Legge, Ibis, 1875, p. 285; Hume, Str. Feath. 1877, p. 219.

Zanclostomus sirkee, Jerdon, Cat. B. S. India, Madr. Journ. 1840, xi. p. 223; Blyth, J. A. S. B. 1842, xi. p. 98.

The Southern Sirkeer, Jerdon.

Jungli totah, Hind.; *Adavi chilluka* and *Potu chilluka*, lit. "Jungle-Parrot" and "Ant-hill Parrakeet," Telugu (apud Jerdon).

Adult male and female. Length 15·5 to 16·0 inches; wing 5·9 to 6·25; tail 8·2 to 9·0; tarsus 1·6 to 1·7; outer anterior toe 1·0, its claw (straight) 0·35; bill to gape 1·4 to 1·55. Weight 5¼ oz. Longest upper tail-covert feather 4·5 (about). The bill is very variable in size.

Iris reddish, with a brown inner circle and sometimes a yellowish exterior edge; bill cherry-red, with the tips yellowish and an angular black marginal patch continued along the edge to the gape; legs and feet bluish plumbeous or plumbeous, claws blackish; orbital skin blackish (?).

Above olivaceous brown, with a strong greyish-green lustre on the back, scapulars, and wings; the shafts of the head, neck, interscapular region, as well as the throat bristly and blackish in colour; tail metallic brownish green, becoming much darker towards the sides, the two outer pairs of feathers being deep brown above; all but the centre pair deeply tipped white, increasing towards the outer feathers.

Orbital bristles or eyelash black, with white bases; feathers of the lores and round the orbital skin whitish; chin and upper part of throat whitish, passing into pale brownish on the fore neck and chest; beneath this the under surface is rufous, deepest on the lower parts and tinged with yellowish on the breast; vent and under tail-coverts grey-brown, the feathers of the latter tinged with rufous at their extremities; rectrices dark brown beneath.

Examples vary in the depth of the rufous of the under surface, and in those which have it deep the throat is pervaded with a fulvous hue.

Young. Birds of the year have the wing-coverts, tertials, and scapulars tipped strongly with fulvous.

Obs. Ceylonese examples all belong to the dark-backed race, considered to be the typical *leschenaulti*. Four species have been recognized of this genus, two of which were separated by Mr. Bligh from Lesson's and Gray's types (*T. leschenaulti* and *T. sirkee*) and styled by him *T. infuscata* and *T. affinis*. All four are very closely allied; and

Mr. Hume, who appears to have now a larger series than has ever been before got together, writes that he can only satisfy himself of the existence of two forms—the present, with the dark olive-brown back, and *T. sirkee* with the pale sandy or satiny-brown upper surface. From an examination of a small series in the British Museum from different localities, I think that his conclusions are likely to prove correct. Three examples from Capt. Pinwill's collection, now in the British Museum, measure in the wing 6·4, 5·9, 6·3 inches, two exceeding my maximum dimension; these are the dark-backed race; but they differ slightly from the Ceylonese bird in the forehead being somewhat rufous and in the rufescent hue of the breast ascending up the throat: the island race is characterized, on the contrary, by its darker or grey fore neck and whitish chin, and the forehead is concolorous with the crown. Typical specimens of *T. sirkee*, the Bengal species, have the back, scapulars, and wing-coverts very pale sandy yellowish above, and the throat and fore neck very pale rufous.

Distribution.—The first specimen of this curious Cuckoo procured in the island was killed by Mr. Forbes Laurie in Dumbara, whither the species ascends from the low country at Bintenne. It is not a rare bird, but, being very shy and inhabiting the densest thickets, appears to have escaped the researches of ornithologists previously working in Ceylon. Its head-quarters, I consider, are the hot jungle-clad districts lying to the south of Haputale and stretching thence from the eastern slopes of the southern ranges through the Bootala and Maha Vedda Ratas to the country lying between Bintenne and the east coast. Thence it ascends the mountain-slopes—on the south, those of the Badulla and Haputale ranges; on the east, those extending from Hewa Elliya past Maturata to Medamahanuwara. Although I have not met with it north of the latter region, it is most probable that it inhabits the whole of the Vedda country round the "Gunner's Coin" mountain almost to the Virgel river, for this is precisely similar in character to that about Kattregama, where I first saw it and where it is common. I have procured it in the Wellaway Korale, and Mr. Bligh has killed several specimens above Lemastota at about 2500 feet elevation. It is pretty common near Nilgulla, inhabiting the open jungle on the elevated ebennas between Keloday and Bibile. Here I saw three or four specimens in a single day. The most elevated region in which it has as yet been observed is the Uva patna-district, in which I have met with it near Wellemade on a hill about 3500 feet in altitude. This portion of the Central Province, consisting of steep patnas and deep wooded ravines, is little known to naturalists, or, in fact, to any but occasional sportsmen, who descend to it from the neighbouring coffee-estates either for Snipe- or Partridge-shooting. It attains an altitude near Bandarawella of about 4000 feet, and on the north-east slopes away to Badulla, and thence into the low country at Teldeniya, where the Sirkeer is found, and whence it ascends into the patnas, very probably inhabiting the whole region. On the mainland this species is found in Southern India. Jerdon writes that he procured it on the Eastern Ghâts, in the Deccan, and on the Nilghiris, finding it in grassy slopes from 5000 to 6000 feet elevation. Its range, however, would appear to extend to the north of India. I have seen specimens from the N.W. Provinces; and Mr. Hume has it from Dehra Doon and Kumaon Bhabur (still further north), as also from Sumbhulpoor, Raipoor, Khandala, and other places in Central India. The Bengal species is found in the Sambhur, Guzerat, Kutch, and other western districts, as well as in other parts of the Presidency.

Habits.—The Sirkeer is a shy bird, frequenting dry jungle in open grassy country, low scrub, tangled thickets, and bushy patna-tracts in the Central Province. It feeds almost entirely on the ground in long grass, never straying far from its native fastnesses, and, as far as I have been able to observe, only issuing from them in the morning and evening, at which times it principally feeds. It is found by the sides of jungle-roads and on patches of ground under native cultivation which are surrounded by dense scrub. I have, following the winding native track, more than once entered these enclosures, generally from 5 to 10 acres in extent, and immediately on my emerging from the wood into the open have espied one of these birds at the far end making off instantly for the cover; on alighting at the edge of the jungle they quickly thread their way, like a *Centropus* ("Jungle-Crow"), from branch to branch, and are not many seconds before they disappear into the impenetrable thicket around them.

Its diet principally consists of grasshoppers, Mantidæ, and other insects, which it captures in long grass and with which it crams itself to excess. Mr. Bligh writes me that since I left the island he shot one near Lemastota with a freshly killed brown lizard in its stomach; it was very thick and about 8 inches long, and

2 м 2

was coiled away neatly, even to the tip of its tail ; it had one deep cut across the brain-region nearly severing its head in two.

Jerdon writes that in India it is seen much about white ants' nests, whence its Telugu name, "the appellation of Parrot being given to it from its red bill."

I know nothing of its nesting ; but Mr. Bligh writes me that a female killed during the S.W. monsoon showed signs in its breast-plumage of having lately incubated, which points to the breeding-season being in June and July.

Fam. TROGONIDÆ.

Bill short, stout, very wide at the gape, and consequently somewhat triangular ; culmen curved. Nostrils concealed by bristles. Tail of twelve feathers, long and broad, sometimes surmounted by a long caudal train. Tarsus very short ; feet small and zygodactyle.

TROGONIDÆ.

Genus HARPACTES.

Bill very short and broad; upper mandible deep; culmen much curved, the tip with a small notch; gonys short, deep, and much ascending. Nostrils basal, narrow, situated in a membrane, which is protected by bristles. Chin furnished with weak bristles. Wings short, the primaries much decurved; the 4th and 5th quills the longest, the 1st rather short. Tail broad, much graduated, even at the tip. Tarsus half-feathered; inner anterior toe slightly longer than the outer; inner posterior toe much longer than the outer one.

Eye surrounded by a naked skin.

HARPACTES FASCIATUS.

(THE CEYLONESE TROGON.)

Trogon fasciatus, Forster, Ind. Zool. p. 34, pl. 5 (1781); Gm. ed. Linn. Syst. Nat. i. pt. 1, p. 405 (1788).

Trogon ceylonensis, Briss. Orn. vol. ii. p. 19 (1763).

Trogon malabaricus, Gould, Mon. Trogonidæ, 1st ed. pl. 31 (1838).

Harpactes malabaricus, Sw. Classif. Birds, vol. ii. p. 337 (1839); Jerd. Cat. B. S. India, Madr. Journ. 1840, xi. p. 232.

Harpactes fasciatus, Blyth, Cat. B. Mus. A. S. B. p. 80 (1849); Kelaart, Prodromus, Cat. p. 118 (1852); Layard, Ann. & Mag. Nat. Hist. 1853, xii. p. 171; Horsf. & Moore, Cat. B. Mus. E. I. Co. ii. p. 714 (1856); Legge, J. A. S. (Ceylon Br.), 1870–71, p. 35; Holdsworth, P. Z. S. 1872, p. 422; Hume, Str. Feath. 1873, p. 432, et 1876, p. 498; Ball, ibid. 1874, p. 385, et 1876, p. 231; Legge, Ibis, 1874, p. 13, et 1875, p. 281; Bourdillon, Str. Feath. 1876, p. 382; Fairbank, ibid. 1877, p. 393.

Pyrotrogon fasciatus, Cab. et Heine, Mus. Hein. iv. p. 156 (1862).

Der Band-Kuruku, Forster; *The Fasciated Curucui*, Gmelin; *The Fasciated Trogon*, Kelaart; *Red Flycatcher*, Europeans in Ceylon.

Kufui churi, Hind.; *Karra*, Mahrattas; *Kakarne hakki*, Canarese.

Nawa nila kurulla, Ranwan kondea, Ginni kurulla, Sinhalese.

Adult male and female. Length 10·5 to 11·2 inches; wing 4·4 to 5·0 (average about 4·7); tail 5·4 to 6·0, outermost feather 3·0 shorter; tarsus 0·55 to 0·7; inner anterior toe 0·6, its claw (straight) 0·3; inner posterior toe 0·4; bill to gape 0·9 to 1·05.

Females slightly the smaller of the sexes.

Iris hazel-brown or reddish brown, in some with a pale outer circle; bill, orbital skin, eyelid, and gape French blue, the orbital skin being the palest; culmen and tips of mandibles black; legs and feet delicate greyish blue, claws bluish horn.

Male. Head, nape, face, and chin dull black, paling gradually to dark slate on the fore neck and upper part of the chest; hind neck partially denuded of feathers; back and scapulars yellowish olive-brown, paling into rufescent fulvous on the rump and upper tail-coverts; least wing-coverts concolorous with the back, the remainder of the wing black, crossed with narrow bars of white on the wing-coverts, tertials, outer webs and tips of secondaries; all but the first primary with a clearly-defined white outer edge; the three centre pairs of tail-feathers cinnamon-rufous, the central pair almost entirely so, with a fine black tip; the next two black at the tip and on the terminal portion of the inner web; the next two with almost all the inner webs black; the three outer pure white on the terminal half, black on the basal, and with a rufous edge except on the outermost.

Female. Has the back and rump as in the male; but the head and hind neck are brown, darker than the back: the throat and fore neck light olive-brown and the chin blackish; the wing-coverts, outer webs of secondaries, and the tertials are barred with bands of fulvescent-rufous, broader than the white bars of the male: breast and under surface fulvous, the white pectoral band wanting.

Young male. Bill and orbital skin duller than in the adult. In nest-plumage the male has the head and face slaty black, back and tail as in the adult: the median wing-coverts with narrow bars of fulvous, and the outer webs of the secondaries with broader bars of the same, slightly paler than these markings in the female; the chin is black and the fore neck slate-colour: the under surface is paler fulvous than the adult female, and the white pectoral band is present. An individual shot in January, in the Northern Province, has the wing-coverts with white-and-rufous barred feathers, and the under surface with fulvous and scarlet ones.

Obs. Mr. Hume has called attention (*loc. cit.*) to the fact that Ceylonese examples are smaller than Indian; and he points out the following difference in the tail of the island race:—"Instead of the central tail-feathers being entirely chestnut with moderately black tips, and the next pair entirely black, they have all the four central tail-feathers black on the inner webs and on the outer webs for about one inch, the rest of the outer webs being chestnut." As a matter of fact the pair adjacent to the central one have the black only on the inner web, at least in a good series I have obtained, so that these feathers may be said to be almost entirely rufous, which is a great dissimilarity to the same in the Indian bird. I have not been able to examine any South-Indian specimens, and cannot express an opinion as to whether it is the rule to find them with such black tails. If the Indian species is to be separated, it must bear another name, as it is the Ceylonese bird which is *fasciatus*, it having been described by Forster, in his 'Indische Zoologie,' from Ceylon.

Mr. Fairbank gives the following measurements of specimens killed in the Palanis:— ♂, length 12·5 inches, wing 5·0, expanse 16·0, tail 7·0, bill from gape 1·1: ♀, length 12·0, wing 5·0, expanse 15·75, tail 7·0, bill from gape 1·0. An individual shot in Sambalpur by Mr. Ball measures—length 11·5, wing 5·0, tail 7·0. From these dimensions it would appear that Indian examples differ chiefly in the length of the tail, but do not much exceed Ceylonese ones in the wing.

Forster's plate of this species is a good representation of it: the figure is that of a male bird lying on the stump of a tree.

Distribution.—This very handsome bird is widely diffused throughout Ceylon, and is by no means uncommon, although, being entirely a denizen of the forest, it is not much known among Europeans. In all parts of the island it is found wherever there is lofty jungle, which it frequents by choice. It is met with near Colombo, at Atturugeria and Ikkade Barawe, and inhabits the forests in the interior of the Western Province. In the south it is found in the timber-jungles near the Gindurah, those throughout the Hinedun Pattu, and in the Kukkul and Morowak Korales. The Singha-Rajah forest is a great stronghold of this species; its gloomy ravines clothed with fine timber-jungle, entwined in many places with enormous ratan-canes, which flourish on the incessant rains of that region, afford it a paradise. In the Eastern Province I found it common in the Friars-Hood hills, in the Nilgalla district, and other localities clothed with heavy jungle. In the north it is locally distributed, being confined to heavy forest, in which I have procured it about 15 miles from Trincomalie. At the northern base of the Matale ranges it is common, and is diffused throughout all the coffee-districts, ascending to the upper ranges in the dry season. Mr. Holdsworth met with it at Nuwara Elliya in February, and I have seen it at Kandapolla in January.

In India Jerdon found it in the forests of Malabar, from the extreme south up to about north latitude 17°, reaching up the Ghâts and hill-ranges to at least 3000 feet. Referring to 'Stray Feathers,' we find Mr. Fairbank procuring it first on the Palani hills at an elevation of 3500 feet, and finding it up to 5000 feet elevation. Mr. Bourdillon records it as a common bird in heavy jungle on the Travancore hills above 1000 feet; north of this region the former gentleman notices it as found in the woods of Sawant Wade, in the Khandala district. In the Central Provinces Mr. Thompson has procured it in the Ahiri forests, in lat. 19° 30'; Mr. Ball at

Jaipur, and also at Rehrakole in 21° N. lat.; and Mr. Blanford has obtained it further to the east in the Godaveri valley. Rehrakole appears to be the most northerly locality to which its range has as yet been traced.

Habits.—The gloomy recesses of the forests this Trogon inhabits serve to bring out its beautiful plumage in striking relief; nothing can form a greater contrast than its brilliantly-coloured breast does with the sombre trunks and subdued foliage of the timber-jungles in the south of Ceylon. Were it not for its shyness in taking wing at the sight of man, it would seldom be observed; for it loves to perch across some horizontal limb, many feet from the ground, and there remains utterly motionless, with its head sunk between its shoulders, until the sight of a passing moth rouses it into activity, and it launches itself out with a loud fluttering of its wings, seizes the prey, and starts off to another branch not far distant from its first. It sits bolt upright, and when viewed from behind appears to have no neck and but very little head! The natives of India have named it *Kufni churi*, from this singular appearance, as if dressed in a fakir's "kufni." I have usually found it in pairs, and not solitary, although the two birds are seldom seen close together; but if one be shot the other will almost sure to be seen close at hand. It is this bird which makes the curious monosyllabic note *chok*, which is often heard in the Ceylon forests; for many years I was unable to identify this sound with any species, until I saw a Trogon in the act of uttering it in some dense forest near Ambepussa. It has another purring call, which it commonly utters; but I am not aware that the Ceylonese birds have any querulous note like the mewing of a cat. Mr. Bourdillon says that it gives this out continuously in the Travancore forests. In the recesses of the timber-jungles in the south of Ceylon, considerable tracts of forest may be traversed without seeing or hearing a single bird; as the naturalist is perhaps commenting on the dearth of bird-life, he suddenly comes on a sociable little troop of his feathered friends, who seem to have collected together in these lonely solitudes for companionship's sake: several Forest-Bulbuls (*Criniger ictericus*) and some Black-headed Bulbuls (*Rubigula melanictera*) are sure to be among the assembly, the rest of which is made up with one or two Azure Flycatchers (*Myiagra azurea*) and a casual *Pomatorhinus* leisurely uttering its melodious call as it clings to the mossy bark of some giant trunk, while, lastly, at a little distance from the sociable gathering, sits aloof a solitary Trogon, as if it had come to see what was the matter, but scorned to associate with its lively neighbours. Jerdon remarks that he has sometimes seen four or five of these birds together.

The food of this species consists chiefly of coleopterous insects, bugs (Hemiptera), moths, &c., which it catches on the wing like a Flycatcher; and hence its ordinary name with gentlemen in the Survey Department, and others who frequent the jungle and have made its acquaintance. It is peculiar for the extraordinarily delicate nature of its skin and consequent looseness of the body-feathers, which fall out in abundance on the bird striking the ground when shot. It is on this account that the Trogon is the most difficult of all Ceylonese species to preserve for the cabinet.

I know nothing certain as to its nidification; but a gentleman in the Survey Department assured me that he found a nest with two young ones in a Kitool-palm during the month of May. It was situated in a hole in the trunk of the palm which stood near his hut in the Three Korales, and the young were lying on the bard wood of the nest-cavity.

Fam. BUCEROTIDÆ.

Bill very large, curved from the base, with or without a casque on the upper mandible. Nostrils small, pierced in the bill, without a membrane, at the junction of the casque with the upper mandible or near the ridge. Wings short. Tail long, of ten feathers. Tarsus short. Feet syndactyle ; three toes in front.

Tongue short and heart-shaped. Sternum wider at the posterior edge than in front, and with a shallow emargination on each side.

Genus ANTHRACOCEROS.

Bill enormous, curved from the base to the tip ; the upper mandible surmounted by a long, high, and sharp casque, its anterior edge projecting forward. Nostrils narrow, situated at the base of the casque ; orbital and gular skin nude. Wings short and rounded, the 1st three quills evenly graduated ; the 1st short and the 5th and 6th the longest ; tertials reaching beyond the primaries. Tail very long, of ten feathers. Legs and feet stout, covered with broad, prominent, transverse scales. Tarsus longer than the middle toe ; toes syndactyle, the outer connected with the middle as far as the last joint ; sole very broad, claws short and stout.

ANTHRACOCEROS CORONATUS.
(THE CROWNED HORNBILL.)

Buceros coronatus, Bodd. Tabl. Pl. Enl. p. 53 (1783) ; Blyth, Ibis, 1860, p. 352.

Buceros violaceus, Shaw, Gen. Zool. viii. p. 19 (1811) ; Blyth, J. A. S. B. 1849, p. 803 ; Kelaart, Prodromus, Cat. p. 126 (1852).

Buceros malabaricus, Tickell, J. A. S. B. 1853, ii. p. 579 ; Jerdon, Cat. B. S. India, Madr. Journ. 1840, xi. p. 38 ; Layard, Ann. & Mag. Nat. Hist. 1854, xiii. p. 260.

Anthracoceros coronata, Reich. Syst. Av. pl. 49 (1849).

Hydrocissa coronata, Horsf. & Moore, Cat. B. Mus. E. I. Co. ii. p. 588 (1856) ; Cab. et Heine, Mus. Hein. ii. p. 170 (1860) ; Jerdon, B. of Ind. i. p. 245 (1862) ; Holdsworth, P. Z. S. 1872, p. 425 ; Ball, Str. Feath. 1874, p. 387.

Anthracoceros coronatus, Elliot, Mon. Bucerotidæ, pt. iv. (1877).

The Large Hornbill, Kelaart ; *The Malabar Pied Hornbill*, Jerdon ; *Toucan, Double-billed Bird*, Europeans in Ceylon ; *Danchuri*, Hind. ; *Bagma-dunes*, Bengal. ; *Wayera*, Mahrattas ; *Peshta-ganda*, Gonds ; *Suliman murghi*, lit. " Solomon's Fowl," Musselmen in South India ; *Kuchla-kha* in Goomsoor (Jerdon).

Porowa kǣndetta, lit. " Axe Hornbill " (from the shape of the bill), Sinhalese ; *Atta-kǣndetta, apud* Layard ; *Errana-chundoo-kuruvi*, Ceylonese Tamils, lit. " double-billed bird " (*apud* Layard).

Adult male. Length 36·0 inches ; wing 13·0 to 13·3 ; tail 13·0 : tarsus 2·5 ; middle toe 2·1, its claw (straight) 0·75 ; hind toe 1·1, its claw (straight) 0·8 ; bill from gape to tip across the arc 7·0, casque along ridge 7·5 to 9·5, height of bill with casque 4·0.

Adult female. Length 34·5 inches ; wing 12·75 : tail 14·0 : bill from gape to tip across are 6·8, casque along the ridge 7·0 to 8·5.

The casque projects back over the crown and gradually becomes compressed to a sharp edge at its anterior part, which recedes downwards to the mandible, joining it about 2¼ inches from the tip. The size of the projection forward beyond the point of contact and the consequent angle of connexion depend on age.

Iris crimson ; eyelid black ; orbital skin and gular region "fleshy;" bill and casque fleshy white ; above and beneath the gape, the posterior face of the casque and its anterior three fourths black, the colour *never descending onto the mandible*, and not reaching quite to the anterior edge of the casque : legs and feet blackish leaden colour ; edges of tarsal scales whitish, soles yellowish.

In the female the black at the gape does not extend to the upper mandible, nor is the posterior edge of the casque black.

Entirely glossy green-black, except the under surface from the chest downwards, the terminal portion of the secondaries and all but the first two primaries, the three outer tail-feathers, and terminal half of next pair, all of which parts are pure white ; base of primaries whitish.

In some examples the tips of some of the tertials are white, as also those of the centre tail-feathers ; while the 4th tail-feather is sometimes entirely white, and the corresponding one perhaps of the normal colour.

Young. *In the bird of the year* the casque is partly undeveloped, the posterior edge is perpendicular, and the anterior portion grades into the ridge being continuous with that of the tip. In the second year the anterior projection of the casque begins to develop. The bill is devoid of the black, there being merely a dusky patch at the gape and a slight dark wash near the anterior portion of the casque.

A male shot at Jaffna measures :—wing 12·3 inches : tail 13·2 ; tarsus 2·5 : bill across are, gape to tip 5·4, along gape 5·5.

A female :—wing 11·8 inches : tail 11·5 ; tarsus 2·5 ; bill, gape to tip across are, 4·85.

The terminal 2 inches of the primaries only are white, while in the adult this colour extends to 3 inches from the tip ; on the secondaries the white diminishes to ¼ inch on the innermost feather.

Obs. Ceylonese individuals are quite as fine as those from India. Mr. Ball gives the wing of a Chota-Nagpur male as only 11·25 inches, and the bill from gape 6·2. The present species is closely allied to *A. malabaricus*, which has been described under the names of *Hydrocissa albirostris* and *H. affinis*, and frequently referred to by these titles in the writings of Indian naturalists. It differs from the present species in the slightly smaller casque, which has the black patch *extending onto the upper mandible*, and in the coloration of the tail-feathers, the three outer pairs of which have the terminal portions only white instead of being entirely so, as in *A. coronatus*.

Distribution.—This fine Hornbill frequents the wild *dry* jungle-districts of the low country, perhaps ascending into the Haputale range and up the eastern slopes of Madulsima, Medamahanuwara, and the Knuckles to some elevation during the N.E. monsoon. Commencing in the south, its range begins in the Hambantota district, where it is numerous, and, taking in all the forest-country up to Lunustota, extends northward through the eastern and northern portions of the island of Jaffna. Down the west coast it is found as far south as Chilaw and the Seven Korales ; but near Kurunegala itself I was unable to detect its presence, although I searched diligently for it. I have seen specimens from the Kurunegala district ; but I imagine they must have been killed nearer Puttalam than that place, for Mr. Parker tells me that it is found at Uswewa, but probably does not extend further inland than Nikerawettiya. It occurs throughout the interior of the north-central part of the island, but not so commonly as near the coast, along which it is always more abundant than further inland.

Layard speaks of a second species of Pied Hornbill which he said he saw twice in the hills ; he supposed it to be the *Buceros albirostris*, above referred to. On one occasion his collector "Muttu" saw it at Gillymally in forest. As will be seen, the slight differences existing in this species are not such as could ensure its identification on the wing ; and I am therefore of opinion that Layard must have met with the immature of the present bird, the peculiar bill of which might have led to the supposed identification of a

2 N

new species. I was told by a native superintendent that a large black-and-white Hornbill is seen sometimes in the jungles at the eastern end of the Haputale range; but I have no doubt that it is the present species, which ascends from the low country to the higher jungles during the N.E. monsoon.

Jerdon remarks that the Malabar Pied Hornbill is found in all the heavy jungles of Southern India, and that he met with it in Malabar, Goomsoor, and Central India. It does not seem to be an inhabitant of the hills in the extreme south of the peninsula, where, however, the Great Hornbill (*Homraius bicornis*) is found. Mr. Fairbank records it from Ratnaghiri, near Bombay, and Mr. Ball from Chota Nagpur; the latter writes, "The Malabar Pied Hornbill affects certain localities in Chota Nagpur, where it may generally be found in a flock numbering from 6 to 10 individuals. I have shot it in Manbhum, Singhbhum, and Sirguja, and seen it in the fine jungles which border the Ghât from the Ranchi plateau to Purulia."

Habits.—The Crowned Hornbill lives in small parties, frequenting the tops of trees and feeding on the many fruits with which the Ceylon jungles abound. These it swallows whole, whether large or small. Layard says that to procure its food, "when attached to a branch, it resorts to an odd expedient—the coveted morsel is seized in its powerful bill, and the bird throws itself from its perch, twisting and flapping its wings until the fruit is detached; on this the wings are extended, the descent arrested, and the bird regains its footing."

An individual which Layard kept in captivity was observed to use its bill in recovering its perch in the same manner that a Parrot would do, except that instead of the upper mandible only it employed the whole of the bill to hook itself on by. It is a shy bird, taking wing at once on seeing itself approached; but it usually does not take long flights; when it does the momentum of its huge bill and heavy neck are such as to cause it on alighting to topple forward before gaining its equilibrium. When flying it proceeds with rather quick flapping of the wings, and then sails along with them outstretched, its long tail and motionless primaries giving it a singular aspect. It has a loud harsh note, and is very noisy in the morning and evening, three or four together without much difficulty making themselves heard far and wide. In the jungles of the eastern side of the island it is partial to the tall forest-trees growing on the margins of the rivers, as in the less fertile tracts away from the influence of the water there is not so much means of subsistence for it, except where the iron-wood tree is to be found, the luscious fruit of which attracts to it every fruit-eating bird in the forest. It is likewise very fond of the banyan fruit. Layard remarks that they are often to be seen feeding on the ground; but this I have never been fortunate enough to see myself.

Nidification.—This bird breeds in the cavity of a tree, and the male, as is the case with other species, closes up the entrance while the female is incubating her eggs, leaving a small hole only sufficiently large to admit of his feeding his imprisoned partner. After the young are hatched the mud wall is broken down either by the male or the female, and both assist in feeding their offspring. In the case of the present species we have nothing but native evidence in support of this extraordinary habit; but I think it may well be credited, in the face of what has been seen by reliable witnesses of the nesting of other Hornbills. The natives attribute the cause of this strange proceeding to the birds' fear of the monkeys, which inhabit the Ceylon forests in such numbers: be this as it may, I doubt not that the incarceration actually does take place; and it would be very interesting if some undeniable proof of it could be obtained by observation on the part of some of my readers in the Ceylon Civil Service or the Public Works Department, who, by offering a reward for the finding of a nest in the forests surrounding their Station, might perhaps succeed in making some valuable notes on the subject. I have no information concerning the eggs of the Crowned Hornbill, for they do not appear, as yet, to have been procured.

Genus TOCKUS.

Bill much smaller than in *Anthracoceros*, without the casque, but with the ridge of the culmen sharp and slightly elevated, and with the sides of the upper mandible vertical at the base; the cutting-edge serrated. Nostril basal and round ; orbital region wide. Eyelids furnished with stiff lashes. Wings, tail, and feet as in the last genus.

TOCKUS GINGALENSIS.

(THE CEYLONESE HORNBILL.)

(Peculiar to Ceylon.)

Buceros gingala, Wilkes, Encycl. Lond. iii. p. 480 (1808).
Buceros gingalensis, Shaw, Gen. Zool. viii. p. 37 (1811); Temm. Pl. Col. ii. p. 17 (1824);
    Blyth, Cat. B. Mus. A. S. B. p. 44 (1849); Kelaart, Prodromus, Cat. p. 126 (1852);
    Layard, Ann. & Mag. Nat. Hist. 1854, xiii. p. 260 ; Schlegel, Mus. P.-B. 1862, p. 12.
Buceros pyrrhopygus, Wagl. Syst. Av. (1827).
Tockus gingalensis, Bonap. Consp. Gen. Av. 1850, p. 91; Jerdon, B. of Ind. p. 250 (1862,
    in part); Blyth, Ibis, 1867, p. 296 ; Jerd. Ibis, 1872, p. 5 ; Holdsworth, P. Z. S. 1872,
    p. 425 ; Legge, Ibis, 1874, p. 14, et 1875, p. 282 ; Elliot, Mon. Bucerotidæ, pt. iv. (1877).
Rhinoplax gingalensis, Bonap. Consp. Vol. Anisod. 1854, p. 3.
Buceros (Penelopides) gingalensis, Von Mart. Journ. für Ornith. 1866, p. 18.
The Small Hornbill, Kelaart ; *Toucan*, Europeans in Ceylon ; *The Grey Hornbill* or *Jungle Grey Hornbill*.
Kœndetta, Sinhalese.

♂ *ad.* suprà sordidè cinerascens, pilei plumis vix brunnescentioribus medialiter albido obscurè lineatis : tectricibus alarum purius cineraceis, nigro limbatis : remigibus nigris vix viridi lavatis, primariis ad basin extremam albis, primariis medialiter albo extùs marginatis et albo latè terminatis, secundariis extùs cineraceis angustè albo limbatis, intimis dorso concoloribus ; rectricibus centralibus cineraceis, reliquis viridi-nigris, basaliter cineraceis, exterioribus latè albo terminatis : regione parotica nigricanti-brunneâ, albido angustè striolatA : genis et corpore subtùs toto albidis, crisso rufescente : tibiis cineraceis : rostro albido, frontem versùs rufescente, culmine et mandibulâ nigricantibus : pedibus cinereis, unguibus nigris ; iride rubrâ.

♀ haud a mare distinguenda.

Adult male and female. Length 22 to 23 inches; wing 8·0 to 8·3 ; tail 0·0 to 0·5 ; tarsus 1·6 to 1·7 ; middle toe 1·3, claw (straight) 0·55 ; bill, gape to tip, straight 3·9 to 4·3, along culmen 4·3 to 4·4. Expanse 27·0.
Iris red ; orbital skin and eyelash black ; bill fleshy white in some, with a reddish tinge adjacent to the forehead, the vertical part of the upper mandible black, lower mandible with a blackish patch beneath ; legs and feet slaty bluish or greenish plumbeous, claws blackish.
Head and nape reddish cinereous brown, each feather with a pale mesial stripe; ear-coverts blackish brown, with pale centres ; back and upper tail-coverts cinereous brown, paling to slaty on the hind neck, and with a slightly rufous cast on the back in some ; wing-coverts greyish slate, the feathers margined with blackish ; quills black, the outer webs of secondaries mostly slaty, with a still paler edge ; terminal portion of 3rd, 4th, 5th, and 6th primaries white ; tail greenish black, the central feathers pervaded with a cinereous hue, and the terminal portion of the remainder white ; normally this extends to half the feather on the two outer pairs, and decreases on each

2 X 2

succeeding pair ; beneath, including the sides of the neck, greyish white, the vent and under tail-coverts rufescent yellowish, and the thighs bluish cinereous.

Young. Birds of the year have a total length of about 20 to 21 inches ; wing 7·7 to 7·0 ; bill from gape to tip (straight) 3·2 to 3·6.

The bill is shaped somewhat differently from the adult, inasmuch as the perpendicular lateral portion extends forward until it meets the margin ; with age the upper edge of this " wall " disappears, leaving only about ½ inch of this part at the base of the mandible.

Iris red ; bill black, usually a white stripe of greater or less extent on the wall of the bill, and in some with patches of the same on the lower mandible ; legs and feet bluish brown.

Head and hind neck darker than in the adult ; under tail-coverts perhaps, as a rule, more rufous.

Obs. The amount of black even on the bill of the adult varies slightly. In the young stage this bird was thought by Layard to be perhaps a different race : he had only procured specimens in one district, viz. the south, which coincidence, I suppose, strengthened his belief as to there being two species in the island. I was, at one time, inclined to think that he might perhaps be correct in his supposition, basing my ideas, however, on a difference of size in the bill ; but a good series, afterwards collected by me, demonstrated both the cause of the black bill and the variability in size of that of the adult. The development of white in some specimens is more than in others ; in certain individuals the penultimates may both be entirely white, while one of the primaries in others may be similarly coloured.

Tockus gingalensis is allied to the South-Indian *T. griseus*, which Jerdon confounded with it in his notice of the Indian bird (*loc. cit.*). The latter has the plumage more of a brownish grey than a slate-colour ; the bill is reddish at the base, paling to yellowish at the tip ; the orbital skin is purplish.

Tockus griseus has not yet been detected in Ceylon.

Distribution.—This Hornbill, commonly known, as is also the last, by the name of Toucan, is an inhabitant of most of the tall forests and heavy jungles of the low country, ascending the mountains of both the Central and Southern Province, in the former of which I have met with it at an elevation of 4000 feet. It is plentiful throughout the northern forests, and Mr. Holdsworth found it inhabiting the scrub-country round Aripu.

I do not know that it has been detected in the Jaffna peninsula, but it may possibly be found in the jungles near Elephant Pass. Passing over the Seven Korales and the Puttalam district, in which it is tolerably plentiful, we find it in the forests about Ambepussa and Avisawella, in the Raygam and Hewagam Korales, in Saffragam, the Pasdun and Kukkul Korales, and in the jungles between Galle and the "Haycock." In the forest of Koltowe I never failed to notice it whenever I visited that place. In the Wellaway Korale and the Friars-Hood Hills it is likewise tolerably frequent. As regards the Kandyan Province I think it is commoner in Uva than elsewhere ; I have seen it from the Knuckles district, and have been told that it has occurred in the main range at Kandapolla ; to such an elevated region, however, I should say it could only be a straggler during the dry season, unless, indeed, it be a resident in Udu-pusselawa, from which it would naturally extend to the jungles above the Elephant Plains.

Habits.—The Ceylonese Hornbill is a shy bird, frequenting the tops of tall trees, and rarely descending into the low jungle beneath them. In the lofty timber-forests of the south and west, therefore, it is difficult to procure ; but in the north, where the jungle is of altogether a different character (thick, with rather low trees), it may easily be shot, as the dense wood conceals the sportsman, and the distance of the bird from him is much less than when it is feeding in the top of some noble Keena-tree, or *kaing* in the upper branches of a gigantic Hora. It generally consorts in troops of five or six and is very noisy, its note being a loud laugh, commencing with the syllables *kū-kā-kā*, slowly uttered, and then quickening into *kakakaka*. In the early morning it roams about a good deal in search of fruit, but after feeding is not much on the wing. Its flight, like that of the last species, is laboured and slow ; it is a combination of flapping of the pinions and quick dips, particularly when descending to alight on a tree. Its diet consists mainly of fruit, that of the Banyan, Bo, wild cinnamon, and Dawata (*Carallia integerrima*) being much in favour with it ; it also devours reptiles and insects, for I have found green lizards and scorpions in the stomachs of some individuals. Its flesh is tender and not distasteful, and when subjected to the usual jungle-test (curry), makes a meal which the hungry hunter is far from despising ; on such occasions it is always in great demand with one's Cingalese and Tamil servants.

I have never been able to procure any information concerning its nesting beyond the native assertion that it breeds in hollow trees like the last species.

The figures in the Plate represent an adult in the foreground, and an immature bird (placed by the artist, failing a knowledge of its habits, upon a cocoanut-tree) in the background. The feet and legs, I regret to say, have been coloured much too dark.

PICARIÆ.

Fam. UPUPIDÆ.

Bill very slender, long, and curved from the base. Wings rounded. Tail moderately long, even or rounded at the tip. Tarsi short. Feet with three toes in front and one behind. Tongue small and heart-shaped. Sternum with either a notch on each side of the posterior edge or a foramen in place of a notch.

Subfam. UPUPINÆ.

Bill more slender and longer than in *Irrisorinæ*. Wings with ten primaries. Tail with ten feathers. Tarsus shielded in front with broad transverse scales.

Sternum with an open notch on each side of the posterior edge. Head crested.

Of terrestrial habits.

Genus UPUPA.

Bill typically long and slender, much compressed; gape rather wide. Nostrils round, partially concealed by the plumes. Wings with the 4th quill the longest, and the 1st a little more than half the length of the 4th. Tail even. Tarsus equal to the middle toe without the claw. Outer toe joined to the middle one at the base, and considerably longer than the inner; hind toe equal to the inner one, its claw long and straight.

Crest very large and deep.

UPUPA NIGRIPENNIS.

(THE SOUTH-INDIAN HOOPOE.)

Upupa nigripennis, Gould, MS.; Horsf. & Moore, Cat. B. Mus. E. I. Co. ii. p. 725 (1856);
    Jerdon, B. of Ind. i. p. 392 (1862); Holdsworth, P. Z. S. 1872, p. 435; Hume, Nests
    and Eggs, p. 163 (1873); Legge, Ibis, 1875, p. 286; Hume, Str. Feath. 1876, p. 458.
Upupa senegalensis, Blyth, Cat. B. Mus. A. S. B. p. 46 (1849); Kelaart, Prodromus, Cat.
    p. 119 (1852); Layard, Ann. & Mag. Nat. Hist. 1853, xii. p. 174.
Upupa ceylonensis, Reich. Handb. Scansoriæ, p. 320. no. 753, tab. 595. fig. 4036 (1851).
Upupa indica, Sharpe & Dresser, B. of Europe, pt. vii. *U. epops*, p. 6 (1871).
Hudhud, Hind.; *Kondeh pitta*, lit. " Crested Bird," also *Kukudeu guwa*, Telugu.
Chaval kuruvi, lit. " Cock Bird," Tamils in Ceylon.

Adult male. Length 10·9 to 11·75 inches; wing 5·1 to 5·5; tail 3·5 to 4·0; tarsus 0·85 to 0·0; middle toe and claw
0·85 to 0·95; bill from gape (straight) 2·1 to 2·56.

Female. Length 10·25 to 10·8 inches; wing 4·7 to 5·0; bill to gape 2·0 to 2·2.

Iris brown; bill black, pale brown at the base of upper mandible, fleshy red at the base of lower; legs and foot pale
slate-blue or plumbeous, in some tinged with brown.

General hue of head, crest, hind neck, and throat fine cinnamon-brown, becoming smoky brownish on the interscapular
region, and pale vinaceous on the fore neck and chest; crest-feathers, which are about 2 inches in length, with a
terminal black bar and *occasionally* a *pale* adjacent patch; back, upper tail-coverts, tail, and wings black; the
lower part of back crossed with white, and the rump entirely so; an angular bar across the centre of the tail, a
broad band across the terminal portion of primaries (the first excepted), three on the secondaries, and another on
the median coverts and scapulars white; 1st primary sometimes with a white spot, at other times without; tertials
with white edges, an oblique streak across the inner webs and another down the centre from the base, the light
parts often deeply tinged with buff; the point of the wing concolorous with the hind neck; beneath, from the
upper breast, white, dashed on the belly in some, and in others on the sides only, with blackish mesial streaks;
under wing pale cinnamon-red.

Young. The nestling is covered at first with pure white down, which is quickly interspersed with feathers of the
normal colour, the crest showing at once.

Obs. In Ceylonese examples of this Hoopoe, a great variety in the depth of coloration is met with; this is particularly
noticeable on the head and hind neck; again, scarcely any two specimens have the lower parts striated alike or
the tertials similarly marked: the spot on the 1st primary is sometimes absent, and may perhaps be a character of
nonage. I have noticed that the largest individuals that I have met with are the palest in colour and always
have the white spot on the 1st primary. It is the exception to find an example with the *whitish* or *pallid* bar
anterior to the black tips of the crest-feathers; but notwithstanding it does exist, though it is not so white as in
examples from northern parts of India—the race, *U. indica*, of the European bird; and it is in the form of a
marginal spot at each side of the shaft, the web next to which is of the same colour as the rest of the feather.
The length and shape of bill cannot be relied upon at all as a characteristic of this Hoopoe; some are tolerably
straight, others much curved; some long, others short.

The North-Indian variety (*Upupa indica* of Hodgson), if it be considered distinct from the present, has more white
(and has it more constantly) at the edge of the black crest-bar; Hodgson's type was collected in Nepal, and the
race it represents seems to me worthy of being considered intermediate between the present species and *U. epops*.
Specimens from " North Bengal," in the British Museum, have the pale heads of the European bird; but they are
longer in the bill than the generality of the latter, and the light patch anterior to the black tip is not so white:
two examples have the wings 5·6 and 5·1 inches, and the bill to gape 2·4 and 2·3 respectively. In *U. epops* the
bill is variable in length, but its pale plumage and white covert-bar make it very distinct from the North-Indian bird,
than which also it has a longer wing.

As to the Ceylonese bird, it is *identical*, in all respects, with specimens I have examined from Mysore, which represent the true *nigripennis* of Gould.

The Burmese form (*U. longirostris*, Jerdon) has not got a longer bill than Ceylonese specimens often have; it has the white spot on the quill which I have shown to exist in the latter, although this is a worthless character in the present species, its absence in specimens which Jerdon handled probably causing him to err in saying that the species wanted it; this, however, was afterwards corrected by him in the 'Ibis,' 1872, p. 22. Both species want the *white* on the hinder crest-feathers; and examples of each may, I think, be found equally dark as to their rufous coloration; I therefore imagine that the two races are scarcely separable.

Distribution.—The Indian Hoopoe is an inhabitant of many of the dry districts in Ceylon. It is very common both in the north and south-east of the island. In the former district it spreads from the Jaffna peninsula down the west coast as far as the neighbourhood of Puttalam. I have seen it in the island of Manaar; and Mr. Holdsworth says that it is very abundant at Aripu during the winter months, its numbers being largely increased about October. In the south-east it is common throughout the year between Hambantota and Yâla, and likewise in portions of the Park country and the Eastern Province. I found it in August on the patnas near Bibile, at the foot of the Madulsima range. It is not unfrequent in Uva, and occurs occasionally on the Elephant and Kandapolla plains and at high elevations in Maturata. I am indebted to Col. Watson for the possession of an example which he shot at Kandapolla in May at an elevation of 6300 feet; and he informs me that he has often seen it in that locality. It is sometimes found in Dumbara, straying thither, in all probability, up the valley of the Mahawelliganga from the low country of Binteune. Near this locality I have met with it at Minery Lake; but I never saw it nearer Trincomalie than this, although it may possibly visit the plains in the delta of the Mahawelliganga.

Layard writes that he procured a solitary specimen at Colombo; but any occurrence of it in that neighbourhood, or anywhere south of Chilaw, must be looked upon as that of a straggler down the west coast. It has never been found in the south-west.

Jerdon writes of this species that it "is found throughout Southern India, extending through part of Central India to the North-west Provinces and the Dehra Doon." Whether the examples from the latter locality really belong to this species or to the race *U. indica*, I am unable to say. In the Khandala district Mr. Fairbank says it is common, and Burgess writes of it as the same in the upper portion of the Deccan. Mr. Adam speaks of it as "not common" in the Sambhur-Lake district, and Captain Butler writes the same of it in the Guzerat region; but these birds, I imagine, probably pertain to the intermediate form. From Sindh, Mr. Hume remarks that he has never seen it. In the extreme south of India it appears to be chiefly restricted to the east coast; for it is found in the island of Ramisserum, and Mr. Fairbank observed it in the lower Palanis, whereas I find no record of it in Mr. Bourdillon's list of the birds of Travancore.

The Burmese race, *U. longirostris*, is common in the plains of Pegu throughout the year, but is, according to Mr. Hume, most numerous in February and March. In the Irrawaddy delta, Dr. Armstrong found it very abundant in open country. Swinhoe found it at Hainan, in China, and records it from Siam.

Habits.—This charming bird frequents, in the island of Ceylon, open sparsely-timbered ground, scrub-dotted plains, cultivated fields, dry grazing-land in the jungles of the interior, and patnas in the Central Province. In its nature it is a tame bird, and when scratching for insects, with its handsome crest depressed, allows a near approach before taking flight; when flushed it does not usually fly far, but takes refuge in a neighbouring tree, where it will sit quietly, giving out its soft and melodious call, *hoo-poo*, *hoo-poo*, accompanied by a movement of its handsome crest and an oscillation to and fro of its head at each note. In Jaffna it may be seen close to the houses of the English residents, and I have known it breed in the garden of a bungalow within a few yards of the verandah. It feeds entirely on the ground, strutting about with an easy gait, and scratching vigorously for insects in dry soil. It often scrutinizes the odure of cattle, beneath which it finds an abundance of food.

In India Jerdon remarks that it frequents "old deserted buildings, such as mosques, tombs, and large mud walls;" he found its food to consist of ants, Coleoptera, and small grasshoppers. Burgess says that in the Deccan it affects sandy plots of ground outside the walls of villages, where the ground is perforated with the conical holes of the ant-lion, on the larvæ of which it feeds.

There is something very striking in the soft tone of this bird's note when heard amidst the chatter and chirping of the numerous Passerine birds which inhabit the Ceylon coast-jungles. Though perhaps uttered tolerably close to the listener it seems to be wafted on the mild sea-breezes from afar off, and tends to rivet the sportsman's attention as he is returning to his bivouac beneath the already burning rays of an 8 o'clock sun, after a long morning's shooting in the parched-up scrubs of the northern coast. The flight of this Hoopoe is buoyant but undulating, and when pressed it is able to show considerable powers of wing, for in India a trained Hawk is said generally to fail in seizing it.

Nidification.—The breeding-season in the north of Ceylon lasts from November until April, and possibly a second brood may be reared later on in the year, as Layard mentions the shooting of young birds in August. It breeds in holes of trees, showing, in this respect, as well as in points of anatomy, its affinity to the last family, the Hornbills. It sometimes, however, chooses a hole in a wall, in which I have known it to nest in the garden of an English residence in the Jaffna fort. Burgess writes, with reference to its habit of building in walls in India, "it breeds in the middle of April and May, constructing its nest in holes in the mud walls which surround the towns and villages in the Deccan." The nests are composed of grass, hemp, and feathers. In the same district a nest made of soft pieces of hemp was found in a fort wall. Miss Cockburn, again, tells us that at Kotagherry it selects holes in stone walls and in earthern banks to build in, making a mere apology for a nest of a few hairs and leaves, which in a short time has a most offensive smell. This, it is asserted, arises from the oily matter secreted by the sebaceous gland on the tail-bone, which in the female at the breeding-time assumes an intolerable stench, whence obtains the idea, according to Jerdon, that the bird constructs its nest of cowdung.

Mr. Holdsworth found one in a hole in a small mustard-tree (*Salvadora persica*) at Aripu; the young were reposing on the bare wood at the bottom of the cavity. The same fact has been noticed by Indian observers, viz. that when holes in trees are resorted to no nest whatever is constructed.

The eggs vary from three to seven, five or six being the usual number. Mr. Hume writes that they "are commonly a very lengthened oval, almost always a good deal pointed towards one end, and sometimes showing a tendency to be pointed at the other end too—a most remarkable form of egg, which I cannot recall having observed in any other species . . . . When quite fresh they are of a pale greyish-blue tint, but many are of a pale olive-brown or dingy olive-green, and every intermediate shade of colour is observable. As a rule they have scarcely any gloss at all, and of course are devoid of markings. In length they vary from 0·9 to 1·05 inch, and in breadth from 0·65 to 0·73 inch."

Fam. CORACIIDÆ.

Bill large, wide at the base, more or less curved and the tip hooked. Legs and toes covered with strong scuta.

Sternum with two emarginations of variable depth in the posterior margin.

Plumage gay, especially on the wing ; feathers of the body with an axillary plume.

PICARIÆ.

CORACIIDÆ.

Subfam. CORACIINÆ.

Bill variable in length and width. Wings moderately long. Tarsus shorter than the middle toe.

Genus CORACIAS.

Bill long, broad, and high at the base, from which the culmen is gradually curved to the tip, which is bent down. Nostrils basal, oval, and oblique; gape armed with short strong bristles. Wings long, the 3rd and 4th quills subequal and longest, the 1st longer than the 6th. Tail moderately even. Legs and feet robust. Tarsus subequal to the outer toe, and covered, as well as the toes, with strong transverse scales; inner toe much shorter than the outer, and slightly exceeding the hind one; claws strong, moderately straight.

CORACIAS INDICA.

(THE INDIAN ROLLER.)

Coracias indica, Linn. Syst. Nat. i. p. 159 (1766); Gmelin, Syst. Nat. i. p. 378 (1788); Sykes, Cat. Birds Deccan, J. A. S. B. 1834, iii. p. 541; Blyth, Cat. B. Mus. A. S. B. p. 51 (1849); Kelaart, Prodromus, Cat. p. 118 (1852); Layard, Ann. & Mag. Nat. Hist. 1853, xii. p. 171; Horsf. & Moore, Cat. B. Mus. E. I. Co. p. 571 (1856–8); Jerdon, B. of Ind. i. p. 214 (1862); Holdsworth, P. Z. S. 1872, p. 423; Hume, Nests and Eggs, p. 103 (1873); id. Str. Feath. 1873, p. 167; Butler, ibid. 1875, p. 456; Morgan, Ibis, 1875, p. 314; Bourdillon, Str. F. 1876, p. 382; Fairbank, ibid. 1877, p. 394.

Coracias bengalensis, Linn. Syst. Nat. i. p. 159 (1766).

Garrulus nævius, Vieill. N. Dict. d'Hist. Nat. xxix. p. 431 (1819).

Rollier de Mindano, Buffon, Pl. Enl. pl. 285; *Blue Jay from East Indies*, Edwards, Gleap. pl. 326 (1764); *Blue Jay* or *Jay*, Europeans in India and Ceylon.

Subzak, lit. "Greenish bird;" also *Nilkant*, lit. "Blue Throat," Hind.; *Tas*, Mahratta; *Pálú pitta*, lit. "Milk-bird," Tel.; *Katta-kade*, Tamul; *Towe*, Mahri (Jerdon).

Doong-kowluwa, lit. "Smoke-bird," Sinhalese; *Panang karda*, Tamils, North Ceylon; also *Kotta-killi*, lit. "Palmyra-Parrot," *apud* Layard.

Adult male and female. Length 12·5 to 13·2 inches; wing 6·9 to 7·1; tail 4·6 to 4·9; tarsus 0·9 to 1·0; middle toe 1·0, its claw (straight) 0·38; bill to gape 1·8 to 2·1. The bill appears to vary in length without regard to size.

Male. Iris grey or yellowish grey, with a rufescent brown inner circle, orbital skin and eyelid dull orange-yellow; bill black or blackish, paling to reddish at base beneath; legs and feet olivaceous yellow or smoky yellow, claws brownish.

2 o

Head dusky bluish green, brightening above and behind the eye to turquoise-blue; above the nostril the forehead is greyish yellow, with a tinge of violet in some; lower hind neck, interscapular region, and scapulars dull brownish green, separated from the blue of the nape by a vinous collar; lower back cerulean blue; upper tail-coverts, base and terminal portion of all but centre rectrices, least wing-coverts, greater part of primaries, and terminal half of secondaries deep violet-blue, with a brilliant cobalt lustre close to the shafts and at the edge of the wing-coverts: central rectrices dusky green, with a blue wash at base: a broad band across the remaining rectrices, another across the six outer primaries, primary-coverts, and bases of secondaries pale cerulean blue.

Lores tawny brown; beneath the eye and the ear-coverts vinous-brown, with whitish mesial streaks; throat and chest pale greyish vinous, the feathers with mesial buff lines, and broadly margined on the fore neck and upper part of chest with purple-violet; beneath, from the chest, with the under wing, pale greenish blue.

Young. Iris brown, the grey outer portion in the adult reduced to a narrow ring; this latter increases with age very gradually, imparting considerable variation to the eye; bill blackish brown, pale or reddish at the base beneath; tarsus slightly tinted with olivaceous: gape yellowish.

Head and back duskier than in the adult; forehead with more of the pale colour; band across hind neck fawn; lesser wing-coverts (in the nestling) almost concolorous with the back; chin and throat paler than in the adult, the purplish lilac on the latter faint.

Obs. Ceylonese examples average, I think, smaller than Indian. Two of the latter from Kamptee measure 7·1 inches in the wing; another 7·4—the former being the maximum limit (according to my experience) of the insular bird. The lilac tints show considerable variation in continental as well as in Ceylonese specimens, the depth of tint depending on age.

Distribution.—The Roller has a peculiarly local distribution in Ceylon, dwelling in the dry portions of the island, and migrating to the damp district of the west chiefly during the dry season (N.E monsoon). Its head-quarters may be said to be the Jaffna peninsula, the open portions of the northern sea-board, and certain parts of the interior of the Northern and N.W. Provinces. In these districts it is common in many places and absent from others. Neither Mr. Holdsworth nor myself observed it in the Aripu district, but on the adjacent island of Manaar it occurs. To the south of the jungles bordering the coast of the Bay of Kalpentyn it is not uncommon. I have seen it in the Kalpentyn peninsula itself, and about Puttalam and Chilaw it is a well-known bird. It is a resident as far south as Madampe, and likewise in the region between that and Kurunegala; but below this line it occurs chiefly as a straggler between the months of October and March. In this season it may often be seen about Vonagodde and Ambepussa, and I have procured it in the Hewagam Korale, a little to the south of Colombo, in July. I doubt, however, if it resides in that district. I have never seen or heard of it to the south of the Kaluganga, nor did I meet with it in the very likely country between Haputale and Hambantota. It may occur in the Eastern Province, but I have no information to that effect. In the Trincomalie district it is now and then seen from December to February; but a little inland, about Ratmalie, it is common enough. Eastward of this point, through the centre of the island, it musters, as above remarked, strongly, confining itself, of course, to open districts, fields surrounded by the village tanks, and dried-up paddy-land. Even here, however, it is local; for although it is common near Hurullé, I have never seen it about Haborenna, which is separated from the former place by a tract of forest.

It has been found now and then in the valley of Dumbara, but I do not know that it occurs elsewhere in the Kandyan country.

On the mainland this species is found throughout nearly all India, from the extreme south to the Himalayas; it does not extend into Burmah, being there replaced by the closely-allied race *C. affinis*. The two forms blend into one another in such a gradual manner that it is difficult to say where *indica* ends and *affinis* begins. Mr. Hume remarks of Mr. Inglis's specimens from N.E. Cachar, that "they are not very typical, but that they are nearer to typical *affinis* than to *indica*." Its range is not by any means so limited towards the north-west; for in that direction it extends through Persia to Asia Minor, mingling thus with its European ally *C. garrula*. Mr. Danford observed it in Asia Minor at the base of the Aladagh mountains; and Messrs. Sclater and Taylor have seen a specimen in the Museum of the American College at Constantinople which was shot on the Asiatic side of the Bosphorus.

Returning to India we find that it is a seasonal visitant to some parts of the country, perhaps avoiding the extreme temperature of the hot season. It is said to leave the Deccan about the middle of April; and Captain Butler notices that it quits the hills at Aboo during the hot season, although the singular fact is testified to that it remains in the plains at that time. In the wooded and cultivated portions of Sindh, Mr. Hume observes that it is common, but absent from the desert tracts; he further remarks that it is in the Terai between Darjiling and in Eastern Bengal that the two races *indica* and *affinis* first commence to intermix.

Habits.—In Ceylon the "Jay" is found in open compounds, cocoanut-groves, tobacco-fields, waste scrubby land, grass-fields near the borders of tanks, and also newly cleared spaces in the forest. It perches on some bare tree, fence, or other prominent object, and sallies out after insects, which it captures cleverly on the wing, either returning to its original post or taking up another close by to devour its quarry. It is fond of perching on cocoanut-fronds, and in the Jaffna district often selects the lofty well-whips used to draw the water for irrigating the native tobacco, and presents a striking appearance with its head drawn into its shoulders and its bright plumage glistening in the sun. It is generally difficult of approach, flying from one fence or stump to another before one can get within shot of it; and when fired at, if not hit, flies off, mounting above the tree-tops and rolling from side to side in its course as if it had a difficulty in balancing itself on the wing. However much it is alarmed it generally returns to the field from which it has been chased, making a wide detour and reappearing perhaps at the opposite end from that at which it left. When the ripe paddy has been cut in the fields round the village tanks the Roller is sure to be seen taking his part in the harvest-making, which consists in consuming as many of the newly exposed terrestrial insects as it can, and flying in the meanwhile from one haycock to the other. Grasshoppers and beetles at such times form its chief diet. Its harsh cry is often uttered when it has been shot at and wounded, it being one of the few birds I have ever met possessed of this singular habit.

Its flight is performed with vigorous flappings of the wings, the points of which appear almost to meet beneath its body while it turns or rolls about in that strange manner which has acquired for it its peculiar name. It varies its course in the air by darting off sometimes at right angles to the original direction and then almost tumbling over in rapidly descending to the ground. These extraordinary evolutions it performs to some purpose when flown at by the Turumti, or Red-headed Merlin, mention of which I have already made at page 112.

Jerdon has some interesting notes on this handsome bird which I subjoin here. He writes, in his 'Birds of India':—"It is often caught by a contrivance called the *chou-gaddi*. This consists of two thin pieces of cane or bamboo bent down at right angles to each other to form a semicircle and tied in the centre. To the middle of this the bait is tied, usually a mole-cricket, sometimes a small field-mouse (*Mus lepidus*). The bait is just allowed tether enough to move about in a small circle. The cane is previously smeared with bird-lime, and it is placed on the ground not far from the tree where the bird is perched. On spying the insect moving about down swoops the Roller, seizes the bait, and on raising its wings to start back one or both are certain to be caught by the viscid bird-lime. By means of this very simple contrivance many birds that descend to the ground to capture insects are taken, as the King-Crows (*Dicruri*), Common Shrikes, some Thrushes, Flycatchers, and even large Kingfishers (*Halcyon*) . . . .

"The *Nilkant* is sacred to Siva, who assumed its form; and at the feast of the Dasserugh, at Nagpore, one or more used to be liberated by the Rajah, amidst the firing of cannon and musketry, at a grand parade attended by all the officers of the station.

"Buchanan Hamilton states that before the Durga Puja the Hindoos of Calcutta purchased one of these birds, and at the time when they threw the image of Durga into the river, set the *Nilkant* at liberty. It is considered propitious to see it on this day, and those who cannot afford to buy one discharge their matchlocks to put it on the wing. The Telugu name of the Roller, signifying Milk-bird, is given because it is supposed that when a cow gives little milk if a few of the feathers of this bird are chopped up and given along with grass to the cow the quantity will greatly increase. It is one of the birds on whose movements many omens depend. If it cross a traveller just after shooting it is considered a bad omen."

The Roller is very tenacious of life, requiring a large amount of hitting before coming to earth.

Nidification.—In Ceylon the Roller breeds from January until June, chiefly rearing its young about March. It nests in holes in trees, one which Mr. Parker found being situated in a palm-tree, and contained 3 white eggs, much resembling those of *Halcyon smyrnensis*. Mr. Hume writes :—"They build in hollow trees, in old walls, in roofs, or under the eaves of bungalows ; they sometimes make a good deal of a nest of feathers, grass, &c., especially when the site they choose is not well closed in ; but when they build in a small-mouthed hole there is usually a very scanty lining. I *have* found the nest in a large niche in an old wall, in which the birds had contracted the entrance with masses of torn vegetable fibre and old rags ; but this is quite exceptional ; and, again, I have taken the eggs from a hole in a Siris-tree, in which there was not the slightest lining beyond a few fragments of decayed wood. I have never found more than five eggs in any nest, and four I take to be the normal number . . . . The eggs are very broad ovals, in some instances almost spherical and like those of the Bee-eater's ; they are of the purest china-white and highly glossy. The average of a large series of measurements is 1·3 by 1·06 inch."

Genus EURYSTOMUS.

Bill very broad at the gape, shorter than the last, much curved, abruptly so at the tip. Nostrils oblique and narrow ; rictal bristles absent. Wings longer than in the last genus ; 2nd quill the longest, the 1st slightly shorter. Feet differing from those of *Coracias* by having the outer toe slightly joined at the base to the middle one.

EURYSTOMUS ORIENTALIS.

(THE INDIAN BROAD-BILLED ROLLER.)

Coracias orientalis, Linn. Syst. Nat. i. p. 159 (1766).

Eurystomus orientalis, Steph. Gen. Zool. xiii. p. 99 (1826); Blyth, Cat. B. Mus. A. S. B. no. 220, p. 51 (1849); Layard, Ann. & Mag. Nat. Hist. 1853, xii. p. 171; Horsf. & Moore, Cat. B. Mus. E. I. Co. no. 148, p. 121 (1854); Jerdon, B. of Ind. i. p. 219 (1862); Holdsworth, P. Z. S. 1872, p. 423; Hume, Str. Feath. 1874, p. 164; Morgan, ibid. p. 531; Bourdillon, ibid. 1876, p. 382.

Eurystomus cyanicollis, Vieill. N. Dict. d'Hist. Nat. xxix. p. 425 (1816).

The Oriental Roller (Horsfield); *Tiong Batu*, Sumatra (Raffles); *Tihong Lampay*, Malay.

Adult male and female. Length (from skin) 11·0 to 12·0 inches; wing 7·2 to 7·5; tail 3·7 to 3·9; tarsus 0·75; middle toe 0·85, claw (straight) 0·35; bill to gape 1·5.

The above are from 3 Ceylonese examples. A Nepaul bird in the British Museum measures, wing 7·4 inches; another from Labuan, wing 7·3 inches.

Iris hazel-brown; bill deep orange-red, the tip of the upper mandible red; orbital skin red; tarsi and feet orange-red; feet duskier than the tarsus.

Head, face, and chin brown, darker in some adults than in others, and slightly suffused with greenish on the nape, which passes into the opaque leaf-green of the hind neck, back, least wing-coverts, tertials, and rump: median and greater wing-coverts greenish blue, blending into the duller hue of the lesser coverts; primary-coverts, primaries, and secondaries black, washed on the outer webs and on the inner just inside the shaft with ultramarine: a broad band of pallid cerulean blue extending from the inner web of the 1st primary to the outer web of the 7th, and tinging the surrounding ultramarine at the point of contact; tail black, the feathers washed with ultramarine at the edges, and the reverse part beneath, except near the tip, blue; centre of the throat cerulean blue, blending into the obscure greenish blue of the fore neck and under surface; the centre of the breast and abdomen verditer-blue; under wing-coverts concolorous with the breast.

The above description is from Ceylonese examples. One from Nepal has the head and hind neck darker, and the blue colour of the breast not so bright; another is very similar to the Ceylonese birds, but has the back and wings more sombre, the wing-bar smaller (its hue spreading down the outer edges of the quills in the form of an edging), and the under surface much greener.

Young. Mr. Hume writes of the immature bird that the bill, which is much smaller than in the adult, is almost black, with the gonys pale orange, which gradually deepens in colour with the age of the bird and spreads over the whole mandible, the upper mandible becoming reddish black, after which the orange hue spreads from the gape over the whole upper mandible except the tip.

An example which I have examined from the Andamans is paler on the head and neck than an adult; the feathers of the upper surface are slightly pale-edged; chin and along the base of the under mandible brown: a portion of the throat tinged with hyacinth-blue, the rest greenish blue, and the feathers pale-tipped, with a faint tinge of the hyacinth hue on the centres of many. The under parts are paler than in the adult, and the feathers of the chest tipped with a light colour.

Obs. This is a variable species in colour, which character is no doubt due to the age of respective individuals: one example from Labuan corresponds with Ceylonese and Indian ones; it is slightly more nigrescent on the hind neck and interscapulars, and the blue of the throat is more extensive. Another from the island of Negros and one from Java are also not to be separated.

Eurystomus pacificus, of which I have examined specimens in the national collection from Ceram and the Sula Islands, is closely allied to *E. orientalis*. The wings of three specimens measure 7·8, 7·8, and 7·5 inches respectively. The upper surface is greener, the under parts paler, and the throat less coloured with blue than in *E. orientalis*: the basal outer margins of the tail-feathers are tinged with greenish blue. A Sula-Island individual, however, has the throat quite as blue as a Ceylonese ; and a Pinang example has a slight inclination towards the greenish edging of the caudal feathers. It would seem that there are connecting-links between the two species.

Distribution.—This handsome Roller is, almost without exception, the rarest resident form in Ceylon. I conclude that it *is* resident, as the only two specimens I have ever met with, and both of which I failed in shooting, were seen during the south-west monsoon. One was at Maha-oya, on the new Batticaloa Road, and the other in Mr. Chas. de Soyza's timber-forest at Kuruwite, near Ratnapura. Layard remarks that but three specimens fell under his notice, one of which he killed in the Pasdun Korale, and the other two near Gillymally. In the British Museum is an example from the collection of Mr. Cuming ; but the precise locality is not stated. Another example was shot some years ago near Kandy, and preserved by Messrs. Whyte and Co. In addition to these instances of its capture I am indebted to Mr. Delancy, of the Kirimattie Estate, near Kadugannawa, for an account of three or four birds which visited the neighbourhood of his bungalow for several days at the close of 1875, and after remaining about some tall trees, disappeared again; from his description of these visitors, and observations which he made on their habits, they must have belonged to the present species.

In Southern India it appears to make its appearance in certain localities and then disappear again. Mr. R. W. Morgan says that it is by no means rare in the Malabar forests, and he procured several specimens at Nellumbore. Captain Vipan observed it near the foot of the Carcoor Ghât of the Nilghiris ; Mr. Bourdillon remarks that it is nowhere abundant in the Travancore hills, and that it is, he thinks, only a visitor ; he has observed it "in August, during the winter months, in April, and as late as May." Regarding its distribution in the northern parts of India and elsewhere, Jerdon writes (*loc. cit.*) that "it is found at the base of the Himalayas in Lower Bengal, Assam, and the Burmese countries, extending to Malayana and China;" and he further remarks that it is said to visit Central India in the cold weather. In Cachar, Mr. Inglis says it is not uncommon and is a resident in that district; Mr. Oates records one specimen as being brought to him from the Arrakan hills, and remarks that it occurs rarely in Pegu. From Tenasserim Mr. Hume notices it as procured by Mr. Davison ; this gentleman writes to Mr. Hume that in the Andamans it is comparatively common about Fort Mouat, Mount Harriet, and other wooded places; it has also been procured about Port Blair from December until April.

I have already remarked that specimens are in the British Museum from Pinang, Java, and Labuan, and there is likewise an example from Negros, in the Philippines, which I cannot separate from Indian. It is said to occur in Sumatra. Concerning Chinese individuals, the late Mr. Swinhoe writes ("Cat. Chinese Birds," P. Z. S. 1871, p. 347) that they do not agree quite with specimens from Java, India, and Lombok ; and therefore they are, as suggested by Blyth, referable to the nearly allied *E. pacificus*.

Habits.—On both occasions that I met with this species, it was frequenting lofty dead trees, on the outermost branches of which it was perched. On the Maha-oya, the individual which I attempted to shoot flew out of the tree and returned at once to its perch, which, being at the top of an enormous tree, was beyond the range of my shot, and on my firing a second time it flew off into the forest. In the distance it has the appearance of a short-tailed Nightjar when perched, its short neck and broad bill giving it a curious outline. Its flight has the same peculiar swerving or rolling character as that of the last genus, but in a modified degree. Layard shot all his specimens in the act of tearing away the decayed wood round holes in trees; they clung to the bark after the manner of Woodpeckers, and were probably seeking a situation to nest in; he found their stomachs full of wood-boring Coleoptera, swallowed whole, and he observed that they beat their food against the bark before swallowing it. It is entirely a forest species, and is only found in regions which are well-wooded throughout. Mr. Morgan writes that in the Malabar forests it may frequently be seen perched on a lofty bamboo in the neighbourhood of some forest-stream, and that it is an exceedingly silent bird, sitting for hours together on a twig, occasionally taking a short flight after some passing insect, but almost invariably, unless disturbed, returning to the same perch. Blyth had one, which he kept in confinement for some time, and which displayed the somewhat abnormal propensity of eating plantains ; it devoured them eagerly, and would fly to him for one when he had it in his hand. The experience of Messrs. Motley and Dillwyn of it in the Malay Islands was that it is a most active and lively bird, haunting very tall jungle in parties of five or six together; these fly rapidly in large circles with quick strokes of the wing, like Woodpeckers, frequently swooping down upon one another with loud chattering. When perched, their note is a single, full, deep-toned whistle, or something between that and the sound "*you*" when uttered with forcible expulsion of the breath. Its mode of flight, when executing these circular manœuvres, must be somewhat abnormal, for any thing less like those of a Woodpecker than its actions when ordinarily on the wing cannot be imagined!

Nidification.—Mr. Bourdillon has lately had the good fortune to discover this interesting bird breeding in Travancore. Mention is made of this occurrence by him in his interesting paper on the birds of the Travancore Hills; and I am indebted to Mr. Hume for the following account written to him by Mr. Bourdillon for publication in the revised edition of 'Nests and Eggs of Indian Birds':—

"On March 17th I was attracted by hearing the chattering of a pair of these Rollers. On going to the spot I found them engaged in ejecting from a hole in a Vedu-plā stump (*Cullenia excelsa*), about 40 feet from the ground, a pair of our Hill-Mynahs (*Eulabes religiosa*). One of the Rollers was in the mouth of the hole, and enlarging it by tearing away with its beak the soft rotten wood. The other Roller, seated on a tree close by, was doing most of the chattering, making an occasional swoop at the Mynahs whenever they ventured too close. I watched the birds for some time, until the Mynahs went off and there and then began building in a 'Pinney'-tree (*Calophyllum elatum*) within the distance of 100 yards. Ten days after I sent for some hillmen ('Khanirs,' we call them here), who managed to ascend by tying-up sticks with strips of cane, in the way that they erect ladders to obtain the wild honey from the tallest trees in the forest. It was past six o'clock in the evening before the Khanir reached the hole in which the birds had bred. He found not the slightest vestige of a nest, but a few chips of rotten wood, upon which were laid the three eggs. These I found to be slightly set. While the man was climbing the tree, the birds behaved in a very ridiculous and excited manner. Seated side by side on a bough, they alternately jerked head and tail, keeping up an incessant noisy chatter, and as the crisis approached, and the man drew nearer their property, they dashed repeatedly at his head.

"After the eggs were taken, the birds disappeared for about a fortnight, but returned, and, I believe, laid again in the same position. I did not molest them this time, wishing to get the young. Unfortunately I had to leave home, and on my return I found the birds, old and young, had disappeared."

Mr. Hume writes:—"Eggs of this species, sent me from Mynall by Mr. Bourdillon, closely resemble those of the Indian Roller, but are somewhat larger, though not quite so large as those of the European Roller. They are very broad ovals, pure white, and faintly glossy.

"The specimens I have vary in length from 1·34 to 1·42 inch, and in width from 1·14 to 1·16."

Fam. ALCEDINIDÆ.

Bill long, straight, conical, and very acute at the tip; gape wide and smooth. Wings with 10 primaries. Tail short. Legs and feet small; the toes syndactyle, the inner one sometimes wanting; soles broad and flat. Sternum with two emarginations on the posterior edge. Head large. Tongue diminutive.

PICARIÆ.

ALCEDINIDÆ.

Subfam. ALCEDININÆ.

Bill long, compressed, with the culmen keeled and the gonys straight. Wings reaching, when closed, beyond the middle of the tail; the 1st quill longer than the 5th. Tail moderate or short. Legs and feet small. Tarsus hardly longer than the inner toe; outer toe nearly as long as the middle, and united to it as far as the last joint; inner toe united to middle as far as the first joint.

Genus CERYLE.

Bill typically long, very straight, the culmen scarcely bent towards the tip, flattened above, with a well-pronounced groove adjacent to it; gonys very long and straight; gape angulated. Nostrils linear and oblique. Wings moderate; the 2nd and 3rd quills subequal and longest, and the 4th considerably longer than the 1st. Tail moderately long, about equal to the bill from tip to the gape, even at the tip. Tibia bare above the knee. Tarsus smooth, very short, much less than the middle toe; feet with a broad sole. Outer toe nearly as long as the middle, and joined to it as far as the last joint; inner toe much shorter, and joined to the middle as far as the 1st joint; hind toe very short.

CERYLE RUDIS.

(THE PIED KINGFISHER.)

Alcedo rudis, Linn. Syst. Nat. i. p. 181 (1766); Sykes, P. Z. S. 1832, p. 84; Gould, B. of Eur. pl. 62 (1837).

Ceryle rudis, Boie, Isis, 1828, p. 316; Blyth, Cat. B. Mus. A. S. B. p. 49 (1849); Kelaart, Prodromus, Cat. p. 119 (1852); Layard, Ann. & Mag. Nat. Hist. 1853, xii. p. 172; Horsf. & Moore, Cat. B. Mus. E. I. Co. p. 131 (1854); Jerdon, B. of Ind. i. p. 232 (1862); Layard, B. of S. Africa, p. 67 (1867); Tristram, Ibis, 1866, p. 84; Sharpe, Monog. Alced. pl. 19 (1868–71); Holdsworth, P. Z. S. 1872, p. 424; Shelley, B. of Egypt, p. 167 (1872); Hume, Nests and Eggs, p. 109 (1873); Legge, Ibis, 1874, p. 14, et 1875, p. 282; Hume, Str. Feath. 1875, p. 52.

Ispida rudis, Jerd. Madr. Journ. 1840, p. 232.

Ceryle varia, Strickl. Ann. & Mag. Nat. Hist. 1837, vi. p. 418.

Ispida bitorquata, Swains. Classif. B. p. 336 (1827).

Ceryle leucomelanura, Reichenbach, Handb. Alced. p. 21, pl. 309. fig. 3488 (1851).

The Black-and-white Kingfisher, Edwards, pl. 9; Kelaart, Prodromus. *Martin Pêcheur noir et blanc de Sénégal*, Buffon, Pl. Enl. 62.

Korayala kilkila, lit. "Spotted Kingfisher," Hind.; *Phutka match-ranga* and *Karikata*,
Beng. (Jerdon); *To-he-haw*, lit. "Fishing Tiger" (Swinhoe).
Pelihuduwa, Waturanuwa, Gomera pelihuduwa, Sinhalese.

Adult male and female. Length 11·5 to 11·75 inches; wing 5·3 to 5·6; tail 3·0; tarsus 0·4 to 0·45; middle toe 0·6, its
claw (straight) 0·35; hind toe 0·25; bill to gape 2·8 to 3·0, at front 0·23. Females average slightly larger than
males.

Iris brown; bill black, the tip somewhat pale; legs and feet blackish, soles paler.

Adult male. Head, nape, terminal portions of the back, rump, and wing-covert feathers, primaries, and secondaries,
central portion of tail, cheeks, a broad band across the chest (sometimes complete, at others interrupted in the
centre), and another narrower one across the breast black; a broad patch above the lores continued as a supercilium to the nape, basal half and tip of tail, basal portion of the primaries and secondaries, the inner webs and
tips of the latter, lateral margins of the crown and nape-feathers, the tips of the back, scapular, and wing-covert
feathers, the major portion of the median wing-coverts, and the entire under surface with the under tail- and under
wing-coverts pure white; edge of 1st primary likewise white; the lower plumage with a silky texture; the forehead more or less uniform black; a few fine black streaks on the white of the lower part of cheeks; a patch of
feathers at each side of the belly, with large black subterminal markings.

Female. Differs from the male in wanting the lower or breast-band of black, and in having the upper broad chest-
band interrupted in the centre.

The extent of the white edgings on the upper surface *is* variable in both sexes, and the older the bird the greater the
gap in the breast-band of the female.

Young. Iris pale brown; bill reddish black, with a considerable portion of the tips yellowish white; legs and feet
brown.

Very similar to adults, but with more white perhaps about the back of the neck; the feathers of the back more deeply
tipped, and the wing-coverts and outer webs of the secondaries more marked with white. In the female, the
chest-band is rather narrow and *complete*, dividing the centre more and more as the bird grows older; in the
male it is very broad, and likewise uninterrupted in the centre; more of the feathers of the lower flanks are
spotted with black than in the adult. As considerable confusion has existed concerning the pectoral bands in the
two sexes, I have noted the above peculiarities from a male and female nestling, able to fly, taken from the same nest.

Obs. Mr. Hume observes that in India females are larger than males; Ceylonese examples correspond in size with
those from the mainland. Four females in the national collection measure as follows:—(1) wing 5·6, bill to gape
3·0 (Assam); (2) wing 5·4, bill to gape 2·8 (Kamptee); (3) wing 5·6, bill to gape 2·85; (4) wing 5·5, bill to gape
2·55. Four males:—(1) wing 5·2, bill to gape 2·7; (2) wing 5·4, bill to gape 2·05; (3) wing 5·5, bill to gape 2·85;
(4) wing 5·45, bill to gape 2·85. The fourth female example is exceptionally short in the bill. The white of the
primaries appears, as a rule, to approach nearer the tips of the feathers than in Ceylonese specimens that I have
examined; in one Indian example it is 1·3 from the tip of the first quill, while in Ceylonese it varies from 2·0 to
1·5 inch from it. I also observe that the heads of the specimens above enumerated are more conspicuously
striated with white; but this, as I have remarked with regard to Ceylonese examples, is variable. Reichenbach
separated the Ceylon *Ceryle* as *C. leucomelanura*, on account of what he stated to be a large roundish spot under the
shoulder, and of the band on the outer tail-feathers being divided into two parts: the first characteristic is nothing
more than the incomplete breast-band in the *female*; and with regard to the second feature, this band will be found
to be more or less divided in specimens from all districts; in scarcely any two examples are these feathers the same.
A Mesopotamian female example measures 5·7 in the wing, another from Kuyam, South Africa, the same, and
one from Egypt 5·65. Western-Assam and African birds would seem, therefore, to be larger than Ceylonese.

Distribution.—The Black-and-White Kingfisher is more or less common throughout the whole sea-board,
and in the northern half of the island its range extends inland to the great tanks, such as Kauthelai, Minery,
Toparc, &c., where it is tolerably frequent. In the Western Province it is found on the Kaluganga, and on the
Bolgodde and Pantura lakes, the Negombo and Puttalam Canal, and other waters which are surrounded with
open land. It is likewise common on the Gindurah and other large rivers in the south, keeping chiefly to
those parts which flow through cultivated districts. On all the leways and salt lakes of the south-east and

2 F

round the whole of the east and north coasts it is common; on the Batticaloa lakes it is especially numerous. I have not observed it on any waters near the base of the hill-zone, nor have I any testimony of its having ever been procured in Dumbara or in other valleys in the upland.

This is the most widely-distributed of any Kingfisher, being found throughout the greater part of the continents of Asia and Africa. Commencing with India, we find it recorded by all observers as common in all open and well-watered districts, be they inland or skirting the shores of the peninsula. It is plentiful in the south, in the Deccan, in Chota Nagpur, and lower Bengal, but locally rare about the Sambhur Lake and in Rajpootana, though very abundant further east in Sindh; it extends to the base of the Himalayas, but does not ascend above the low country, as is the case in South India. Eastward of India it is found throughout Burmah and Tenasserim, extending thence into Siam and northwards into China, in some parts of which it is plentiful and in others rare. Of the latter localities Mr. Swinhoe cites Ningpo as one; on the Yangtsze, according to him, it does not occur below Szechuen, and this river seems to be its northernmost limit in China. Capt. Blakiston, however, records it from Hakodadi in Northern Japan. Turning westwards from India, we find Canon Tristram speaking of it as the commonest and most conspicuous Kingfisher in Palestine, being particularly abundant about Tyre and Sidon, along the shore to Mount Carmel, on the Jordan, and on the lake of Gennesaret. In Asia Minor, Mr. Duruford observed it at the waterfalls of the Cydnus.

On the sister continent of Africa it is equally well distributed: Captain Shelley and Mr. E. C. Taylor have it as common in Nubia and Egypt; but Mr. Drake does not seem to have observed it in Morocco. On the Gold Coast, again, Captain Shelley with Mr. Buckley met with it, and Governor Ussher writes of it as very common in Fantee generally, and it literally swarms on the river Volta. Messrs. Layard, Shelley, and Buckley all record it from South Africa—the latter gentleman mentioning it as pretty common in Natal, but much more so in the north of the Transvaal.

As regards Europe, Degland recorded it from Spain; but Mr. Saunders says that he has no authentic information of its occurrence there. Malherbe records it from Sicily. Lindermayer, as quoted by Mr. Sharpe, observes, in his 'Birds of Greece,' that it is found on the islands of Thermia and Mykone, and that Erhardt includes it as a summer visitant to the Cyclades. Demidoff says that it is confined, as regards the Black Sea, to the Sea of Marmora, not being found on the northern coast of the Euxine.

Habits.—This interesting Kingfisher is not particular in its choice of position, provided a plentiful supply of fish exists to tempt its clever fishing-powers; it certainly avoids rivers and water in forest-country, but otherwise it is equally at home on freshwater tanks or lakes, the half-dried lewny, the broad and brackish estuary, the meadow-lined river or winding canal, the salt lagoon or land-locked bay, or even, in some parts of the world, the foaming shore. Although found in all such situations in Ceylon, it is, I think, most partial to brackish lagoons or backwaters, whereon it is a most persevering fisher, perching on stakes driven in to assist in laying nets or to mark the road across the shallows, or seating itself on some outstanding rock; thus it is to be seen flying about in the blazing noonday heat when scarcely another bird is abroad, and patiently hovering with downward-pointed bill about 30 feet in the air over some "fishy" spot, until with a sudden plunge it captures its well-earned prey and makes off to its favourite perch. It is generally in pairs and is most wary and watchful in its nature, starting off long before it is observed, and flying straight away to a place of safety; but when not alarmed it is constantly on the wing, flying up and down in a restless manner, and uttering its querulous quick-repeated note generally while on the wing. In addition to being so shy, it is a bird which is exceedingly tenacious of life, flying away more or less no matter how hard it is hit, and even when picked up exhausted from its wounds is hard to deprive of life. It darts invariably on its food from the wing, and descends perpendicularly and not in a slanting direction like other Kingfishers. Governor Ussher has seen them "hawking over the surf, and picking up waifs and strays brought in by the rollers, or now and then pouncing on an unwary fish." On the shores of the Holy Land, to which these birds resort in immense numbers in winter, Canon Tristram observed them "hovering by dozens over the sea about a hundred yards from the land, and occasionally perching with loud cries on an outlying rock. . . . . . During the most stormy gales of winter they continue, regardless of the weather, to hover over the breakers, ever and anon dashing down into the surf, and apparently diving to the bottom for their prey." I have observed them hover three successive times without flying back to their perch; but they usually settle down again after making a plunge, from which they do not often return empty-mouthed.

Nidification.—Throughout the northern countries included in its geographical range this Kingfisher breeds in March, April, and May. In the former month I found it nesting in Ceylon in the earthy or alluvial banks of the Gindurah : the nest was situated about 3 feet from the entrance of the hole, which was about 4 inches in diameter; the eggs were deposited in a cavity of some 7 or 8 inches in diameter. As a rule grass is found on the floor of the chamber; and Canon Tristram speaks of finding an "abundantly heaped nest of grass and weeds" in all that he dug out in Palestine; bones, however, do not seem to be used, although by the time the young leave the nest it is a mass of such, the refuse of the large supply of food brought for their sustenance. Captain Marshall, as quoted by Mr. Hume (*loc. cit.*), notices a singular feature in this bird's economy, viz. that it is sometimes a gregarious breeder; he speaks of finding a hole leading to a sort of cavern about 3 feet across which was plentifully strewn with grass and rubbish and contained eggs in different corners. The number of eggs is usually four, but sometimes six; they are, of course, white and glossy, sometimes nearly spheroid and at others pointed at one end; they average, says Mr. Hume, 1·15 by 0·92 inch.

Mr. Blewitt witnessed these birds constructing the hole leading to their nest, and writes as follows :— "They alternately relieved each other at the work, and when tired sat together some short distance off for a few minutes." When the young first leave the nest they sit together on the bank near at hand, while the old birds bring them food; this I have observed in the meadows bordering the Gindurah river.

In South Africa, where the seasons are opposite to ours, it breeds at the end of the year. Layard found its nest in November, and says that it was composed entirely of fish-bones and scales.

Genus ALCEDO.

Bill not so straight as in the last, the culmen perceptibly curved from the base, not flattened above, compressed throughout; the groove slightly developed. Nostrils oblique and nearer the commissure than in *Ceryle*. Wings moderately rounded; the 2nd and 3rd quills subequal and longest; the 4th shorter and slightly exceeding the 1st. Tail very short and rounded at the tip. Legs and feet as in *Ceryle*, the hind toe longer in proportion to the inner.

ALCEDO BENGALENSIS.

(LITTLE INDIAN KINGFISHER.)

Alcedo bengalensis, Gm. Syst. Nat. i. p. 450 (1788); Kittl. Kupf. Vög. pl. 29 (1832); Sykes,
P. Z. S. 1832, p. 84; Jerd. Madr. Journ. 1840, p. 231; Blyth, Cat. B. Mus. A. S. B.
p. 49 (1849); Kelaart, Prodromus, Cat. p. 119 (1852); Layard, Ann. & Mag.
Nat. Hist. 1853, xii. p. 172; Horsf. & Moore, Cat. B. Mus. E. I. Co. p. 129 (1854);
Temm. & Schl. Faun. Jap. Av. pl. 38 (1850); Jerdon, B. of Ind. i. p. 230 (1862);
Sharpe, Mon. Alced. pl. 2, p. 11 (1868-71); Holdsworth, P. Z. S. 1872, p. 424;
Hume, Nests and Eggs, p. 107 (1873); id. Str. Feath. 1873, p. 168, et 1875, p. 173;
Ball, ibid. p. 387; Legge, Ibis, 1874, p. 14; Oates, Str. Feath. 1875, p. 52; Butler,
ibid. p. 456; Armstrong, ibid. 1876, p. 307; Inglis, ibid. 1877, p. 19.

Alcedo minor, Schl. Mus. P.-B. *Alced.* p. 7 (1863).

Alcedo japonica, Bonap. Consp. Vol. Anis. p. 10 (1854).

The Little Blue Kingfisher of some; *The Common Indian Kingfisher*, Jerdon; "*King of
the Shrimps*," China (Swinhoe).

Chota kilkila, Hind.; *Chota match-ranga*, Beng.; *Nila buché gadu*, Telugu; *Ung-chim-pho*,
Lepch.; *Gârun*, natives in Himalayas; *To ho-âng*, lit. "Fishing Reverence," or "the
old gentleman that fishes!" Chinese of Amoy (Swinhoe).

Mal-pelihuduwa, lit. "Flower-Kingfisher," from its bright colours; also *Diya pelihuduwa*,
Sinhalese.

Adult male and female. Length 6·0 to 6·3 inches; wing 2·7 to 2·82; tail 1·2 to 1·4; tarsus 0·3 to 0·4; middle toe and
claw 0·67; bill to gape 1·72 to 1·95, average length 1·8.

Iris deep brown; bill, upper mandible blackish brown, lower yellow or reddish yellow; legs and feet coral-red, claws
dusky.

Some male specimens which I have shot, and which seem fully adult, have the under mandible black, from which it
appears that the coloration of this is uncertain; Mr. Armstrong notes it in some Irrawaddy examples as brownish
white.

Basal portion of feathers of the head, hind neck, and a broad stripe leading from the lower mandible down the sides
of the neck blackish brown; the terminal parts of these feathers, together with the tips of the wing-coverts,
French blue; scapulars, ground-colour of the wing-coverts, outer webs of the quills, and the tail-feathers duller
blue; back, rump, and upper tail-coverts bright cerulean blue (this colour becomes a shining green if hold away
from the light); inner webs of the primaries and secondaries, and terminal portions of the latter, dark hair-
brown; lateral feathers of the rump and upper tail-coverts cobalt-blue.

Lower part of loral region black; upper part of the same, a broad streak passing over the ears, chest, and under
surface, with the under tail- and under wing-coverts orange-rufous; chin, throat, and a continuation of the ear-
stripe white, the latter separated from the throat by the blue cheek- and side-neck stripe; bases of the under-
surface feathers white, imparting a non-uniform appearance to the plumage.

Young. Bill in some examples (males) with the under mandible black, like the upper, and tipped with whitish; in a
female example which, from the green hue of the blue parts and the state of the organs, appears to be immature
it is yellowish.

The distribution of the colours in the nestling is the same as in the adult, but the blue tints are greener than when
older. This greenish blue is an individual peculiarity, as some immature examples are quite as blue as old birds.

Obs. This species is a small race of *A. ispida*, the European Kingfisher, differing from it in its proportionally longer
bill and much less bulky body, although it measures very nearly as much as the latter in the wing. Ceylonese and

Indian specimens of *A. bengalensis* correspond very fairly in size, the balance perhaps being in favour of the latter. The measurements of several from different parts of India, which I have examined in the British Museum, are as follows:—(1) wing 2·95 inches, bill to gape 1·72; (2) wing 2·9, bill to gape 1·82; (3) wing 2·8, bill to gape 1·85 (Assam); (4) wing 2·85 (Kamptee). The dimensions of four specimens from the Irrawaddy delta, recorded in 'Stray Feathers,' are:—wing 2·75 to 2·8 inches, bill to gape 1·8 to 2·0, the latter measurement exceeding any that I have note of from Ceylon. Mr. Sharpe, in his exhaustive article in the 'Monograph of the Alcedinidæ,' gives the wing of Central-Asian and Philippine birds as 2·9 inches; and one I have examined from Celebes measures 2·7, bill to gape 1·07, and very stout. Compared with the above dimensions, Mr. Sharpe notes the average size of the wing in *A. ispida* as from 2·95 to 3·1. An example from Belgium, examined by myself in the national collection, has the wing 2·95, and the bill to gape 1·6; another from England, wing 3·05, and bill to gape 1·95. A Cairo specimen of *A. bengalensis* has the wing 2·8, bill to gape 2·0, and is referred to this species by Mr. Sharpe purely on account of its length of bill. In fact the two species grade into one another at the north-west confines of India and throughout the west of Asia to the borders of Europe in such a manner that it would be difficult, from a mere perusal of dimensions, to arrive at a proper identification; typical specimens of the Indian form are found far to the west and out of its usual habitat, but no typical examples of the European form are found further within the habitat of *A. bengalensis* than Sindh. In this latter region Mr. Hume considers the race to be an intermediate one, which averages as large as *A. ispida*, while the bills are, as a rule, shorter than in either species. He also notes that the birds from the Andamans and Pegu have very short bills.

Distribution.—The present species inhabits the whole island of Ceylon, from the sea-coast to the level of the Nuwara-Elliya plain. Wherever there is water, be it the tiny pond resorted to by buffaloes and wild animals in the midst of a parched-up district, or the flooded paddy-field, the lonely tank or forest river, the brackish lagoon, or even the rocky sea-shore, the Little Kingfisher is sure to be found. In the wet districts of the west and south its numbers are greater than in the north and east, but nevertheless in these it congregates in great numbers in those few spots where water is to be found.

Every forest-lined river has its pair of Kingfishers at every quarter of a mile, which dwell in the out-spreading branches of the Koombook and Mee-trees, and ever and anon plunge into the trickling stream beneath them. It is common enough in the Central-Province valleys drained by the Mahawelliganga and its affluents, but above 3500 feet becomes tolerably scarce. It finds its way to the Nuwara-Elliya lake up the streams from the Fort-Macdonald patnas; but I have not seen it on the streams between there and the Horton Plains, nor on the source of the Maha Elliya in the plain itself, the rise through forest from Galagama of the latter stream to the level of the plain (about 5600 feet) being too great for the explorations of the Little "Fisher."

This bird is found all over India, being in nearly all parts the most numerous of its family in the peninsula. It is not frequent in some of the hill-districts of the south, for I observe that neither Mr. Bourdillon nor Mr. Fairbank met with it on the Travancore and Palani hills. It is, however, not uncommon in the Nilghiris, and has been found nesting as high as Ootacamund. It is noted as being very common in the Kandhala district and also in Chota Nagpur. Turning to the north-west we find it rare at the Sambhur Lake, common at Mount Aboo and in the Guzerat plains, and very rare again in Sindh, where it is replaced by a larger race as above noticed. It extends north of India into Central Asia and the Amoorland, where Schrenk procured it; and to the westward Mr. Sharpe notes it from Cairo, the Sinaitic peninsula, and Nubia. Canon Tristram, however, only met with *A. ispida* in Palestine. To the east and south-east of India it has an extensive range, being found in Burmah, Tenasserim, Malacca, the Andamans and Nicobars, Java, Sumatra, Labuan, Borneo, and Celebes, extending northwards again to Formosa, the Loochoo Islands, Eastern China, and Japan. Swinhoe received it from Hakodadi, Northern Japan, which is its most northerly observed limit on the eastern bounds of Asia. The only locality in Sumatra from where I can find it recorded is Lampong, on the south-east coast; but when this vast island has been more explored it will doubtless be found in its western portions.

Habits.—This tame and watchful little bird passes the entire day in the constant search for its prey; no bird in Ceylon is more diligent in seeking for the means of existence than this pretty little Kingfisher, which takes up its post on any object over water, and while calling to its mate, who is generally close at hand, executes its curious little gesture of frequently jerking up its head with a combined similar movement of its tail, and darts with an unerring aim on the tiny inhabitants of the pool. It is bold and regardless of man to a degree, not hesitating to seize a fish close to a bystander; and, indeed, I have more than once seen it take up its quarters over my head while camped on the sandy bed of a forest river, and dash over and over again into the water at my feet. It is possessed of the keenest sight, pouncing often on its prey from a very considerable height above the water. It usually lives in pairs, which dwell together on terms of the greatest sociability; on one joining its companion the two become quite garrulous for some minutes, uttering in consort their clear piercing little whistle, accompanied by a vigorous bobbing up and down of heads and sundry spasmodic up-jerkings of their tails. The flight of this species is very swift; it flashes past like an arrow, its blue plumage gleaming against the sombre green of the forest, and its clear note often rousing the tired sportsman from his reverie. I have more than once observed it hovering for an instant close to the water, it having suddenly checked itself in its flight, perhaps to observe some fish too deep at the moment to pounce upon. Swinhoe notes the same habit, remarking that it is done close to the surface of the water and not high up after the manner of the last species. Concerning this little bird's temerity in seizing fish, there is an interesting note in 'Stray Feathers,' 1873, by Mr. H. J. Rainey, which shows likewise the occasional rapacity of the Brahminy Kite. This gentleman writes :—"I observed a Brahminy Kite make a rather leisurely swoop at a fish swimming on the surface of the stream; but when almost within its grasp a King-fisher (*A. bengalensis*), which had darted down swiftly, carried off the prey. This appears to have infuriated the Kite, and it immediately followed in hot pursuit of the Kingfisher, and after a long and 'stern' chase, it eventually succeeded in seizing its unresisting quarry; holding the screeching bird securely in its talons it bore it to the shore, and after complacently plucking the feathers of its (then still living) victim it set about devouring its flesh with evident satisfaction. On my approaching the spot, soon after the Kite had commenced its savage repast, it flew away, leaving little else than a few bare bones of the Kingfisher" (and, as I should have added, me vowing vengeance against the whole race of Brahminies). Layard speaks of this little Kingfisher being caught in Ceylon by Moormen, who export the skins to China, where they are used for embellishing fans. This trade does not seem to be carried on now.

Nidification.—In South, West, and Central Ceylon the breeding-season of this species is from February until June; but in the north I have known it to nest in November. It excavates a hole about 2 feet 6 inches or 3 feet deep in the soft or upper earth of a stream or river-bank, or, in fact, in any situation where such soil exists, for I have found its nest in the sides of the cavities excavated by coolies in making roads and far away from any water. At the end of the hole the little miners scoop a cavity about 6 inches in diameter and deposit frequently a layer of small fish-bones on the earth, on which the eggs are laid. In this its habits are one with those of its European representative. The eggs are sold in India to be usually five to seven in number; three are, however, sometimes laid, as Dr. Holden writes me of finding a nest with three young in Hewahette. They are very round and glossy, and pinky white when unblown, averaging 0·8 by 0·68 inch. One specimen brought to me as the egg of this species, from Baddegama, measured 0·81 by 0·76 inch.

PICARIÆ.

ALCEDINIDÆ.

Subfam. HALCYONINÆ.

Bill shorter, much broader at the base, and less compressed than in the last subfamily; lower mandible very deep at the gonys-angle, with the gonys ascending in a curve to the tip. Wings more rounded, the 1st quill shorter, and the tips of the primaries not reaching, when closed, to half the length of the tail.

Mostly of large size, and, to a great extent, reptile feeders.

Genus PELARGOPSIS.

Bill very large, stout, the culmen flat and perfectly straight to the tip; groove pronounced and parallel to the ridge. Nostrils slightly advanced, gape angulated; gonys deep and ascending in a curve to the tip. Wings with the 3rd quill the longest, and the 1st much shorter than the 5th. Tail rather long and even; tibia bare in front above the knee; tarsus stout; toes scutellate, the outer and middle subequal, but the middle claw much longer than the outer; claws deep and expanded at the sides.

Of large size.

PELARGOPSIS GURIAL.

(THE INDIAN STORK-BILLED KINGFISHER.)

Halcyon gurial, Pearson, J. A. S. B. 1841, x. p. 633; Blyth, Cat. B. Mus. A. S. B. p. 47. no. 200 (1849), et Ibis, 1865, p. 30.

Halcyon capensis, Jerd. Madr. Journ. 1840, Cat. no. 245, p. 231; Kelaart, Prodromus, Cat. p. 118 (1851); Layard, Ann. & Mag. Nat. Hist. 1853, xii. p. 177; Legge, Ibis, 1874, p. 14.

Halcyon brunniceps, Jerdon, 2nd Suppl. Cat. Madr. Journ. 1844, p. 143.

Halcyon leucocephalus, Horsf. & Moore, Cat. B. Mus. E. I. Co. p. 123 (1854); Jerdon, Birds of Ind. i. p. 222.

Pelargopsis gurial, Cab. & Heine, Mus. Hein. ii. p. 156 (1860); Sharpe, P. Z. S. 1870, p. 66; id. Mon. Alced. pl. 34 (1868–71); Holdsworth, P. Z. S. 1872, p. 423; Legge, Ibis, 1875, p. 275; Hume, Nests and Eggs, p. 105 (1873); Ball, Str. Feath. 1874, p. 386.

The Gurial Kingfisher, Latham, Hist. iv. p. 12; The Cape Kingfisher, Kelaart; Brown-headed Kingfisher, Jerdon; Gurial, Beng.; Alcyone, Portuguese in Ceylon.

Maha pelihuduwa, lit. "Great Kingfisher," also Waturanuwa, Sinhalese.

Adult male. Length 14·75 to 15·3 inches; wing 5·75 to 5·9; tail 3·75 to 4·0; tarsus 0·7; middle toe 1·0, its claw (straight) 0·12; outer toe 1·0, its claw (straight) 0·3; bill to gape 3·6 to 3·75, depth at gonys-angle 0·8.

Female. Length 15·0 to 15·3 inches; wing 6·0 to 6·3; tail 4·0 to 4·4; tarsus 0·8; bill to gape 3·8.

Iris brown, chestnut-brown in some; eyelid dull red; bill arterial blood-red, dusky at tips of both mandibles; inside of mouth coral-red; legs and feet coral-red, claws dusky.

Head and hind neck, including the face and ear-coverts, dull brown, tinged with greenish on the crown and hind neck, which is most perceptible when the feathers are new; forehead and lores slightly paler; interscapular region and scapulars dingy bluish green; lesser secondary wing-coverts almost concolorous with the scapulars, while the greater wing-coverts, outer webs and tips of secondaries, and tertials are dull greenish blue; primaries and inner webs of secondaries dark glossy brown, the basal portions of the outer webs of the primaries concolorous with the blue of the secondaries, and the terminal portions faintly tinged with blue; back and rump brilliant pale azure-blue, with a silky lustre; upper tail-coverts bluish green; tail greenish blue, with the inner webs changing into French blue; shafts deep black.

Entire under surface, sides of neck, and a broad nuchal collar just below the lower cap orange-buff, paling to albescent buff on the gorge and chin, and deepest on the flanks and under wing-coverts; under surface, quills, and tail pale brown.

Females have the head scarcely tinged with greenish, and the brown in old feathers paler than in new.

Young. Bill darker at the tips than in the adult; eyelid yellowish red; legs dusky red. *Birds of the year* have the chin almost quite white, the buff of the under surface overcast with a brownish hue, particularly on the chest, and the feathers of the fore neck, chest, nuchal collar, breast, and flanks with crescentic margins of brown, coalescing on the sides of the chest, just beneath the point of the wing when closed, into a narrow band, which joins the green of the interscapular region; lores and forehead darker than in the adult; least wing-coverts faintly edged with fulvous; ground-colour of the scapulars darker than in the adult.

With age the dark pencillings on the under surface disappear from the chest and remain only on the sides of the breast, from which they do not vanish until the bird is fully aged.

Obs. The Ceylon race of this Kingfisher appears to be, as a rule, more tinged with green on the "cap" than Indian birds, and resembles, in this respect, *Pelargopsis malaccensis*, Sharpe, differing from this, in the adult stage, in the less dark mantle, although I must say young birds are very like the latter species: this is, however, a smaller bird, the wings of two specimens measuring 5·55 and 5·65 inches. Indian examples of *P. gurial* from Madras measure 5·7 to 5·95 inches in the wing, and 3·7 to 3·9 in the bill from the gape. Mr. Ball gives the following dimensions, *loc. cit.* :—(Rahunchal) wing 6·15 inches, bill from gape 3·9; (Calcutta) wing 5·95, bill from gape 3·7; (Satpuras), ♂, wing 6·1, bill from gape 3·55. The Indian and Ceylonese bird comes very near to *P. fraseri* from Java and *P. burmanica* from Burmah, two other closely allied species: the former has the back and wings of that peculiar blue tint considered to be characteristic of *P. malaccensis*, and the brown cap is sometimes absent. The wing of an example which I have examined is 6·1, bill to gape 3·5; the latter has the cap very pale and the back greyer than in *P. gurial*, being simply a pale form of this bird. All these species are so nearly allied that they appear to me to be merely races of *P. gurial*; and I observe Mr. Hume remarks to the same effect, 'Stray Feathers,' 1877, p. 19. Mr. Holdsworth was the first to rectify the synonymy of this species as a Ceylonese bird, Kelaart and Layard having followed Jerdon's name of *capensis*, bestowed on it in the Madras Journal, 1840.

Distribution.—This large and noisy Kingfisher is found more or less on all the rivers and wild streams throughout the island, frequenting likewise the brackish lagoons and backwaters round the eastern and northern coasts, and the large sea-board lakes of the Western Province; in the latter district it is found in large jungle-begirt paddy-fields, and on the Gindurah, Kalugauga, Kelaniganga, and Maha-oya rivers. It is also an inhabitant of the Ikkude-Barawe forest and other large jungles not far from Colombo which are traversed by streams. It is pretty generally diffused through the hill-country near Galle, in which there are numerous isolated paddy-fields lying between hills, and generally drained by a stream fairly stocked with fish. The lonely tanks, particularly the smaller sheets of water surrounded by large trees which are scattered throughout the northern half of the island, and the romantic rivers which flow both east and west through that region from the hill-zone, are its favourite abode; along the whole course of the Mahawelliganga from Kottiar to the base of the hills it is common, and, I believe, ascends this river into Dumbara, though it is not of very frequent occurrence in that valley.

Jerdon remarks of this Kingfisher that it is found over all India, from the extreme south to Bengal, chiefly where there is much jungle or forest or where the banks of rivers are well wooded—precisely the same conditions which regulate its habitat in Ceylon. Mr. Fairbank saw it at the base of the Palani hills, and once near Mahabaleshwar, and Jerdon remarks that it is rarely seen in the Carnatic or the tableland. It is common in Bengal, but has been met but rarely in the north-west. At the Sambhur Lake and in Sindh it does not appear to be found. In the contrary direction, in Chota Nagpur, Mr. Ball says it is met with occasionally, as also on the Rajmehal and Satpura hills. It is not uncommon about Calcutta, and Dr. Hamilton observed that it bred in mud walls in that neighbourhood; it extends to the lower Himalayas. In Burmah it is replaced by the paler race *P. burmanica*, and even in Cachar Mr. Hume says the Stork-billed Kingfisher belongs more to the latter than to the present species.

Habits.—The Stork-billed Kingfisher always frequents the vicinity of water, and, as far as my experience goes, feeds entirely on fish and frogs. It is solitary in habit and rather sluggish, taking up its post on the branches of forest-trees overhanging water, or in the mangroves lining brackish lagoons, and at long intervals plunges headlong down on its prey, splashing up the water in its descent. Every now and then it gives out its loud discordant cries, and generally moves on to some other likely spot with a straight-on-end and powerful flight. It is very early astir in the morning, awakening with its far-sounding laugh the traveller who has halted for the night on the borders of the forest-lined river, or welcoming the sportsman on the termination of his long and early morning drive to some lonely Snipe-ground. I have seen it, when disturbed by gun-shots, take long flights across extensive paddy-fields, and after reaching a place of safety shout vociferously for a quarter of an hour. When wounded it is capable of inflicting a severe blow with its huge bill; and a Mr. Smith, in his MS. notes quoted by Dr. Horsfield, mentions an instance in which he "once observed a contest between one of these birds and a Hawk of considerable size, in which the Hawk was worsted and obliged to leave his hold, from the effects of a severe blow which the other administered to him on the breast." Mr. Ball remarks that he has only once seen it plunge into water for the purpose of capturing a fish. I have been more fortunate than this; for I have seen it several times in the act of seizing its prey; but it certainly is a far less active fisher than other members of its family that have come under my notice. Layard found this bird feeding on crabs and small Mollusca, as well as on fish.

Nidification.—Breeds in secluded spots, excavating a deep hole in the side of a river-bank or in the bund of a tank beneath shady trees. The nesting-time in Ceylon is during the first three or four months in the year. Mr. Edward Creasey, Ceylon Survey Dept., found a nest in the Jaffna district which was situated 7 feet from the entrance to the hole; it contained two eggs, which were spherical in shape, pure white, and measured 1·45 by 1·23 inch. Mr. Thompson found it breeding in May on the streams debouching from the Himalayas, and speaks of a nest containing five young ones, near which there were some deserted habitations, each having the appearance of having served its turn as a breeding-place in former years. Another writer, Mr. Theobald, notes its laying in the fourth week in June.

2 Q

Bill differing from the last in having the culmen sharply keeled and curved slightly near the tip, and the upper mandible suddenly compressed. Nostrils more oblique, less advanced ; gape less angulated. Wings with the 2nd quill subequal to the 3rd. Tibia feathered in front to the knee. Of smaller size than *Pelargopsis*.

HALCYON SMYRNENSIS.

(THE WHITE-BREASTED KINGFISHER.)

Alcedo smyrnensis, Linn. Syst. Nat. i. p. 181 (1766).

Halcyon smyrnensis, Steph. Gen. Zool. xiii. p. 99 (1826); Sykes, P. Z. S. 1852, p. 84; Blyth, Cat. B. Mus. A. S. B. p. 47 (1849); Kelaart, Prodromus, Cat. p. 118 (1852); Layard, Ann. & Mag. Nat. Hist. 1853, xii. p. 172; Horsf. & Moore, Cat. B. Mus. E. I. Co. p. 125 (1854); Tristram, Ibis, 1866, p. 86; Sharpe, Mon. Alced. pl. 59 (1868–71); Holdsworth, P. Z. S. 1872, p. 424; Hume, Nests and Eggs, p. 105 (1873); Adam, Str. Feath. 1873, p. 372; Hume, ibid. 1874, p. 167; Legge, Ibis, 1874, p. 14.

Alcedo fusca, Bodd. Tab. Pl. Enl. 54 (1783).

Halcyon fuscus (Bodd.), Jerdon, B. of Ind. i. p. 224; G. R. Gray, Gen. of Birds, i. p. 79 (1849).

The Smyrna Kingfisher, Latham; *The Indian Kingfisher*, Horsfield; *Blue Kingfisher*, Europeans in Ceylon.

Kikila, Hind.; *Sade-buk match ranga*, Beng.; *Lak nuka*, Tel.; *Vichuli*, Tam. (Jerdon); *Matsya-ranga*, Sanscrit; *Fei-tsuy*, China (Swinh.).

Pelihuduwa, Sinhalese; *Kalari kuruvi*, lit. "Wide-mouthed Bird," Ceylon Tamils (Layard).

Adult male and female. Length 10·8 to 11·1 inches; wing 4·4 to 4·6; tail 3·2 to 3·4; tarsus 0·5; middle toe 0·7 to 0·75, claw (straight) 0·37; bill to gape 2·5 to 2·7; depth at gonys-angle 4·9 to 5·7.

Iris sepia-brown; bill deep arterial red; inside of mouth vermilion; anterior portion of legs and feet dark brownish red; posterior portion and soles of feet orange-red; claws blackish.

Head, cheeks, back, and sides of neck, sides of chest, and all the lower parts from the breast downwards with the under wing-coverts deep chestnut-brown or reddish chocolate-colour, darkest on the head, hind neck, and sides of chest: back, scapulars, rump, upper tail-coverts, tail, secondaries, and basal portion of outer webs of all but the first primary, when viewed against the light, turquoise-blue, brightest on the back, rump, and secondaries, and when viewed with the light malachite-green; tertials and margins of the tail-feathers with a decided greenish hue; first primary, terminal portion of the rest, tips of secondaries, and inner half of the inner webs blackish brown; least wing-coverts lighter chestnut than the head, the median secondary coverts coal-black; shafts of tail-feathers black; a fine line just beneath the lower eyelid, chin, fore neck, centre of the chest, edge of the wing, and basal portion of the inner webs of the primaries white, increasing on the latter towards the inner feather, on which it approaches close to the tip.

Some examples have a brownish wash on the forehead and crown, and, in fact, the chestnut portions of the plumage are, as a rule, variable, some birds being darker in this respect than others. It is worthy of remark that if this Kingfisher be held away from the light, the *white chest assumes a greenish hue*.

Young. The nestling has the bill red at the base, paling to yellowish towards the tip, which is black.

The feathers of the head and hind neck are pale-tipped, mostly so on the forehead; the least wing-coverts are tinged with black.

Obs. Although the chestnut colour in this species is variable, I doubt not that, if a large series of Ceylonese examples were compared with a good many from most parts of India, they would be found to be, as a rule, darker than the latter; and I do not think they attain the same size as some Indian specimens. Mr. Hume gives the largest of forty birds as 4·85 in the wing, and remarks that his extreme southern specimens from Anjaugo are the darkest and smallest, and therefore correspond best with ours. The least wing-dimension in Anjaugo birds is 4·4. I have, however, a specimen from Rauisserum Island with the wing 4·5 inches, but with a very small bill, measuring 2·35 to gape and 0·48 in depth at gonys-angle. It likewise has the chest very strongly tinged with green. As regards the head and hind-neck hues, Ceylonese birds resemble those from the Andamans; but these latter, in addition to being darker than those from *any other part of Asia*, are larger, and have therefore been separated as *H. saturatior* by Mr. Hume. The greenish-blue tint on the white chest is observable in Nepal, Kamptee, and Beloochistan specimens, also in one from Jericho; but they must be held *from* the light, with the bill pointed towards the eye, in order to produce this colour to the greatest extent. The Jericho specimen is somewhat paler on the head than one from Colombo; but the under parts and sides of the chest are darker if any thing: it has the wing 5·1 inches; bill to gape 2·7. Another from Beloochistan is slightly greener in all lights than Ceylonese individuals, and has a white stripe above the lores; wing 4·95, bill to gape 2·7. An example from Bagdad is pale on the head and has a white superciliary line. For purposes of comparison, I will add that an Andaman example of *H. saturatior* measures 5·1 inches in the wing, but the bill to gape is only 2·75.

Distribution.—This handsome Kingfisher is extremely common in Ceylon, and is spread over the whole island, inhabiting the Kandyan Province up to the altitude of Nuwara Elliya, at which place it has made its appearance since the lake was found. It is more plentiful in the Western and Southern Provinces and in the cultivated portions of the northern district than in the jungle-covered country of the interior, for though it occurs on the forest-rivers it is not so abundant as the Stork-billed or little Blue Kingfishers. It is fairly numerous in the islands of the Jaffna district and in Mannar, and Mr. Holdsworth says it is not uncommon at Aripu. In the northern forests it is more often found near village tanks and on new clearings than elsewhere. In the Kandyan Province it is chiefly an inhabitant of the terraced paddy-fields, and is tolerably numerous in the well-cultivated valleys.

Out of Ceylon it has a very wide range, being found all over India, extending eastward to China and westward to Palestine and Asia Minor. As regards India it has been recorded as a common bird from all parts of the low-lying districts which have been worked out; but though Mr. Bourdillon found it plentiful at the foot of the Travancore hills, it did not ascend there to any height. Mr. Fairbank likewise only observed it in the lower Palanis.

From the low districts of Bengal, where it is very common, it extends to the base of the Himalayas, and westward through Sindh into Persia and Palestine, where Canon Tristram found it in the Jordan valley up to the sources of the river; beyond this Russel recorded it, in the last century, in his ‘Natural History of Aleppo,’ to be an inhabitant of Asia Minor. Captain Graves met with it in the same locality after the lapse of a century, during which time it had escaped the observation of naturalists. Canon Tristram and Mr. Sharpe note it as a doubtful straggler to Europe. From Burmah it extends into Tenasserim and the Malay peninsula. In many parts of China it is common, and resident, according to Mr. Swinhoe, from Canton to the river Yangtsze; he likewise procured it in Formosa.

Habits.—Although this Kingfisher frequents paddy-fields, streams, rivers, swamps, and fresh water in all situations, it is almost as often found affecting clearings in the jungle, dried-up fields, cultivated gardens, and the edges of open wastes, and in such places subsists on lizards, grasshoppers, crickets, locusts, and even small snakes. It invariably resorts to new clearings in the forest after they have been burnt off, and takes up its position on stumps or branches of charred trees, and therefrom flies down on the lizards and insects which it espies on the blackened soil. Mr. Inglis, in his ‘List of Birds of Cachar,’ mentions seeing one so occupied for half an hour, and on shooting it found its stomach crammed with crickets. Mr. Ball has seen it dive for fish on one occasion; but this must be an occurrence of extreme rarity; he writes that in Chota Nagpur it is snared, and the flattened-out skins disposed of to merchants, who sell them to Burmese traders for ornamenting court-dresses. In Ceylon it is best known to those who do not penetrate into the wilds as an inhabitant of the paddy-fields, of which it is one of the chief ornaments in the way of bird-life, and is the first bird which attracts the attention of the new arrival in the island as he trudges through his first hot December-day's Snipe-shooting.

It is, perhaps, the first bird astir at daybreak, and when there is scarcely enough light to discern it, flies up to the top of the highest tree near at hand and pipes out its plaintive trilling note for a considerable time, and then makes off to some favourite outlook, uttering its loud harsh call, very different from that which it has just indulged in. This latter is always uttered when the bird is on the wing, while the former is only heard when it is perched. When a lizard, which is a favourite meal, is captured, it is hammered against a stone or branch of a tree until dead, and then devoured whole, and crabs and mollusks are treated in the same way when the bird has taken up its quarters by a stream. I have observed one launch out from a high tree, in the manner described by Layard, on a butterfly; but this writer records an evil deed against the lovely bird, which is worthy only of such a cannibal as the *Koloruwa* (*Megalæna zeylanica*). He relates that one which was "unluckily introduced into an aviary, destroyed most of the lesser captives ere he was detected as the culprit; he was at last caught in the act of seizing a small bird in his powerful bill; he beat it for a moment against his perch, and then swallowed it whole!" The habits of this species as observed in Palestine by Canon Tristram are somewhat different to those which obtain with it in India and Ceylon. He writes:—"It loves to sit moodily for hours on a slender bough overhanging a swamp or pool, where the foliage helps to conceal its brilliant plumage, and where, with cast-down eyes and bill leaning on its breast, it seems benumbed or sleepy, until the motions of some lizard or frog in the marsh beneath rouse it to a temporary activity. When disturbed, it rather slinks away under the cover of the overhanging oleanders than trusts for safety to direct flight." In one example he found a snake 18 inches long, entire. In the Holy Land it is solitary in habit as in Ceylon, where two birds are scarcely ever seen together.

Nidification.—In the west and south of Ceylon this species breeds from January till April, and in the north I have found its nest as late as July. It nests in a bank generally near water or in the bund of a tank, penetrating from 2 to 4 feet, and then excavating a large vault, sometimes 9 inches in width, in which it lays its eggs, which are usually four in number, though sometimes six. In a nest which I took in the breach in the great "bund" of Hurullé tank there were no bones, nor any thing used for a lining to the nest; the passage and egg-chamber, however, frequently contain remains of frogs, lizards, &c., which have been taken in by the old birds for feeding their young. The eggs are pure white, round in shape, and those that I have seen from Ceylon vary from 1·14 to 1·2 inch in length by 1·0 to 1·04 inch in breadth. In India this bird often nests in mud walls and sometimes in open wells, Mr. Hume recording an instance of one building in a hole in the side of a well 100 feet below the surface of the ground. The eggs, when first laid, have, it is said, a beautiful gloss; but they rapidly lose this, as those I have taken were rather dull than otherwise. Some attain a size of 1·27 by 1·12 inch, or as large (as Mr. Hume remarks) as a Roller's egg.

HALCYON PILEATA.

(THE BLACK-CAPPED PURPLE KINGFISHER.)

Alcedo pileata, Bodd. Tabl. Pl. Enl. 41 (1783).

Halcyon pileata, Gray & Mitchell, Gen. of Birds, i. p. 79 (1844); Sharpe, Mon. Alced.
p. 169, pl. 62 (1868–70); Holdsworth, P. Z. S. 1872, p. 424; Hume, Str. Feath.
1875, p. 51; Armstrong, ibid. 1876, p. 306; Sharpe, Ibis, 1876, p. 33.

Alcedo atricapilla, Gm. Syst. Nat. i. p. 453 (1788).

Dacelo pileata, Schl. Mus. P.-B. *Alced.* p. 27 (1863); id. Vog. Ned. Ind. *Alced.* pp. 22, 54,
pl. 9 (1864).

Halcyon atricapillus, Blyth, Cat. B. Mus. A. S. B. no. 204, p. 47 (1849); Layard, Ann. &
Mag. Nat. Hist. 1853, xii. p. 171; Horsf. & Moore, Cat. B. Mus. E. I. Co. p. 124
(1854); Gould, B. of Asia, pt. xii. (1860); Jerdon, B. of Ind. i. p. 226 (1862); Hume,
Str. Feath. 1874, p. 168.

Entomobia pileata, Salvad. Ucc. di Borneo, p. 102 (1874).

Martin Pêcheur de la Chine, Buff. Pl. Enl. 673 (1770); *The Black-capped Kingfisher*; *Black-
winged Kingfisher*.

Udang, Malay; *Burong udang*, Sumatra (Raffles).

Adult male and female (Burmah). "Length 11·7 to 12·5 inches; wing 4·9 to 5·3, expanse 18·0 to 18·75; tail from
vent 3·3 to 3·75; tarsus 0·6 to 0·7; bill to gape 2·9 to 3·15" (*Armstrong*).

Layard's Ceylonese specimen measures 5·4, a male shot by Mr. Oates 5·3, and two examples in my own collection
4·8 and 5·1 inches (the former is an immature bird).

Iris reddish brown, dark brown, or olive-brown; bill deep coral-red; legs and feet dull red, brownish on the front of
tarsus; claws "horny brown."

Head, face, ear-coverts, nape, and wing-coverts coal-black; back, scapulars, upper surface of tail, primary-coverts, and
the outer webs of the secondaries and tertials ultramarine-blue, very brilliant on the interscapular region, and
changing into a lustrous smalt-blue on the upper tail-coverts; a broad band of white across the hind neck,
immediately beneath which the blue of the back is shaded with black; terminal half of primaries and tips and
inner webs of secondaries dull black, the basal half of the former delicate bluish, or bluish white on the outer
webs and pure white on the inner.

Chin, fore neck, centre of chest, and upper breast white; sides of chest and fore neck, flanks, lower breast, abdomen,
under tail-, and under wing-coverts fine tawny rufous, blending into the white of the fore neck, and often tinging
the hind-neck collar; under surface of tail blackish.

Young. Birds of the year have the black of the upper parts and the blue of the back and rump less pure, and the
sides of the chest and breast, as also the feathers of the hind-neck collar, marked with crescentic tippings of
blackish brown; but in some examples the latter part is striated with brown instead of barred. These crescentic
markings appear to remain until the bird is fully aged, as they are present in many specimens which have the upper
surface in beautiful adult feather.

Distribution.—This lovely Kingfisher has been only once recorded from Ceylon. Layard speaks of
one specimen having been shot in the island of Valenny, near Jaffna. This bird, which must have been a
straggler driven to the coasts of Ceylon by the northerly winds of December, is now in the Poole collection
and is in a fair state of preservation. Its occurrence in Ceylon is very interesting, as it is a rare bird in
India, and particularly so in the south. Jerdon shot a specimen at Tellichery, on the Malabar coast, and
saw others from the same locality; he speaks of it having been procured as high up the Ganges as Monghyr,
although it is rare in Bengal. It affects wooded country near the sea, and consequently is more common in

the Sunderbunds than elsewhere in India. It has not, I believe, been found anywhere to the west of Lower Bengal. In Burmah it is common near the sea, though rare up at Thayetmyo. Mr. Armstrong writes :— "This beautiful Kingfisher formed a marked characteristic of the avifauna belonging to the Irrawaddy delta. It was to be seen everywhere. It was abundant among the mangroves on each side of every creek and nullah ; the shore-jungle along the coast from Elephant Point to China-Ba-keer resounded with its discordant cry." It is found in Tenasserim and throughout the Malay peninsula, where it is far from uncommon, inhabiting likewise the islands in the Bay. In these, however, it is rare, both as regards the Nicobars and the Andamans. Mr. Davison saw it at Triukut and Kondul in the former, and Mr. Hume has received it from Port Blair, Andamans. It is known from both Java and Sumatra, and Count Salvadori includes it in his ' Birds of Borneo,' where also Mr. Alfred Everett has of late years procured it. Further north it is an inhabitant of China, in which country, Mr. Swinhoe remarks, it is found from Canton to the Yangtsze, and is rare in the neighbourhood of Amoy. Dr. Zelebor, who accompanied the ' Novara ' Expedition, found it at Hong Kong.

Habits.—This species loves thickly wooded estuaries and brackish creeks such as are found in the great Sunderbunds near Calcutta, in the delta of the Irrawaddy, and other similar localities, in the impenetrable jungle of which it passes a generally unmolested existence, feeding on the crabs which abound in the muddy creeks and nullahs. These crustaceans form its favourite food. Mr. Armstrong says that in the Irrawaddy delta "under every little projecting twig along the mud shore a quantity of white excreta and the remains of the legs and bodies of small crabs showed where one of these birds had been making its dinner and indulging in its siesta. Each bird appears to have its own favourite watch-tower, and when disturbed flies away with a shrill cry, taking a semicircular stoop to some dry twig on ahead, and as soon as it thinks that the danger is passed by returns again to the post from which it has been dislodged." Captain Wimberley, who shot this bird at Port Blair, says it is excessively shy and wary, and that he had to go out day after day before he could procure it. It has a harsh crowing call according to Jerdon, and which is described by other writers as discordant. Dr. Zelebor likens it to the cry of the European Great Spotted Woodpecker.

The Chinese, with their usual admiration for the feathers of Kingfishers, put those of this species also to ornamental purposes, using them for the manufacture of their fans.

I am unable to give any information concerning the nesting of this species.

Genus CEYX.

Bill much as in *Halcyon*, the culmen less keeled. Wings with the 1st quill as long as in *Alcedo*, and the 4th not much shorter than the 3rd. Tail short and broad at the base, rounded at the tip. Tarsus much shorter than the anterior toes ; *inner toe wanting* ; claw of outer toe very short.

CEYX TRIDACTYLA.

(THE INDIAN THREE-TOED KINGFISHER.)

Alcedo tridactyla, Pall. Spic. Zool. vii. p. 10, pl. 11. fig. 1 (1769).

Ceyx tridactyla, Sykes, P. Z. S. 1832, p. 84; Jerdon, Ill. Ind. Orn. pl. 25 (1847); Kelaart, Prodromus, Cat. p. 118 (1852); Layard, Ann. & Mag. Nat. Hist. 1853, xiii. p. 172: Jerdon, B. of Ind. i. p. 229 (1862); Sharpe, P. Z. S. 1868, p. 270; id. Mon. Alced. pl. 40, p. 119 (1868–71); Holdsworth, P. Z. S. 1872, p. 424; Hume, Str. Feath. 1874, p. 173, et 1875, p. 51, et 1876, p. 287; Inglis, ibid. 1877, p. 19.

Alcedo erythaca, Gm. Syst. Nat. i. p. 449 (1788).

Ceyx erythaca, Blyth, Cat. B. Mus. A. S. B. no. 220, p. 50 (1849).

Ceyx microsoma, Jerd. Cat. B. S. India, Madr. Journ. 1840, xi. p. 231.

Martin Pêcheur de Pondicherry, Buff. Pl. Enl. 778. fig. 2.

The Three-toed Kingfisher, Europeans in Ceylon; *The Pinang Kingfisher*, Sharpe, Mon. Alced.

Dein-nyoeen, Arracan; *Raja whodan*, Malay (Blyth).

Punchi Mal-pelihuduwa, Sinhalese.

Adult male and female. Length 5·25 to 5·4 inches; wing 2·1 to 2·3; tail 0·9; tarsus 0·35; innermost toe and claw 0·05; hind toe and claw 0·3; bill to gape 1·44 to 1·6, at front 1·3. Expanse 8·3.

Iris brown; bill coral-red; legs and feet coral-red, slightly paler than bill; claws yellowish.

Head, hind neck, face, lower back, rump, and tail with the least wing-coverts and under wing rufous, overlaid on the back, upper tail-coverts, and behind the eye with delicate shining lilac, and tinged with the same on the head: upper back black, overlaid with a patch of brilliant cobalt-blue; wings blackish brown; a spot at the side of the nape, a wash over the back and tertials, and edges of wing-coverts fine deep violet-blue; beneath the nuchal spot a white streak; forehead edged with deep blue at the bill; eyelid and a spot in front of eye black; outer web of 1st primary and edge of winglet, inner margins of quills, and base of secondaries pale cinnamon; chin, throat, and centre of abdomen flavescent whitish; rest of under surface saffron-yellow, shaded with rufous on the flanks. In some specimens the centre of the head wants the violet tinge, this part being plain rufous; others, probably not adult, have the tail tipped dark.

Obs. Ceylonese examples are identical in character with Indian and Malaccan. A Pegu specimen, recorded in 'Stray Feathers,' measures 2·2 in the wing and 1·55 in the bill from gape: another I have seen from Malacca, 2·3 in the wing and 1·6 in the bill from gape. A male example, with a similarly large bill, I procured at Kaotholai; but the average size of the bill in Ceylon specimens is about 1·45. Mr. Sharpe figures an example in his plate ('Monog. Alced.') with a brown tail, and remarks that it may be sexual or a sign of immaturity; it certainly is not the former, as I have sexed males and females without any trace of dark colour in the tail: and as to the latter the nearest approach to a dark tail in what appeared to be a young bird, from the state of the organs, was a dark tip of about $\frac{1}{16}$ inch in depth to the centre tail-feathers. It seems not unreasonable to infer that the coloration in the specimen figured by Mr. Sharpe was abnormal, and at the same time very remarkable.

To many of my readers who are not well acquainted with this beautiful genus of Kingfishers, it may not be uninteresting to peruse a short *résumé* of its members, taken from Mr. Sharpe's magnificent 'Monograph of the Kingfishers,' which I here give. Commencing with the species which ranges next to ours in its habitat, we have :—

Ceyx rufidorsa, Strickland, P. Z. S. 1846, p. 99; Sharpe, Mon. Alced. pl. 41. Indo-Malayan region.

Differs chiefly from *C. tridactyla* in having the back and wing-coverts of the same hue as the head, rump, and tail, which are lilac-rufous. Wing 2·2.

Ceyx dillwynni, Sharpe, P. Z. S. 1868, p. 591; id. Mon. Alced. pl. 43. Labuan.

Larger than the above; head, back, rump, and tail lilac-rufous; scapulars black, washed with blue. Wing 2·45.

Ceyx sharpii, Salvad. Atti R. Accad. Tor. 1869, p. 463; Sharpe, Mon. Alced. pl. 42. Borneo.

Nearly all the upper surface brilliant lilac-rufous, with a portion of the scapulars black, and the wing-coverts tipped with blue. Wing 2·3.

Ceyx solitaria, Temm. Pl. Col. 595 ; Sharpe, Mon. Alced. pl. 38. New Guinea and adjacent isles.

Back rich ultramarine ; the head, tail, and wings chiefly black ; bill black. Wing 2·1.

Ceyx cajeli, Wall. P. Z. S. 1863, p. 25, pl. v. ; Sharpe, Mon. Alced. pl. 44. Bouru Island.

Chiefly black above, with the back and rump silvery blue : head and wing-coverts spotted with silvery blue. Wing 2·5.

Ceyx wallacei, Sharpe, P. Z. S. 1868, p. 270 ; id. Mon. Alced. pl. 45. Sula Islands.

A large species, chiefly black above, with the back very rich shining cobalt : distinguished by its black scapulars from the next. Wing 2·5.

Ceyx lepida, Temm. Pl. Col. 595 ; Sharpe, Mon. Alced. pl. 46. Ceram, Amboina, south-west coast of New Guinea.

Likewise a large species. Chief characteristics of upper plumage black, spotted with rich ultramarine on the head and hind neck : back "rich ultramarine." Wing 2·5.

Ceyx uropygialis, Gray, P. Z. S. 1860, p. 318 ; Sharpe, Mon. Alced. pl. 47.

Smaller than the above. Upper surface chiefly black, spotted minutely and striped with ultramarine on the head ; back ultramarine ; rump silvery blue. Wing 2·4.

Ceyx melanura, Kaup ; Sharpe, Mon. Alced. pl. 90. Philippine Islands.

Above chiefly lilac-rufous, with a patch of feathers on each side of the neck blue, under which is another white patch : head spotted with lilac-blue. Wing 2·1.

Ceyx philippinensis, Gould, P. Z. S. 1868, p. 404 ; Sharpe, Mon. Alced. pl. 37. Philippine Islands.

Chiefly indigo-blue above, banded with light cobalt on the head and face : under surface deep rufous. Resembles the Indian Kingfisher somewhat in general appearance. Wing 2·3.

Distribution.—This diminutive and beautiful little Kingfisher is the rarest of the indigenous species of the family in Ceylon, occurring here and there in localities few and far between throughout the low country, and inhabiting the upland valley of the Mahawelliganga and its affluents to an elevation of about 2000 feet. I have procured it in forest on the Trincomalie and Anaradjapura road, near Kanthelai tank, and at Devilane in the Friars-Hood district. In 1875, while residing at Harellé tank, Mr. Cotteril, C.E., met with a little flock of four, and it has been seen in the Mullaittivu district. Layard speaks of meeting with it at Galle, Trincomalie, Anaradjapura, Matale, Puttalam, and Ratnapura. I closely scrutinized the rocky streams and rivers during two years' wanderings in the jungles of the south-west, but never saw it, nor did I ever encounter it in any of the humid districts of the island, and am therefore convinced that it is chiefly to be found in the dry portions only. It is not uncommon in Dumbara ; but is chiefly located, I imagine, down the valley, from Kandy towards the bend of the Mahawelliganga. Mr. Holdsworth "at various times obtained three specimens, which were killed in the central district ;" and it has been described to me (whether correctly identified or not I cannot say) as inhabiting the tributaries of the Kelani in Lower Dickoya.

It is scattered all over India, but nowhere, says Jerdon, common. He procured it in the south of India, and remarks that it seems to be a coast-bird for the most part. Col. Sykes got it in the Deccan ; but Mr. Fairbank does not appear to have met with it in that part. In the north-west of India it has not, that I am aware of, ever been found, its distribution being decidedly eastern. Mr. Ball does not even record it from Chota Nagpur or the Satpura jungles, and we next find it in the Sikhim Terai, and thence eastward in Cachar and Burmah. In Pegu Mr. Oates only found it on the eastern slope of the Pegu-Yuma hills, where the country is covered with evergreen forest, in the deep-wooded nullahs of which it was not uncommon. In Northern Tenasserim Mr. Davison found it between Tavoy and Meeta Myo, at Kavope, and near Ye. In the peninsula and the island of Pinang it is well known, and it has been procured at Ross Island, Andamans, and at Kondul, a small islet adjoining the Great Nicobar Island. It has been found in Java and Sumatra and some of the Indo-Malayan Islands, and Mr. Sharpe instances it as having been procured in the Philippines ; but the last-named locality requires confirmation.

Habits.—The Three-toed Kingfisher, which is the loveliest of all Ceylon birds, is a shy and usually solitary species, delighting in the gloom of the forest, where it frequents the edges of tiny brooks and damp or swampy spots containing small water-holes, subsisting on diminutive fish and small aqueous insects. It is so small that it is next to impossible for the collector, however keen-eyed he be, to detect it on its little perch before it is alarmed and takes wing with a shrill piping note, glancing instantaneously round the nearest tree to a place of safety. It is consequently very difficult to procure ; but in the evening, just as darkness is setting in and the jungle becomes gradually enshrouded in gloom, it becomes restless and noisy, continuing to whistle and fly from place to place round its diurnal position until dark, and may then be watched and easily shot. Unless when breeding it is always found alone; and though it frequents the banks of streams and rivers in the jungle, it evidently prefers the interior of the forest to the vicinity of exposed water. We find Mr. Inglis noting it, in Cachar, as affecting thick jungle with small streams running through it; and at Devilane I procured one of my specimens frequenting the jungle through which the sluice-stream ran, and rejecting completely the open water of the tank which abounded with fish. Mr. Inglis observes that they sit very close, and that he has more than once attempted to catch them with his hand. This is an illustration of the many instances in which the habits of different species vary entirely according to the district or country which they inhabit, for, as I have just remarked, this is a very shy bird in Ceylon. I have been told that the Singhalese occasionally catch it on the Mahawelliganga, but in what manner I do not know.

No information appears as yet to have been acquired concerning the nidification of this little Kingfisher.

PICARIÆ.

Fam. MEROPIDÆ.

Of small size. Bill long, slender, curved, both mandibles much pointed. Wings long and pointed. Tail with the central feathers often elongated. Legs and feet feeble.

Sternum with two emarginations on the posterior edge.

Genus MEROPS.

Bill much lengthened, slender, acute, compressed from the nostrils to the tip; both mandibles curved gently throughout. Nostrils oval, basal, placed midway between the margin and the culmen, partially protected by short bristles; rictal bristles short and stiff. Wings long and pointed; 1st quill minute, 2nd the longest. Tail of 12 feathers, even at the tip, or with the two central rectrices prolonged beyond the rest and much attenuated. Tarsus short, covered in front with transverse scales. Feet with the lateral toes joined to the middle, the outer beyond, and the inner as far as, the last joint; claws curved and hollowed beneath.

MEROPS PHILIPPINUS.

(THE BLUE-TAILED BEE-EATER.)

Merops philippinus, Linn. Syst. Nat. ed. xiii. tom. i. p. 183 (1767) ; Lath. Ind. Orn. tom. i.
p. 271 (1790) ; Blyth, Cat. B. Mus. A. S. B. p. 52 (1849) ; Kelaurt, Prodromus, Cat.
p. 118 (1852) ; Layard, Ann. & Mag. Nat. Hist. 1853, xii. p. 173 ; Horsf. & Moore, Cat.
B. Mus. E. I. Co. p. 87 (1854) ; Gould, B. of Asia, pt. vii. (1855) ; Holdsworth, P. Z. S.
1872, p. 422 ; Legge, Ibis, 1875, p. 281 ; Hume, Nests and Eggs, i. p. 101 ; id. Str. Feath.
1876, p. 287.
Merops javanicus, Horsf. Trans. Linn. Soc. xiii. p. 294 (1820).
Merops daudinii, Cuv. Règn. An. 1829, t. i. p. 442.
Merops philippensis, Jerd. B. of Ind. i. p. 207 ; Blyth, Comm. Jerd. B. of Ind., Ibis, 1866,
p. 344 ; Legge, Ibis, 1874, p. 13.
Grand Guêpier des Philippines (juv.), Buffon, Pl. Enl. 57 ; *Le Guêpier Daudin* (juv.), Levaill.
pl. 14, p. 49 ; " *Flycatcher* " of Europeans in India and Ceylon.
Boro-putringa, Beng. ; *Burra-putringa,* Hind. ; *Komu passeriki,* Tel. (Jerdon) ; *Kachangan,*
Java (Horsf.) ; *Berray Berray,* Malay ; *Shale,* Nicobarese (Davison).
Kurumenne kurulla, lit. " Beetle-bird," Sinhalese ; *Kattalan kuruvi,* lit. " Aloe-bird " [*], Tam. ;
Pappugai de Champ, Portug., lit. " Ground-Parrot " (*apud* Layard).

Adult male and female. Length 12·0 inches ; wing 5·0 to 5·4 ; tail 5·9, central feathers 2·3 longer than the rest ; tarsus 0·45 to 0·5 ; mid toe and claw 0·85 ; bill to gape (straight) 2·0 to 2·1. Expanse 10·75.
Iris scarlet ; bill black ; legs and feet blackish, hinder part of tarsus paler.
Head, back, and sides of neck, back, scapulars, and wing-coverts shining brownish green, brownest on the head and hind neck, and passing into the bright green-blue of the rump and upper tail-coverts : external edges of the primaries and secondaries greenish blue, the remaining portion of the feathers pervaded with brown, which changes at the basal part of the inner webs into cinnamon-rufous ; tips of the shorter primaries and of all the secondaries blackish brown ; terminal portion of the tertials and the tail (with the exception of the blackish elongated tips of the central feathers) bright greenish blue, the rectrices brownish internally.
A broad black streak from the gape over the eye and ear-coverts, above it a faint line from the forehead to the posterior corner of the eye, and beneath it a broader stripe of bright greenish blue, the latter very pale at the termination : chin and upper part of throat yellowish ; fore neck chestnut-colour, gradually changing into the faded greenish of the breast, which brightens into cerulean blue on the under tail-coverts ; the basal portions of the under-surface feathers light brownish, showing more or less throughout ; under wing concolorous with the cinnamon bases of the quills ; shafts of the quills and rectrices white beneath.

Young. Iris dull red or brownish red, changing into the hue of the adult during the first year.
Above greener than the adult ; the bases of the feathers brownish green ; rump and upper tail-coverts not so bright as in the adult ; central rectrices not elongated, but slightly exceeding the rest and more pointed at the tips. The blue loral and cheek-stripes less conspicuous, and the chin not so yellow as in the adult ; under tail-coverts paling at their lateral margins.

The above is the plumage of the young birds arriving in Ceylon in September ; they quickly acquire the adult tail, and meanwhile the normal yellowish feathers of the chin and the chestnut ones of the throat make their appearance, the latter part in the quite young bird being much paler than in the adult.

Obs. I have examined some examples from Sumatra, and one or two from India, in the British Museum, which have

---

* According to Layard from a fancied resemblance in the tail of this bird to the aloe-plant.

the blue cheek-stripe broader than in any I have procured in Ceylon. Philippine specimens are identical with Ceylonese in plumage, but they are a smaller race; the wing of a Negros example is 4·0 inches, another 5·0. A Sumatran example measures 5·2 inches; two from Japan 5·15 and 5·25 respectively.

Distribution.—This fine Bee-eater, migratory to Ceylon, arrives in the north of the island about the beginning of September, and rapidly spreads more or less through all parts of it before the end of the month. It seems to find its way to the south-west corner, or Galle district, almost as soon as to any part of the island, and collects there in greater numbers than elsewhere on the western side. I have met with it in the interior of the country, between Galle and Akurresse, as early as the 8th of September. It locates itself in great numbers in the Jaffna peninsula, and on the north-west coast as far south as Puttalam, and spreads in tolerable numbers into the interior, passing over the forest-clad portions, however, to a great extent, and ascending to the patnas and open hills of the Kandyan Province. In Uva and Pusselawa and on the Agra, Lindula, and Ropatalawa patnas, at an elevation of 5000 feet, it is common; but I have never seen it on the "plains" of the Nuwara-Elliya plateau. In the Eastern Province it confines itself mostly to the sea-board, being less numerous in the Park country and the south-eastern "jungle-plain" than the next resident species. Its departure from the island is as sudden as it is regular, in proof of which I may state that at Galle, in two successive seasons, I observed it collect in large flocks between the 29th and 31st March, and disappear entirely on the 1st April. Mr. Holdsworth, who writes that at Aripu it was so abundant that the common resident species (*M. viridis*) was scarce in comparison with it, states that it left about the beginning of April; and by the end of that month I believe it has quitted the island entirely. In the neighbourhood of Colombo it is chiefly located in large tracts of paddy-ground and about the great swamp between there and Negombo. It is now and then met with in the cinnamon-gardens.

The Blue-tailed Bee-eater is found throughout most of the empire of India, being very generally distributed throughout the central and eastern portions of the peninsula during the cool season, while in the breeding-time it locates itself in those parts which furnish it with localities suitable to its nesting-habits. In some places it is rare: Mr. Fairbank met with it but once in the Khandala district; and it is mostly replaced by the Egyptian Bee-eater in the north-west, for though it "often occurs," according to Mr. Hume, in the Mount-Aboo district, it is neither found in Northern Guzerat nor in Sindh. In Chota Nagpur it appears to be local; but Mr. Ball writes, in 'Stray Feathers,' 1875, that he met with large numbers in the vicinity of a river in that region in April, and that he infers that they were breeding there. To the eastward of India this species is found in Tenasserim and Burmah, and likewise in the Malay peninsula, taking into its range the Nicobars and South Audamans. Further south still it is found in Java, Sumatra, Flores and Timor, and the Philippine Islands, and has been met with in China and Formosa by Mr. Swinhoe.

Habits.—In Ceylon this species prefers to frequent open lands, plains studded with bushes near the sea-shore, esplanades, paddy-fields, swamps, and the patnas of the hill-region. It passes a great part of its existence on the wing in pursuit of insects, after which it dashes with a very rapid flight, constantly uttering meanwhile its loud notes. When reposing from its labours, it rests on low objects, such as stumps of trees, fences, low projecting branches, little eminences on the ground, and often on the level earth itself. It is tame in its nature, allowing a near approach before it takes wing. On rainy evenings in November and December, when the air is swarming with insects, and particularly with winged termites, which issue forth from their nests on such occasions, the Blue-tailed Bee-eater congregates in large flocks on the wing, dashes to and fro for hours together, ascending to a great height in pursuit of its prey, and keeping up its not unpleasant notes without intermission. When exhausted with these exertions, they settle on walls, trees, or the ground in little parties, and when rested resume their flight. I have seen such flocks as these night after night on the Galle esplanade, and often observed them flying round and round high above the fort before finally moving off for the night to some distant and common roosting-place. When its prey consists of beetles, dragonflies, or other large insects, which it espies from its perch, it is captured after a sometimes prolonged flight, brought back, and killed before being swallowed by being repeatedly struck against whatever object the bird is seated on. This may often be witnessed when the bird is perched on telegraph-wires, which are a very favourite look-out with it. I have seen it dash on to the surface of ponds and rivers, and seize insects which were passing over the

2 R 2

water. Mr. Holdsworth has observed it hunting close to the surface of the sea, at a distance of a quarter of a mile from the shore. Jerdon notices its habit of congregating together, and writes that on one occasion he saw an "immense flock of them, probably many thousands, at Caroor, on the road from Trinchinopoly to the Nilghiris." They were sallying out from the trees lining the road for half an hour or so, capturing insects, and then returning to them again. As a rule they do not consort in close company, but live in scattered flocks of about half a dozen, and often one or two birds constantly frequent the same locality. The note is difficult to describe. Jerdon not inaptly speaks of it as "a full mellow rolling whistle." This Bee-eater retires late to roost, collecting to one spot from many miles round, and forming a large colony which pass the night in thickly foliaged trees or bushes. On Karativoe Island I discovered one of these roosting-places; the birds were flying over from the mainland some miles distant, and continued to arrive from various points on the opposite coast until it was too dark to distinguish them on the wing. They resorted to the borders of a small back-water beneath the high sand hills of the island, which was lined with mangrove-trees, the thick branches of which afforded them a safe refuge.

Nidification.—Mr. Hume writes, in 'Nests and Eggs' (Rough Draft), that "the Blue-tailed Bee-eater breeds from March until June pretty well all over continental India, in well-cultivated and open country. Like all the rest of the family it breeds in holes in banks, and lays usually four or five eggs. The holes are rarely less than four feet deep, and I have known them to extend to seven feet. At the far extremity a rounded chamber, as a rule not less than six inches in diameter, is hollowed out for the eggs, and at times this chamber has a thin lining of grass and feathers, which I have never yet met with in the nests of the other species." The banks of the Nerbudda, Mahanuddee, Ganges, a stream near Barnich, and localities at Lahore, Nujgeebulud, and Mirzapore are cited as breeding-places of the species; and Mr. Hume himself found a colony established in a railway-cutting at Agra, where the engines "passed twenty times a day within two feet of the mouths of the holes." The eggs are white, highly glossed, and very spherical ovals, averaging 0·88 by 0·76 inch.

MEROPS VIRIDIS.

(THE GREEN BEE-EATER.)

Merops viridis, Linn. Syst. Nat. i. p. 182 (1766); Bonn. Enc. Méth. Orn. pt. i. p. 273,
pl. 105. fig. 3 (1790); Sykes, Cat. no. 23, J. A. S. B. iii. (1834); Blyth, Cat. B. Mus.
A. S. B. no. 236, p. 53 (1849); Kelaart, Prodromus, Cat. p. 119 (1852); Layard,
Ann. & Mag. Nat. Hist. 1853, xii. p. 173; Horsf. & Moore, Cat. B. Mus. E. I. Co.
p. 84 (1851); Gould, B. of Asia, pt. vii. (1855); Jerdon, B. of Ind. i. p. 205 (1862);
Holdsworth, P. Z. S. 1872, p. 422; Hume, Nests and Eggs, p. 99 (1873); Adam,
Str. Feath. 1873, p. 371; Hume, ibid. 1875, p. 49; Legge, Ibis, 1875, p. 281; Oates,
Str. Feath. 1876, p. 304; Dresser, B. of Europe, pt. 51 (1876).
Merops orientalis, Lath. Ind. Orn. Suppl. p. 33 (1801).
Merops indicus, Jerd. Madr. Journ. xi. p. 227 (1840).
Merops torquatus, Hodgs. Gray's Zool. Misc. 1844, p. 82.
The Indian Bee-eater, Edwards, pl. 183.
Le Guêpier à gorge bleue, Levaill. pl. 10. p. 39.
The Common Indian Bee-eater (Jerdon); *Flycatcher*, Europeans in India and Ceylon; *Hurrial*,
Patringa, Hind.; *Bansputtee*, lit. "Bamboo-leaf," Bengal; *Chinna passeriki*, Tel., lit.
"Small green bird" (Jerd.); *Mo-na-gyee*, Arracan (Blyth).
Kurumenne kurulla, Sinhalese; *Kattalan kuruvi*, Tamils in Ceylon.

Adult male and female. Length 9·5 to 10·5 inches, according to length of tail; wing 3·6 to 3·8; tail 5·1, central
feathers 2·0 to 2·3 longer than rest; tarsus 0·4; middle toe and claw 0·6; bill to gape 1·4 to 1·55.
Iris scarlet; bill black; legs and feet brown, the edges of scales whitish.
Above leaf-green with a bronze lustre, paling to bluish green on the tertials, rump, and upper tail-coverts; basal or
concealed portion of the head- and nape-feathers golden fulvous, showing on the surface at the occiput and nape:
quills deeply tipped with blackish; inner webs of secondaries and borders of those of primaries pale cinnamon,
which is likewise the colour of the under wing; tail green, with the tips of the shorter and elongated portion of
the central feathers blackish.
A broad black stripe from nostril and gape over the eye and ear-coverts; above it a narrow yellowish-green supercilium; chin and throat greenish turquoise-blue, deepening into brownish green on the upper breast, and paling
into bluish green on the lower parts and under tail-coverts; across the throat a conspicuous black band, edged
above and beneath with bright yellow-green; vent whitish.

Birds in old plumage have the nape and occiput much yellower than those in good feather, the paler colour being the
result of abrasion; this must not, however, be confounded with the fine aureous lustre observable in some
specimens, particularly those from N.E. India and Burmah.

Young. Iris light red or yellowish red; bill generally pale at the base beneath; legs and feet blackish slate. Central
tail-feathers not lengthened.
Above green, the feathers edged with bluish; aural stripe blackish brown; throat, neck, and chest greenish blue,
palest on the chin; lower breast and belly albescent; under tail-coverts bluish green. Some nestlings have the
throat tinted with yellowish. The black throat-bar is acquired at a very early age, but is narrow and ill-defined,
and in some edged with blue; the long central tail-feathers are likewise acquired, about the same time, by a
"nestling" moult, although tolerably old yearlings may now and then be seen without them.

Obs. Ceylonese specimens of this Bee-eater vary, as above mentioned, in the golden hue of the nape and hind neck,
but do not exhibit the brilliant hue of birds from Cachar and Burmah, to which Hodgson gave his name of

ferrugineiceps: they are typical *M. viridis*, like birds from Central and Southern India; but it must be remarked that occasionally very rufous-headed specimens are procured in Madras. That the species is variable in this character throughout its entire habitat may be gathered from the fact, demonstrated by Mr. Hume, of the Sindh race almost wanting the rusty golden tinge. In Ceylon I have observed that nestling birds vary in the extent of the brighter colours of their plumage when these are first put on, the development of such tints depending perhaps on the physical vigour of the individual. I once shot a pair of young green Bee-eaters together, which were, of course, out of the same nest—one with the normal plain green throat and short tail of the nestling, the other with the blue throat-band appearing and the central tail-feathers half-grown. Perhaps the latter would always have been a more brilliantly plumaged bird than the former; for the difference in age, at most 24 hours, could scarcely have accounted for the backwardness of the plainer specimen in acquiring its adult character. As regards the relative size of Indian and Ceylonese birds, I find that the wings in 8 specimens from Pegu (as given in 'Stray Feathers') vary from 3·0 to 3·8 inches, precisely the measurements given above for Ceylonese birds. Some Indian examples have the central tail-feathers longer than any I have seen in Ceylon; one specimen from Kamptee in the British Museum has them 2·0 inches beyond the adjacent pair, 2·3 being my limit. The dimensions given by Mr. Armstrong of the wings of several Burmese specimens, viz. 4·6 to 5·2 inches, are most probably those of some other species entered by a printer's error in his note on *M. viridis*.

Concerning the species in North Africa, Mr. Dresser writes that examples from Egypt, India, and Abyssinia all have the throat markedly green and the head but slightly tinged with rufous. This is, of course, to be expected, in continuation of the characters displayed by the westernmost of Indian birds, viz. those from Sindh. He further remarks that, according to his experience, Indian specimens have, as a rule, the throat tinged with verditer-blue, and that those from Ceylon exhibit this character to a still greater extent; this, however, is with us somewhat variable, as I have demonstrated above.

Distribution.—The Green Bee-eater is a resident species and very numerous in all the dry parts of the low country. It is most abundant about open scrubby land near the sea-coast round the north of the island and along the south-east and eastern sea-boards. Its habitat seems to be restricted to a nicety by the influence of climate. It is common in the interior of the northern half of the island, as well as in the maritime regions, and can be traced along the foot of the western slopes of the Matale ranges from Dambulla to Kurunegala, and thence across the dry country on the north of the Polgahawella and Ambepussa hills to Chilaw and Madampe, near which it stops, not being found south of Nattande. So much does it avoid a moist atmosphere that it extends for a few miles south of Kurunegala, on the high road to Polgahawella, and suddenly vanishes on the road entering the hills. South of these limits it is unknown throughout the Western Province and the south-west hill-region, reappearing again just to the eastward of Tangalla, where the climate again becomes dry; beyond this all round the coast it is common, being particularly numerous in the Hambantota and Yála districts. I have traced it through the interior to the foot of the Haputale hills, but it is much scarcer there than at the sea-coast. In the Eastern Province it inhabits the high cheenas in the neighbourhood of Bibile, which attain an altitude of 1000 feet, and which is the highest point I have found it to attain in Ceylon. Mr. Holdsworth remarks, *loc. cit.*, that it occurs about Colombo. I conclude that the evidence on which this place is included in its range must be that of a stray bird; for I have never observed it anywhere nearer to it than the above limits, neither has Mr. MacVicar nor the taxidermist of the Colombo Museum, both of whom have collected for many years in that part.

This species is spread all over India, extending into Burmah, Tenasserim, Arrakan, and the Indo-Chinese countries. It is common in the south of the Peninsula and ascends the hills. Mr. Fairbank procured it at the base of the Palanis, and Mr. Davison has shot it at an elevation of 6000 feet above the sea in the Nilghiris and found it breeding at about 5000 feet. In the Deccan and Khandala district it is common according to Mr. Fairbank, and the same is true as regards the north-west of India; for Mr. Adam records it as very plentiful about the Sambhur Lake, and Mr. Hume found it pretty common all the year round in Upper Sindh, though comparatively rare in Lower Sindh. It is found along the base of the Himalayas, but does not extend to any elevation. In Chota Nagpur it is one of the "most abundant of birds." In Cachar Mr. Inglis says it is common between August and April, in which latter month a large number migrate. In Pegu it is extremely numerous in the low country, but not in the hills. In Tenasserim it is generally distributed; but it is absent from the islands of the Bay of Bengal, where our other two species are found. It appears to be a seasonal visitant to the neighbourhood of Calcutta, for Capt. Beavan records that it arrives at Barrackpore in October. Westward of India it extends through Beluchistan and Persia to Northern Africa, and there is not uncommonly found in Egypt, Nubia, and Abyssinia.

Habits.—This is one of the most charmingly fearless little birds in Ceylon; unlike the last it is very terrestrial in its habits, perching all day on some little bush or low stick near the ground, and sallying out like a Flycatcher after its food, when it at once returns to its perch or sweeps off to another close by. It is generally found in pairs, or three or four in scattered company, which frequent roadsides and dry open ground of all description where they can find objects to take up their watch upon. About Trincomalie, and, in fact, anywhere on the sea-coast of the eastern side of the island, it is very fond of the sandy scrubby wastes lining the sea-beach, and is so tame that it may be almost knocked down with a stick, so near an approach will it allow before taking wing. In the interior a favourite locality with it is the dried-up paddy-fields in the neighbourhood of the village tanks. It roosts in little colonies, retiring early to rest and congregating in close company ; it resorts usually to the same tree, round which much noisy preparation goes on—flying up and wheeling round, alighting on a neighbouring tree-top and then returning, after which the little flock will start out again from the branches and make another little detour, keeping up all the while a continuous clamour. Its note is a sweet little chirrup, unlike the loud voice of the last species. It is either uttered when the bird is perched or when it is sailing along in pursuit of an insect, which it seizes with an audible snap of its bill. It usually preys on small flies or minute Coleoptera, avoiding large dragonflies and other giants of the insect kingdom, upon which the last species feasts and beats to death in the manner aforementioned. Jerdon says that he has seen one occasionally pick an insect off a branch or a stalk of grain or grass ; and Blyth has seen them assembled round a small tank seizing objects from the surface of the water, after the manner of a Kingfisher. I have also observed them about rushy jheels and small tanks, but they are not particularly partial to the vicinity of water.

Nidification.—This Bee-eater breeds in the sand hills at Hambantota and other similar localities in Ceylon. I found the young fledged, on the south-east coast, in June, but did not succeed in finding any nests. The nesting-time is in April and May. Mr. Hume says that it prefers to breed in sandy banks or cliffs, but that he has found the nest in a mud wall, and once in a perfectly level barren plain. It cuts the hole, after the manner of the last species, with its bill, scraping away the loose earth with its little feet, and sometimes excavates to a depth of 5 feet, the passage increasing in width and often, according to Mr. Adam, declining at an angle of 30° from the entrance to the egg-cavity, which is about 3½ inches in width. No nesting-materials are used, the eggs, which vary from three or four (the usual number) to seven, being laid on the bare ground. The eggs are nearly spherical in shape, milky white, and "brilliantly glossy." The average size of a large series is 0·78 by 0·7 inch.

MEROPS SWINHOII.

(THE CHESTNUT-HEADED BEE-EATER.)

Merops quinticolor, Vieillot, N. Dict. xiv. p. 81 (1817); Kelaart, Prodromus, Cat. p. 119 (1852); Layard, Ann. & Mag. Nat. Hist. 1853, xii. p. 174; Horsf. & Moore, Cat. B. Mus. E. I. Co. p. 88 (1854); Jerdon, B. of Ind. i. p. 208; Holdsworth, P. Z. S. 1872, p. 423; Walden, Ibis, 1873, p. 301; Legge, Ibis, 1874, p. 13.

Merops erythrocephalus, Brisson, Av. iv. p. 563; Blyth, Cat. B. Mus. A. S. B. p. 53 (1849); Swinhoe, P. Z. S. 1871, p. 348.

Merops swinhoei, Hume, Nests and Eggs (Rough Draft), p. 102; id. Str. Feath. 1874, p. 163; Ball, ibid. p. 386; Armstrong, ibid. 1876, p. 305.

Le Guêpier quinticolor, Levaillant, Hist. Nat. Guêpiers, p. 51, pl. 15 (ex Ceylon).

The Five-coloured Bee-eater, Kelaart, Prodromus; "*Flycatcher*" of Europeans in India and Ceylon.

Kurumenne kurulla, Sinhalese, Southern Province; *Pook-kira*, Sinh., N.W. Province.

Adult male and female. Length 8·4 to 8·6 inches; wing 4·2 to 4·3; tail 3·3; tarsus 0·45, middle toe and claw 0·65: bill to gape 1·6 to 1·8.

(In this species the tail-feathers are not elongated, but the tail is somewhat sinuated, the central pair being rounded at the tips and longer than those adjacent, though shorter than the laterals.)

Iris scarlet; bill black; legs and feet dark vinous brown or purplish brown.

Head, hind neck, sides of the same, interscapular region, and upper edge of black throat-band bright chestnut; wings and tail dull green, edges of wing-coverts, terminal portion of tertials, and edges of rectrices bluish; rump and upper coverts pale cerulean blue, tips of the longer-coverts darker; tips of quills and rectrices, with the exception of the centrals, brownish black; inner webs of secondaries, borders of those of primaries, and under wing cinnamon-red as in the other species.

A black facial stripe, narrower than in the last, passing from the gape beneath the eye; chin and throat rich saffron-yellow; black throat-band bordered beneath with golden yellow; beneath this the underparts are green, passing into pale greenish blue on the lower breast, abdomen, and under tail-coverts.

Young. *Birds of the year* have the chestnut of the upper surface paler, the throat whitish, the black band ill-defined and slightly edged with yellow beneath, the wing-coverts and secondaries margined with blue, and the chest greenish blue like the lower parts. The nestlings, which are blind for the first few days, quickly acquire the feathers of their first plumage as here described.

Obs. This species was first made known from Ceylon—that is to say, specimens were sent to Levaillant from there, and the bird was named by him, in his work on the ' Guêpiers,' the *Guêpier quinticolor*; but by some oversight he gave a plate of the species inhabiting Java, and accompanied it by a description, in which he stated the colour of the throat to be "d'un jaune jonquille, lequel jaune est terminé au bas par un collier noir," making no mention of the triangular chestnut throat-patch above the black mark, which character is wanting in the Javan bird, as it likewise is in his plate. His plate and description did not therefore apply to the Ceylon bird, nor can Vieillot's name, which was founded on the plate. *Merops quinticolor* accordingly is the Javan bird, and not the Indian. The matter has been referred to by the late Mr. Swinhoe and Lord Tweeddale in the references above given, and Mr. Hume gave the Indian bird its present title in his notice of it in 'Nests and Eggs,' as it was without a name. Ceylonese examples correspond with Indian and Burmese in size and likewise in coloration of the throat, though individuals from any district will differ *inter se* in this latter respect. One specimen I have examined in the British Museum from Madras has a wider black throat-band than any I have seen from Ceylon. Pinang specimens correspond with Ceylonese.

Distribution.—This handsome Bee-eater is sparingly dispersed over the island, inhabiting some localities in considerable numbers, while in other districts mere stragglers are met with. In the south it is common on the Gindurah river, commencing above Baddegama and extending up into the hills of the Hinedun Pattu; it likewise frequents the banks of the Kaluganga, Kelaniganga, and Maha-oya in the Western Province, and is found here and there through Saffragam. To the north of these localities it is located about Kurunegala, on the Deduru-oya, in the Puttalam district, and in isolated spots in the neighbourhood of Dambulla. Mr. Parker has met with it in the Anaradjapura district, and it occurs sparingly throughout the northern forests. I have seen it between Trincomalie and Mullaittivu, but I do not think it is to be found much to the north of the latter place. In the Kandyan Province it is much more common than in most parts of the low country, inhabiting the vale of Dumbara, Deltota, Nilambe, Maturatta, and Uva generally. It does not ascend to the Nuwara-Elliya plateau.

This species is found in most of the forest-districts of India, Burmah, and Tenasserim, inhabiting the Andamans and extending to Pinang. Jerdon writes that it occurs in the Malabar forests and adjoining mountains, and is not uncommon in the Wynaad and other elevated wooded districts. I notice that Mr. Bourdillon did not procure it in the Travancore hills, nor Mr. Fairbank in the Palanis. The latter gentleman found it on the sides and base of the Goa and Savant-Wade hills, and records it as an inhabitant of the entire west coast as far north as Guzerat, whence, however, I do not observe that it has been procured. Capt. Marshall writes, in 'The Ibis,' 1872, that it is found in the Doon and the Terai, and along the whole of the southern skirts of the Himalayas to the valley of the Brahmapootra. In Chota Nagpur it is rare, Mr. Ball recording the occurrence of a single pair only; in Cachar it is migratory, being common during April and May: in Southern Pegu it occurs very sparingly; Mr. Armstrong met with it there in the month of February: at Thayetmyo Captain Feilden says it is rare, and in the plains of Pegu Mr. Oates did not meet with it at all. Mr. Davison found it throughout Northern Tenasserim, and in the Andamans he procured many specimens, meeting with it in Port Blair, Great and Little Cocos Islands, &c., but in the Nicobars it was not found.

Habits.—The banks of rivers which flow through forest or the borders of jungle-begirt tanks are the favourite localities of this bird in the low country. In the Central Province I have seen it principally in the vicinity of rivers in the deep valleys leading to the Mahawelliganga, on roads leading through jungle, and in spots studded with high trees on the sides of steep ravines. It is usually in pairs, and is very arboreal in its habits, sitting on the topmost or most outstretching branches of high trees overhanging water, and darting thence on its prey, much after the manner of a Flycatcher. It takes short flights, and often returns to the same perch again. It is a very pretty object, with its bright green plumage and glistening rufous head, as it darts from the fine old trees lining the forest-rivers down to the edge of the sparkling stream, and glides over the sandy bed, quickly catching up some passing insect. A pair may sometimes be seen seated on a dead twig, touching one another, so very sociable is it in its disposition. It has a soft note, differing from that of either of the foregoing species, which it generally utters from its perch.

Nidification.—I found the nest of this bird on the banks of the Gindurah in the month of April. The hole was excavated in the soft mould near the top of the bank, went in about 2 feet, with an average diameter of 2 inches, and at the end widened into a cavity 4 or 5 inches in height and nearly double that in width. There were four young ones lying on the bare ground, which was swarming with living maggots, ants, and flies, brought in for their food by the old birds. The nestlings showed a marked difference in age; two were perhaps not three days old, and the others had the green scapular feathers already sprouting. Layard found the nest in the same month, and says the eggs are two in number.

Mr. Davison writes that the hole is sometimes 6 feet in depth when excavated in sand, and that some turn off at a right angle, while others take a circular direction. The eggs are stated to vary from four to six in number, and to be pure white, very glossy, and nearly spherical in shape; they average 0·87 by 0·76 inch. The old birds are said to sit very close, allowing themselves to be dug out.

2 s

PICARIÆ.

Fam. CYPSELIDÆ.

Bill very small, but with the gape enormous, unfurnished with rictal bristles; tip hooked. Wings very long and pointed, with ten primaries. Tail variable, short and even, or long and much forked, of ten feathers. Legs and feet small and feeble; hind toe either directed forward or more or less reversible to the front.

Sternum with the keel very deep. Humerus very short. Throat furnished with large salivary glands.

Genus CHÆTURA.

Bill very small, triangular, the gape receding far back and very wide; culmen curved, flattened at the base, the tip hooked. Nostrils exposed. Wings very long and pointed. The humerus and ulna very short; the 1st quill the longest; the inner very short, imparting a sickle-shape to the wing. Tail short, even or rounded at the tip; the shafts rigid, very acute, and projecting some distance from the web. Tarsus short, stout, feathered just below the knee, and the rest covered with a naked skin. The three front toes nearly equal, the hallux directed backward but reversible to the front; claws stout, deep, and much curved.

CHÆTURA GIGANTEA.

(THE BROWN-NECKED SPINE-TAIL.)

Cypselus giganteus (V. Hasselt), Temm. Pl. Col. 364 (1825).
Acanthylis caudacuta, Blyth, Cat. B. Mus. A. S. B. p. 84 (1849); Kelaart, Prodromus, Cat. p. 118 (1852); Layard, Ann. & Mag. Nat. Hist. 1853, xii. p. 170.
Acanthylis gigantea, Jerd. B. of Ind. i. p. 172 (1862); Holdsworth, P. Z. S. 1872, p. 419; Ball, Str. Feath. 1873, p. 55; Legge, Ibis, 1875, p. 280; Tweeddale, Blyth, B. Burmah, ext. no. 1875, p. 84. no. 183.
Chætura gigantea, Sclater, P. Z. S. 1865, p. 608.
Chætura indica, Hume, Str. Feath, 1873, p. 471, et 1876, p. 286.
Hirundinapus giganteus, Walden, Ibis, 1874, p. 131.
The Needle-tailed Swallow, Lath. Gen. Synopsis; *The Spiny-tailed Swift*, Kelaart. *Wæhælaniya*, Sinhalese.

Adult (from three Ceylonese specimens). Total length, estimated from skins, 9·5 inches; wing 7·8 to 7·95, reaching 1·5 beyond tail when closed; tail 2·7 to 2·9, bare shafts of central feathers 0·35 to 0·4; tarsus 0·65 to 0·7; middle toe 0·5, its claw (straight) 0·35; bill to gape 1·0.

Iris brown; bill blackish or dark brown; legs and feet livid brown or fleshy purple, claws blackish brown.

Head, back and sides of neck, upper part of back, anterior scapular feathers, wings, sides of rump, and upper tail-coverts shining green-black, glossed on the wing-coverts, secondaries, and sides of rump more or less with blue; back whity brown, of variable paleness, blending into the surrounding green; inner margins of quills and tertials light mauve-brown, palest on the latter; shafts of tail-feathers blackish brown.

Lores intense black, between which and the nostril there is a whitish or whity-brown spot : throat a corresponding pale colour—that is, lightest in those birds which have the palest frontal spots; beneath umber-brown, glossed obscurely with green, and blending gradually on the throat into the pale hue of the chin; under tail-coverts and a broad streak leading from them above the flank to opposite the centre of the back white : shafts of under tail-coverts black ; under wing-coverts pale mouse-brown.

Young. Immature birds have the frontal patches scarcely discernible, the head browner than the adult, the back darker, and the under surface less suffused with green.

Obs. This Swift is variable in the pale markings about the face and chin, in the light hue of the back, and in the extent of the blue gloss on the upper plumage. I have examined a series from Labuan, Malacca, Singapore, and South India, and I find that the dark-backed birds, which are evidently not fully aged, have the chin and loral spots of a correspondingly dark hue. Mr. Hume has separated the Indian birds as *C. indica*, on account of their more pronounced white chin and frontal patches, as distinguished from what he considers to be true *C. gigantea* from Java, without the white chin. If the type from this island had not the whitish markings it must have been, in all probability, an immature bird. Temminck's plate shows no white nostril-patches; but in those days artists were not particular.

I am not conversant enough with Indian specimens to say whether they never show an absence of the white patches either as young birds or as individuals; but those from all other quarters, as I have just remarked, vary in this respect. Birds from each end of the geographical limit of the species, viz. from India and Celebes, have the white spots alike, which argues in favour of there being but one species. Two examples from Labuan measure 8·1 and 8·2 inches in the wing: one is a dark-backed bird, the other a light one, and the chin and forehead tally with the back in each : two from Malacca measure 8·1 and 7·9 inches in the wing; one has a dark back and no loral spot, the other is slightly paler and has an indication of the light patches. One from Singapore measures 7·9 inches, has a very dark back, no frontal patches, and a dull brown under surface : it is evidently a young bird. Another from the Nilghiris is entirely a pale bird, with light chin- and nostril-spots. Lord Tweeddale finds that adolescent examples from the Andamans agree with Malaccan ones in his collection.

Distribution.—The Brown-necked Spine-tail is a resident in the Ceylon hills, wandering at uncertain times during its day's peregrinations over the whole island. In the upper ranges it is most often seen frequenting the Horton, Nuwara-Elliya, Kandapolla, and Elephant Plains, over which it dashes at one moment, while at the next it sweeps round the adjacent hills in its headlong course. It is frequently noticed in the coffee-estates in the surrounding districts. Mr. Elwes writes that it is often seen in Dimbulla; and Mr. Bligh, who observes it yearly in the Haputale gorges, tells me that it comes into that district to breed usually about the month of April. It inhabits the Morowak-Korale and Kukkul-Korale hills, in which I have seen it in various months, and I have no doubt it breeds there in sequestered places. I have seen it in large flocks on the sea-coast at Tangalla, and Capt. Wade has met with it at Yála. On one occasion, too, I encountered it in the north of the island. It hawks, as I have seen *C. caudacuta* in Australia, at an enormous height, and when rained on by a monsoon shower descends to earth, and is thus seen for a few minutes in the low country, vanishing again on the return of sunshine. Layard knew it principally from Nuwara Elliya.

In India Jerdon observed it chiefly in the south of the peninsula, specifying the Nilghiris, Malabar, and the Wynaad as the localities where he met with it. Mr. Carter found it during the S.W. monsoon at Coimbatore, Salem, and on the Anamully hills at various elevations up to 6000 feet. The species does not seem to extend into the north of India, where its Australian and Chinese congener, *C. caudacuta*, singularly enough, is found in considerable numbers. Our bird inhabits Tenasserim, and Mr. Inglis obtained it in Cachar. It is common in the Andamans, but has not been procured in the Nicobars. It extends down the Malay peninsula (taking in Pinang) to Singapore, and thence to Java, Labuan, Borneo, and Celebes, to the south-east of which latter group it has not yet been observed.

Habits.—This magnificent Swift and its Australian ally are the swiftest creatures in existence, excelling all other living beings to such an extent in their powers of locomotion that they cannot fail, as the per-

2 s 2

fection of an all-wise Creator's handiwork, to excite wonder and admiration in the mind of the naturalist and true lover of nature. That any bird can sustain an aerial course of such rapidity for 12 or 14 hours at a time, without any cessation from its exertions, must of necessity excite the astonishment of the most careless thinker, while to the inquiring mind it amply demonstrates what a marvel of strength and perfection of structure are exhibited in this wonderful bird. A casual glance at one of these Swifts will show that it is entirely formed for speed. The pointed aspect of its face and bill, with the thick lores and stiff superciliary feathers to protect the eye from the rush of air, its broad body, gradually tapering from the rump to the acute tip of the tail, give it the form of a feathered *projectile* constructed to acquire immense velocity, which, in truth, its rigid sickle-shaped wings, with their specially lengthened metacarpal bones imparting so much power to the downward stroke, cannot fail to give it. It is this peculiar outward form which imparts to it a so much higher power of speed than exists in other Swifts, such as the next species, for the structure of the sternum is not so very much superior to that of the Alpine Swift. Dr. Sclater writes, in explanation of the drawing of the sternum of this Spine-tail which is contained in the P. Z. S. 1865, that it is broader in proportion and less elongated than in *Cypselus*, and that the anterior point or apex of the keel is not carried so far forward. Apparently these slight differences would not give the Spine-tail the superiority over the ordinary Swift which it possesses were it not for its admirable external shape and greater length of metacarpus.

This Spine-tail haunts the vicinity of rocky precipices and steep hill-sides, dividing its time between careering round them and up and down adjacent valleys and sweeping over the surrounding country, especially where there exist open tracts, in search of food. When hawking in a large flock its flight is not unlike that of the Alpine Swift; but it is varied by vast circles and detours made with astonishing swiftness, as if merely for exercise, returning in a moment to its place in the flock. It is not in this manner, however, that its great powers of flight are put forth; it is in returning at nights from its day's labours to its far-distant roosting-place that these are brought out, and then its flight is as swift as the momentary rush on its quarry of the Peregrine Falcon. I have experienced this on more than one occasion in the Ceylon hills, where a whiz just over my head, like that of a bullet, has brought my attention to the onward course of one of these birds, which the next moment had disappeared far away in the gloom of the tropical evening. Mr. Carter writes, concerning a flock that he fired at, "I should not like to say how many I missed; but some idea of their rate of speed may be formed when I say that in seeing one coming towards me and turning sharp round, by the time I sighted it it was too far . . . . The two I got I killed passing over me, making great allowance and firing far in front. One, although quite dead when I came up to it, had managed to clutch a stone, which remained tight in its claws." Mr. Davison observed that they hawked very high in the air, betraying their almost invisible presence by a sharp clear whistle. At nights they were found, in company with other Swifts, about ponds or tanks. Concerning the roosting of this Swift, which is one of the most interesting points in its economy, very little seems to be known. Its spinous tail is evidently a provision of nature to afford it support against the rock to which it clings at night. It most probably, as suggested by Jerdon, has some fixed roosting-places, to which large flocks resort from immense distances, arriving no doubt at a late hour, and thus preventing the possibility of their haunt being discovered from observations of the birds on their way thither. He observed that they flew towards the coast, and on one occasion witnessed an enormous flock passing him on their way towards the sea some time after sunset, although there was no situation on the west coast where they could have roosted; consequently the idea suggests itself that they make for the sea-shore and then travel along it to their nightly rendezvous.

Layard was informed by the natives that this species nested in rhododendron-trees, which, it is scarcely necessary to remark, is an erroneous idea. It breeds, as its near ally the White-necked Spine-tail, in lofty cliffs. Mr. Bligh informs me that they yearly resort to some inaccessible precipices in the Haputale ranges for the purpose of breeding, but he has been unable to find their nests or procure their eggs.

Genus CYPSELUS.

Bill slightly stouter and more curved from the base than in *Chætura*. Wings equally long, the metacarpus shorter in proportion; the 2nd quill equal to, or longer than, the first. Tail variable in length, emarginate or deeply forked. Tarsus very short, feathered; all four toes directed forward, but the two inner reversible, shorter than in the last.

CYPSELUS MELBA.

(THE ALPINE SWIFT.)

Hirundo melba, Linn. Syst. Nat. i. p. 345 (1766).
Hirundo alpina, Scop. Ann. i. Hist. Nat. p. 166 (1769).
Cypselus melba, Illig. Prod. Syst. Mamm. et Av. p. 230 (1811); Gould, B. of Europe, pl. 53 (1837); Blyth, Cat. B. Mus. A. S. B. no. 421. p. 85 (1849); Kelaart, Prodromus, Cat. p. 117 (1852); Layard, Ann. & Mag. Nat. Hist. 1853, xii. p. 167; Jerdon, B. of Ind. i. p. 175 (1862); Tristram, P. Z. S. 1864, p. 431; Sclater, P. Z. S. 1865, p. 598; Holdsworth, P. Z. S. 1872, p. 419; Severtzoff, Faun. Turkestan, pp. 67, 145 (1873); Dresser, B. of Europe, pt. 31 (1874); Butler & Hume, Str. Feath. 1875, p. 453.
Le Grand Martinet à ventre blanc, Mont. Hist. Ois. vii. p. 316 (1783).
Le Martinet à gorge blanche, Levaillant, Ois. d'Afr. (1806).
Andorinhão gaivão, Portuguese; *Avion*, Spanish; *Alpensegler*, German.
The Common Large Swift, Kelaart, Prodromus.
Wæhælaniya, Sinhalese.

Adult male and female. Length 8·5 inches; wing 8·0 to 8·25; tail 3·0 to 3·5 tarsus 0·55; middle toe 0·35, its claw (straight) 0·32 to 0·35; bill to gape 0·85 to 0·9.
The tail is slightly forked in this species.

Obs. These measurements are taken from three Ceylon examples, and are below those of birds from Europe and Africa, some of which, from Switzerland, range as high as 8·7 in the wing. Possibly these Ceylonese specimens were bred in the island, and would almost of necessity be smaller than those from cold countries.
Iris brown; bill blackish, darkest at the tip; feet livid brown, claws black.
Head, all the upper surface with the wings and tail glossy earth-brown, passing over the chest and down the flanks to the under tail-coverts; on the wings and tail a strong brownish-green lustre is often present: feathers of the back, rump, and upper tail-coverts with the shafts perceptibly darker than the web; quills and rectrices darker than the back; lores black, surmounted by a thin whitish line; chin, throat, breast, and abdomen white; the feathers above and below the brown pectoral band and those of the flanks more or less tipped with the same; thighs and tarsal feathers concolorous with the flanks; under wing-coverts dark brown, some of the feathers tipped with white; edge of the wing more or less narrowly margined with white.

Young. Birds of the year have the feathers of the head, sides of the neck, and all the upper surface with fine whitish terminal margins, external edge of wing-lining with conspicuous white edgings, the white throat-patch more extensive, reducing the extent of the brown pectoral band; under tail-coverts tipped with white.

Distribution.—The Alpine Swift takes up its quarters almost exclusively in the upper regions of the Kandyan Province; but, being a bird of such immense powers of flight, it wanders with ease, in the course of a day's hawking, over all parts of the island. Hence Layard observed it at Dambulla and Ratnapura, and I have seen it at Topare tank. Mr. Holdsworth records it as frequenting Nuwara Elliya throughout the cool season, and Mr. Bligh has noticed it both there and in Haputale at various times of the year. In May I found it in great numbers congregated about the high cliffs of Ragalla, which rises above the Elephant Plains, where, as Mr. E. Watson informs me, it is often to be seen. It probably frequents the Gongalla range, in the southern coffee-district, in common with the last species.

Ceylon appears to be the most southerly point of this Swift's range in Asia. It is found all through India, more particularly in the Ghâts, Nilghiris, and Cashmere hills, from which it extends through Western Asia to Europe, which may be most properly styled its head-quarters, and where it is well known in the Alps, Pyrenees, and other groups of mountains. Through Africa it wanders as far as Cape Colony, whence it is recorded by Layard, Andersson, Ayres, Shelley, and others, but in the tropical region south of the Atlas it has not as yet been observed ; in the northern parts of the continent it is common, wandering over Egypt and Algeria in the summer, and the same may be said of the northern sea-board of the Mediterranean. Mr. G. C. Taylor records it as plentiful in the Crimea and at Constantinople ; Mr. Danford noticed it as a summer visitant to parts of Asia Minor, and Severtzoff found it breeding in scattered localities in Central Asia (Turkestan). To parts of India it is a cold-weather visitant ; at Mount Aboo it arrives, according to Captain Butler, in large numbers about the beginning of September, and remains throughout the season. It has not been found to the eastward of the Bay of Bengal, being replaced in Burmah and Tenasserim by *C. pacificus*.

Habits.—This splendid Swift, which, next to the larger species of the foregoing genus, is the swiftest bird in existence, loves to haunt the vicinity of great mountain declivities, towering precipices, ravines, or great river-gorges, about which it dashes at tremendous speed, either in search of its insect-prey, or, as would appear to an eye-witness, from some normal habit of exercising its marvellous muscular power. It is most active, like other Swifts, before rain, when the atmosphere teems with life, or on still evenings, when it may be seen varying its headlong flight with extensive curves and vast swoops, from which it will rise with renewed swiftness and redoubled beatings of its long, sickle-shaped wings. It hawks late in the evening, and it is generally nearly dusk before it directs its course towards the far-off roosting-place which it left in the morning, and the reaching of which will perhaps add some hundreds of miles to the immense distance which it has traversed during the day. Dr. Jerdou, who, to judge by his writings, took much interest in this family, observed them in the south of India flying towards the sea-coast about sunset, and was of opinion that it was their habit to make for the seaside and then follow the coast-line, "picking up stragglers from other regions on their way to the cliffs of Gairsoppa," where he discovered that they roosted. Tickell, as quoted by the same author, noticed these Swifts assembled "of an evening near large ponds in the jungle, dashing into the water with loud screams," like the Common Swift of Europe. They assemble in very large flocks, and, as I noticed at Polanarua, suddenly appear in a locality, and, after hawking it well, as quickly disappear again. It has a shrill, tremulous cry, which has a curious sound as the bird rapidly approaches the spectator, and, instantly passing overhead, is again quickly out of hearing. It is said to roost against cliffs, clinging to the rock in an upright position, for which its powerful and much-curved claws are well adapted.

Nidification.—As regards Ceylon, little or nothing is known of this Swift's breeding. Mr. Bligh is of opinion that it nests in April and May near Nuwara Elliya and on the southern slopes of the Haputale range, and it is not impossible that some of the birds observed by me at Ragalla were breeding in the great precipice there. It does not confine itself to cliffs and rock-faces, but will nest in churches and other large buildings. Mr. Hume describes nests sent to him by Miss Cockburn from the Nilghiris as being made of "feathers firmly cemented together with saliva ; but vegetable fibre of different kinds and dry grass formed part of the structure, which was a coarse felt-like mass of about 5 inches in diameter, with walls 1 inch thick ; and several nests appear to have been grouped together. The eggs are four or five in number, pure white."

CYPSELUS AFFINIS.

(THE INDIAN SWIFT.)

Cypselus affinis, J. E. Gray, Ill. Ind. Zool. i. pl. 35. fig. 2 (1832); Sykes, P. Z. S. 1832,
p. 83; Jerdon, Cat. B. S. India, Madr. Journ. 1840, xi. p. 235. no. 255; Blyth, Cat.
B. Mus. A. S. B. p. 86 (1849); Kelaart, Prodromus, Cat. p. 117 (1852); Layard, Ann.
& Mag. Nat. Hist. 1853, xii. p. 167; Horsf. & Moore, Cat. B. Mus. E. I. Co. p. 106
(1854); Jerdon, B. of Ind. i. p. 177 (1862); Sclater, Ibis, 1865, p. 235; id. P. Z. S.
1865, p. 603; Blyth, Ibis, 1866, p. 339; Holdsworth, P. Z. S. 1872, p. 419; Hume,
Str. Feath. 1873, p. 166; Ball, Str. Feath. 1873, p. 370; Hume, Nests and Eggs, i.
p. 85 (1873); Dresser, B. of Eur. pt. 33 (1874); Aitken, Str. Feath. 1875, p. 214.

Cypselus nipalensis, Hodgs. J. A. S. B. v. p. 780 (1836).

C. galilejensis, Antinori, Cat. Collez. di Uccelli, p. 24 (1864).

C. galilæensis, Tristram, Ibis, 1865, p. 76.

C. abyssinicus, Streubel, Isis, 1848, p. 354.

The Allied Swift, Gray; *White-rumped Swift*, Jerdon.

Ababil or *Babila*, Hind.; *Huwa bil-bil*, Natives at Saharunpore (Jerdon).

Wæhælaniya, Læniya, Sinhalese.

Adult male and female. Length 5·1 to 5·5 inches; wing 5·1 to 5·3; tail 1·8; tarsus 0·4; mid toe and claw 0·5; bill
to gape 0·65 to 0·7.

In this species the tail is short, slightly forked, but the feathers not pointed.

Iris deep brown; bill black; feet vinous-brown, claws black.

Head, hind neck, wings, and tail blackish brown, with a slight greenish lustre, and the forehead paler than the crown;
back and scapulars glossy green-black, blending into the hue of the hind neck; primaries pale on the inner webs,
the tertials and the feathers along the metacarpal joint with fine light edges; rump and its sides, with the chin
and centre of the throat, white, some of the feathers of the former region generally with dark shafts; under surface
glossy black, paler on the under tail-coverts; under wing brownish black.

Young. Immature birds have the feathers of the under wing-coverts margined with whitish, and the rump more
lineated than in the adult; the breast and lower parts are likewise more or less finely edged with whitish.

Obs. This Swift varies considerably in size in different portions of its habitat. In India Mr. Hume has found it
varying in the wing from 4·8 to 5·5 inches; and Dr. Finsch gives the wing of specimens from the Blue Nile as
high as 5·6 inches; he likewise remarks that a more or less visible superciliary stripe is occasionally visible. I
have found the amount of white on the throat to be variable in some examples; it does not quite extend to the
chin; probably such are mature birds.

Distribution.—The common Indian Swift is not migratory to Ceylon, as was supposed by Layard, but is
merely a wanderer throughout the low country, its movements appearing to be regulated by the weather and
monsoon winds. In the south-west of the island I have noticed it at the seaside only during the first three
months of the year, although I have seen it in the hilly parts of the interior during the S.W. monsoon, at
which season Mr. Parker, of the Ceylon Public Works Department, has observed it at Puttalam. In the
north-east I have seen it at both seasons of the year, but am of opinion that it is no more than a straggler
over that flat region, traversing it in the course of a day's wandering from its head-quarters in the hills. In
the Kandyan Province it is a common bird and a permanent resident there. It appears to prefer the dry
climate of Uva to other parts, although I have noticed it in most of the coffee-districts. It is sometimes

met with about Nuwara Elliya and on the Horton Plains, but in all probability does not roost in such high regions.

It is a bird of very extensive range, for besides inhabiting the whole of India and Western Asia as far as Palestine, where it is the *C. galilaensis* of Antinori, it extends through Africa to the extreme south. Although found throughout India from the south to the Himalayas, Jerdon remarks that large tracts of country may be traversed at times without seeing a single individual, and Mr. Hume has likewise found it to be very local. In many parts of Sindh he met with it commonly, but throughout Upper Sindh to Schwan he did not see it. At Mount Aboo and the plains of the surrounding country it is common, breeding in the celebrated Dilwarra temples. It is rare in the Deccan; and Col. Sykes remarks that though found in all districts in India, it is often confined to a small tract in the neighbourhood of some fine large pagodas and other buildings. In the central regions of Nepal it is said by Hodgson to remain throughout the year. In Palestine Canon Tristram records that it is a permanent resident in the Jordan valley, while every other species of its genus is migratory there. In the portions of Africa which are inhabited by it it is likewise non-migratory. With regard to this peculiarity in its economy, it is singular that the same is true of its representative on the eastern side of the Bay of Bengal, the *Cypselus subfurcatus* of Blyth, which Mr. Swinhoe recorded as "resident on the Chinese coast" as far north as Amoy.

Habits.—In the mountains of Ceylon this stout little Swift is usually seen coursing over coffee-estates, steep patnas, or the so-called "Plains" in the upper ranges, while in the low country it affects every variety of open situation, particularly on sultry rainy evenings, when the damp tropical air is teeming with an abundance of insect-food. It congregates in large flocks, and hawks about with a rapid powerful flight, careering round and round at a great height, and then suddenly descending, will fly as low as the Common Swallow, picking up its evening meal right and left with no apparent exertion. In the hills it consorts with the Swiftlet, and may often be seen late in the evening flying with that species in some given direction on its way to a distant roosting-place, probably some inaccessible cliff where it has been bred. It is not usually a noisy bird, its note being a weak scream, resembling that of the European Swift, but not so soft in tone, and which Blyth styled a "shivering" cry. In the breeding-season, however, its cries are incessant; packing in small troops like the common Swift of Europe, it dashes round the spot where its nests are swarming with young, alighting for an instant to convey to the hungry mouths the food which it carries in its bill, and then sweeping off in a body, separates in search of a fresh store or continues its circular peregrinations. Jerdon, who remarks that its flight is fluttering and irregular in the morning and evening, writes that "small parties at these times may be seen flying close together, rather high up in the air and slowly, with much fluttering of the wings and a good deal of twittering talk; and after a short period of this intercourse all of a sudden they separate at once and take a rapid downward plunge, again to unite after a longer or shorter interval." They may occasionally be seen flying beneath culverts and road-bridges like a Swallow, evidently feeding on the insects which congregate about the water in such places. Mr. Blyth, it may be remarked, has stated that he has seen this Swift rise from off the ground.

Nidification.—This species breeds either in large colonies or in company with a few of its fellows, and rears its young at various periods between the months of March and July. It builds in the verandahs of outhouses, beneath bridges and culverts, under overhanging rocks, or in caves, in all of which situations I have known its nest to be found. Layard found them breeding at Dambulla in April about the rocks there, and at Tangalla beneath a bridge. I met with a large colony nesting in March in a salt-store at Kirinde, and another in May under the celebrated wooden bridge at Wellemade in Uva. In the month of April several pairs used to breed annually in a small seaside cave near Trincomalie. Mr. Holdsworth found it nesting "under the rocks overhanging the entrance to the famous temple at Dambulla." The nest is constructed of feathers, straw, grass, and at times pieces of rag, wool, twine, or any miscellaneous material which the bird can find and which will assort well with the rest of the structure. The whole mass is firmly cemented together with the saliva of the bird, and is shaped in accordance with the situation in which it is built, which likewise determines the position of the aperture. The interior is spacious, and sometimes several nests are fastened together. Nests which I have seen in caves or beneath bridges have had the entrance at the top, and others fixed under tiles have been very long structures with the opening at the end. My correspondent, Mr. Parker, writes me of a pair which took possession of a Red-bellied Swallow's nest under a road-bridge near Kurunegala. To get possession of the eggs a hole had to be made in the side of the nest, which the bird used afterwards as an outlet. On a second visit a piece of the side came out, which the bird clumsily repaired the third year with feathers and leaves, making up a piece of patchwork which reminded one of a "hole in a window-pane stuffed with a piece of cloth!" The number of eggs is generally three; they are long ovals in shape, smooth in texture, and pure white in colour; they vary from 0·8 to 1·0 inch in length by 0·55 to 0·65 inch in breadth.

From what has been written of its nidification in India, it appears that there its nest varies in character, as in Ceylon, according to its situation. Mr. Aitken, writing of its breeding at Berar, remarks that when the nest is attached to the roof of a building and not supported in any way, the straws of which it is composed are so firmly agglutinated that it tears like a piece of matting.

CYPSELUS BATASSIENSIS.

(THE PALM-SWIFT.)

Cypselus batassiensis, Gray, Griff. An. Kingd. ii. p. 60 (1829); Horsf. & Moore, Cat. B. Mus.
    E. I. Co. p. 128 (1854); Jerdon, B. of Ind. i. p. 180; Blyth, Ibis, 1866, p. 340; Sclater,
    P. Z. S. 1865, p. 602; Holdsworth, P. Z. S. 1872, p. 420.

Cypselus balasiensis, Blyth, Cat. B. Mus. A. S. B. p. 86 (1849); Layard, Ann. & Mag. Nat.
    Hist. 1853, xii. p. 167.

Cypselus balisiensis, Kelaart, Prodromus, Cat. p. 117.

Cypselus palmarum, Gray & Hardwicke, Ill. Ind. Zool. i. pl. 35 (1832); Hume, Nests and
    Eggs, i. p. 87; Ball, Str. Feath. 1874, p. 384.

Putta-deuli and *Tari ababil,* Hind.; *Tal-chatta,* Bengal, lit. "Palm-Swallow;" *Batassia,*
    Bengal (Jerdon); *Chamchiki,* Beng., a name also applied to Bats (Blyth).

Wœhœlaniya, Sinhalese.

Adult male and female. Length 5·1 to 5·3 inches; wing 4·3 to 4·7; tail 2·4 to 2·8, outer feather 1·0 longer than the
    middle; tarsus 0·4; middle toe and claw 0·32; bill to gape 0·5. The wings reach 0·5 beyond tail, which is deeply
    forked, with the feathers pointed at the tips.
Iris sepia-brown; bill black; legs and feet vinous-brown; claws blackish.
Above glossy ash-brown, darkest on the head and tail; the lower back and rump paler than the interscapular region
    and with dark shafts to the feathers; quills blackish brown, with the internal margins slightly paler than the rest.
Bases of the loral feathers white; beneath mouse-grey; the under tail-coverts with dark shafts, and the flanks darker
    than the breast.

Young. On leaving the nest the young bird is clothed like the adult, but the upper surface is not so glossy.

Distribution.—The little Palm-Swift is the most numerous of its genus in Ceylon, and is found throughout
the entire low country and sub-hill region. It is seen now and then in the Kandy district; but is not a
permanent resident there, and on the Uva side of the Central Province it ascends from the plains in fine
weather to a considerable altitude, Mr. Bligh informing me that he has seen it in Haputale as high as 4000 feet.
It is a common bird in the south and west of the island, and more numerous on the sea-board than in the
interior. In the palmyra-districts, on the northern coasts, it is very abundant, and is the only Swift, as far as
I can ascertain, which commonly affects the Jaffna peninsula and adjacent islands.

As regards the Palm-Swift's distribution in India, Jerdon informs us that it is abundant in all districts
where palmyra- and cocoanut-palms are found, and that it is common on the Malabar coast, the Carnatic, the
northern Circars, and Bengal, but rare in the central tableland and North-west Provinces. In Chota Nagpur
Mr. Ball says it is found in abundance where its favourite trees are common, and so local is it that he has
observed a small colony settled in a single tree, where, perhaps, for many miles around not another tree or
Swift could be found. It is said to extend into Assam and Burmah; but this can only be as a straggler, as it
is not recorded in 'Stray Feathers' from either Pegu or Tenasserim; it is replaced in these provinces by
Cypselus infumatus, the Sooty, or, as called by some, the "Palm Roof-Swift." It has not as yet been procured
in Sindh.

Habits.—The localities preferred by this Swift are fields and open lands in the vicinity of cocoanut- and
palmyra-groves. In the northern parts of the island it is seen much about the sea-shore, which is, in many
places, completely lined with the widely spread *Borassus* palm, its favourite tree all over India; indeed Jerdon
remarks that it is seldom found at any distance from where this palm grows. This, however, is not its
habit in Ceylon; for it abounds in many parts of the Western Province, where the tree is unknown, but
where its place is supplied by the cocoanut, and particularly the areca-palm, around which latter it careers

in little flocks with lively screams in just the same manner. These remain about the place of their birth throughout life, roosting in the trees which contain the nests in which they were reared, and to which they return early in the evening, flying up to the fronds and again darting off in search of their evening meal. It associates in parties of considerable numbers, and may often be seen, in company with the Swiftlet, hawking at evening time over the paddy-fields in the Western Province. Its flight is swift and regular at times and fluttering at others, particularly when hawking in a flock; it flies late at nights, and, as Dr. Jerdon remarks, it is not uncommon to see Bats and these Swifts hawking together at dusk, a circumstance which perhaps has given rise to the belief that it is nocturnal in its habits, and is also doubtless the origin of its Bengal appellation "*Chamchiki*." I have seen it flying rather leisurely about, taking winged termites, at sunset. Its note, which it constantly utters, is likened by Blyth to the sound *titteya*, which is a very correct rendering of it, although there is a pretty shrillness in the cry that cannot be well expressed in words. Dr. Hamilton considered this bird to be nocturnal in Bengal, appearing at sunset and going to rest at sunrise! It certainly hawks very late; but it is difficult to understand what became of those that were seen at sunrise, and whose disappearance must have given rise to this strange belief.

Nidification.—This species breeds from October until April, probably rearing two broods in the season, as I have found eggs and young of the same colony during both these months. Although it invariably nests in the palmyra-palm wherever these trees are to be found, I am of opinion that it takes to the areca in the south of the island, as I have seen them thronging around these trees at Galle during the breeding-season. It very often selects an isolated palmyra, and sometimes one situated in a most public spot, to breed in—to wit, the solitary tree which stands on the shore in front of Fort Frederick at Trincomalie, and in which there is always a little colony to be found. The nest is built on the under surface of the hanging fronds, which droop round the head of the trunk beneath the cluster of more vital and horizontal ones; it is attached principally to the ribs of the leaf, and situated high up where these lie at a convenient distance from one another. If, however, it is placed low down, near the tip, it is firmly fixed to the hollow portions as well as the ribs. In shape it resembles a little open pocket, with a shallow interior of about 1 inch in depth and 1¾ in width; the back part, adjoining the leaf, which is thin, is continued for some distance up, affording an additional support, and often a partial foundation, for another nest built immediately above it. The materials consist of "wild cotton," the down from the pod of the cotton-tree, mixed with feathers which are placed in regular layers round the front and firmly incorporated with the cotton, which is agglutinated with the saliva of the bird. Sundevall, remarks Jerdon, shot these birds with their mouths slimy and filled with the down of some syngenesious plant which they appeared to catch during their flight. Mr. Hume finds the nests in India to be constructed of the fine down of the *Argemone mexicana* and similar plants. The eggs are two or three in number, much elongated and smooth in texture, pure white, and the shell very thin; they measure from 0·65 to 0·7 inch in length, and from 0·43 to 0·46 inch in breadth. The young, when able to use their feet, cling to the leaf above the nest, supporting themselves in an upright position; the old birds, when feeding them or entering their nest, alight at the bottom of the palm-leaf and run nimbly up the ribs.

Genus COLLOCALIA.

Bill smaller and more hooked than in *Cypselus*. Wings with the 1st quill considerably shorter than the 2nd. Tail slightly forked, and the tips of the feathers rounded. Tarsi and feet very small and feeble; tarsus naked, the hind toe directed backward and only partially reversible.

COLLOCALIA FRANCICA.

(THE INDIAN SWIFTLET.)

Collocalia francica, Gm. Syst. Nat. i. p. 1017. no. 15 (1788); Walden, Ibis, 1874, p. 132.
Hirundo brevirostris, M'Clelland, P. Z. S. 1839, p. 155.
Hirundo unicolor, Jerdon, Madr. Journ. Sc. xi. p. 238 (1840).
Collocalia nidifica, G. R. Gray, Gen. Birds, i. p. 55. no. 1 (1844); Blyth, Cat. B. Mus.
    A. S. B. p. 86 (1849); Horsf. & Moore, Cat. B. Mus. E. I. Co. i. p. 98 (1854);
    Bernstein, J. f. O. 1859, p. 118; Jerd. B. of Ind. i. p. 182 (1862); Legge, Ibis,
    1874, p. 13.
Collocalia brevirostris, Kelaart, Prodromus, Cat. p. 118; Layard, Ann. & Mag. Nat. Hist. 1853,
    xii. p. 168.
Collocalia fuciphaga, Holdsworth, P. Z. S. 1872, p. 420.
Collocalia unicolor, Bourdillon and Hume, Str. Feath. 1876, pp. 374, 375.
Esculent Swallow of Latham and Stephens; *Indian Edible-nest Swiftlet*.
Wœhœlaniya, Sinhalese.

Adult male and female. Length 4·5 to 4·8 inches; wing 4·1 to 4·6, reaching 0·8 to 1·1 beyond the tail; tail 1·0 to 2·1; tarsus 0·4; middle toe and claw about 0·4; bill to gape 0·4.

Iris brown; bill black, vinous-brown at base; legs and foot dusky fleshy reddish, in some fleshy brown.

Above uniform dark smoke-brown, with a green lustre on the back, wings, and tail; primaries and tail deep glossy brown; the feathers of the rump albescent at the margins near the base, the light portions concealed beneath the overlying feathers; lores whitish at the base and tipped black; beneath glossy mouse-grey, palest on the neck and chest; the under tail-coverts with a slightly greenish gloss.

Young. The nestling is plumaged like the adult as soon as fledged; the tips of the quills finely margined with albescent.

The skin of the unfeathered chick is dark brown; and the head becomes quite feathered before the body commences, the scapulars following next.

Obs. No little confusion has existed in the synonymy of this and the Javan Swiftlet, *C. fuciphaga* of Thunberg; and ornithologists are therefore much indebted to Lord Tweeddale for his note on these species in the 'Ibis,' 1874, in which Indian, Ceylonese, and Andaman specimens of the species are shown to be identical with those from Mauritius and Seychelles. His lordship writes me that a specimen which I have lately forwarded him for examination is identical with birds from the Nilghiris, Darjiling, Andamans, and Malacca. The peculiarity of this species is that the tips of the concealed basal parts of the webs of the dorsal feathers are albescent, which increases in paleness towards the rump, showing in some specimens on the surface of the plumage and imparting a light appearance to that region. I regret that I did not collect more examples of this Swiftlet while in Ceylon: in the several that I have examined this latter degree of paleness has not been perceptible on the surface of the rump-plumage, although the basal portions of the feathers exhibit the above-mentioned character. Layard, in his correspondence with Blyth on the subject of this Swiftlet's nesting, writes of it as *C. nidifica*, but styled it *C. brevirostris* in his published notes, an older title bestowed by M'Clelland on a specimen from Assam, but which Mr. Hume is of opinion in reality applies to *Cypselus infumatus*, Sclater. Gmelin's title has not yet come into use in the pages of 'Stray Feathers,' as Mr. Hume still applies Jerdon's name of *C. unicolor* (bestowed in the Madras Journal on the first specimens he received from the Nilghiris) to examples from Southern and Northern India.

C. fuciphaga from the Andamans as well as Java is a much smaller bird than *C. francica*. Total length about 3·5 inches; wing 3·8, reaching 1·5 beyond tail; tail 1·5.

Above glossy black-green, with a very strong lustre on rump and upper tail-coverts and tail; throat and sides of fore neck dark brownish grey, chest-feathers edged with whitish; breast and abdomen white, the feathers with mesial brown lines; under tail-coverts concolorous with the back, the shorter feathers broadly margined with white.

Distribution.—The little Swiftlet of Ceylon is spread over the whole island, taking up its quarters in the low country near the many isolated rocky hills which abound therein, and wandering thence over the surrounding districts, while in the Kandyan Province, full of precipices and caves, it everywhere finds a home. There are consequently many parts of the low-lying forest districts where it may always be found, such as the rocky ranges in the Eastern Province, the hilly Pattus and Korales in the south-west, from which it strays to the neighbourhood of Galle, the vicinity of the curious rock-ridges stretching from Kurunegala to Dambulla and northwards to the isolated and singular mountain of Rittagalla, whence it overruns all the Vanni to the extreme north; in these localities, as also about sundry rocks on the north-east coast to the south of Tirei, the precipices of Yakhahatua near Avisawella, and other crags in the Raygam Korale, I have invariably noticed the Swiftlet. It is occasionally seen, on fine mornings, about the cinnamon-gardens of Colombo, but not so often as round the southern port. It is abundant in the higher parts of Uva, round Nuwara Elliya and Hakgala, and similar spots in the main range.

This species is found throughout the south of the Indian peninsula, and is said to be more abundant on the Travancore and Nilghiri hills than in the low country. In the north of India it is found in Sikhim and in the neighbourhood of Darjiling. Southward it extends into Malacca and to the Andamans, where a nearly allied species, *C. spodiopygia*, Peale, with a paler rump, is found. In the opposite direction it reappears in the Mauritius and Seychelle group of islands.

Habits.—This Swift generally affects the crags and rocky hills in which it has been bred, wandering great distances during the day over the surrounding country. At early morn, when sallying out from its roosting-places, the caves of its birth, it flies about the vicinity with a rather tardy, uncertain flight, and then starts off for distant questing-grounds, when numbers may be met with, all making for the same direction, whence they doubtless spread outwards in search of food. In the afternoon they return in great numbers and pack into a large flock, dashing about their native rocks in close company, uttering their low, hissing cries. They commonly associate with the Palm-Swift, and when questing with these on open ground, such as the "cinnamon," fly very low and may easily be shot. They can always be recognized from *C. batassiensis*, on the wing, by the short tail and the absence of the well-known note of this latter species. I have noticed them hawking about the bunds of large tanks, flying close to the water and keeping up their evening meal until quite dark. Jerdon mentions them returning to the caves in Pigeon Island, off Honore, as late as 9 P.M., and comments on the vast distance they must have flown to arrive at their roosting-place three hours after dark! Their powers of flight are certainly very great, their progress being much more rapid than that of the Palm-Swift. The food of this species consists of gnats, mosquitos, and other small flies. It appears, like other Swifts, to be constantly in the act of catching its food; even late at night, when sitting on a lofty cliff overlooking one of the magnificent prospects of the splendid province of Uva, I have watched them picking off insects in their rapid progress homeward.

Nidification.—The breeding-season of this little Swiftlet in Ceylon lasts from March until June. It nests in large colonies in various caves in the hills and mountains of the central and southern parts of the island. Many of these are known from seeing the birds haunt the vicinity of certain precipitous hills; but few have been visited and examined, on account of the general inaccessibility of these resorts. Among those which are known are:—two situated on the rocky hills of Diagallagoolawa, near Pittegalla, on the banks of the Bentota river, and which are referred to in the extract given below from Layard's notes; several occupied by large and small colonies on the Dambetenne and Piteratualic estates on the south face of the Haputale range; one on Pedrotallagalla, spoken of by Kelaart; and another which I was informed of in a hill called Maha-ellagala, near the "Haycock" mountain, as also another in the Nitre-cave district. Besides these there are, I believe, colonies in the "Friars-Hood" or some of the surrounding rock-hills and in Rittagalla, the above-mentioned mountain situated between the Central and Trincomalie roads. The celebrated cave in the Haputale range, and the only one which I have had the good fortune to visit, is situated in a bold peak standing out above and towering over the Dambetenne and adjoining estates, which form one of the finest sweeps of coffee-ground in Ceylon. On a sultry day in May 1876, my friend Mr. Bligh and myself set out from Catton bungalow to see the Swifts' cave. A long tramp round the adjacent spur brought us to the gorge in which lies the fine estate of Mousakelia, up which we toiled, gradually winding our way up the zigzag paths, and at last reached the inviting shade of the tall forest crowning the top of the ridge. Here our journey was enlivened by the notes of the usual denizens of these belts of fine jungle; and as we trudged along, listening to the clear, strong whistle of the Grey-headed Flycatcher, the churr of the handsome Trogon, and the twittering of the brilliant "Sultan-bird" (*Pericrocotus flammeus*), we congratulated ourselves that we had reached the highest point of our journey (6000 feet), and that we had but a short and immediate descent to our destination. Another half-mile and we had passed over the ridge and came into sudden view of the glorious prospect beneath, such a one as only can be witnessed in the higher ranges of the beautiful Central Province. Before us lay a magnificent amphitheatre, the top of it a dark sweep of forest, and the middle a splendid basin of coffee, consisting of the Dambetenne and Piteratmalie estates, in luxuriant growth, between which and ourselves a narrow ravine ran down from the range on our right and suddenly opened out into an abysmal gorge, the wooded slopes of which stretched up to the foot of the coffee. In these woods Mr. Bligh, some years previous, had discovered the handsome Whistling Thrush (*Arrenga blighi*). At a point where the great gorge suddenly commenced, by a sheer precipice dropping down about 1000 feet into the lower estate, stood the fine bungalow occupied by the gentleman, Mr. Imray, who was to be our kind host for the night; and at the back of this, at the top of a rich slope of coffee, towered up a rocky buttress, in which the Swiftlets of Haputale propagate their species. In this precipice a vast boulder, about 70 feet in height and 50 in breadth, has at some period slipped away from the face of the mountain, and leans against it at an angle of about 30°, forming a lofty narrow cavern. Here about 300 pairs of birds have their nests built against the inner side of the boulder, which is convex and corresponds with the concave face of the main mass. There are no nests on this latter, down which there is doubtless a considerable amount of drainage; and the instinct of the little birds is here wonderfully displayed in rejecting the wet side of the cavern, which would seriously impair the stability of their gelatinous nests. These are placed in tiers, one above the other, about 15 feet from the guano at the bottom of the cave; in places three or four were joined together, the back part of the under nest being prolonged up to the bottom of the one above it. The little structures were by no means edible, being constructed of moss and fine tendrils, arranged in layers and cemented with the inspissated saliva of the bird, the back part attaching the nest to the rock, as well as the interior of the cup, being, however, entirely of this material. I have seen one or two nests from Pittegalla almost wholly made of this substance; but even these were mixed, to a certain extent, with foreign or vegetable material. The interior of these Dambetenne nests was in most cases oval, the longest diameter, which varied from 2 to 2½ inches, being parallel to the rock. In depth the egg-cup was, on the average, about 1 inch. At the date of my visit, the 22nd May, nearly all the nests contained young, two being the average number. A series of eggs procured at another time, and which I have examined, were of various shapes, long ovals being the predominant; they are pure white, and varied from 0·81 to 0·83 inch in length by 0·51 to 0·54 in breadth. It is noteworthy that the partially-fledged young which were procured for me on this occasion, and which I kept for the night, scrambled out on to the exterior of the nests and slept in an upright position, with the bill pointed straight up. This is evidently the normal mode of roosting resorted to by the species.

The interior of this cave, with its numbers of active tenants, presented a singular appearance. The bottom was filled with a vast deposit of liquid guano, reaching, I was informed, to a depth of 30 feet, and composed of droppings, old nests, and dead young fallen from above, the whole mingled into a loathsome mass with the water lodged in the crevice, and causing an awful stench, which would have been intolerable for a moment even had not the hundreds of frightened little birds, as they screamed and whirred in and out of the gloomy cave with a hum like a storm in a ship's rigging, powerfully excited my interest and induced a prolonged examination of the colony. This guano-deposit is a source of considerable profit to the estate, the hospitable manager of which informed us that he had manured 100 acres of coffee with it during that season. Besides this colony, there are two other smaller offshoots on the adjoining estate, in one of which, Mr. Bligh tells me, the birds have to pass through a cloud of spray in order to gain access to their nests.

Concerning the large breeding-station on the Bentota river, Mr. Layard writes, in the 'Annals and Magazine of Natural History' for 1853, xii. p. 168, as follows :—" Having fully described my acquaintance with these birds in a letter to my friend Mr. Blyth, I cannot do better than copy what I then wrote :—'The cave is situated at a place called Havissay, about thirty-five miles from the sea and twenty from the river, and about 500 feet up a fine wood-clad hill called Diagallagoolawa or Hoonoomooloocota. Its dimensions are as follows—length between 50 and 60 feet, about 25 broad and 20 high. It is a mass of limestone rock, which has cracked off the hill-side and slipped down on to some boulders below its original position, forming a hollow triangle. There are three entrances to the cave, one at each end, and one very small in the centre. The floor consists of large boulders, covered, to the depth of 2 or 3 inches, with the droppings of the birds, old and young, and the bits of grass they bring in to fabricate their nests. The only light which penetrates the cavern from the entrances above mentioned is very dim; when my eyes, however, got accustomed to the light, I could see many hundreds of nests glued to the side of the fallen rock, but none to the other side, or hill itself. This I attribute to the fact of the face of the main rock being evidently subject to the influence of the weather, and perhaps even to the heavy dews off the trees; but for this the side in question would have been far more convenient for the birds to have built on, as it sloped gently outward, whereas the other was much overhung and caused the birds to build their nests of an awkward shape, besides taking up more substance. I was at the spot a few days before Christmas, and fancy that must be about the time to see the nests in perfection. This is corroborated by the fact of my finding young birds in all the nests taken by me, and by what the old Chinaman said, that the 'take' came on in October. I find that they have three different qualities of nests, and send two for your inspection; the best is very clean, white as snow and thin, and is also very expensive. The most inferior are composed of dry grasses, hair, &c.; but I could not detect any thing like the bloody secretion as described (though only under peculiar circumstances of exhaustion) by Mr. Barbe, even in a fresh nest. I was in the cave late (after 5 P.M.) in the evening of a day which threatened rain; but the old birds were still flying round the summit of the mountain at a vast altitude, occasionally dashing down into the cave with food for their nestlings.'"

Genus DENDROCHELIDON.

Bill much as in *Collocalia*, but smaller, if any thing, and much hooked at the tip. Wings with the 1st quill the longest. Tail very long and deeply forked, the lateral feathers much attenuated. Tarsus very short, much less than the middle toe; three anterior toes subequal; hallux long, directed backwards, and not reversible.

DENDROCHELIDON CORONATUS.

(THE INDIAN CRESTED SWIFT.)

Hirundo coronatus, Tickell, J. A. S. ii. p. 580, xv. p. 21 (1833).
Macropteryx coronatus, Blyth, Cat. B. Mus. A. S. B. p. 87 (1849); Kelaart, Prodromus,
    Cat. p. 117 (1852); Layard, Ann. & Mag. Nat. Hist. 1853, xii. p. 167.
Macropteryx longipennis (Swainson), Jerdon, Cat. B. S. India, Madr. Journ. 1840, xi. p. 236.
Dendrochelidon schisticolor, Bonap. Consp. Av. i. p. 66 (1850).
Dendrochelidon coronatus, Gould, B. of Asia, pt. xi. (1859); Jerdon, B. of Ind. i. p. 185
    (1862); Holdsworth, P. Z. S. 1872, p. 420; Hume, Nests and Eggs, i. p. 92 (1873);
    Legge, Ibis, 1874, p. 13; Ball, Str. Feath. 1874, p. 384; Oates, Str. Feath. 1875,
    p. 45; Fairbank, ibid. 1877, p. 393.

Adult male and female. Length 9·3 to 9·75 inches; wing 6·1 to 6·3; tail 5·3 to 5·7 (outer feathers 3·25 to 3·6
    longer than the middle); tarsus 0·35; middle toe and claw 0·65; bill to gape 0·75. Female slightly the larger.
Iris deep brown; bill black, inside of mouth slate; legs and feet vinous-brown, claws blackish.
Eye very large for the size of the bird. Coronal feathers elongated and capable of being erected.

Adult male. Head, back and sides of neck, back, scapulars, and rump bluish ashy, palest on the back, and with a
    greenish gloss on the head and upper tail-coverts, continued in some birds to the back; wing-coverts deep though
    obscure lustrous green; quills and tail obscure metallic green, with a steel-bluish lustre about the tips of the
    shorter primaries; terminal portions of the tertials greyish.
Lores black, surmounted by a thin white supercilium; a blackish orbital circle; chin, cheeks, and ear-coverts glossy
    chestnut, palest on the chin; throat and chest ashy, blending into the deeper hue of the sides of the neck and
    passing down the flanks, the lower parts of which are concolorous with the rump: lower breast, abdomen, and
    under tail-coverts white, blending into the surrounding grey hue; under wing-coverts dusky bluish ashy.

Female. Wants the rufous chin and cheeks of the male, the face and behind the eye being black, as are the lores;
    beneath the cheeks a whitish line: chin concolorous with the throat; white of the lower parts less pure than in
    the male; under tail-coverts with dark shafts.

Young. On leaving the nest the nestling has the head, back, and rump fulvescent greyish, the feathers rounded at
    the tips and with silky white edges and the basal portions metallic green; the lower scapulars and rump paler
    than the back; wing-coverts, quills, and tail deep metallic green, the terminal portions of the coverts of the same
    hue as the back; under surface delicate ash-grey, with finer white edges than on the back, which are separated
    from the grey by a fine dark border.

Bird of the year. Upper surface less glossy than in the adult, the feathers of the hind neck, rump, and tail-coverts
    terminally margined with white; tertials deeply tipped with the same and brownish on their terminal portions;
    shorter primaries tipped with white; beneath pale bluish grey, paling to albescent on the centre of the breast
    and under tail-coverts, which parts have the feathers tipped with brown, most extensively on the latter.

Obs. Indian examples of this Swift correspond in size to Ceylonese. Mr. Ball gives the wing of two males as 6·05
    and 6·1 inches, and that of two females as 6·15 and 6·35; a female from Pegu measured 6·3.

Distribution.—The Crested Swift is diffused throughout the whole island of Ceylon, extending into all
parts of the Kandyan Province and the mountain-ranges of the south. In the low country it is more common
as a resident in some districts than in others; but wandering about in its powerful flight as all Swifts must do,
it is liable to be met with anywhere as a straggler. I have never found it so numerous on the sea-board of
the Western Province during the S.W. monsoon as at the opposite time of the year, although I have occasionally

met with it about the Slave-Island lake in the height of the boisterous weather. In the Galle district it is a very common bird, and follows the coast round to the Hambantota district in fair numbers, and thence spreads throughout the flat, jungle-clad country to the Hapatale hills. I have noticed it in all parts of the Eastern Province that I have visited, and in the jungles of the northern half of the island have found it chiefly confining itself to the vicinity of the grand old tanks, such as Minery, Topare, Kanthelai, &c., and likewise affecting any large clearings which may exist in the forests of the Vanni. In the Kandyan Province it is common enough in the coffee-districts, and in fine weather may be seen about the Elephant, Kandapolla, and Horton Plains. Mr. Holdsworth does not record it from Nuwara Elliya ; but I have seen it a few miles from that place, about which it no doubt flies in the course of its day's wanderings.

On the mainland this fine Swift is found, according to Jerdon, throughout Southern and Central India, but "most abundant on the Malabar coast and the Wynaad, extending up the slopes of the Nilghiris to 4000 feet or thereabouts." Mr. Fairbank only observed one example in the Palani hills. It is recorded by other observers to inhabit the sub-Himalayan districts ; and Mr. Hume says, in 'Nests and Eggs,' that it breeds " below Kumaon and Gurwhal." Mr. Ball says it is found in most parts of Chota Nagpur, but nowhere abundant ; he also obtained it in the Satpura range and Rajmehal hills. It extends into Burmah. Mr. Oates found it common throughout the year in Upper Pegu ; Mr. Davison procured it in the pine forests north of Kollidoo in Northern Tenasserim ; but in the south he did not meet with it, as it appears to be replaced there by *D. comatus* and *D. klecho,* which two species, in common with other Malayan forms, do not seem to extend much to the north of Mergui.

Habits.—This species is strikingly arboreal in its habits, haunting open hill-sides or clearings in the jungle studded with dead trees, on which it perches almost as freely as a Passerine bird. In such localities little colonies may often be met with, the majority of the birds in which will be seen dashing about with great velocity in quest of insects, while half a dozen or less are perched on the topmost branches of some tall dead tree standing among a group of rocks, where it has escaped the woodman's axe, but has been charred and killed by the fire which has swept his clearing. Here it sits elevating and depressing its crest and constantly uttering its loud call, until it dashes forth and commences to hawk round the adjacent tract with its companions, who, in their turn, settle for a while and join in the noisy cries. When thus perched the Crested Swift presents a singular appearance, its long wings, crossed widely over its attenuated tail, forming a broad arrow, the striking aspect of which is increased by the long body in continuation of it, and the crest erected as fiercely as that of a Cockatoo. When wandering about from place to place, it has a very swift flight, performed with quick and powerful strokes of the wing, varied with wide sweeps and downward plunges, from which it gracefully rises on its rapid course. At times it flies high in the air, but, as a rule, keeps a short distance above the trees of the forest or the wooded tank over which it is hawking. Jerdon remarks that " should there be a tank or pool of water or river near, it is fond of descending suddenly, just touching the water, and then rising again with unrivalled grace and speed." It utters its loud cry when flying, as do other members of the family, but not so repeatedly as when perched, at which time it appears to call to its companions on the wing, and is then very tame, allowing a near approach without taking flight. Its food consists of small flies, of which it consumes quantities, its stomach being very capacious for the size of its body.

Some Indian writers speak of the great velocity with which this Swift flies. This has never struck me as any thing very extraordinary if only compared with that of the lightning-like speed of the Spine-tail. Mr. Oates speaks of it flying over a certain bungalow in Pegu, and "dipping with incredible velocity to the surface of the Irrawaddy." It certainly has, as I have remarked above, a great speed when thus launching itself downward from its course.

Nidification.—Nothing authentic has ever been discovered of the breeding of this bird in Ceylon. The natives assured Layard that it built in old *Euphorbia*-trees ; possibly it may ; but this tree is not well suited to its habits, and I have never myself seen the bird about it. The inhabitants of the Malabar coast informed Jerdon that it bred in holes in trees ; this certainly is erroneous, for, as a matter of fact, its curious nest, which has several times been found in India, is, according to Mr. Hume, "a little, shallow, saucer-shaped structure, composed of thin flakes of bark, gummed, probably by the bird's own saliva, against the side of a

2 U

tiny horizontal branch. The nest is nowhere more than ¼ inch in thickness, is at most ½ inch deep in the deepest part, and can be exactly covered by half-a-crown." Mr. Thompson writes "that it is entirely filled by the solitary, rather largish, white, oval egg. The bird looks for all the world as if she were sitting on the branch, and no amount of looking from underneath would show you that there was a nest under her." The flakes of bark of which the nest is composed are sometimes mixed with a few feathers, which, cemented with the inspissated saliva of the bird, serve to bind the whole together. Mr. Hume gives the measurements of an egg in his possession as 0·85 by 0·55 inch.

PICARIÆ.

Fam. CAPRIMULGIDÆ.

Bill with the culmen short and curved and the gape very wide, receding below the eyes, and furnished, in some, with stout bristles. Wings moderate, or long and pointed. Tail of ten feathers. Legs and feet very small.

Sternum short, deeply keeled, the posterior edge emarginated.

Plumage soft and mottled. Eyes very large. Of nocturnal habit.

Subfam. STEATORNINÆ.

Bill large, inflated, the margin curved and receding beyond the posterior corner of the eye: gape enormously wide; base of upper mandible clothed with bristly feathers. Wings, when closed, scarcely reaching beyond the middle of the tail. Feet very small, the middle claw not pectinated.

Genus BATRACHOSTOMUS.

Of small size.

Bill short and enormously wide, both mandibles inflated at the sides and suddenly compressed at the tips; culmen much curved and the tip of the upper mandible hooked. Nostrils horizontal, linear, placed in a membrane, which is completely covered by the frontal plumes; gape smooth; a series of erect branching plumes in front of the eyes. Wings short, rounded; the 1st quill about two thirds of the length of the 4th and 5th, which are subequal and longest. Tail long, even at the tip much graduated, the lateral feathers very short. Legs and feet small; the tarsus shorter than the middle toe, feathered more or less in front, the bare portion scutellate. Middle toe considerably longer than the lateral toes, both of which are joined to it at the base by a membrane; hind toe short.

Sternum small, with a shallow keel, with two deep emarginations in each half of the posterior edge.

A tuft of long hair-tipped feathers springing from above the ears.

BATRACHOSTOMUS MONILIGER.

(THE CEYLONESE FROG-MOUTH.)

Batrachostomus moniliger (Layard), Blyth, J. A. S. B. 1849, xviii. p. 806, ♀; Kelaart, Pro-
dromus, Cat. p. 117 (1852); Layard, Ann. & Mag. Nat. Hist. 1853, xii. p. 165; Jerdon,
B. of Ind. i. p. 189 (1862); Nevill, J. A. S. B. (Ceylon B.) 1870–71, p. 33; Holdsworth,
P. Z. S. 1872, p. 420; Legge, Ibis, 1874, p. 12; id. Str. Feath. 1875, p. 198;
Walden, J. A. S. B. 1875, pt. ii. Extr. No. p. 84; Hume, Str. Feath. 1876, p. 376;
Tweeddale, P. Z. S. 1877, p. 439, pls. 48 & 49, et Ibis, 1877, p. 391; Hume, Ibis, 1878,
p. 122.

? *Podargus javanensis* (Horsf.), Jerdon, Madr. J. L. Sc. (*nec* Horsf.).

Batrachostomus punctatus, Hume, Str. Feath. 1874, p. 354; Blanford, Ibis, 1877, p. 251;
Tweeddale, Ibis, 1877, p. 391; Hume, Ibis, 1878, p. 122.

The Ceylon Oil-bird, Kelaart; *The Wynaad Frog-mouth* of Jerdon.

CEYLON.

Collection.	Length.	Wing.	Tail.	Tarsus.	Middle toe.	Claw (straight).	Width of bill at gape.	Bill to gape.
	in.	in.	in.	in.	in.	in.	in.	in.
(1) ♀. Brit. Mus. (Whyte)	9·4 (from skin)	5·0	4·7	0·5	0·6	0·25	1·3	1·35
(2) ♂. „ „ („)	8·0 („ „)	4·6	4·5	0·6	0·6	0·25	1·3	1·3
(3) ♂. Legge	9·1	4·6	4·2	0·5	0·6	0·2	1·4	1·4
(4) ♀. „	9·0 (from skin)	mutilated	4·35	0·50	0·6	0·21	1·4	1·45
(5) Juv. ♂?. Legge	..	4·8	mutilated	0·55	0·6	0·2	1·3	1·4
(6) ♀. Lord Tweeddale	(not recorded)	4·6	4·6	0·5	0·63 (with claw)	..	1·25	..
(7) ♀?. „ „	(„ „)	4·6	4·4	0·6	0·65 („ „)	..	1·25	..
(8) ♂. „ „ .. (Nevill) .	(„ „)	4·68	4·5	0·5	0·75 („ „) ..		1·2	..
(9) ♂. „ „	(„ „)	4·75	4·75	0·5	0·7 („ „) ..		1·2	..
(10) ♂. Hume (type of *B. punctatus*).	7·75 (from skin)	4·3	4·0	0·5	0·75 („ „) ..		1·2	1·3
(11) ♀?. Layard (Poole Collection) ..	..	4·65	4·5	0·6	..	..	1·2	1·4

TRAVANCORE.

(12) ♀. Hume (Bourdillon)	9·0	4·75	4·0	0·6			1·37	1·35
(13) ♂. „ („)	9·0	4·75	4·5	0·50			1·4	1·4

Iris yellow; bill olive-brown or greenish brown, the under mandible paler than the upper; lower part of tarsus and
feet fleshy grey, darkest on the toes; claws dark brown, inside of mouth dull greenish.

Male, nearly adult (Galle). General aspect above brown, mingled with rufous and grey, the head and upper part of
hind neck being the darkest and the wing-coverts the most rufous portions. Most of the feathers of the head, hind
neck, and back with a terminal spot of black; the extreme tips of the head and neck-feathers rufous, and the spots
bordered above by the same, while the rest of the web is mottled with fulvous and whitish; superciliary feathers
creamy white, tipped with black; feathers across the lower part of the hind neck much mottled with pale fulvous,
imparting the appearance of a lightish band; on the back and rump the ground-colour is chiefly rufous and the
mottlings black; the upper tail-coverts with an angular white terminal spot; outer webs of the longer lateral
scapulars and the entire portion of the external shorter feathers dull white, mottled with blackish; wing-coverts
boldly marked with black, and the innermost secondary wing-coverts with terminal white feathers on the outer
webs; primary wing-coverts blackish brown, mottled with rufous; primaries and secondaries deep neutral brown,

2 U 2

more or less mottled with fulvous-grey at the tips; the outer webs of the primaries with deep rufous and buff indentations, and those of the secondaries with mottled bar-like markings of the same; tertials silvery grey, mottled chiefly with blackish and slightly with rufous, the feathers with terminal black spots; tail crossed with five mottled and vermiculated grey bands, each with an anterior irregular black border, increasing in width towards the tip, the extreme tip whitish, preceded by a black spot, outer web of lateral tail-feather chiefly buff-white with black bars.

Loral and frontal plumes tipped and barred with black; cheeks and throat pale rusty fulvous, with numerous black cross pencillings, the feathers just beneath the gape darker than the rest; feathers across the lower part of the fore neck with large white terminal spots, in some of which the portions of the extreme tips and an anterior border are black; beneath from the chest pale fulvous, mottled with blackish, many of the feathers at the sides of the breast with the terminal portions pure white, conspicuously marked with black cross pencillings; abdomen, thigh-plumes, and under tail-coverts more fulvous than the breast and less mottled, and crossed with regular black lines; longer under tail-coverts with terminal white spots; under wing-coverts pallid rust-colour with blackish markings; tarsal plumes pale fulvous, marked with blackish brown.

Male (British Museum, *ex* Whyte). Slightly less rufous above than the foregoing, with the black terminal markings deeper, and the whitish stipplings across the hind neck clearly defined; the scapulars with more white, and the terminal spots larger, which is likewise the case with the tertials and outer and median wing-coverts, the spots on the latter preceded by bold black markings: the light indentations on the outer webs of the primaries larger and more albescent; tail paler in the ground-colour, the bands the same; throat and chest less rufous, the white breast-spots larger and the black anterior edges bolder.

This specimen, I conclude, is somewhat older than the foregoing. An adult (?) male, described by Lord Tweeddale in his excellent monographic notice (P. Z. S. 1877, p. 442), is similar to the above examples, with the exception of a white collar across the hind neck, thus portrayed:—" Nuchal plumes with a subterminal white band confined between an upper and a terminal dark brown transverse line: a well-defined nuchal collar is thus formed." The outer rectrices have pure white marginal spots; the margins of the white throat- and breast-spots are described as dark brown instead of black; the wing-coverts are terminated with very bold white spots; the breast and flanks appear to be more rufous than in my specimens; but the greater development of the white markings pronounce it, I think, to be one of the oldest birds yet procured.

Young male? It is probable that the young of this sex have a grey or cinereous character from the nest, though the plumage may be much mixed with rufous. I have an example (No. 5 in the table of measurements) of a dark brown general aspect, with the wing-coverts rufous and with most of the upper plumage mottled with fulvous, which I take to be a male, on account of its very pale scapulars, the black tippings of the hind neck and inter-scapular region, and the spotted appearance of the breast above the lower pectoral region where the white black-bordered spots exist. In the females I have examined the breast is of a uniform hue from the white necklace down to the ventral spottings; there is a pale supercilium, the feathers being tipped with black; the narial plumes have the terminal portions black, as are also the bases of the feathers round the gape; the tips of the auricular plumes are blackish brown; the lateral scapulars are fulvous-white, with black terminal spots, and the greater wing-coverts have large terminal white spots; some of the long nuchal feathers are barred with black, with a light edge, but there is no further indication of any collar; the outer webs of the primaries are rusty fulvous; the under surface is rufescent fulvous, the breast mottled with black and tipped with the same; the white throat-spots have an anterior border of black, the white extending to the tip, and the same is true of the ventral and lower flank-feathers. I regret to say that the rectrices are wanting in this interesting specimen (No. 5), it having knocked itself about while in confinement in Kandy and lost the entire tail. Both in this and the first example described in this article there is a very remarkable tuft of downy undeveloped feathers, similar to that well known in the Herons, springing from the side just above the femur; it lies so close to the skin that it is with difficulty detected.

A presumed immature male described by Lord Tweeddale (*loc. cit.*) has a greyish-brown general aspect; the super-ciliary plumes are rusty fulvous on the outer webs and brown on the inner; a few feathers on the nape slightly tipped with white, some with fulvous, forming "a rudimentary uncompleted nuchal collar;" rest of the upper plumage somewhat similar to the first specimen above described, but more rufous, and the margins of the white spots brown; the chin and throat are rusty, and the white necklace well developed; upper pectoral plumes rusty, the lower fulvous-grey, broadly tipped with white. The wing in this specimen (which is No. 9 in the above Table) is 4·75, a very large dimension indeed for an immature male; and I have no doubt that some of these rufous males are older than has been supposed, and that adults will be found, when a large series has been got together, to vary somewhat in the character of their plumage.

Adult female (British Museum, *ex* Whyte; No. 1 in table of measurements). General hue of head, hind neck, back, rump, and wing-coverts dull rufous, darkest on the back and lesser coverts, and very pale on the tertials and outer scapulars, the whole very closely and finely stippled with dark cinereous brown, deepening into a blackish hue on the wing-coverts; the brown hue of these mottlings overcomes the rufous on the head and hind neck, which parts have an ashen tint; feathers of the forehead with a small black subterminal spot succeeded by a fine white tip; across the lower hind neck the feathers are terminated with broad white bars, bounded above and beneath with a black border; the median row of wing-coverts with the same, without the terminal black border; greater secondary coverts, an indistinct pale terminal spot, secondaries, and the primaries with their coverts dark ashen brown, mottled at the tips with fulvescent rufous; the outer webs of the primaries with a deep wavy rufous edge; tail rufous, crossed with six indistinctly defined mottled bands; the outer feathers indented outwardly with white, and crossed with black markings on the outer webs; the extreme tips of the feathers fulvous, preceded by a black edge.

Loral plumes blackish, rufous at the base; over the eye a pale undefined streak; under surface very similar to the upper in hue, as far as the middle of the breast, when the ground-colour becomes fulvous and extends thus to the abdomen and under tail-coverts; across the throat a band of large terminal white spots bordered by a black edging; terminal portions of the lower-breast, flank, and abdominal feathers whitish, with transverse black verniculations and a border of the same at the tips; under tail-coverts with a black-bordered terminal white spot; under wing-coverts rufous, faintly mottled with blackish brown.

Another example in my collection has the upper surface almost uniform rufous; there are indistinct mottlings on the scapulars and upper tail-coverts; the latter have the tips slightly paler than the rest of the feathers; the crown is rufous-brown, with the long auricular feathers rufous. The nuchal feathers have the white black-bordered bars forming the collar somewhat distant from the tips of the feathers; the scapulars have a terminal black white-tipped spot, the wing-coverts a very large white anteriorly bordered spot on the tips of the outer webs; the tail is crossed with mottlings of blackish brown gathered into the form of bars, and the penultimate or second lateral feather has the outer web indented with fulvous-white and crossed with black; superciliun pale fulvous; the chin is rufescent fulvous; the chest bright uniform rufous, with the normal white necklace; on the lower breast the feathers change abruptly on their terminal portions into white, pencilled with black; the under tail-coverts have a white black-bordered tip; the abdominal feathers have the middle portions pale fulvous.

Another in the Poole collection (the specimen mentioned by Layard in his Notes) is very similar to this; the nuchal collar and necklace are the same, but there are a few black terminal spots on the frontal feathers just behind the long plumes; the tertials and scapulars are both marked with white terminal spots surrounded by a deep black border: the chest and breast are very rufous as in the above. I am of opinion that these last two specimens are fully aged females; they agree in having the upper surfaces almost uniform in their hue, the mottlings being almost obsolete.

A similar bird (No. 6 in the table of measurements) is described by Lord Tweeddale, *loc. cit.* pl. 49. Judging from the minute description of this bird given by his lordship, it corresponds almost feather for feather with the rufous example treated of above; the tail is perhaps slightly less uniform than in mine, but the upper surface and chest have the same unmarked rufous ground-colour, and the wing-covert spots, nuchal collar, and necklace are likewise to all intents similar.

In a second female example noticed by his lordship the rufous is still deeper in tone than in the latter; the distribution of the white markings the same, but the white bars and terminal rufous-brown fringes of the nuchal-collar plumes are more pronounced.

Young. The nestling procured by Mr. Bourdillon and noticed below was probably a female, and is described by Mr. Hume as "a curious little rufous-brown ball with the characteristic bill of the species, and with distinct traces of black terminal bars to the feathers of the upper back and scapular region."

Obs. At the risk of being wearisome to the most scientific of my readers I have given as complete a series of observations on the plumage of this most remarkable of Ceylonese birds as it was in my power to do. This was, I think, necessary, as there has been so much controversy on the subject of Mr. Hume's presumed new species from Ceylon, *B. punctatus,* which I have thought expedient to unite with the present. This bird, as will be seen at a glance at the above table of measurements, is most variable in size, irrespective of sex. Some males exceed the average of females; but the latter will, I think, be found to contain larger birds in their ranks than the former. Blyth's type specimen, sent to him by Layard, was evidently a large female; while Mr. Hume's (No. 10 in the above list), sent to him, I believe, by Mr. Nevill from the south of Ceylon (where I have obtained a similar example), was a very small male. It would take up too much room in my pages to recapitulate Mr. Hume's description of his specimen, and it will suffice to say that it is a grey bird, corresponding exactly in plumage with those above described. Mr. Hume remarks, at page 122 of 'The Ibis' for 1878, "In no adult *B. moniliger* does the wing fall short of 4·7 inches ; in *B. punctatus*, on the other hand, of which several specimens have now, Mr. Whyte informs me, been obtained, the wing appears to be always under 4·5 (in the type it is only 4·3)"*. With regard to the size of *B. moniliger*, I refer my readers to the above table : in respect to that of Mr. Hume's *B. punctatus*, I have only to remark that the second male specimen described in this article (a strictly *punctatus* type of bird) has lately been sent home by Messrs. Whyte and Co. with "*p. inceps.*" written on the label (!)—proof evident that these naturalists do not know which phase of plumage represents *B. punctatus*. The female of Mr. Bourdillon's pair, sent from Travancore to Mr. Hume, is similar to the example from Ceylon described by Blyth, and the male, as described by Mr. Hume, corresponds exactly with my Southern Ceylon one. If, however, these Travancore birds prove different from Ceylonese, a new title should be bestowed upon them : and then *B. moniliger* will stand as one of the peculiar Ceylonese forms. That there should be two species of this rare and remarkable genus in Ceylon is most unlikely.

Distribution.—The Frog-mouth is widely spread throughout Ceylon, but is very seldom procured, as it is strictly nocturnal and an inhabitant of the inmost recesses of the jungle. Two examples were brought to Layard from the Western-Province jungles round Avisawella and Ratnapura, and a pair were met with near the latter place by Mr. Mitford. One was obtained by Mr. Nevill near Amblangoda, in the south of Ceylon, and another by myself at Wackwella near Galle. A third was shot in the Chilaw district in 1868 by the taxidermist of the Colombo Museum. Two more were captured on their flying into houses in Kandy at the latter end of 1875 and the beginning of 1876. In March of the former year Major Sandford, of the Royal Engineers, came upon one seated asleep on a branch in low jungle near the Peria-Kulam tank, Trincomalie, and described to me its toad-like and inanimate appearance as it sat with its bill pointed upwards. In February 1875 Mr. Edwin Watson met with another, under similar circumstances, in jungle above Ragalla Estate, at an elevation of 5600 feet. Besides the above-mentioned examples, there are the male and female sent home to the British Museum by Messrs. Whyte and Co., both of which were procured near Kandy, the latter at the end of last year, and the former on the 30th of January of the present year. The south-eastern, eastern, and north-western portions of the island are therefore the districts in which the Frog-mouth has not yet been procured or observed ; and of all parts in which it has been found the neighbourhood of Kandy is that in which it has proved most numerous. It is, however, in my private opinion, much more common than is supposed ; for throughout much of the northern forest-tract, as well as in many of the bamboo-districts between Colombo and Ratnapura, I have heard a singular note which I firmly believe is that of this bird. It is uttered just about sunset, and from that until about 10 o'clock, and is renewed again at daybreak on the following morning. It always proceeds from dense jungle, and all my efforts, which were many, proved fruitless in getting a sight of the mysterious bird.

The Frog-mouth of South India, which is presumably the same as the Ceylonese, has been found in the Wynaad and in the Travancore hills up to an elevation of 2100 feet ; north of these districts it has not as yet been traced.

Habits.—This singular nocturnal bird frequents thick bamboo-jungle, dense thickets, low umbrageous jungle, and such-like localities. It does not appear, as a rule, to sally forth before dark into the open, as I know of the testimony of but one person who has ever seen them out of jungle. Mr. Mitford, as quoted by Layard in his Notes, says that he observed a pair "frequenting a tree in full flower and capturing the beetles which flew about it." Its position of rest during the day is seated across a branch with the bill pointed straight upwards and its eyes of course fast closed. The bird which I was fortunate enough to meet with was perched on a low bamboo in a dense thicket through which I was creeping. I was close to it when I first saw it, but it was not awake ; and struck with its extreme likeness in general aspect to *Podargus cuvieri*, with which I was acquainted in a state of nature, I at once identified it as the much sought after Frog-mouth, and crept away to a convenient distance for shooting it. While moving off a slight crackling of the sticks beneath my feet awoke it, and it slowly turned its head round in my direction, but I do not think it saw me. Its stomach contained beetles, which form, to a considerable extent, the food of the *Podargi* in Australia. In the stomach of one of these latter I once found a green stone of considerable size, from which I infer that

---

* 5·5 and 5·3 in 'The Ibis'; presumably a printer's error (*vide* Str. Feath. 1874, p. 354).

perhaps these birds pick their food from the trunks of trees as well as capture it on the wing; in so doing the *Podargus* most likely took in the stone, which easily descended its capacious throat. Layard writes of one of the two examples he met with, that "it lived three days with me, but refused all food; during the day it slept, squatting on the ground with its head sunk between its shoulders: on being alarmed it sprang upwards with a sudden jerk, and after executing a rapid summersault in its confined cage, it would again alight and settle down like the *Caprimulgi*." Mr. Whyte, of Kandy, kept one, which was taken in a room into which it had flown, for some few days, during which I saw it; it perched on the bottom of the cage with its head up, and when approached and awakened would open its eyes and mouth wide, glare at me and then commence slowly closing its mandibles, which finally came together with a sudden jerk. I made it repeat this gesture several times, and the mouth always closed in the same curious manner.

Mr. Bourdillon speaks of the peculiar note which I have alluded to above, and likens it to "a loud chuckling cry, with something of the tone of a Goatsucker and not unlike the laugh of a Kingfisher." It is, as he remarks, a difficult call to describe; but his representation of it is, I think, the best that could be given. I will leave it to my numerous ornithological friends and acquaintances in Ceylon, who will, I hope, be interested in this bird, to prove whether or not this is its note. On one or two occasions I heard it in an isolated bamboo-thicket in the Ratnapura district; but I was too hurried to halt for the night and search the copse in the morning, which would probably have resulted in my finding the Frog-mouth had it been there.

Nidification.—The members of this remarkable subfamily of the Caprimulgidæ differ from their allies the true Goatsuckers as much in their nesting-habits as in their anatomy. The *Podargi* of Australia construct a nest which they fix on the limb of a tree, and the smaller *Batrachostomi* of Asia nest in a similar manner. Nothing has been discovered concerning the Ceylonese Frog-mouth's nesting; but I will subjoin the following interesting account of the nidification of the Travancore bird, inasmuch as I think it applies to the Ceylonese one as well. Mr. Bourdillon is the only person who has been fortunate enough to discover any particulars concerning the nidification of any of the *Batrachostomi*. The nest to which he refers in the following notes, which Mr. Hume kindly sends me, was found on the 24th of February 1876, in rather open jungle at an elevation of 2100 feet on the Travancore hills.

Mr. Hume says, *in epist.*, Bourdillon's account of the nest of *B. moniliger* is as follows :—

"The nest was brought to me one evening by a coolie who had been working in the jungle.

"It was composed of vegetable down neatly and compactly interwoven with pieces of dead leaves, fragments of bark and dry wood, and one or two pieces of lichen. In shape it is a sort of disk about 2¼ inches broad and 1¼ inch deep, the upper surface being slightly hollowed out. The young one, partially fledged, was unmistakably a Frog-mouth, from the colour of his plumage, bill, and huge gape. On receiving the nest I at once went with the man, and restoring it to its original position, sat down to watch.

"The chick (I quote from my notes) was much pleased at finding himself in his old quarters, and repeatedly shook himself, as if he could not at first settle down into a comfortable position, this shaking being attended with some danger, as once or twice the bird seemed within an ace of rolling out of the nest. At intervals of about ten minutes it uttered a feeble chirruping call, not unlike an "Ice"-bird at a distance. As darkness increased its cry was more frequent and became a single chirp. I watched till night closed in and it became pitch dark without seeing any thing of the old bird, though once something which might have been either bird or bat flitted past.

"Next morning I returned some time before sunrise, and in the moonlight had a good view of one of the old birds seated on the nest.

"It was in a very peculiar position, more lying down than sitting, with its head well up in the air. The nest was not 15 feet from the ground, in a fork of a sapling, apparently without any attempt at conceal-ment, so that I was able to approach very close to the bird, which, without moving, merely opened its large eyes to stare at me. Now comes the worst part of the story. I was so anxious to secure the specimen that I determined to shoot it on the nest; accordingly I retired as far as possible and fired, the result, owing to intervening bushes, being that, to my great disappointment, the bird went off into the jungle hard hit and was lost.

"Thinking at first the bird could not possibly have escaped I searched about for it, and at the foot of the small tree where the nest was I found the remains of an egg. These I have kept and will send with the nest, as I at least have no doubt that they originally enclosed the young Frog-mouth. You will see, from these fragments, that the egg of the bird is probably pure white, almost round, of thin texture, and with a smooth, glossless surface."

PICARIÆ.

CAPRIMULGIDÆ.

Subfam. CAPRIMULGINÆ.

Bill short, very weak and flexible, the tip hooked slightly. Nostrils tubular; gape furnished with stout bristles. Wings and tail long. Middle toe with the claw pectinated.

Genus CAPRIMULGUS.

Bill short, very wide at base, suddenly compressed towards the tip, which is gently curved and grooved parallel to the culmen. Gape enormous, and protected by long stout bristles. Wings long and pointed; the 3rd quill the longest, and the 1st shorter than the 4th. Tail moderately long and expanding slightly towards the tip, which is even. Tarsus short, more or less feathered, the bare portion in front covered with transverse scales; lateral toes short and united to the middle at the base by a membrane; claws straight, the middle one with the inner edge strongly pectinated.

CAPRIMULGUS KELAARTI.

(KELAART'S NIGHTJAR.)

Caprimulgus indicus, Blyth, J. A. S. B. 1845, xiv. p. 208.
Caprimulgus kelaarti, Blyth, J. A. S. B. 1851, xx. p. 175; Kelaart, Prodromus, Cat. p. 117
(1852); Layard, Ann. & Mag. Nat. Hist. 1853, xii. p. 167; Jerdon, B. of Ind. i. p. 193;
Holdsworth, P. Z. S. 1872, p. 421; Hume, Nests and Eggs, i. p. 97 (1873); Morgan,
Ibis, 1875, p. 314; Bourdillon et Hume, Str. Feath. 1876, p. 381.
Caprimulgus indicus, Jerdon, Cat. Madr. Journ. no. 251; Ill. Ind. Orn. pl. 24 (1847).
The Nilgherry Nightjar, Jerdon; *The Newara-Elliya Goatsucker*, Kelaart; *Night-Hawk*,
Europeans in Central Province.
Bim-bassa, Sinhalese, lit. "Ground-Owl"; *Pay-marrettai*, Tam., lit. "Devil-bird."

Adult male. Length 10·0 to 10·6 inches; wing 7·0 to 7·5; tail 4·1 to 5·0; tarsus 0·6; middle toe and claw 0·8 to
0·85; bill to gape 1·2 to 1·3. Expanse 22·4.

Female. Length 9·5 to 10·0; wing 6·9; tail 4·0.

Iris deep brown; eyelid brownish yellow; bill vinous brown, paler at the gape, the tip black; legs and feet vinous
brown, darker on the toes; soles pale, claws blackish.

Male. Light portions of head, back, and wings pale cinereous, finely pencilled with dark brown, and mottled on the
hind neck, wing-coverts, and scapulars with white: over the centre of the forehead and crown a broad black
stripe; feathers of the back, rump, and upper tail-coverts crossed with wavy marks of black; the scapulars with
velvety black centres and tips of au arrow-shaped or bar-like form set off by pale buff margins; wing-coverts
blackish brown, mottled on the inner webs with cinereous, and with a conspicuous terminal buff, dark-mottled spot
on the outer webs; tertials with black mesial portions and boldly pencilled with dark brown; quills blackish
brown, clouded with cinereous at the tips, and with a round white spot on the inner web of 1st primary and a
broad bar across the next three, generally interrupted on the 2nd; the outer quill indented with buff-white:
tail black, the central feathers with mottled cinereous transverse spaces, the remainder with mottled, distinctly
separated bars, and the four outer feathers with a large subterminal white spot.
A white stripe from gape to beneath the ear-coverts; across the throat a white band, interrupted in the centre and
edged below with rich ferruginous buff, which reappears on the sides of the neck, and is continued as a white
tracing round to the centre of the hind neck; throat, chest, and upper breast light cinereous, crossed with blackish
pencillings, which on the lower parts take the form of dark bands on a whitish ground; belly and under tail-
coverts whitish buff, the latter with a few brown bars.

Female. Darker above and also on the chest than the male; spots on the quills buff, of smaller size than in the male,
and that on the 2nd quill interrupted in the centre: the four outer tail-feathers wanting the white terminal spots,
and merely having a pale bar at the tips mottled with brown.

Note. The group to which this species and *C. indicus* belong is distinguished by having the tail, in the male, with
the four outer feathers on each side terminated with a white spot and the tarsus feathered.

Obs. Jerdon first pointed out the differences between Southern Indian examples of this species and *C. indicus*. Blyth
afterwards noticed them in 1845, *loc. cit.*, in an example from the Nilghiris, which he, however, still recognized
under the latter name. Subsequently, in 1851, he described the species from specimens sent from Ceylon by
Dr. Kelaart as *C. kelaarti*, finding these identical with his Nilghiri bird. It differs from *C. indicus* in its more
cinereous or albescent hue compared with the rufous tint of the latter, and also in the more mottled black markings,
which give it altogether a darker shade. It is likewise, at least so Blyth considered, a smaller bird. Of late years,
however, Hume, from the evidence afforded by a large number of examples from different parts of India, finds
that neither of these distinctions will hold good as regards peninsular birds, and remarks that every intermediate
link between the two typical forms occurs over all India. Some of the very smallest birds are rufous ones from

2 x

Mahabaleswar and Ahmednuggur, and also from Raipore, Sankra, and Etawah, while silver-grey and black-mottled birds are found near Simla, altogether out of the accepted range of *C. kelaarti*. Moreover in Travancore Mr. Bourdillon has procured both grey and rufous birds, the latter being quite as much so as North-Indian specimens. There is no reason, however, that the two species should not inhabit the same regions; and if we extend the limits of the range of each, this difficulty will be got over. As regards the Ceylonese birds, it is necessary to remark that they are all grey and like typical *C. kelaarti*, which militates against the possibility of suppressing the species in Ceylon, whatever may be done in future as regards India, where it seems difficult to draw the line of separation between it and *C. indicus*. Two of Mr. Bourdillon's specimens from Travancore measure—♂, wing 6·75 inches ; ♀, wing 7·25. In Ceylon the females are much the smaller of the sexes.

Distribution.—This very handsome Nightjar, first noticed in the island by Dr. Kelaart, and named by Blyth from specimens sent him by the Doctor, is almost entirely confined to the mountain-zone, and therein inhabits chiefly the upper ranges and the higher parts of Uva. I have seen it in great numbers about Nuwara Elliya, where its discoverer remarks, in his 'Prodromus,' that it swarms in the dusk of the evening in the marshy plains. It is, however, equally abundant during the S.W. monsoon in all the higher parts of the main range which are open and favourable to its habits, such as the Kandapolla and Elephant Plains and similar localities as far south as the Horton Plain. It appears to leave these high regions for warmer districts during the cold nights of the opposite season, as I found it rare in all the above districts in December, and did not meet with it at all on the elevated plateau between Totapolla and Kirigalpotta. In Haputale and other parts of Uva, as well as in most of the coffee-districts of about 3000 to 4000 feet in altitude, it is common enough throughout the year; but it is almost unknown in Dumbara, its usual limit being the neighbourhood of Deltota and Hewahette on the south of the valley, and Kalebokka on the north. It does not appear to have been hitherto known from any portion of the low country, although Mr. Holdsworth records as his opinion the probability of its leaving Nuwara Elliya during the cold season; but in August, 1875, I met with it in one locality of the Eastern Province which is at the sea-level, and where it was not at all to be expected. This was in the forest-region at the base of the Friars-Hood group of isolated hills, which form so prominent an object in the Batticaloa country. This tract is connected with the eastern slopes of Uva by detached groups of hills; but they spring from a low base, and are not situated in such a manner as to favourably foster a migration from the mountains to such a remote part as the Devilane district; and I therefore am inclined to think that the species must be resident in portions of the Eastern Province, particularly as I found it there at the season when it flocks to the upper hills. In corresponding parts of the Western Province, which lie much higher than the Friars Hood, it does not appear ever to be found ; nor have I any evidence of its inhabiting the Morowak-Korale mountains, although it doubtless does do so, but has been overlooked by gentlemen collecting in that part of the island.

On the continent of India, Kelaart's Nightjar is found in the Nilghiris and the wooded Ghâts of the Central Provinces, all over which latter hills Mr. R. Thompson records it as being common. Mr. Bourdillon notes it as a winter visitor to the Travancore hills, occurring rather abundantly from November until March. It must, in this case, ascend the range from the low country, which is the very opposite of its habit in Ceylon. I observe that Mr. Fairbank did not meet with it on either of his trips to the Palani hills, which does not augur in favour of its being widely spread in the mountains of South India.

Habits.—This Nightjar affects stony patnas, open glades in the forest, and all the confines of the Downs or so-called Plains which are such a singular feature of the fine jungle-clad ranges of Ceylon. It hides during the day among rocks near the edge of the jungle or among coffee-bushes, and from such places of concealment sallies out early in the evening and on all sides simultaneously is heard its curious call-note, *chump-pud, chump-pud,* repeated for several minutes and then suddenly stopped on the bird moving out to some conspicuous perch, such as a stump or huge rock, from which it recommences to utter its call. It is a very noisy bird in the breeding-season, but in the cold weather is almost silent, a peculiarity which was curiously noticeable in the birds I met with at Devilane tank, which, on three consecutive evenings before I shot them, were observed silently hawking on the bund of the tank. This species has a bold and dashing flight, rapidly and noiselessly performed, with frequent dexterous turns in the air, as it seizes its prey, and when disturbed in the daytime it quickly darts off and realights on the ground. It is, however, more rarely flushed during the day than either of the two following species, as it lies very close and does not repose in open spots like the Common Nightjar.

Dr. Jerdon writes, in his 'Birds of India,' that " it is now and then flushed from the woods when beating for game ; and more than one has fallen before the gun of the inexperienced sportsman, its extent of wing and the lazy flapping having caused it to be mistaken for the Woodcock." I have myself observed this peculiarly lazy flapping, which is not the usual mode of progression, at sunset, and several times have heard the strange sound which the bird makes, resembling the beating of an immense fan or wing in the air : whether this is caused by the motion of its pinions, or by the utterance of a guttural note, I am unable to say ; but much as it resembles a mechanical effect, it is doubtless the result of some curious vocal power in the bird. Its food consists almost entirely of beetles, of which it consumes immense numbers, its stomach being crammed with these, one would think, indigestible insects at an early hour in the evening. It is worthy of remark that the majority of specimens procured of this species are *males* : what becomes of the females in the evenings it is hard to say ; but one thing is certain, that they keep out of the way and are seldom shot, except when flushed in the daytime from their nests or in company with a young brood.

Nidification.—Mr. Holdsworth remarks that the breeding-season about Nuwara Elliya commences in March and April. Its eggs appear to be seldom found ; and the only instance of their being taken that ever came under my notice was related to me by a gentleman in Haputale, who informed me that his sons sometimes procured them on the estate. In India they are well known. In the Nilghiris and Central Provinces, according to Mr. Hume's correspondents in 'Nests and Eggs,' it commences to breed in March and continues to lay until August. The eggs are deposited " in a slight depression under a bush or tuft of grass ;" but they have been found, Mr. Davison relates, in a heap of ashes produced by the *Burgas* burning weeds in their fields. The eggs are two in number, and are said to be counterparts of those of the closely allied *C. indicus* ; they are of a pale yellowish or salmon ground-colour, marbled with brown among blotches of a lighter shade, which sometimes resemble a darker tint of the ground-colour ; they are long ovals in shape, and " vary from 1·08 to 1·23 inch in length, and from 0·8 to 0·9 inch in breadth."

Mr. Rhodes Morgan on one occasion found the eggs deposited on a heap of ashes ; he describes them as of a " pinkish buff, blotched with pale violet-brown."

CAPRIMULGUS ATRIPENNIS.

(THE JUNGLE-NIGHTJAR.)

Caprimulgus atripennis, Jerdon, Ill. Ind. Orn. pl. 24, letterpress (1847); id. B. of Ind. i.
p. 196; Holdsworth, P. Z. S. 1872, p. 421; Legge, Ibis, 1874, p. 12.

Caprimulgus spilocercus, Gray, List Fissirostres Brit. Mus. p. 7 (1848); Hume, Stray Feath.
1873, p. 432.

Caprimulgus maharattensis (Sykes), Kelaart, Prodromus, Cat. p. 117 (1852); Layard, Ann.
& Mag. Nat. Hist. 1853, xii. p. 166.

Maharatta Goatsucker, apud Kelaart; *The Spotted-tailed Goatsucker*, Gray.

The Ghât Nightjar, Jerdon; *Goatsucker, Night-Hawk*, Europeans in Ceylon.

Bim-bassa (West Prov.), *Ra-bassa, Omerelliya* (South Province), Sinhalese; *Pathekai*, lit.
" Roadside-bird," Jaffna Tamils, also *Pay-marrettai* (Jerdon).

Adult male. Length 10·6 to 11·0 inches; wing 7·0 to 7·2; tail 4·0 to 5·0; tarsus 0·7; middle toe and claw 0·95 to
1·0; bill to gape 1·3 to 1·4.

Adult female. Length 10·0 to 10·4 inches; wing 6·6 to 6·7.

Iris deep brown; eyelid pale reddish (yellowish in female); bill reddish brown, tip black; legs and feet reddish brown
or pale reddish, claws dusky brown.

Male. Top of the head and upper part of hind neck cinereous brown, very finely stippled with grey, the feathers of
the centre of the crown and nape having broad black mesial stripes, the ground-colour of the latter part passing
with a ferruginous hue on the hind neck into the blackish of the back, rump, and upper tail-coverts; the margins
of the feathers on these parts stippled with fulvescent grey, and the black confined chiefly to a central stripe;
scapulars very handsomely marked with oblique bands and spade-shaped patches of velvety black, the shorter
feathers with oblique external margins of rich buff, the longer feathers being mostly grey near the tips, vermiculated
with blackish: lesser wing-coverts blackish, mottled with ferruginous; anterior feathers of the remaining series
black, marked with mottled spots of buff, and in some examples with white tips to many of the feathers; inner
secondary coverts mostly mottled with grey on a black ground, and with buffy white tips to some of the feathers;
primaries blackish brown, mottled at the tips with pale einereous; the 1st quill with a white spot on the inner
web, and the next three with a white bar in continuation, interrupted on the 2nd quill at the centre; inner
secondaries paler than the primaries, marked in places with ochraceous buff; tertials mottled with cinereous grey
at the tips: tail blackish brown, the four central feathers mottled with dusky fulvous, the two lateral feathers on
each side black, with the terminal third white and the lateral margin tinged with buff, inner margins of all
indented with buff.

Lores and ear-coverts russet-brown, mottled with black; rictal bristles black, with white bases; chin and along the
base of lower mandible mottled black and fulvous; a thin white stripe at the gape; across the throat a broad
white band, its lower edge deeply margined with black, or, in some, barred with this hue and tipped with rufous-
buff; chest and upper breast cinereous, finely stippled with brown, and the latter part washed with a russet hue:
beneath this the under surface is fulvous, crossed with narrow bars of blackish brown, the centre of the breast
being, in many specimens, slightly albescent; under wing-coverts fulvous, cross-marked with brown.

The scapulars vary much in this bird, scarcely any two examples having them marked the same; in some individuals
the broad oblique buff margins are almost entirely wanting; the white tips of the wing-covert feathers likewise are
variable.

Female. Has not the scapulars so conspicuously marked as the male; the wing-spots are buff or buffy white, small
and bar-shaped, that on the 4th quill almost wanting; two lateral rectrices on each side with a buff-white tip,
varying up to half an inch in depth, or with the tip only mottled with buff (such examples being probably young);
white throat-spot smaller than in male, ground-colour of lower parts duller.

Young (male of the year). Wing 6·6.

Bill and feet paler than in adult.

White tail-spots smaller than in adults, the black running out on the outer web much further than on the inner ; the outer margin of the white spot mottled with brown ; throat-bar as in females.

Note. The section to which this species and one or two others in India belong is characterized by having the two outer tail-feathers in the male terminated with white and the tarsus feathered.

Obs. Layard speaks of *C. mahrattensis*, in conjunction with *C. asiaticus*, being very abundant in the vicinity of Colombo and throughout the Southern Province. As there is no other Nightjar besides the latter which is common, or even found, in the districts named, it follows that *C. mahrattensis* was mistaken for the present species, as Mr. Holdsworth (*loc. cit.*) has already suggested. Mr. Hume points out ('Stray Feathers,' 1873) that Ceylon specimens do not agree over well with Nilghiri ones.

Distribution.—This fine Nightjar is a denizen more or less of the entire sea-board of Ceylon, and extends into most of the inland districts, being very numerous in all parts which are clad with forest or are even moderately well-wooded. Mr. Holdsworth does not record it from Aripu on the north-west coast ; but it is abundant in parts of the Jaffna peninsula, and I have met with it on the coast at Illepekadua, north of Mantotte, and at Pomparipu to the south of it ; so that I imagine it is simply locally absent from the open country near the Pearl station, and probably an inhabitant of the adjacent interior. It is very numerous in the northern forest-tract and around Trincomalie, in the wooded districts of the south-west from Kalutura round to Tangalla, and in the jungle-country north of Kattregama. The same may be said of the country north of Kurunegala and many parts of the Western Province, although I found it conspicuously absent from most parts of Saffragam. It ranges into the hills up to an altitude of about 3500 feet, at which elevation I have seen it in Hewahette, and in Dumbara it is common. Mr. Parker does not record it in his letters to me from the Uswewa district ; but I have no doubt that it is found there.

On the mainland, the Ghât Nightjar, as it is styled by Jerdon, is found in various parts of the south of India, to wit, on the Malabar coast and in the Ghâts of the north of the Carnatic. It is tolerably common in the Nilghiris ; but Mr. Bourdillon has not procured it in the Travancore hills, nor Mr. Fairbank in the Palani ranges, which proves that it is a bird of local distribution in the peninsula.

Habits.—This species inhabits dry forest, low jungle, scrub, and wooded tracts in semicultivated country. It is very partial to the "cheena"-woods in the Galle district, and similar secondary jungle in the east and north of the island, such haunts affording it secure shelter whilst it roosts on the ground, and from which it sallies out at dusk, settling in roads, pathways, or any bare spaces in the woods. I have always observed that it avoids localities in which there are not large trees, which habit is exemplified in its locating itself in numbers about the outskirts of the cinnamon-gardens at Colombo, while it does not haunt the open bushy gardens themselves, where the next species is so common. It comes out a little later than the Small Nightjar, first of all flying up to a low stump or branch and uttering its curious call, like the striking of a hammer on a thin plank ; as soon as it is heard this cry is answered by its companions, and in a few minutes these notes resound on all sides and are continued until it is dark enough for the birds to take wing in pursuit of the myriads of beetles and other insects which throng the calm air of a tropical evening. This loud note is generally preceded by a low *grog, grog-grog*, which can only be heard when one is close to the bird. It is a gluttonous feeder, its stomach being generally crammed with beetles or winged termites before dark, which it captures with a powerful swooping flight, often sailing along with very upturned motionless wings. It is just as fond of sitting on roads and paths as the next species; but it is not so tame, and will not suffer itself to be almost kicked as it will. The Tamils in the north of Ceylon call it the "roadside bird" from this habit, and have a strange superstitious notion that it has the power of plucking out the eyes of their cattle ; but they do not seem to be able to account for the fact that there is no ocular testimony of this objectionable habit ever having been put into practice ! It is noteworthy that this Nightjar perches continually on the tops of small dead branches of low trees ; and once I think I saw it sitting in a diagonal manner, though not quite transversely, across a branch, as an ordinary Passerine bird would have done.

Nidification.—In the west of Ceylon the Jungle-Nightjar breeds during the latter part of the dry season and the commencement of the monsoon rains in April and May. It lays two eggs in a slight depression in sandy ground, beneath the shelter of a shrub; they are of a buff ground-colour, and very sparsely spotted with very dark sepia-brown, rather roundish blots. I have seen several eggs, and have not detected any of the streaky markings peculiar to those of other Nightjars. I unfortunately have no data of the size of this Nightjar's eggs, as I omitted to measure those which I examined in Mr. MacVicar's collection; they are, however, considerably larger than those of the next species, measuring, according to Layard, " 14 lines by 11 lines." The dimensions given by Mr. Hume of a pair of eggs from the Nilghiris, viz. 1·13 inch by 0·72, and 1·01 by 0·74, are, I am sure, inferior to those of Ceylonese birds.

CAPRIMULGUS ASIATICUS.

(THE COMMON INDIAN NIGHTJAR.)

Caprimulgus asiaticus, Latham, Ind. Orn. ii. p. 588 (1790); Gray & Hardwicke, Ill. Ind.
Orn. i. pl. 34. fig. 2 (1832); Sykes, Cat. J. A. S. B. iii. no. 30 (1834); Blyth, Cat. B.
Mus. A. S. B. p. 83. no. 415 (1849); Horsf. & Moore, Cat. B. Mus. E. I. Co. p. 115
(1854); Holdsworth, P. Z. S. 1872, p. 419; Hume, Str. Feath. 1873, p. 432; id. Nests
and Eggs, p. 97 (1873); Adam, Str. Feath. 1873, p. 371; Legge, Ibis, 1874, p. 12.
et 1875, p. 281; Ball, Str. Feath. 1874, p. 385; Butler, ibid. 1875, p. 455.

The Indian Goatsucker, Kelaart; *Night-Hawk, Goatsucker, "Ice-bird,"* Europeans in
Ceylon (the latter name from the resemblance in the bird's note to a stone scudding on
ice).

Rim-bassa, Ra-bassa, Sinhalese; *Pathekai*, Tamils in Ceylon.

Adult male. Length 8·0 to 9·1 inches; wing 5·65 to 5·8; tail 4·0 to 4·2; tarsus 0·85; middle toe and claw 0·85 to
0·9; bill to gape 1·2.

Iris deep brown; eyelid light reddish yellow; bill reddish or reddish brown, with the nostril and tips black; legs and
feet brownish red, darker at the ends of the toes, claws dark brown.

Light portion of head and upper surface cinereous ashy, finely and distinctly pencilled with brown, and the scapulars
and wing-coverts richly marked with buff-yellow; centre of the forehead and crown striped with black, the
feathers edged with rufescent yellow; back and upper tail-coverts pale cinereous, most of the feathers with a
narrow mesial black line, and the whole finely pencilled with brown; scapulars with arrow-shaped velvety black
centres, bounded by broad, rich buff margins; secondary wing-coverts with the terminal portions buff, paling to
white at the edges; quills and primary-coverts dark brown, the latter, together with the secondaries, barred with
reddish buff; the primaries mottled with grey near the tips, the first with a white spot on the inner web (in some
with a corresponding external pale edge) and a similar one on both webs of the next three; centre tail-feathers
cinereous, with narrow wavy cross bars; remainder blackish, with wavy cross lines of reddish buff, the two outer
feathers with a terminal white spot (1¾ inch in depth in old birds), the tip of the lateral feather nearly always with
some dark mottling and its outer margin buff.

Ear-coverts dark brown, beneath there is a narrow whitish rictal spot; a white bar across the throat, divided by a
buff-mottled patch in the centre, and continued as a buff collar round the hind neck; chest with the feathers
across the centre deeply tipped with pale buff; breast, flanks, and sides of belly barred with brown on a buff
ground; belly and under tail-coverts whitish buff, unbarred.

Female. Length 8·4 to 8·6 inches; wing 5·6 to 5·8. Bill paler than in the male, brownish olivaceous at the base and
gape; legs and feet brownish olive, claws brown.

Upper surface similar to male; quills paler, edges of primaries greyish near the tips; spots on the outer web of 2nd,
3rd, and 4th quills buff, *in some examples wanting altogether*; tail-spots not so large as in the male, about ⅔ of
an inch in depth, the lateral margin of the outer tail-spot sullied with brown, except in *old* birds.

Young. Iris as in adult; bill dusky olive-brown, the tip dark brown; legs and feet brownish fleshy, palest on the
sides of the tarsi.

Above paler or less marked with dark brown and black than in adults; scapulars in some broadly margined with
buff, in others almost uniform with the back; quills tipped with buff, the primaries apparently darker in the
male than in the female; the white spots on the outer webs of the primaries more or less tinged with buff, as is
also that on the inner web of the 4th quill; outer margin of the terminal tail-spots washed with buff and mottled
with brown; exterior of lateral tail-feathers broadly edged and indented with buff in those birds which have
richly marked scapulars.

Chin and along base of bill whitish in some, this part being, as in the adult, variable in its marking; under surface
in the quite young bird fluffy, and the markings undefined in older examples; the ground-colour is greyer than
in adults; under tail-coverts usually barred with brown.

Note. This species and its allies have the tarsus bare and the tail-feathers as in the last.

Obs. With age the white terminal spots of the rectrices increase in size, and the throat-band develops and becomes whiter. Examples from the dry, hot districts in the south-east and north of the island are more rufous in their tints than those from the west and south; they thus resemble Indian examples of the species, which are, as a rule, as Mr. Holdsworth remarks, *loc. cit.*, much less grey than those from the island. It must, however, be borne in mind that this Nightjar is a very variable bird in its coloration; some individuals seem to have the tendency to buff markings more exaggerated throughout the entire plumage than others, this being particularly noticeable in the scapulars and tail-feathers; the wing-spots vary considerably in character, and while the ground-colour of the primaries is almost black in one bird, it will be a medium brown in another of the same age.

Distribution.—This little Nightjar inhabits, in considerable numbers, all the maritime portions of the island, affecting, by choice, those localities where sandy scrubs or sparsely clothed open lands border the sea-coast; it is consequently less common in the damp wooded district of the south-west than in the hot eastern and northern divisions of the island. It is very abundant in the Batticaloa, Hambantota, and Trincomalie districts, and likewise in the Jaffna peninsula and down the western coast as far south as Kalutara. In the interior it is less numerous, and such wooded tracts as Saffragam, the Pasdun, and lower portion of the Kukkul Korale are haunted but little by it. It ascends into the Kandyan Province, and is by no means uncommon in Dumbara and Deltota and in the low-lying basins drained by the affluents of the Mahawelliganga. In Uva it ranges to a considerable altitude, and I have seen it in May as high as 4000 feet in the Fort-Macdonald district. Higher than this I have no evidence of its occurring.

Elsewhere on the continent this species, which is the commonest of the Indian Nightjars, is found throughout all India, and ranges, according to Mr. Hume, into the Himalayan mountains in the spring and summer, at which season it may be met with as high as 6000 feet. It extends into Burmah, and is common in the British Provinces there, Mr. Oates recording it as numerous in the plains of Pegu, but not in the hills. As regards India proper, I find that it is local in Sindh, having only been met with at Sehwan. In the Sambhur-Lake district Mr. Adam says it is not common, but in northern Guzerat and the surrounding plain country it is so. Mr. Fairbank notes that it is plentiful in the Deccan; it is likewise so in the southern parts of the Madras Presidency, but does not appear to occur in the hills, as Mr. Bourdillon does not record it in his list of Travancore birds, and Mr. Fairbank procured but one example of it at the base of the Palanis.

Habits.—The Common Indian Nightjar affects scrubby waste lands, low sandy jungle-tracts, cinnamon-plantations, and openly wooded country intermingled with small wood. It is a tame and familiar bird, and is better known to most people than the last species. It roosts during the day on bare ground between shrubs and sleeps soundly, suddenly getting up when almost trodden on, and quickly realighting again at a little distance off. The young brood remain with the parents for some time, and thus a little party of three or more may often be surprised roosting in close proximity to native houses. It is a well-known bird in the cinnamon-gardens of Colombo, alighting in the roads just after sunset, and on dull afternoons an hour earlier, and allowing itself to be almost driven over before it rises. Layard well describes the habits of this and the last species when he says that "the belated traveller hurrying homeward ere the last dying gleams of the setting sun fade in the west, is startled by what seems to be a stone flying up with a few rapid querulous notes, and gliding along on noiseless pinions settling again within a few yards of him." It is a very noisy bird at sunset and before daybreak, uttering its notes likewise on moonlight nights, although it is quite silent in darkness. Its well-known note, persistently repeated for a long time together, is wearisome when heard around one's bungalow at midnight, and many liken it, both in India and Ceylon, to the sound made by a stone scudding along ice. It resembles somewhat the sounds *chuk-chuk-chuk chuk-urrr-ruk*; but some liken it, according to Jerdon, to the syllables *tyook-tyook-tyook*. However this may be, the peculiar note has given rise to its name "Ice-bird;" and not unappropriate it is too, notwithstanding that the idea does not assimilate well with a temperature of 81° Fahr.! Its flight is buoyant and skilful, enabling it to capture its coleopterous prey with great ease. It feeds more on beetles than moths, and some say that the singular pectination of the middle claw is adapted by nature for the removal of beetles' claws from its gape. This species usually settles on the ground; but I have several times seen it perched on stumps, like the preceding.

Nidification.—The breeding-season on the western side of the island is during the first three or four months of the year. It lays usually two eggs on the bare ground, often without any depression or nest-formation; but the shelter of a bush or stump is generally chosen. The eggs are ovals in shape and smooth in texture, of a light salmon or reddish-grey ground-colour, marbled slightly and blocked openly throughout the surface with sienna-red over faint clouds of bluish grey. An egg obtained in the cinnamon-gardens measured 1·12 by 0·73 inch; but in 'Nests and Eggs' the average is given at 1·04 by 0·77 inch. The eggs are much more salmon-coloured than those of the last species and smaller. In India this species breeds chiefly in April and May, but its eggs have been taken in July; and Captain Butler is of opinion that it lays twice in the year, he having shot a hen bird, in company with a young one just fledged, on the 20th of July, and found, on dissecting her, that she was about to lay again. It is said not to be so particular in choosing its situation as other Nightjars. Mr. R. Thompson, as quoted by Mr. Hume, says that he has found the eggs " in a quite unsheltered spot in the middle of a dry pebbly *nullah.*"

www.ingramcontent.com/pod-product-compliance
Lightning Source LLC
Chambersburg PA
CBHW021347210326
41599CB00011B/786